HPE Aruba Networking Certified Associate - Network Security

OFFICIAL CERTIFICATION STUDY GUIDE

(EXAM HPE6-A78)

Second Edition

Miriam Allred

HPE Press
660 4th Street, #802
San Francisco, CA 94107

HPE Aruba Networking Certified Associate - Network Security
Official Certification Study Guide (Exam HPE6-A78)
Miriam Allred

Published by:
Hewlett Packard Enterprise Press
660 4th Street, #802
San Francisco, CA 94107

ISBN: 978-1-960058-30-0
Printed in Mexico

WARNING AND DISCLAIMER

This book provides information about the topics covered in the HPE Aruba Networking Certified Associate - Network Security certification exam (HPE6-A78). Every effort has been made to make this book as complete and as accurate as possible, but no warranty or fitness is implied.

The information is provided on an "as is" basis. The author, and Hewlett Packard Enterprise Press, shall have neither liability nor responsibility to any person or entity with respect to any loss or damages arising from the information contained in this book or from the use of the discs or programs that may accompany it.

The opinions expressed in this book belong to the author and are not necessarily those of Hewlett Packard Enterprise Press.

FEEDBACK INFORMATION

At HPE Press, our goal is to create in-depth reference books of the best quality and value. Each book is crafted with care and precision, undergoing rigorous development that involves the expertise of members from the professional technical community.

Readers' feedback is a continuation of the process. If you have any comments regarding how we could improve the quality of this book, or otherwise alter it to better suit your needs, you can contact us through email at hpepress@epac.com. Please make sure to include the book title and ISBN in your message.

We appreciate your feedback.

Publisher: Hewlett Packard Enterprise Press
HPE Press Program Manager: Michael Bishop

About the Author

Miriam Allred has many years' experience configuring, testing, and troubleshooting wired and wireless networks. Miriam combines this wide range of technical expertise with pedagogy and instructional design training, allowing her to create technical training courses for both advanced and entry-level server, storage, and networking professionals. Miriam Allred has a degree from Cleveland State University.

Introduction

This book helps you prepare for the HPE Aruba Networking Certified Associate - Network Security certification exam (HPE6-A78). This certification validates candidates' knowledge and ability to describe common security threats and vulnerabilities. It tests a candidate's knowledge of device hardening, implementing security at the edge with AAA, basic roles and firewall policies, dynamic segmentation, and endpoint classification. It covers basic threat detection technologies, including collecting logs and alarms.

Book topics

Topics covered in this guide include:

- Security threats and HPE Aruba Networking security strategy
- Security technologies
- Hardening AOS-CX switches
- Hardening AOS mobility devices
- Enhancing LAN security
- Network authentication technologies
- Enforcing role-based authentication and access control
- Understanding unified SASE
- Implementing threat detection and forensics
- Troubleshooting, monitoring, and more

To help you learn as you go, Knowledge Check questions are included in each chapter. In addition, Chapter 13 has a 30-question practice exam that simulates the kinds and styles of questions that you can expect to meet in the certification exam (HPE6-A78). Answers to the Knowledge Checks and the sample test can be found in Appendix B.

HPE Certification and Learning

Hewlett Packard Enterprise (HPE) Certification and Learning provides end-to-end continuous learning programs and professional certifications that can help you open doors and accelerate your career. We provide:

- **Professional sales and technical training and certifications** to give you the critical skills needed to design, manage and implement the most sought-after IT disciplines; and
- **Continuous learning activities and job-role based learning plans** to help you keep pace with the demands of the dynamic, fast paced IT industry
- **Advanced training** to help you navigate and seize opportunities within the top IT transformation areas that enable business advantage today.

As a Certification and Learning certified member, your skills, knowledge, and real-world experience are recognized and valued in the marketplace. To continue your professional and career growth, you have access to our large HPE community of world-class IT professionals, trend-makers and decision-makers. Share ideas, best practices, business insights, and challenges as you gain professional connections globally.

To learn more about HPE Certification and Learning certifications and continuous learning programs, please visit: http://certification-learning.hpe.com.

Audience

This guide is designed for network engineers or help desk engineers working in either a customer or partner environment.

Assumed knowledge

The candidate should have worked six months to a year in networking and understand the fundamentals of both wired and wireless networks.

Minimum qualifications

There are no prerequisites for this candidates new to this certification.

Preparing for Exam HPE6-A78

This self-study guide covers the topics that you need to prepare for the HPE Aruba Networking Certified Associate - Network Security certification exam (HPE6-A78). However, it does not guarantee that you will have all the knowledge you need to pass the exam. It is expected that you will also need to draw on real-world experience and would benefit from completing the hands-on lab activities provided in the instructor-led training course.

Recommended HPE Aruba Networking training

Recommended training to prepare for the HPE6-A78 exam is accessible from the exam's page in The Learning Center. See the exam attachment, "Supporting courses," to view and register for the courses.

Additional Resources

At HPE Aruba Networking, we want to help you succeed in your certification journey by providing all the resources you need to pass your certification exams in the technology of your choice. With the *HPE Certification Advantage Program*, you can purchase a comprehensive set of resources to help you prepare and succeed in your certification journey—all at a deep discount, in the technology of your choice. For the price of a typically 5-day course, you get the course, an HPE Press study guide, office hours with instructors, self-directed lab access, and an exam voucher(s), depending on which written or practical exam option you choose. As of the writing of this book, this discount was 45% off the normal price. More information on this program can be found here: https://certification-learning.hpe.com/tr/certifications-advantage.

Obtain hands-on experience

You are not required to take the recommended, supported courses, and completion of training does not guarantee that you will pass the exams. HPE Aruba Networking strongly recommends a combination of training, thorough review of courseware and additional study references, and sufficient on-the-job experience prior to taking a certification exam.

Exam registration

To register for the exam, go to:

https://certification-learning.hpe.com/tr/learn_more_about_exams.html

Contents

Contents

1 Security Threats and Security Strategy

EXAM OBJECTIVES

- ✓ Describe common security threats, including malware, denial of service (DoS), spoofing, man-in-the-middle (MITM), and social engineering
- ✓ Identify the difference between a threat and a vulnerability
- ✓ Describe a kill chain and attack stages

Assumed Knowledge

- Basics of IP networking
- Basics of Ethernet switching
- Basics of 802.11 wireless networks

Overview

This chapter lays the groundwork for the rest of the study guide. You will learn about the threats to organizations' networks and the reasons why organizations require rigorous security more than ever before. You will then apply what you have learned in an activity.

Next you will learn about the specific stages of cyberattacks, from reconnaissance to extracting data onto the hacker's device.

Finally, you will learn why customers need to embrace a zero trust strategy for securing their network. You will examine the HPE Aruba Networking strategy for delivering a zero trust network and get an overview of the pieces of the strategy that this study guide will cover.

Confidentiality, Integrity, and Availability (CIA)

Before you dive into an examination of threats against network services, take a moment to examine what these threats might compromise and what a secure network should deliver.

As defined by widespread industry models, a secure network should provide confidentiality, integrity, and availability (CIA). As shown in Figure 1-1, confidentiality, sometimes called privacy, means that no one can read a message except the intended recipient. Integrity means that the message that the recipient receives matches the message that the actual transmitter originally sent. No one replaced the message with a different one or tampered with the message's contents. Integrity is related to authenticity, which means that the supposed transmitter of the message is the actual transmitter. These two pillars in the CIA model relate to the security of communications. The final pillar, availability, ensures that network services are available for legitimate users.

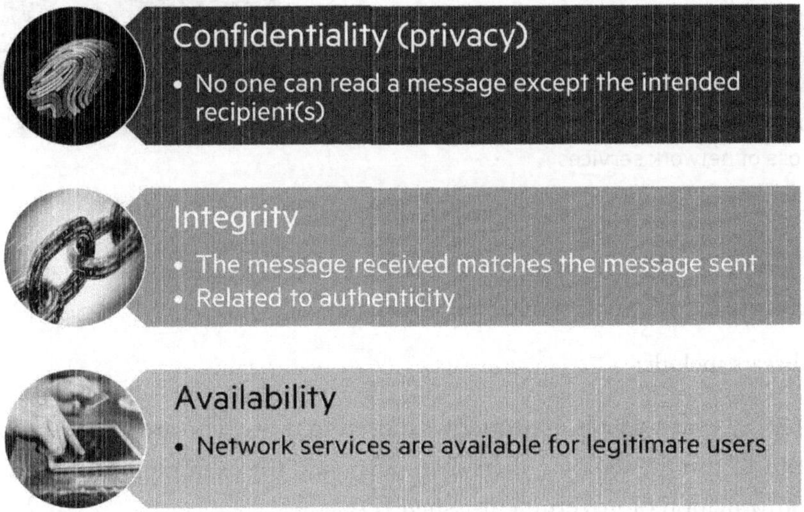

Figure 1-1: Confidentiality, Integrity, and Availability (CIA)

Throughout this chapter, you will look at threats to confidentiality, integrity, and availability. The rest of this study guide will explain how you can use an HPE Aruba Networking network infrastructure to fight those threats.

Threat Overview

You will first learn about the threats that users and organizations face.

Threats

You will begin by looking at common threats and the risks that they pose to your customers' networks and business.

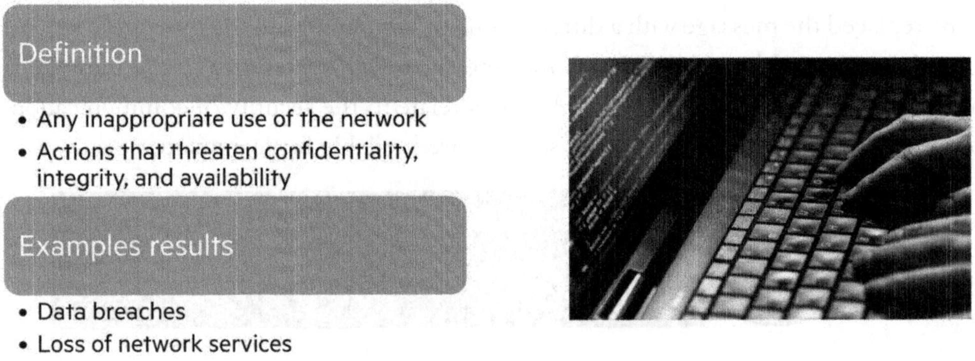

Figure 1-2: Threats

As shown in Figure 1-2, a threat broadly refers to any action executed or initiated by a bad actor who uses or attempts to use the network inappropriately.

The results of threats include:

- Data breaches, in which data is improperly exposed to those who are not authorized to use that data and want to use the data for their own purposes; for example, threats can expose users' credit card information or government-issued IDs.
- Loss of network service; for example, threats can prevent legitimate users from reaching a company's web page.

Vulnerabilities versus Threats

To properly secure your network from threats, you need to understand the relationship between threats and vulnerabilities.

Vulnerability

A vulnerability refers to a state in which a device or network is vulnerable to a threat. A vulnerability does not necessarily mean that an issue has occurred, but the risk of an exploit or attack is present.

As shown in Figure 1-3, for example, a server operating system might have a bug that a hacker could exploit to cause the server to become unstable. Or perhaps software was coded in such a way that unintentionally allows hackers to gain privileged access without the proper credentials. Applications can also have vulnerabilities, as can the operating systems on network infrastructure devices.

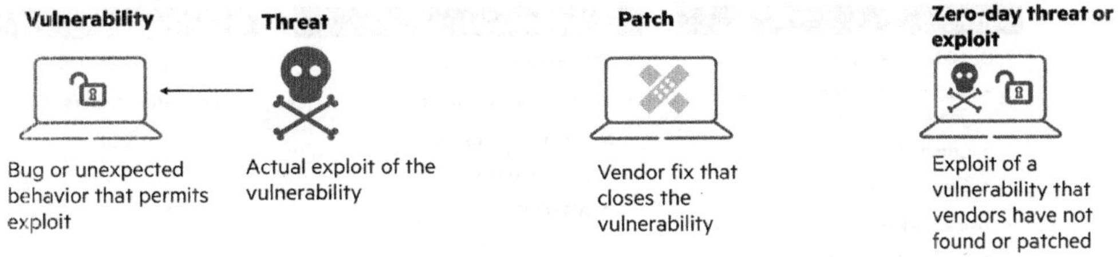

Figure 1-3: Vulnerabilities versus Threats

Patches

When security teams discover vulnerabilities, they inform the device, software, or OS vendor, which can then create a "patch" that closes the vulnerability. Vendors might push the patch to devices automatically—this is often what you are seeing when a window on your computer pops up, telling you that you need to restart your system to apply updates.

In a data center environment, security teams and server admins often work together to test those patches. After they have verified that the patches do not cause unintended changes to their servers' functionality, they deploy the patches. In a network environment, you also typically investigate and test software updates before deploying them in a production network.

Zero-day Vulnerabilities and Exploits

A Zero-day vulnerability means that the vulnerability exists, but the vendor does not yet know about it.

A Zero-day exploit means that unethical hackers have discovered the vulnerability and begun to exploit it before the vendor is aware of the vulnerability or before the vendor is able to create a patch. Because no patch yet exists for the vulnerability, many systems can be susceptible to the threat, making Zero-day exploits highly dangerous.

Actors

Now look a bit more closely at those bad actors, who might launch threats against your network.

The bad actors who come to mind most quickly are outside hackers. As shown in Figure 1-4, in the past those hackers might have been people seeking to cause general mischief. But most often now hackers are criminals seeking some sort of monetary gain. Sometimes hackers work for governments, which spy on each other and sometimes on their own citizens as well. Hackers can also be internal actors such as disgruntled employees or contractors who want to take revenge on the company. Or perhaps the internal actors are seeking monetary gain by abusing their access to the network.

Figure 1-4: Actors

Bad actors can be not so bad—generally well-meaning users who carelessly introduce threats. For example, an inexperienced admin might leak data by accidentally publishing it to the wrong location. Or users might break rules without malicious intent. As just one example, an employee might expose a network to unauthorized access by connecting a wireless router to it and failing to configure any security on that router.

You should also be aware of ethical hackers. These are people who use many of the same tools as other hackers but for ethical purposes. They might be security researchers, who seek to discover vulnerabilities so that vendors can fix them. Companies might also employ or hire "pen testers" to perform penetration testing on their network. In other words, the pen testers try to attack the network and then report any vulnerabilities to the company so that the company can close those vulnerabilities.

Meet Alice, Bob, Eve, and Mallory

As shown in Figure 1-5, discussions of cryptography use standard characters named Alice, Bob, Eve, and Mallory. These characters make discussions a bit more concrete than simply referring to communications between A and B. This study guide will use these characters in discussions of cryptography in the next chapter. For consistency, it will also use these characters in the broader security discussions in this chapter.

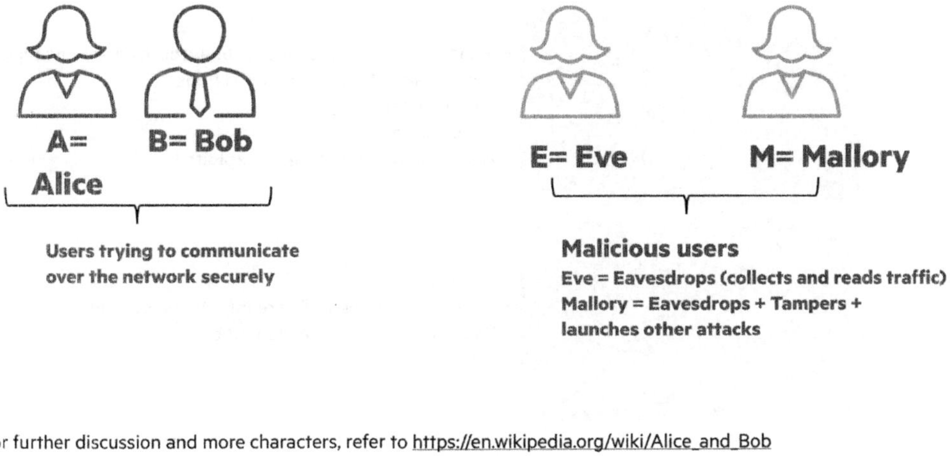

Figure 1-5: Meet Alice, Bob, Eve, and Mallory

Alice and Bob are two users who want to communicate over the network securely. In classic cryptographic discussions, Alice and Bob communicate with each other. In our more generic security examples, Alice and Bob might also be using the network to access network services.

Eve and Mallory are malicious users or hackers. Eve eavesdrops on communications, which means that she collects traffic that does not belong to her and tries to read it. Mallory eavesdrops and also tampers with traffic. For its generic security discussions, this book will extend Mallory's role to that of an all-round hacker, who launches many sorts of attacks.

Full cryptographic discussions have many other characters. If you are interested, you can refer to sources such as https://en.wikipedia.org/wiki/Alice_and_Bob.

What Is Malware?

You are now ready to look at specific types of threats.

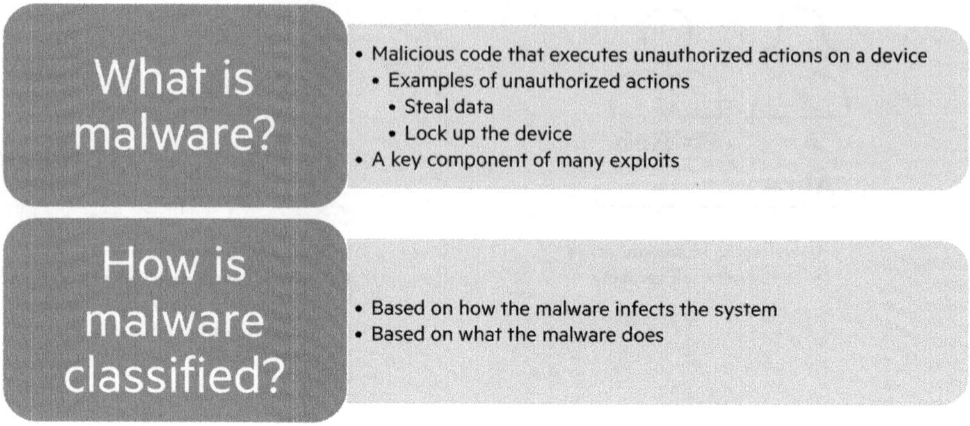

Figure 1-6: What Is Malware?

As shown in Figure 1-6, threats are often related to malware, which refers to any malicious code that executes actions unwanted by the owner of the device. Malware can form a part of many exploits from stealing data to making a device unusable.

You can classify malware based on how it gets on a machine, as well as by what it does once it is on the machine.

Malware: How It Infects a System

First, you will focus on the different ways that malware compromises devices and spreads.

Viruses and Worms

Viruses and worms are closely associated types of malware, which are often more colloquially called viruses (in other words, people might say "virus" when they properly mean "worm"). Both viruses and worms are characterized by their ability to replicate and spread. However, a virus is malicious code that executes from a host program while a worm is malicious code that runs without a host program.

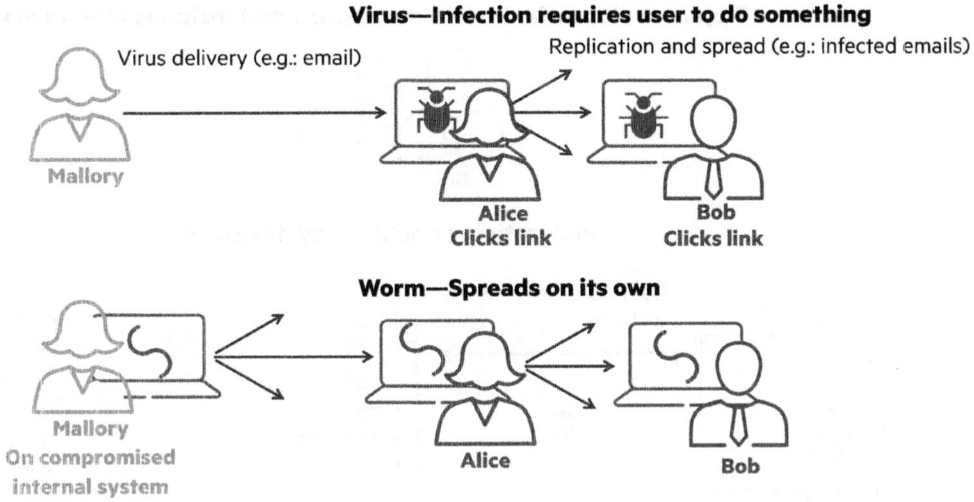

Figure 1-7: Viruses vs Worms

Many years ago, hackers spread viruses through disks. Now viruses often get on a computer when users open an infected file, which might be an email attachment, a file that a user obtained through peer-to-peer sharing, or a file downloaded from an untrusted HTTP or FTP server. A virus requires a human to run the file in order to infect the system. As shown in Figure 1-7, once the virus has infected a system, it will replicate and try to spread. For example, a virus might generate emails to all of the addresses in a compromised device's Outlook address book, attaching itself to each of the emails.

Your system can acquire a worm in the same way as a virus, but worms can also propagate themselves through the network without human interaction. In some cases, they can use network packets to target vulnerable devices' memory directly.

Trojan

In the Iliad, the Greeks pretended to retreat and left the Trojan horse as an ostensible gift to their enemies, the Trojans. Instead of retreating, the Greeks hid in the horse, and after the Trojans brought the horse inside their city walls and trustfully went to sleep, the Greeks emerged to attack the city from within. Similarly, a Trojan is malware that hides itself inside something the user wants, such as a free version of a useful tool. As shown in Figure 1-8, when users install the Trojan, they do indeed get the promised software, but malware also installs and begins working in the background.

If the malware included in the Trojan replicates itself and attempts to spread itself, then the Trojan is a type of virus.

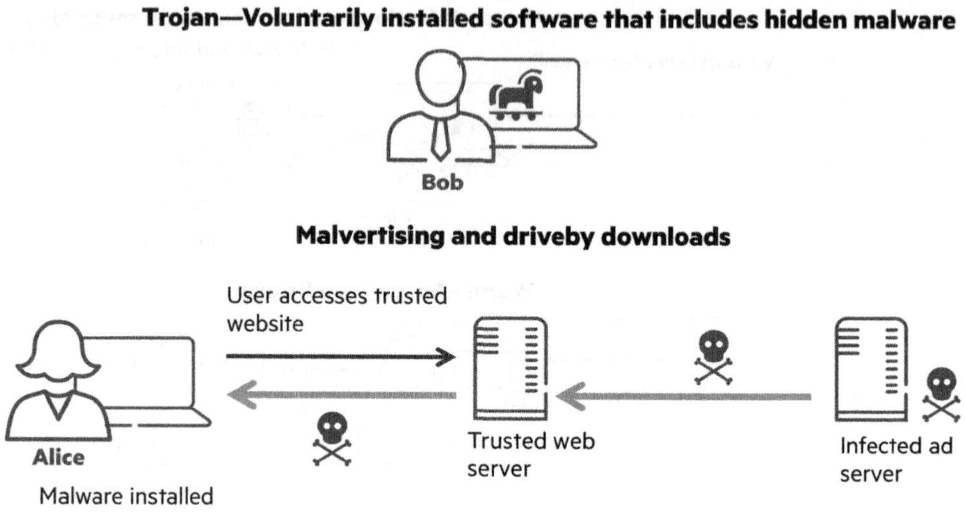

Figure 1-8: How Trojans Infect Systems

Drive-by Downloads and Malvertising

Computers can also acquire malware from a "drive-by download." A drive-by download exploits vulnerabilities in the OS or one of the programs running on a machine to install software on a machine without the user knowing. Users can be exposed to such attacks by clicking links in email or by visiting untrustworthy web sites.

In fact, users can even be infected with malware through ads on *trustworthy* web sites. This is called malvertising. Many web pages bring in code from third-party ad servers. Hackers can disguise their malicious code as ads and pay an ad server to host those ads. The unsuspecting ad server then injects the malicious code into many sites.

Malware: What It Does

You can also classify malware based on what it does. Early viruses typically did something fairly obvious to disrupt the computer. Some early viruses were simple pranks that displayed a message. Others were more malicious and did things like delete files, including system files necessary for the computer to function. Viruses can still do these things. However, now that hacking has become a sophisticated criminal activity, malware has also become more sophisticated and subtle. The following sections describe the categories of malware shown in Figure 1-9.

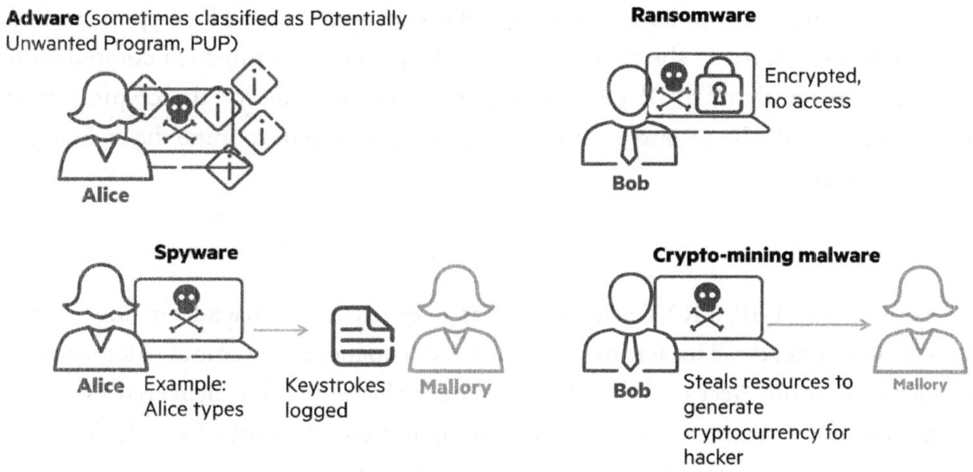

Figure 1-9: Malware classified by what it does

Adware

Adware, often included in Trojans, brings up unwanted ads whenever users open web browsers. Adware is sometimes classified as a Potentially Unwanted Program (PUP) rather than malware because it does not perform actively malicious actions. Some software is monetized through means of adware, which users accept as the price for the freeware.

Spyware

Spyware gathers information, often for use in identify theft or for gaining unauthorized access to other systems. For example, a key logger tracks inputs from the keyboard, typically in order to steal a user's passwords.

Ransomware

Ransomware typically locks a system in some way, often by encrypting all of the files. The hacker demands a payment in exchange for unlocking the system. Alternatively, the hacker might threaten to publish the user's data unless paid a ransom. While ransomware attacks occurred as early as 30 years ago, ransomware has catapulted to prominence in the last decade. Attacks against local governments have been particularly harmful and publicized.

Crypto-mining malware

Hackers might also install malware that steals a system's resources for the hacker's purposes. Crypto currencies have become more prominent in the last several years. These currencies are based on coins that are "mined" by systems running complex computations. The cryptocurrency miner gets to keep the coins that he or she mines. The time that it takes to run such computations limits the amount of currency that can be generated, preserving the value of the cryptocurrency. Crypto-mining malware steals the resources of the compromised system to mine the currency and give the yields to the hacker.

Remote Administration Tool/Remote Access Trojan (RAT) and rootkit

As shown in Figure 1-10, a RAT is malware that gives hackers a "backdoor" into a system to allow the hacker remote access. The acronym is sometimes expanded as Remote Administration Tool and sometimes as Remote Access Trojan because the malware is often delivered as a Trojan. With a RAT, the hacker can extract sensitive information and take control of the device. For example, a RAT can let the hacker search through the machine for cached usernames and passwords or cached credit card information. A rootkit is a particular type of RAT, which gives the hacker elevated, or root, level access to the compromised system's OS. A rootkit is also characterized by actively hiding its presence on the system.

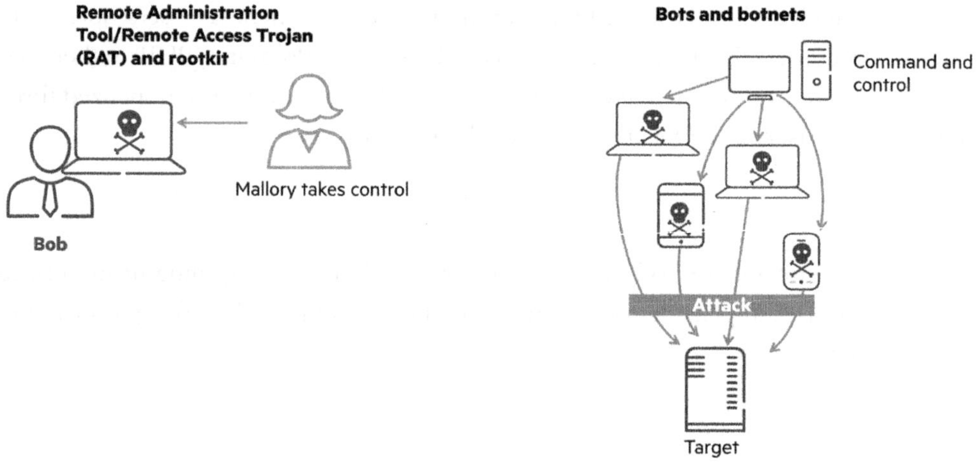

Figure 1-10: RAT, rootkits, bots, and botnets

Bots and Botnets

Malware such as RATs can hijack a device and transform the device into a "bot" in service of the hacker. The compromised device might continue to function more or less as expected so that users do not notice an issue. However, in the background, the device is acting as a bot, doing things like launching attacks on other systems.

The botnet might have a centralized architecture, in which bots check with a central server, called a Command and Control server, for their orders. Or a botnet can use a distributed architecture, in which server functions are distributed across the bots. In either case, the botnet is ultimately under the control of the hackers.

Denial of Service (DoS)

As shown in Figure 1-11, a Denial of Service (DoS) attack is any attack that prevents users from using the network or a network service. Many types of DoS attacks exist, and they can be executed in many ways.

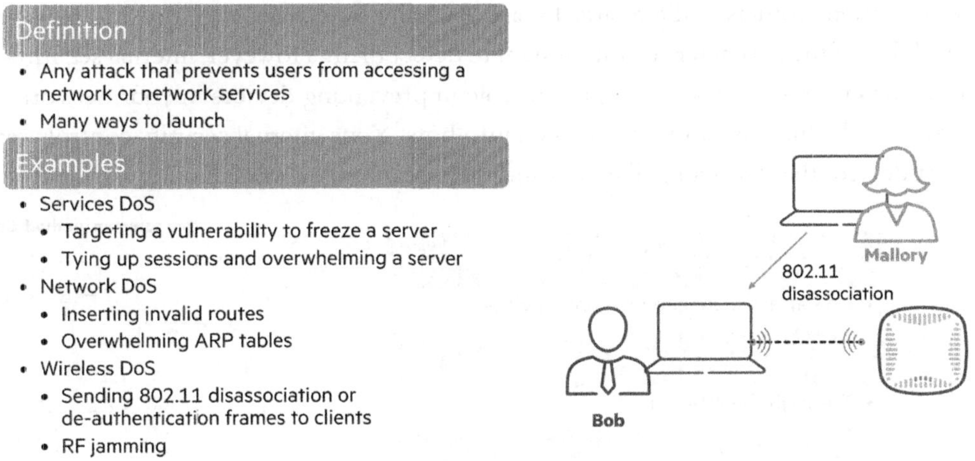

Figure 1-11: Denial of Service (DoS)

DoS attacks can target servers. For example, a hacker could exploit a vulnerability on a server. The resulting attack could make the server freeze, denying users access to the service provided by that server. DoS attacks often involve overwhelming a service so that legitimate users cannot access it. For example, hackers could open so many web sessions on a web server that the web server can no longer form sessions with legitimate users. Similarly, hackers could launch a DoS attack on a DNS server. A DNS server maps user-friendly domain addresses such as hpe.com to IP addresses. By overwhelming a DNS server with requests, hackers can prevent the server from responding to legitimate requests, which can interfere with network connectivity.

Network DoS attacks can target network infrastructure devices such as switches and routers. For example, a hacker could insert a router into a network that uses an insecure routing protocol and configure the router to advertise fake routes, which cause traffic to be misdirected or dropped. Or a hacker could attempt to overwhelm a router's Address Resolution Protocol (ARP) table, which can interfere with traffic delivery.

A wireless DoS attack could involve a hacker sending disassociation or deauthentication frames to clients so that they cannot stay connected to the legitimate access point (AP). In a radio frequency (RF) jamming attack, bad actors disrupt the wireless connections between legitimate users and an AP by drowning the signal out with noise.

These are just a few examples of DoS attacks; many more exist.

Distributed Denial of Service (DDoS)

As shown in Figure 1-12, a Distributed Denial of Service (DDoS) attack amplifies the effects of a DoS attack by using a large number of devices to launch the attack. The devices involved in a DDoS attack are often botnets. DDoS attacks are typically launched over the Internet, so it is the responsibility of the perimeter security system to detect them. However, internal security—the focus of this study guide—can play an important role in preventing devices in your network from being compromised by malware and transformed into bots. Your internal security controls can also help to detect devices that are acting like bots and then quarantine or rate limit those devices.

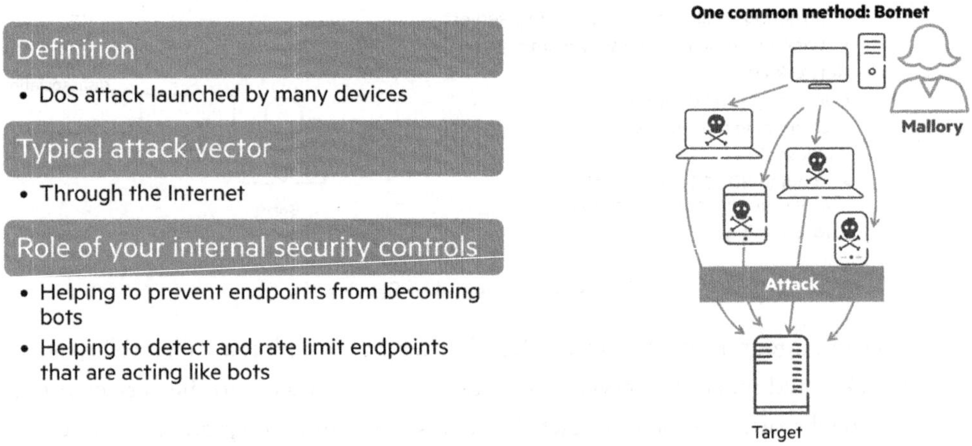

Figure 1-12: Distributed Denial of Service (DDoS)

Eavesdropping

Anytime that hackers obtain data in transit that is not intended for them, the hackers are eavesdropping. A shared medium—such as wireless—makes eavesdropping easy. It is trivial for hackers to configure a NIC in promiscuous mode, which means that the NIC will process packets destined to any MAC address, not just its own. Without security measures, hackers can collect any data sent over the wireless network.

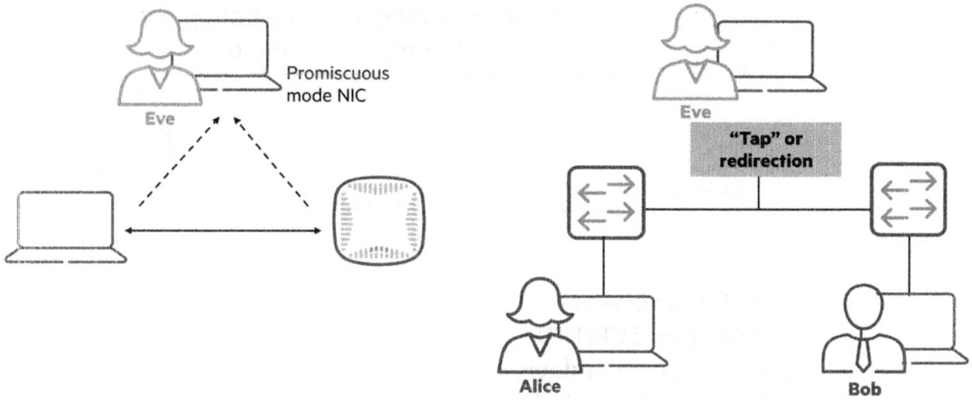

Figure 1-13: Eavesdropping

If hackers want to eavesdrop in a wired network, they have several options. If they have physical access to a network as shown in Figure 1-13, they can install a terminal access point (TAP) between devices for which they want to collect communications.

A network tap is a tool for collecting traffic at the physical layer. The tap has at least three ports. To insert the tap, the hacker—Eve, in the example shown in Figure 1-13—needs access to the physical cable. She can then interrupt the physical link and reconnect it through two ports in the tap. She connects the tap's third port, often called the monitor port, to her own device. She can then collect all traffic that passes over the link.

Eve might not have access to a switch-to-switch cable, but if she can connect to the network, she can collect traffic in other ways. She might try to flood switches' MAC address tables. This will cause the switches to treat all traffic as unknown unicasts and flood the traffic on all ports, including to Eve. Or if she can gain access to network infrastructure devices' management interfaces, she could configure traffic mirroring or other mechanisms to redirect traffic to her device.

Network Reconnaissance

If hackers can succeed in connecting to a network, they can conduct reconnaissance. Network reconnaissance is the process of gathering information about a target network. It can involve passive and active techniques.

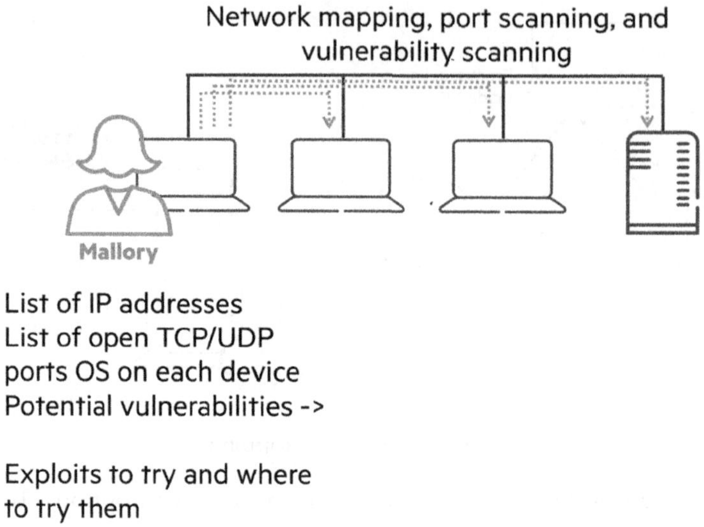

Figure 1-14: Network Reconnaissance

Passive reconnaissance involves eavesdropping on traffic and gathering information in that way.

As shown in Figure 1-14, active techniques include:

- Network mapping, in which hackers seek to discover the topology of the target network
- Port scanning, in which hackers search devices on the target network for open TCP or UDP ports
- Vulnerability scanning, in which the hackers search the target network for known vulnerabilities that the hackers can exploit

While performing active scanning exposes hackers to a higher chance of detection, such scanning yields rich context about the target network.

For example, Nmap, while a legitimate tool used by network admins, can also form part of a hacker's arsenal. Nmap sweeps a network to discover the IP addresses of all the devices on the network. It further probes each IP address with TCP and UDP messages to map out which ports are accepting traffic, or "open." Nmap also features OS fingerprinting capabilities. The program sends out TCP messages and analyzes the devices' responses. Because different operating systems respond in different ways and use different default options within TCP messages, Nmap can determine the OS running on discovered devices. Discovering a device's OS is called device profiling or endpoint identification. It can be a powerful tool for security solutions, but also for hackers, because it can help hackers discover targets for attack.

With this rich collection of data, hackers can use another tool that is valuable to both security teams and hackers: vulnerability scanners. Vulnerability scanners look for unpatched vulnerabilities in systems, which they can target with known exploits. For example, a scanner might attempt to establish an HTTPS session with the scanned device. HTTPS, as you will learn in more detail later, uses TLS, which can be very secure but can also use insecure options. A vulnerability scanner can detect if a system is using an outdated version or algorithm for which known attacks exist.

In short, network reconnaissance forms a critical part of hackers' tool set, enabling them to launch more attacks.

Man-in-the-Middle (MITM) Attack

Next you will examine man-in-the-middle (MITM) attacks. In an MITM attack, a hacker such as Mallory inserts herself between two communicating users (or more properly devices). The hacker can then eavesdrop on communications, or even modify the communications for her own purposes. These purposes might include intercepting credentials or personal information, or sabotaging communications.

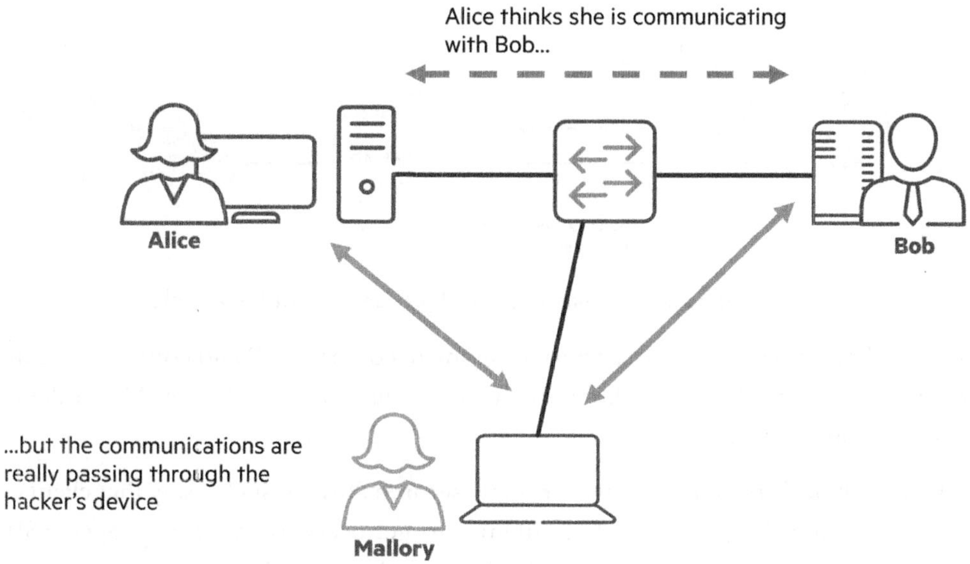

Figure 1-15: Man-in-the-Middle (MITM) Attack

An MITM is illustrated in Figure 1-15. As you see, Alice believes that she is communicating with Bob, but she is really communicating with Mallory. Similarly, Bob believes that he is communicating with Alice, but he is really communicating with Mallory. Mallory might simply eavesdrop on the communications and then pass them on unaltered. If Alice is telling Bob sensitive information such as her credit card number and expiration date (maybe Bob is an online vendor), Mallory has a lot to gain simply from seeing those communications. But Mallory might also alter the communications to suit her purposes, such as soliciting more information from Alice.

The protocols that carry communications can affect how damaging an MITM is. For example, if users are connecting to insecure HTTP websites, the hacker in an MITM can read all of the information that the users send to those websites. The hacker can also add code to the insecure site to infect the user's device with malware. HTTPS, the secure form of HTTP, can help to prevent such issues. However, if users, devices, or websites do not implement HTTPS properly, users can still be subject to such attacks.

Some Methods for Launching MITM Attacks

Hackers can implement wireless MITM attacks with honeypot APs. A honeypot AP lures users to connect to it rather than to a legitimate AP by sending disassociation frames to clients connected to the legitimate AP and by running at a high transmit power. Note that the "AP" might actually be a computer or laptop with a wireless NIC running in AP mode, as shown in Figure 1-16. Such a device is sometimes a software-based or "soft" AP.

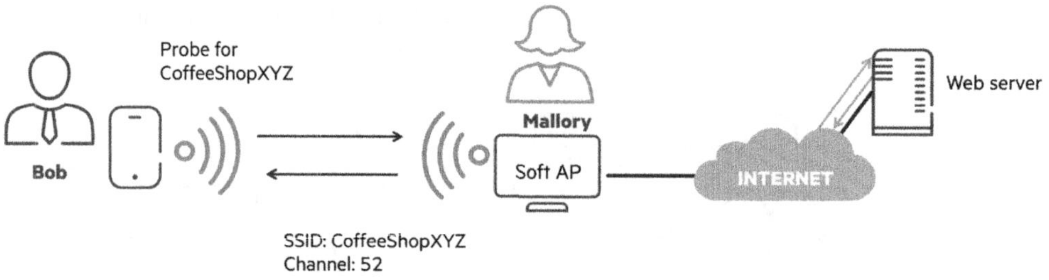

Figure 1-16: Some Methods for Launching MITM Attacks

Hackers might set up a malicious AP near a legitimate company AP and configure the malicious AP to broadcast the same SSID on a different channel. This type of malicious AP is called a honeypot or sometimes an evil twin.

However, it can be difficult to imitate a properly secured SSID in such a way that clients will actually connect to it. Therefore, hackers often attempt to get users to connect to open SSIDs with no security. When a client connects to an SSID, and the user chooses to connect to that SSID automatically, the client will probe for that SSID in the future. Often the client is looking for open SSIDs such as ones from airports, hotels, and coffee shops. The hacker's soft AP examines wireless traffic for the SSIDs for which wireless clients are probing. The hacker then has the honeypot AP broadcast these SSIDs so that wireless devices will connect to them automatically. Because the client is expecting open authentication it simply connects, and often the user notices nothing.

When users connect to the hacker's AP, the hacker can collect their traffic. The hacker typically connects the honeypot AP to a legitimate backend network, so the AP can pass users' traffic to its destination and implement the MITM attacks about which you just learned.

Address Resolution Protocol (ARP) Review

Hackers can also exploit Address Resolution Protocol (ARP) to implement MITM attacks in both wireless and wired settings.

To understand how, you must first review ARP. IP uses ARP to enable devices to resolve IP addresses with MAC addresses so that they can direct their traffic correctly within a wireless or Ethernet network. When a device wants to send traffic to a destination IP address in its subnet, it broadcasts an ARP request for that IP address. The device that owns the IP address sends a response, indicating its MAC address.

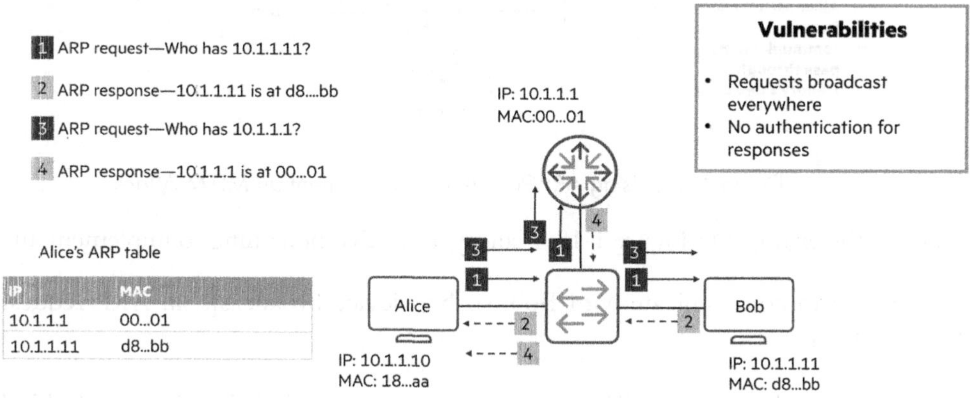

Figure 1-17: Address Resolution Protocol (ARP) Review

As shown in Figure 1-17, Alice's device requests the MAC address for 10.1.1.11, and Bob's device responds. Her device adds this entry to its ARP table, and it can now direct unicasts to Bob's device.

Similarly, if Alice's device needs to transmit routed traffic to another subnet, the device sends an ARP request to the default gateway, which is at 10.1.1.1 in the example shown in Figure 1-17. The router responds, and Alice's device adds this address to the ARP table.

However, ARP has no built-in security, which means that a hacker can just as easily respond to the ARP request as the legitimate device. In addition, most ARP devices update their tables every time they receive an ARP packet even if they did not request the information. These features make ARP vulnerable to attacks.

Using ARP Poisoning to Implement an MITM Attack

Attackers can exploit ARP to "poison" target devices' ARP tables. The attacker tricks the target devices into associating an invalid MAC address—often the hacker's own MAC address—with valid IP addresses. In this way, the hacker can intercept traffic and launch on MITM, as well as block traffic for a DoS attack.

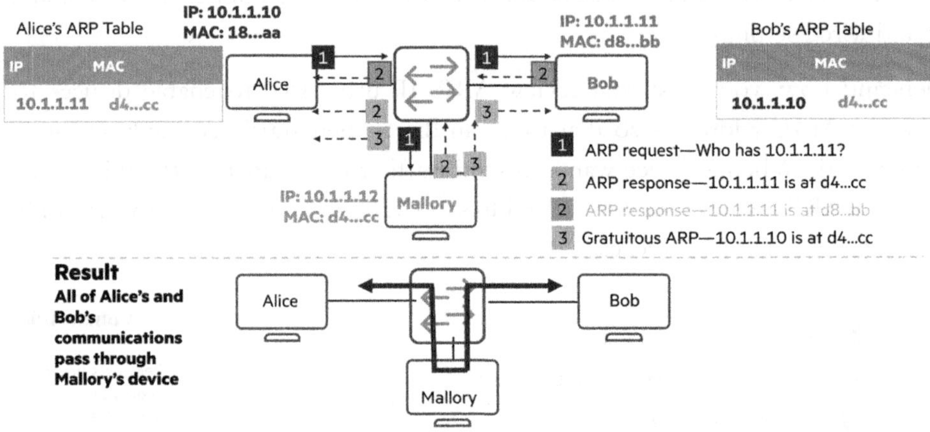

Figure 1-18: Using ARP Poisoning to Implement an MITM Attack

Now look at the example in Figure 1-18: Mallory uses ARP poisoning to implement an MITM:

- Alice wants to communicate with Bob, so her device broadcasts an ARP request for Bob's MAC address.
- Mallory responds with an ARP response that maps her MAC address to Bob's IP address. Because Bob's device would typically also respond to the ARP request, Mallory might also do things like send multiple ARP responses to Alice to override Bob's response.
- Mallory also sends a gratuitous ARP, which is an unrequested broadcast in which she maps her MAC address to Alice's IP address.

When Alice's and Bob's devices update their ARP tables with the spoofed entries, their ARP tables are considered "poisoned." Because the table maps Bob's IP address to Mallory's MAC address, her device sends all IP traffic intended for Bob to Mallory instead. And the same thing occurs when Bob wants to send traffic to Alice.

ARP poisoning only works against devices in the same subnet. However, Mallory can still use ARP poisoning to implement a more widespread MITM attack. As shown in Figure 1-19, she simply uses ARP poisoning to target the default router and one or more of the clients in the network. By implementing ARP poisoning between Alice and the default gateway, Mallory can intercept all traffic that Alice sends into the rest of the LAN and the Internet. This allows Mallory to do all the nefarious actions discussed earlier—eavesdrop on Alice's messages, collect sensitive data, and possibly even install malware.

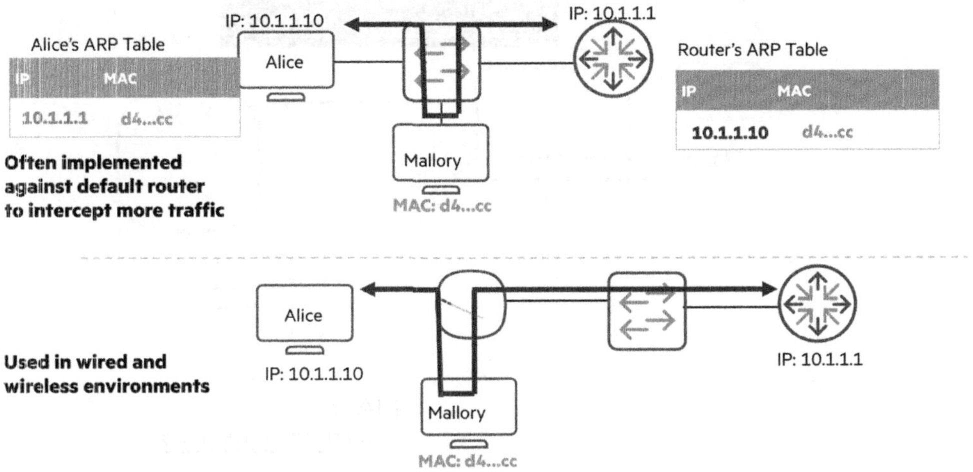

Figure 1-19: Using ARP Poisoning to Implement a More Widespread MITM Attack

These illustrations have shown ARP poisoning in a wired network for simplicity, but ARP poisoning also works in wireless networks. Unsecured wireless networks are most vulnerable to ARP poisoning, of course. However, a malicious authorized user could use ARP poisoning within a secure wireless network as well. This attack is particularly dangerous because, as you see, the traffic passes through the AP. That means that the AP decrypts the protected traffic for the malicious user. The user does need to have the credentials to connect successfully in the first place, though.

MAC and IP Spoofing

When hackers pose in some way as someone or something that they are not, they are "spoofing." Many types of spoofing exist, and hackers can use it as a tool for many purposes.

They can use spoofing for hiding themselves. For example, hackers can spoof the source IP address in their communications to misrepresent the source of an attack.

They can also use spoofing to gain unauthorized access. As shown in Figure 1-20, for example, some networks implement access control based on MAC address. Hackers can change their NIC's MAC address to spoof an authorized device's address to gain access to the network. Spoofing a MAC is trivial; on a Windows machine, for example, you can easily change a NIC's MAC address in the properties.

Hackers can also use spoofing to pose as a legitimate server or infrastructure device as part of eavesdropping or a MITM attack.

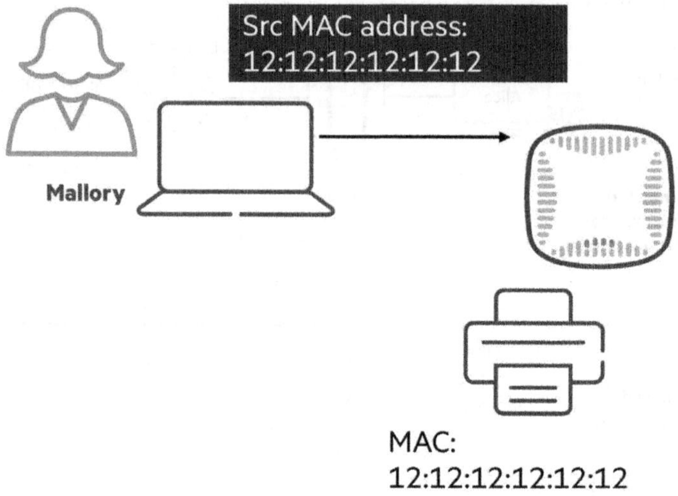

Figure 1-20: MAC and IP Spoofing

Email Spoofing and Phishing

Hackers can also spoof email addresses by specifying a false source email address in an email header. They can use email spoofing to pose as a legitimate institution, trick users into sending information to them, and stealing that information. This is called phishing.

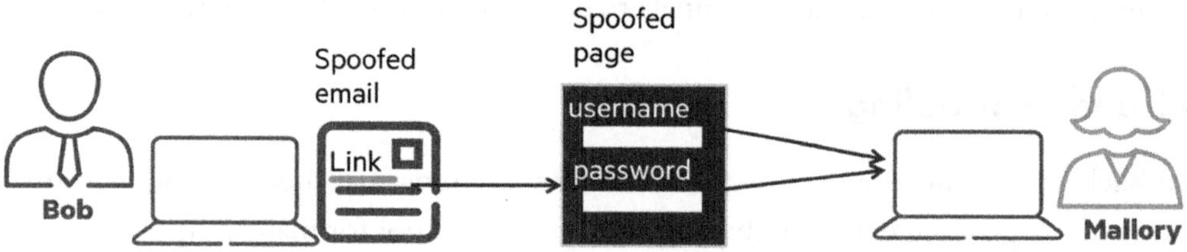

Figure 1-21: Email Spoofing and Phishing

As shown in Figure 1-21, for example, the hackers send an email that seems to come from a legitimate entity like the user's bank. The email tells users that they must do something like log in or reset their password and includes a link for doing so. The link goes to the hackers' site, which impersonates a legitimate login or reset site. But when users enter their credentials, the credentials go to the hacker.

This is just one example of phishing. Hackers might also "phish" for credit card and other sensitive information. Sometimes the links in phishing emails install malware.

Hackers usually send phishing attacks out to large numbers of targets, hoping that a small percentage of targets will fall victim. But hackers can also launch spear phishing attacks, which are phishing attacks that are targeted at a specific individual.

In the past, phishing messages were relatively easy to pick out, but hackers have become very sophisticated at making their phishing messages seem legitimate. Hackers can even apply data analytics and AI to the abundant information available about people on the Internet to make spear phishing attempts more sophisticated. For example, the email might seem to come from an actual business contact whose name the hackers have discovered through a networking site like LinkedIn. If the legitimate contact's email is alice@mycompany.com, the hacker might send an email from a very similar source such as alice@myconpany.com.

Social Engineering

As shown in Figure 1-22, social engineering refers to any form of attack in which hackers use humans to get around defenses. For example, a hacker could call an employee, posing as a member of the IT staff. The hacker might spin a story about problems with the user account and needing the user's password to fix them. Some helpful employees, unaware of the risks, could give up their passwords, allowing the hacker to obtain valid credentials without having to "crack" anything. Hackers can make similar attempts to discover other sensitive information like usernames, security practices, and personal information.

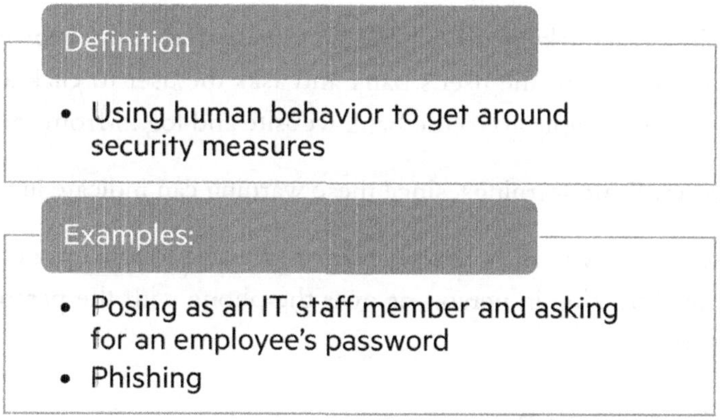

Figure 1-22: Social Engineering

Phishing, which you looked at earlier, has a social engineering component, as it relies on convincing users to click a link and perform an action.

Best Practices for Avoiding Social Engineering Attacks

The best defense against social engineering is good security training. Employees and other users should be trained in best practices. As shown in Figure 1-23, employees should never tell their password to anyone, particularly over a chat; IT staff should typically be able to help them without needing this password.

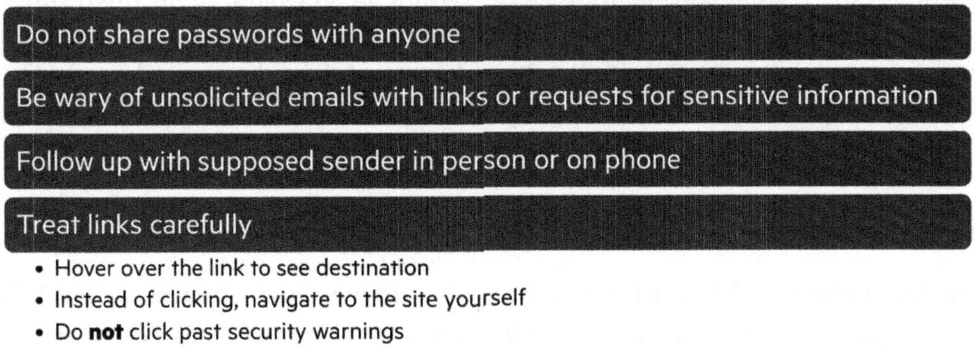

Figure 1-23: Best Practices for Avoiding Social Engineering Attacks

When receiving an email with a link or a request for sensitive information, users should take a few steps:

- Hover over the link to see where it is really going
- Whenever possible, avoid clicking links; instead, navigate to the site themselves; for example, if an email claims to come from the user's bank and asks the user to click a link to log into the account, the user should simply go to the bank website and log in from there
- Do not click past certificate warnings, since these warning can indicate invalid, phishing sites
- Contact the person from whom the email claims to be and follow up; never give up sensitive information without speaking in person or over the phone with the person who claims to be asking for the information

Additional Attacks Against Passwords

Hackers can also attempt to gain unauthorized access to a network, device, or service by attacking the passwords that protect them. Consider, for example, a hacker who has found the email address of a bank employee who has access to sensitive customer information. The hacker wants to hack into this user's email account. In a dictionary attack, the hacker tries a list of common passwords or even every word in the dictionary until one works. The list of passwords to try might also include words with common character replacements such as "@" for "a."

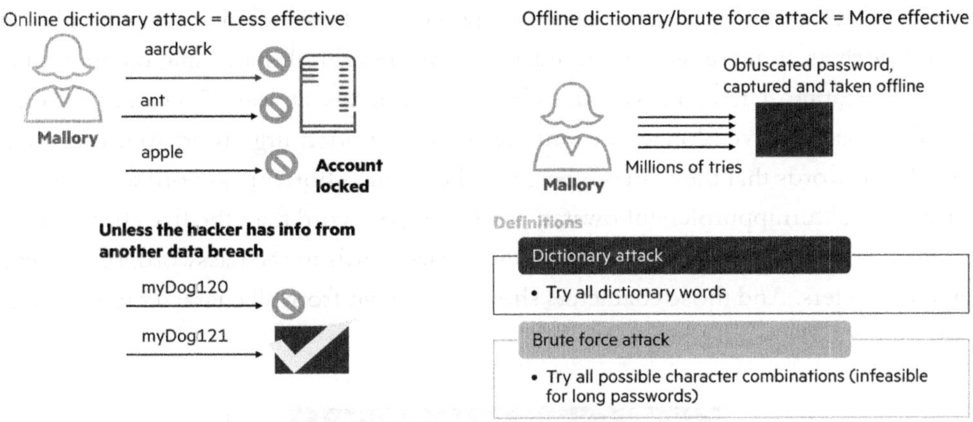

Figure 1-24: Additional Attacks Against Passwords

Hackers might also attempt a brute force attack, in which they simply try every combination of characters until they crack the password. Brute force attacks take longer than dictionary attacks because there are many more combinations to try than words in the dictionary. For longer passwords, too many combinations might exist for a brute force attack to succeed.

However, users often choose dictionary words for passwords, and the passwords are subject to simpler dictionary attacks.

As shown in Figure 1-24, such attacks are less effective if they are online—that is, if the hacker must run the guess against an online entity checking the password. For example, if the hacker must subject every try at a password to an email server, the account will lock, preventing the hacker from making further tries. Dictionary and brute force attacks become much more effective when hackers have data about the password that they can take offline. For example, the hacker might eavesdrop on the legitimate user's submission of the password. The password is almost always obfuscated in some way or another, but hackers can often crack such measures with modern computation power if the password has under 8 characters.

And often hackers do not have to try many words because they can use information from other data breaches to make educated guesses. In other words, if hackers know that a user has one account with password "myDog120," they can try "myDog121," "myDog122!" and so on until they guess the password.

How to Protect against Password-based Attacks?

You are no doubt very accustomed to creating passwords for new accounts. So what makes a good password? Often people think of using complex characters like a mix of upper and lower case and special characters. And certainly, this is a good practice. As shown in Figure 1-25, however, the most critical factor is simply never reusing a password. This includes using simple variations on passwords such as an incrementing number. Also more critical than using complex characters like # and ! is creating a long password. NIST recommends at least 8 characters, but preferably passwords should be even longer. Security experts now often urge users to create passwords that consist of four words that they can remember rather than a shorter password with special characters. In other words, "turnippurplerainbowstar" is a better password than the five character "&!Fa." That said, if it is feasible, it is best not to include dictionary words in the password, but instead a random string of characters. And those characters should be drawn from all character types, including upper and lower case and special characters.

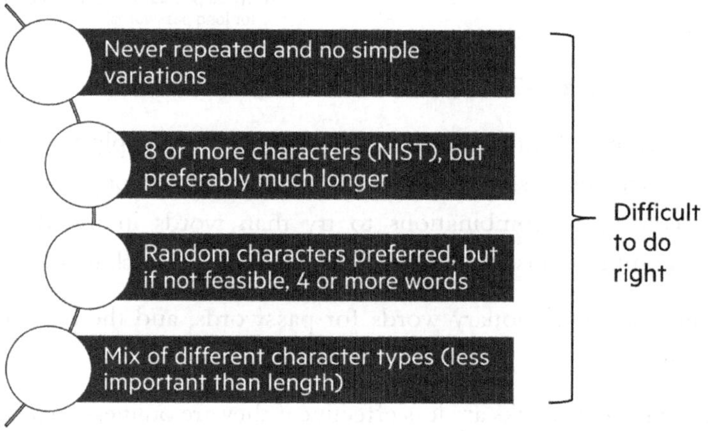

Figure 1-25: How to Protect against Password-based Attacks?

Password management solutions offer one way for users to achieve the goals of using secure passwords. A password management solution stores a unique password for all of a user's accounts. When a user wants to log into any account, the user only needs to remember a single password, which unlocks access to the password management solution. The solution then automatically fills in the correct password in the background. The advantage of such a solution is that users avoid repeating passwords and can use stronger passwords. However, the solution itself becomes a point of vulnerability. You must trust the security of the solution.

Best of all is when you can implement security while avoiding passwords. While things like email account passwords are out of your control, this study guide will teach you how you can achieve this goal within the network, by replacing passwords and shared secrets with something called digital certificates.

Attack Stages

You have looked at various types of bad actions like malware and eavesdropping in isolation, but it can be useful to consider how hackers can put techniques together. You will now focus on attack stages.

Kill Chain

Lockheed Martin originally introduced the concept of a "kill chain" for IT attacks, repurposing a US military concept related to determining the target of an attack and then executing the attack. According to Lockheed Martin, a kill chain consists of seven stages, terminating in Actions on Objective. The objective can be anything, including a denial of service (DoS) attack. But often hackers aim to exfiltrate valuable data such as credit card numbers and PINs, US social security numbers, or the equivalent in other regions. (Exfiltrate means to extract data to the hacker's device.)

Targeted Attacks versus Opportunistic Attacks

The Lockheed Martin kill chain model was designed to describe targeted attacks, in which a hacker or group of hackers are attempting to attack a specific high value organization. For example, hackers might attack a large bank or retailer in an attempt to steal account information for many users.

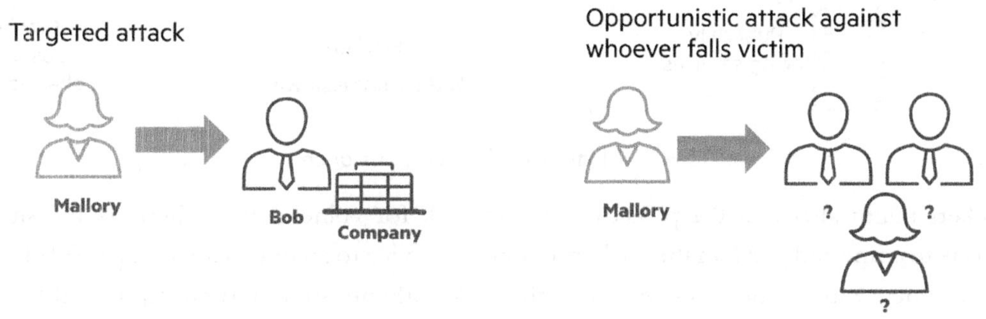

Figure 1-26: Targeted Attacks versus Opportunistic Attacks

As shown in Figure 1-26, however, you can also use some of the kill chain concepts to understand more opportunistic attacks in which hackers cast out wider nets, hoping that just a few users fall victim to the attack. The hacker can then search for anything of value on that user's system, such as credit card numbers.

Targeted Attacks

You will now look at the kill chain for a targeted attack in more detail.

Reconnaissance

In the first stage of the attack, the hackers select a target organization and perform reconnaissance. That is, they gather as much information as possible about the target in order to explore that target's defenses—and to find for vulnerabilities. In this example, the hackers want to gain access to account information stored by Bank XYZ. As shown in Figure 1-27, the hackers start to probe the perimeter defenses, searching for open ports and testing whether the company's public facing servers are unpatched against any known vulnerabilities.

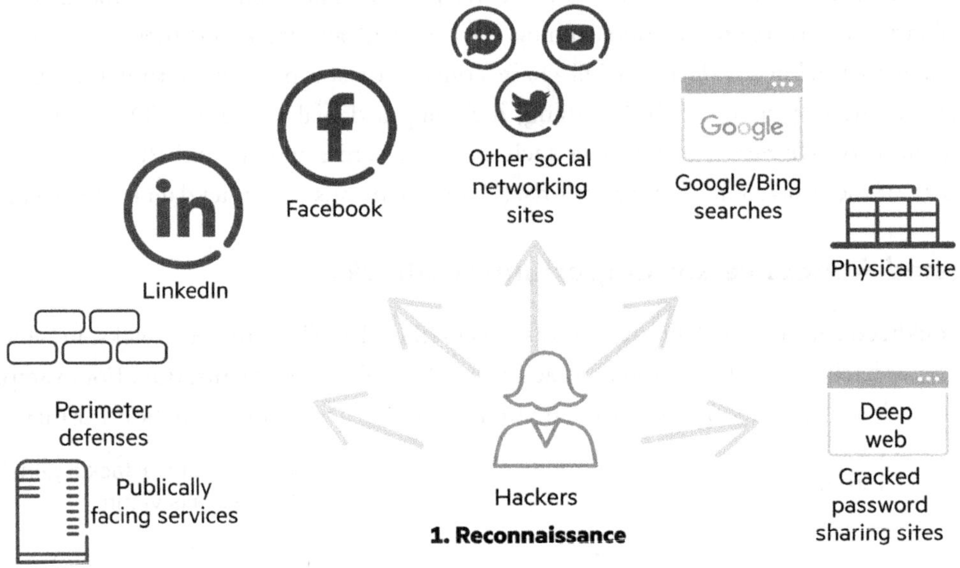

Figure 1-27: Reconnaissance

Hackers might also visit the physical site and look for vulnerabilities there. Is the site open for anyone to come and go? Can the hackers connect a device to an open Ethernet port? If the company has an open wireless network or one with weak authentication and encryption, the hackers will certainly note the vulnerability.

Reconnaissance can extend beyond the technical realm to the realm of social engineering.

The hackers search the company's website and Google for employee names. They use LinkedIn to find more names. They compile a list of information about their top targets, including email addresses, colleagues, and interests. They search social networking sites for the selected employees, collecting information that they can use to make a convincing social engineering attack. For example, say that the hackers have decided to target the CFO, Bob Brown. They find Bob's public profile in Facebook. They discover that he lives in Chicago and is a member of the Chicago Bears' Fan Club Facebook group.

Hackers share usernames and passwords that they have hacked at password sharing sites in the deep web. Because people often use their email accounts for their usernames, the hackers can easily look up the targets' email addresses in these sites. The compromised passwords are usually for accounts with online services, but users often reuse passwords or just make slight variations. Hackers can try the collected passwords and common variations to gain access to the users' business email accounts. They will also store the passwords for use once they have infiltrated the network and potentially have remote access to the users' computers.

If hackers believe that the reward is sufficient, they could follow this exhaustive process for multiple targets at the organization.

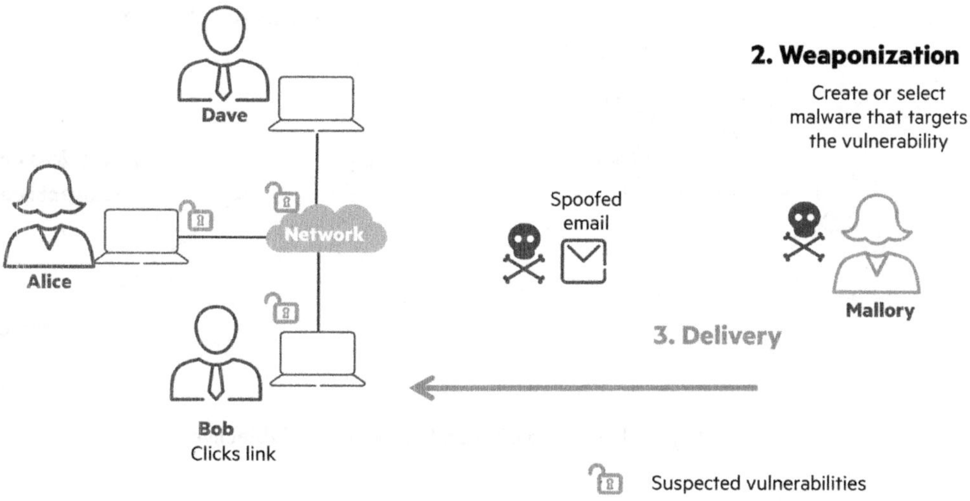

Figure 1-28: Weaponization and Delivery

Weaponization

In this stage, as shown in Figure 1-28, hackers create malware that exploits known vulnerabilities for systems and applications that they know or suspect that the targeted employees use. Often hackers can find readymade malware that exploits vulnerabilities in common operating systems and applications. While these vulnerabilities might have patches, the hacker only needs to find one unpatched system to get a foot in the door.

Delivery

The hacker delivers the malware to the targeted network, as shown in Figure 1-28. You learned about ways that hackers can install malware on devices earlier. Some of these ways include sending malware in emails that entice users to click on them, sending infected USBs to users, using Trojans, and executing "drive-by-installations." In our example, the hackers could use the information collected during reconnaissance to craft a social engineering attack. They send an email to Bob Brown posing as the president of the Chicago Bears Fan Club, promising a chance to win tickets to a game if Bob clicks on the link. Bob might or might not click the link—but the hackers have likely created many such emails targeting Bob and other employees. It takes just one click to give the hackers their chance to deliver the malware.

With the malware delivered, the hacker can execute the next stages of the attack, as shown in Figure 1-29.

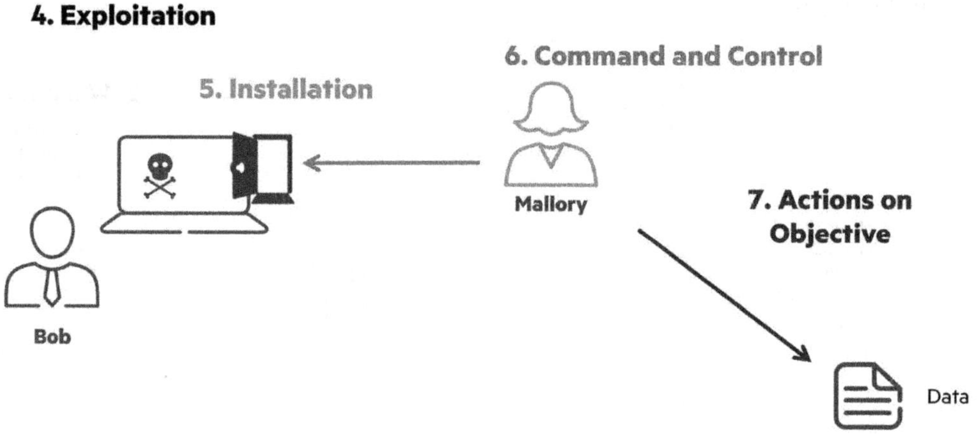

Figure 1-29: Exploitation to Actions on Objective

Exploitation

The malware executes and exploits the vulnerability. Often what this means in practice is that the malware gains a high enough level of access to the system that it can install more malware.

Installation

The malware creates a backdoor into the system through which the hacker can access the system.

Command and Control

Using the backdoor, hackers essentially take over the malware-infected device. They now have "hands on the keyboard," or in other words remote access high enough to move through the OS, start collecting data, and, if necessary, escalate their privileges on the system even more.

Actions on Objective

With this level of control over the system, the hackers can achieve their objectives. In this example, they can exfiltrate the desired data and also cover up their tracks by deleting all logs of their behavior.

Opportunistic Attack Chains

Now look at an opportunistic attack. In this case, hackers are not really targeting anyone in specific, but waiting until someone falls victim to one of the many traps that they have set. As shown in Figure 1-30, they might send out mass emails, embed malware in ad servers, or hide malware in Trojans. The ad server might interact with millions of devices and find just a fraction of devices with a targeted vulnerability on which it can install. Similarly, only a fraction of those exposed to a Trojan will choose to download it on a device with the vulnerability that the Trojan targets. But with so many attempts, the malware finds victims. Once the malware installs on a device, it opens a backdoor and "phones home" to the hacker. The hacker can now investigate the device and decide what data the device has to exploit.

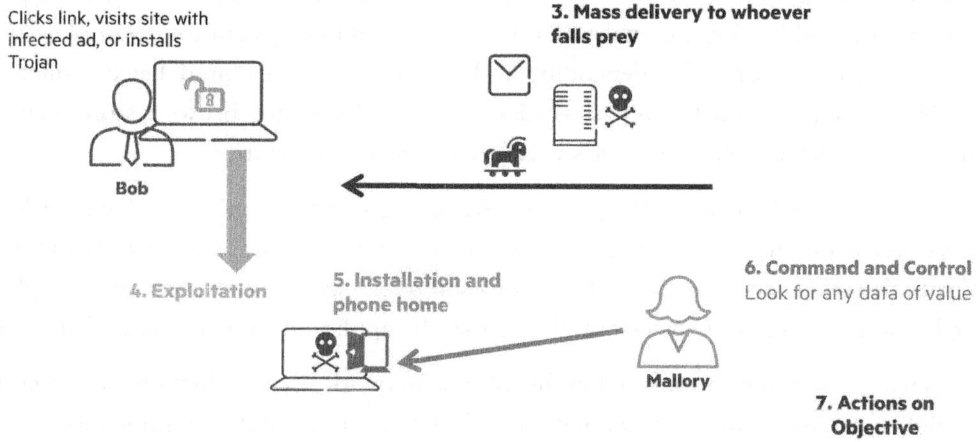

Figure 1-30: Opportunistic Attack Chains

Lateral Movement

Whether hackers are implementing a finely targeted or more scattershot attack, they often target *many* users and devices in the delivery stage of an attack. The system to which the hackers first gain access is usually one of the *less* secure systems, which often means a less valuable target. For this reason, hackers often try to use the access that they gained on one device in a network to gain access to other devices in the network. As shown in Figure 1-31, this is called lateral movement.

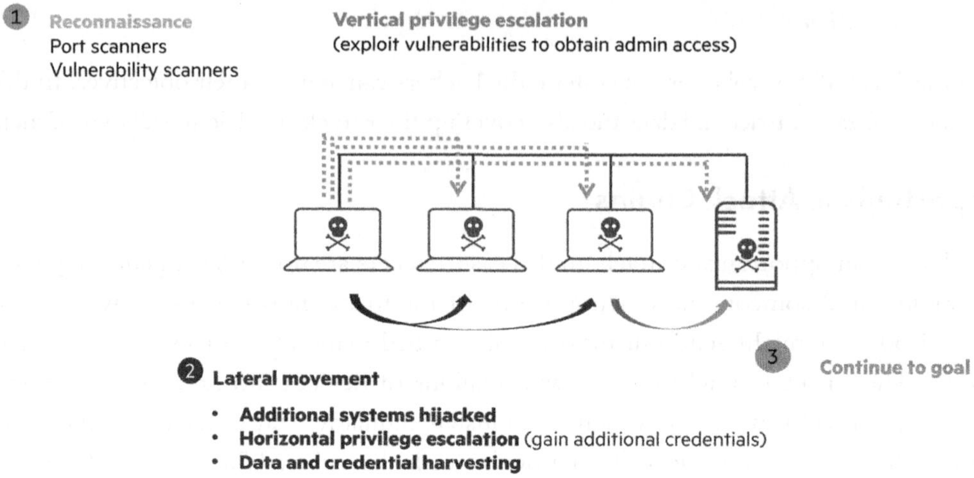

Figure 1-31: Lateral Movement

Privilege escalation often forms a key step in lateral movement. Hackers can implement horizontal privilege escalation, in which they harvest credentials of additional user accounts. For example, the hacker might install key logging malware on the compromised system to capture keystrokes when users enter their passwords. Or, depending on how passwords are stored, hackers might be able to run offline brute force attacks on passwords. Or if users have their browser store their passwords, the hackers only have to crack the password used to secure the store.

Hackers can also implement vertical privilege escalation, also called privilege elevation. With vertical privilege escalation, the hackers attempt to exploit the access that they already have to obtain deeper, administrative-level access to the system. For example, a hacker might be able to exploit a bug in an application or OS to obtain root privileges that the application was not intended to be granted.

Hackers who have gained a foothold in the internal network also conduct network reconnaissance. With tools like Nmap, they map the network, find open ports, and determine discovered devices' operating systems. They can use the vulnerability scanners that security teams use to detect issues for quite another purpose, looking for vulnerabilities to exploit. The hackers can then attempt to gain access to other devices, install malware on them, and open up a backdoor into those systems. Or the hacker can try to use passwords and keys culled from already compromised systems to gain access to other systems without having to first infect them with malware. As the hackers move to more systems, they continue to escalate their privileges, harvest more data, and compromise more devices and user accounts. They continue this lateral movement until they have achieved their goals.

Need for Zero Trust

The classic kill chain model focuses on outside hackers who seek to gain access to devices within a corporate network. However, many security advisors agree that this model no longer holds entirely true. Many threats involve internal attackers. In addition, as you just saw, even if the attacker is external, at some point that attacker could gain internal access and start moving laterally. It is important that the internal network have protections against exploitation.

When companies consider the internal network trusted and only external networks, such as the Internet, as untrusted, they focus on protecting the perimeter at which the external network connects to the internal network. Perimeter protection is still important. But it is not the only form of protection required.

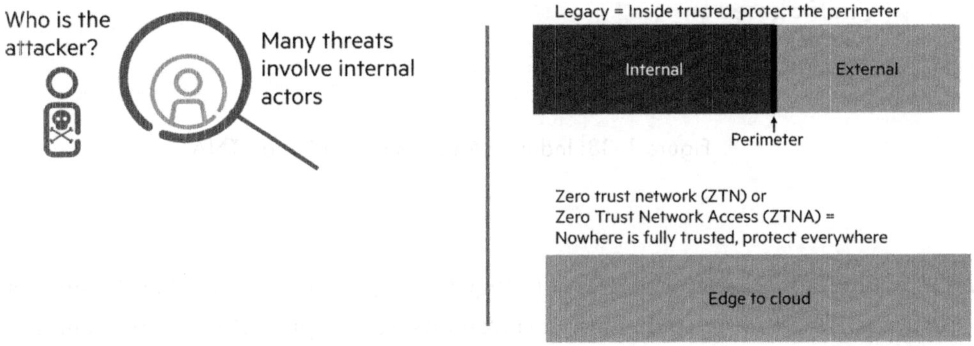

Figure 1-32: Need for Zero Trust

Many security analysts now argue that companies should consider *both* the internal and external network as untrusted. As shown in Figure 1-32, this approach is called a zero trust network (ZTN) or zero trust network access (ZTNA). In ZTN, the assumption is that the internal network could already be compromised and the goal is for the network to remain secure even if that is true. To achieve this goal, companies must take steps to secure *all* communications from the internal edge to the cloud.

The next section of this chapter will explain how the HPE Aruba Networking security strategy for a zero trust world.

HPE Aruba Networking Security Strategy

HPE Aruba Networking has the solutions to support customers in a world in which no part of the network can be fully trusted. This section covers this strategy at a high level and explains which components of the strategy you will learn to implement in this study guide.

Industry Analysts on ZTN and ZTNA

Before you dive into the HPE Aruba Networking security strategy, you will look at what the industry is saying about ZTN.

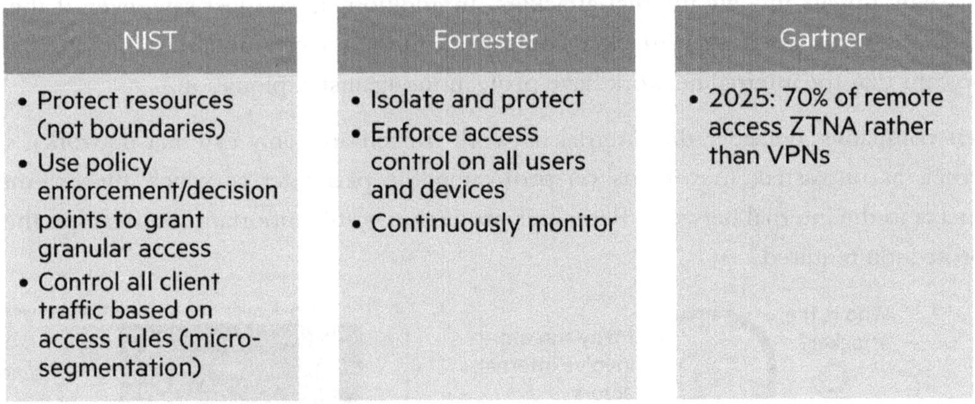

Figure 1-33: Industry Analysts on ZTN and ZTNA

NIST

National Institute of Standards and Technology (NIST) is a US agency that defines many security best practices. NIST is encouraging organizations to adopt a ZTN approach. Its zero trust architecture (ZTA) provides a broad framework in which many technologies can fit. As shown in Figure 1-33, a key feature of the ZTA is a focus on protecting resources rather than protecting boundaries. Policy enforcement and decision points identify each user and device on the network and grants that user or device only the level of access to resources required for that particular entity to do his, her, or its job. This approach is sometimes called micro-segmentation. Rather than control traffic as it passes from one large segment, such as the external network, to another large segment, such as the internal network, micro-segmentation controls all traffic between all devices, applying granular controls based on identity and device type.

For more information, see https://nvlpubs.nist.gov/nistpubs/SpecialPublications/NIST.SP.800-207-draft2.pdf.

Forrester

Forrester, a prominent industry analyst, is similarly guiding organizations towards a ZTN approach. Forrester emphasizes that in a ZTN you cannot trust any data, people, devices, or workloads absolutely. Instead, you must monitor all users, devices, and applications and take appropriate steps to control how they use the network. In this effort, visibility, analytics, and orchestration across multiple security components will be key.

As shown in Figure 1-33, Forrester recommends that customers protect their network by identifying sensitive data that needs to be isolated and protected. To achieve this protection, the company should implement micro-segmentation by applying access controls to *all* clients. Remember: no users are absolutely trusted. Companies must define policies to dictate which authenticated users are allowed access which data. And these policies must apply to users no matter how and where the users access the network, although customers might choose to use this contextual information to adjust the access policies to the situation.

Finally, Forrester recommends that customers continuously monitor and analyze the environment to discover threats and address security incidents in real-time.

(The information about Forrester's ZTN approach is taken from "Five Steps To A Zero Trust Network," Forrester, Stephanie Balaouras, Chase Cunningham, and Peter Cerrato, October 2018.)

Gartner

Organizations are recognizing that ZTN is the future and searching for ways to achieve ZTN. Gartner predicts that by 2025, 70% of remote access will use ZTNA rather than VPNs ("Gartner Identifies Three Factors Influencing Growth in Security Spending," Gartner Press Release, October 13, 2022).

HPE Aruba Networking Security Solutions

To help organizations achieve a ZTN, HPE Aruba Networking offers a framework of security solutions. This comprehensive, integrated framework addresses the unique security needs of mobility, IoT, and cloud to help customers successfully combat new cyber threats and empower their network and security teams.

Create a Trusted Network Infrastructure

HPE Aruba Networking helps customers to reduce their network's vulnerability all the way from the edge. As shown in Figure 1-34, the foundation of the secure network is a trusted network infrastructure.

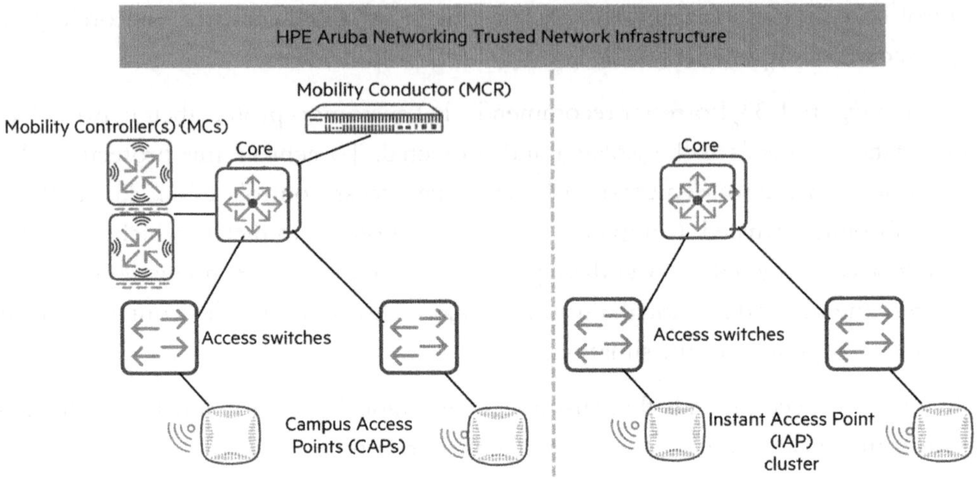

Figure 1-34: Create a Trusted Network Infrastructure

For enterprises, this infrastructure typically consists of HPE Aruba Networking switches at the network core and the access layer. Those switches are typically HPE Aruba Networking CX (AOS-CX) switches. For the mobility solution, the infrastructure includes Campus APs (CAPs), which are controlled by one or more Mobility Controllers (MCs). In other words, the MCs control the CAPs' configuration settings. The CAPs also typically tunnel wireless client traffic to the MCs for application of policies and forwarding.

If the company requires multiple MCs, HPE Aruba Networking typically recommends deploying them in an architecture controlled by a Mobility Conductor (MCR).

Smaller or mid-sized companies might use an Instant AP (IAP)-based architecture instead. IAPs form a cluster that is simple to manage from a single point but does not require a controller. IAPs support many of the same security features as MCs, including the HPE Aruba Networking Wireless Operating System (AOS) firewall. Note that in 2017 HPE Aruba Networking began delivering APs as Unified APs (UAPs), which customers can deploy as either CAPs or IAPs, as they choose.

Throughout this study guide, you will learn how to harden HPE Aruba Networking infrastructure devices against unauthorized access by hackers. You will also learn how the devices can create a secure environment for communications with technologies such as encryption.

Note that this study guide focuses on AOS software version 8.6 and AOS-CX software version 10.10. The study guide occasionally points out AOS-CX features available in version 10.11 or 10.12 but does not focus on those features.

Enforce Micro-Segmentation

HPE Aruba Networking solutions build on the foundation of the trusted network infrastructure with the capability to enforce the micro-segmentation and access controls that are so critical to ZTN. As shown in Figure 1-35, an HPE Aruba Networking infrastructure can identify and begin controlling users' and devices' access from the moment that they connect. HPE Aruba Networking ClearPass Policy Manager (CPPM), working together with the HPE Aruba Networking infrastructure, identifies every user and device as it connects to the network. It can use advanced authentication and device profiling technologies to ensure that users are who they claim to be, as well as determine exactly what type of devices are connecting from laptops to smartphones to Internet of Things (IoT) devices.

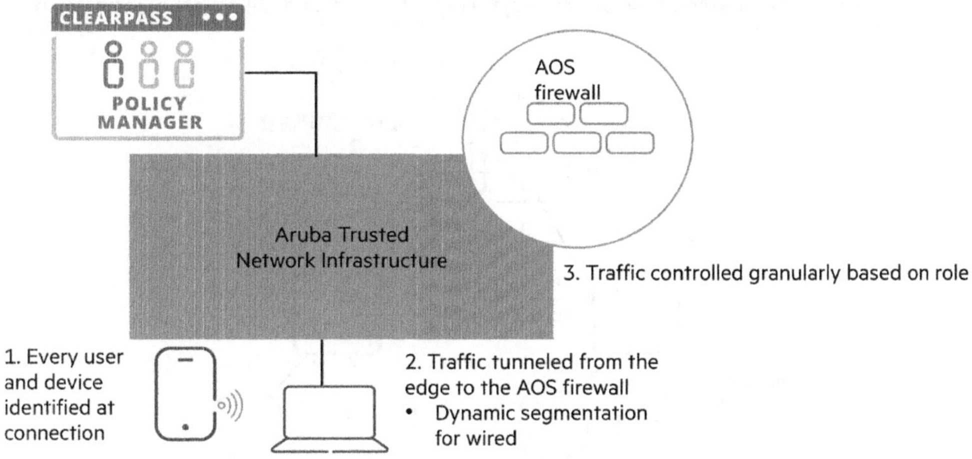

Figure 1-35: Enforce Micro-Segmentation

CPPM uses this information, as well as other optional contextual information such as location, to assign the client to a role, which it communicates to the HPE Aruba Networking infrastructure. The infrastructure can then use that role to apply role and context-based access control policies.

To enforce micro-segmentation of traffic from both wireless and wired users, HPE Aruba Networking infrastructure devices support a technology called dynamic segmentation. Both HPE Aruba Networking APs and AOS-CX switches can tunnel client traffic to MCs, where the AOS firewall implements role-based access control policies before users have the chance to access any sensitive data. The firewall can examine traffic all the way to layer 7 with deep-packet inspection (DPI) and control the traffic accordingly.

If you are not sure exactly what all of these technologies mean, do not worry. You will learn about all of them throughout this study guide.

Log and Inspect within the Trusted Network Infrastructure

To combat modern threats, companies cannot simply permit or deny users access and then stop their security measures there. Network infrastructure solutions must inspect users and devices on an ongoing basis and integrate with other security solutions to mitigate threats. HPE Aruba Networking solutions support a wireless intrusion detection system (WIDS), which detects threats on an ongoing basis. (See the air monitors [AMs] in Figure 1-36, which implement WIDS.) You can also implement logging and auditing to support ongoing network monitoring. You will learn about these features throughout this study guide as well.

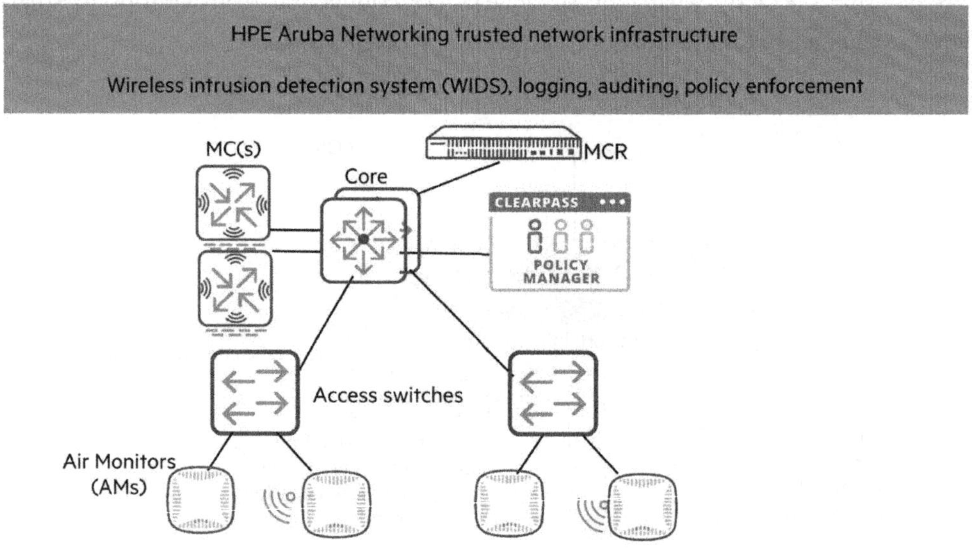

Figure 1-36: Log and Inspect within the Trusted Network Infrastructure

Continuously Monitor, Control, and Orchestrate with the HPE Aruba Networking

HPE Aruba Networking extends the power of its solutions by integrating with a broad range of third-party solutions in the HPE Aruba Networking 360 Security Exchange. These solutions include:

- Next Generation Firewalls, which filter traffic and apply access control policies, as well as sometimes scan traffic for signs of intrusions or attacks
- Endpoint Detection and Response (EDR) systems, which scan devices for issues such as malware

- Enterprise Mobility Management (EMM) solutions, which many companies use to monitor and manage their mobile devices
- Intrusion Detection Systems/Intrusion Prevention Systems (IDS/IPS), which analyze network traffic for indications of malware and attacks and, in the case of an IPS, provide mitigation

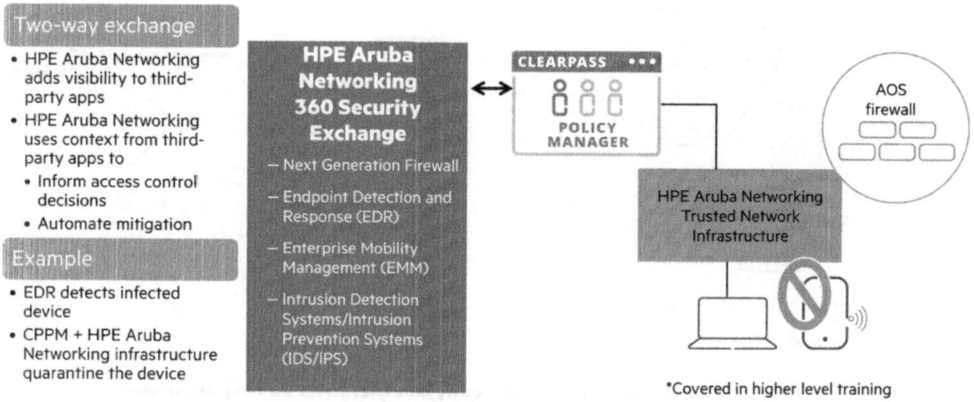

Figure 1-37: Continuously Monitor, Control, and Orchestrate with the HPE Aruba Networking

By integrating with these solutions, CPPM can add visibility to the third-party apps. As shown in Figure 1-37, for example, CPPM can inform a third-party Next Generation firewall of the usernames associated with IP addresses, allowing the firewall to apply more intelligent policies.

CPPM can also accept information from third-party apps to receive the context that it needs to enforce even more sophisticated access control policies. And by working together with EDRs, EMMs, IDS/IPS, and other Security Exchange solutions, CPPM and an HPE Aruba Networking infrastructure can provide the automation and orchestration necessary to detect and mitigate threats more quickly. For example, an EDR system discovers that a device is infected with malware. On its own the EDR can only inform admins of the issue. But CPPM can help the EDR system take action by denying the infected device network access or placing it in a quarantine VLAN.

Similarly, if an IDS/IPS informs CPPM that an endpoint is launching an attack, CPPM can tell HPE Aruba Networking infrastructure devices to lock that endpoint out of the network.

CPPM and the HPE Aruba Networking Security Exchange also help to improve the user experience. For example, if CPPM detects that a user is having issues authenticating, it can automatically log a ticket with help desk software.

You can learn more about the Security Exchange in higher level training in this track.

Implement Same Security Everywhere

Enterprises can implement the security capabilities that you have examined at a main campus. As shown in Figure 1-38, they can also extend them across every branch in a larger solution.

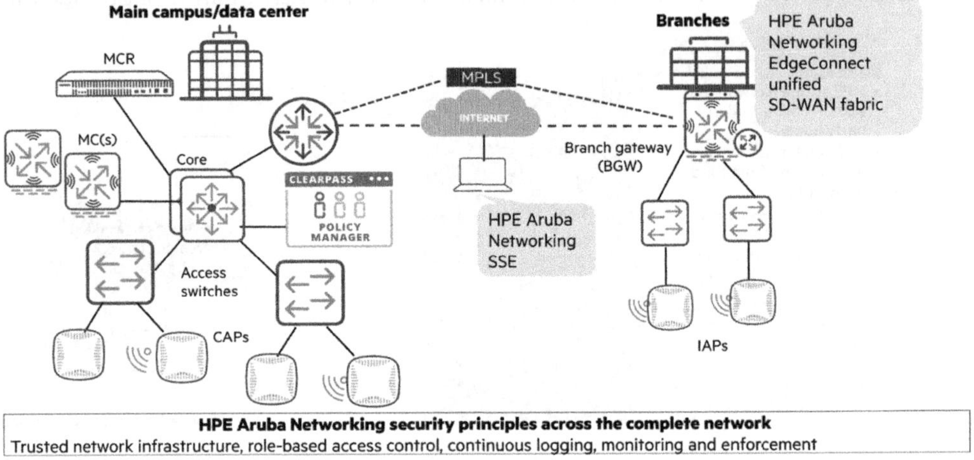

Figure 1-38: Implement Same Security Everywhere

The HPE Aruba Networking EdgeConnect portfolio includes several options for secure software-defined WAN (SD-WAN) solutions, which can micro-segment branch user and device traffic to ensure consistent security, visibility, and experiences across all locations.

You will learn more about the HPE Aruba Networking EdgeConnect unified SD-WAN fabric later in this study guide.

You will also learn how HPE Aruba Networking SSE helps to protect remote access.

Summary

Congratulations! You have completed the first chapter.

This chapter has armed you with the background information that you need to understand how cyberattacks are launched and how HPE Aruba Networking seeks to secure networks against them. In the next chapter, you will learn more about technologies for securing a network.

Learning checks

1. What are ways that you can protect yourself from phishing attempts? (This is an open-ended question.)
2. What is a primary security risk introduced by the fact that ARP has no built-in authentication and integrity mechanisms?
   a. Hackers can add malicious code to ARP messages to install viruses on user machines.
   b. Hackers can send malformed ARP messages that cause user devices to blue screen.
   c. Hackers can poison devices' ARP tables in order to implement MITM attacks.
   d. Hackers can use ARP to change the MAC address on their machines and gain unauthorized network access.
3. What is the correct order for the chain? (Put the steps below in order.)

   __ Delivery

   __ Installation

   __ Reconnaissance

   __ Exploitation

   __ Weaponization

   __ Actions on Objective

   __ Command and Control
4. What is one way an HPE Aruba Networking solution supports micro-segmentation?
   a. It replaces third-party security solutions with a single network-based solution.
   b. It installs a host-based firewall on every device that connects to the network.
   c. It applies granular access controls as external traffic flows into the internal network.
   d. It identifies users and devices at network connection and tunnels their traffic to a firewall that implements role-based access control.

Note

For answers to all learning checks, refer to Appendix B.

---

2 Security Technologies

EXAM OBJECTIVES

✓ Explain common compliance and regulatory requirements

✓ Explain common security protocols and their use cases

✓ Describe Public Key Infrastructure (PKI) components

Assumed Knowledge

- Basics of IP networking
- Basics of Ethernet switching
- Basics of 802.11 wireless networks

Overview

The first part of this chapter presents some examples of regulations that dictate security measures that organizations must take to protect certain types of data. Next you will dive into technologies for securing network communications, including data encryption and authentication. You will learn about the role that certificates can play in establishing trusted communications. Finally, you will look at Transport Layer Security (TLS) and then the authentication, authorization, and accounting (AAA) framework for controlling user access to services.

Regulatory Compliance

Securing the network is not only good for the business; it is also a legal necessity in many cases. Many countries and regions have passed regulations controlling how companies store and transmit various types of data.

Companies must implement specific security measures to comply with these regulations. They must also have the records required to prove to auditors that they comply. This study guide cannot teach you the specifics of proving compliance with specific government standards—which vary from region to region. However, it will introduce you to the types of regulations with which companies might need to comply.

Regulatory Requirements

The table in Figure 2-1 lists some of the most important regulations in several regions. These regulations, including the Privacy Act, Safe Harbor At, PIPEDA, APPI, and GDPR, generally target how companies protect people's personal data. They require organizations to collect and store this information in ways that protect that data from misuse and exposure. GDPR defines personal information as "anything from a name, a photo, an email address, bank details, posts on social networking websites, medical information, or a computer's IP address."

Region	Data protection regulations
US	Privacy Act Safe Harbor Act Health Insurance Portability and Accountability Act (HIPAA) Health Information Technology for Economic and Clinical Health (HITECH) Act Federal Information Processing Standards (FIPS) 140-2 (140-3 is emerging)
Canada	Personal Information Protection and Electronic Documents Act (PIPEDA)
European Union (EU)	General Data Protection Regulation (GDPR)
Japan	Act on the Protection of Personal Information (APPI)
Multiple regions	Payment Card Industry (PCI) Data Security Standard (DSS) 3.2

Applies to all companies that process, store, or manage data about EU citizens

Figure 2-1: Regulatory Requirements

You should be aware of a few additional implications of GDPR, which can even apply to companies based outside of the EU because it applies to all companies that process, store, or manage information about EU citizens. GDPR prohibits companies that offer public access from requiring a user to submit personal information such as an email address or phone number in order to gain access. In addition, GDPR requires users to explicitly consent to their personal information being processed. GDPR also includes the right to be forgotten. To comply with GDPR, customers may need to adjust the clean-up intervals in any network infrastructure solutions that collect data about users, among other measures.

In the US, HIPAA and HITECH specifically regulate how companies must protect personal healthcare-related information.

FIPS is a US regulation that applies to non-military federal government agencies, as well as any companies with contracts with the government. FIPS encompasses a broad range of standards related to how the agencies and contractors handle data. The 140 Series within FIPS specifically regulates the cryptographic standards that these agencies and contractors must use to protect sensitive data. FIPS 140-3 was released on September 22, 2019, and testing for FIPS 140-3 certified hardware begins in September 2020. However, FIPS 140-2 and 140-3 will overlap for several years.

Any company that handles credit card data must comply with the Payment Card Industry (PCI) Data Security Standard (DSS) 3.2, which is discussed in more depth in a moment.

Complying with legal regulations, such as the ones listed in Figure 2-1, not only protects the company from fines, but it also helps the company avoid the risks of data breaches. These risks include a damaged reputation, which can lead to real revenue losses. Lost intellectual property can also lead to revenue losses. And downtime from denial of service (DoS) or ransomware attacks can also affect the company's bottom line. In short, complying with regulations makes good business sense.

PCI DSS 3.2 Requirements

Any organization that transmits or stores credit cardholder data must comply with PCI DSS 3.2. PCI DSS compliance comprises 12 requirements and covers the complete cardholder data environment (CDE), which includes anywhere that cardholder or account data is stored, processed, or transmitted. Cardholder data is defined as the primary account number (PAN), cardholder name, card expiration date, and service code. Account data includes the chip or magnetic stripe data, card validation value, and personal identification number (PIN).

ID	Description
11.1	Detect and respond to rogue APs
1.1.2	Diagram the network
1.2.3	Firewall the network
9.1.3	Impose physical security
2.1.1	Don't use defaults
2.2	Implement config standards
4.1	Use best practice encryption
6.1 and 6.2	Rank vulnerability and get latest patches
7.1 and 7.2	Enforce role-based access control
10.1 and 10.3	Audit and record user access to system
11.4	Use IDS/IPS
12.3	Define proper usage of technologies

PCI DSS certification = A complex, formal process

Figure 2-2: PCI DSS 3.2 Requirements

PCI DSS comprises many requirements; the table in Figure 2-2 focuses on the requirements that are most relevant to the networking solution that transmits cardholder data. You will learn about meeting many of these requirements throughout this study guide. However, you should be aware that achieving PCI DSS certification is a formal process, which involves many steps. If your company needs to achieve this certification, it will typically have a team that works on ensuring compliance and documenting the measures that the company has taken. You would take your instructions from this team.

Understand Your Company's Policies

You will need to work with your company's legal and security teams to determine which regulations apply to your company and what role the network plays in the overall security strategy. Companies often have their own security policies, which are formulated to meet or exceed the requirements of the government regulations that apply to them.

In order to comply with regulations, you not only need to implement best security practices, but you also need to create an audit trail documenting how you have implemented the regulations. For example, in addition to establishing firewall policies to control users' access, you will need to document those policies. And you should also have ways to track and log violations of the policies. While this study guide is not intended to make you an expert in compliance auditing, it will give you the tools that you need to help companies comply with regulations.

Secure Communications: Symmetric Encryption and Hash-Based Authentication

You will now learn about some general security concepts, beginning with methods for ensuring secure communications. In this section, you will concentrate on how technologies called "encryption" and "hashing" can help to protect data in transit.

Goals of Secure Communications

Data in transit over an insecure communication channel is susceptible to eavesdropping and potentially tampering, as shown in Figure 2-3.

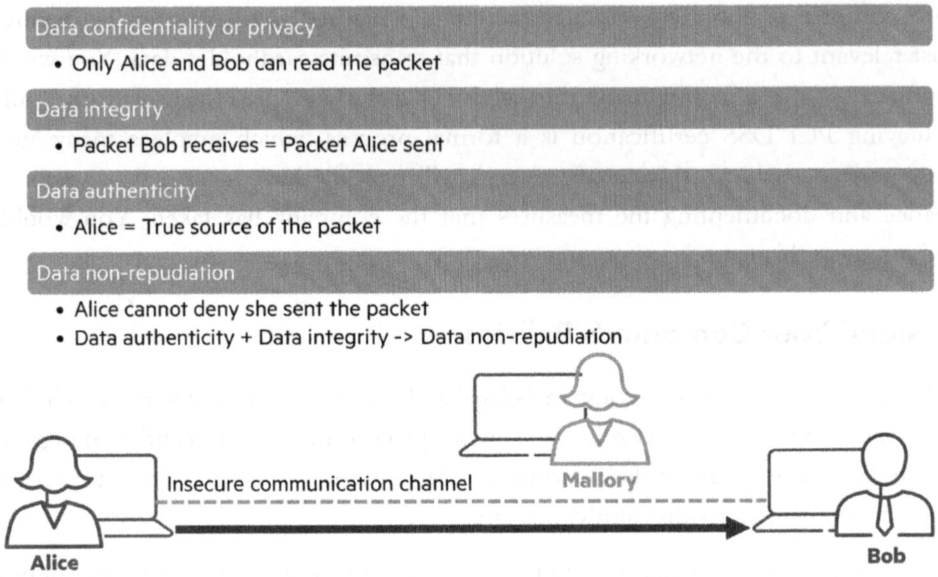

Figure 2-3: Goals of Secure Communications

Goals for securing such communications are:

- **Data confidentiality or privacy**—Only the transmitter and intended receiver or receivers can read the data. Data privacy is important for protecting sensitive data from eavesdropping.
- **Data integrity**—Data remains true to what the transmitter sent and is not altered in transit. Ensuring data integrity is important for many reasons. Of course, communicating parties want to know that what they think the other is saying is actually what the other is saying. But the need for data integrity also goes further. For example without data integrity, a hacker could "replay" packets. A hacker might capture an authorized user's authentication packet, change the timestamp, and resend the packet to gain unauthorized access.
- **Data authenticity**—The transmitter is who he or she claims to be. Data authenticity also helps to prevent man in the middle (MITM) attacks because both sides of the session know with whom they are communicating. Data integrity and authenticity are often ensured together.
- **Data non-repudiation**—The transmitter cannot later claim not to have sent data that he or she sent. For example, if users are signing documents over a network, they should not later be able to claim that they did not do so. If communications provide both data authenticity and data integrity, then data non-repudiation is assured.

Using Encryption for Data Privacy

Encryption can help to provide data privacy.

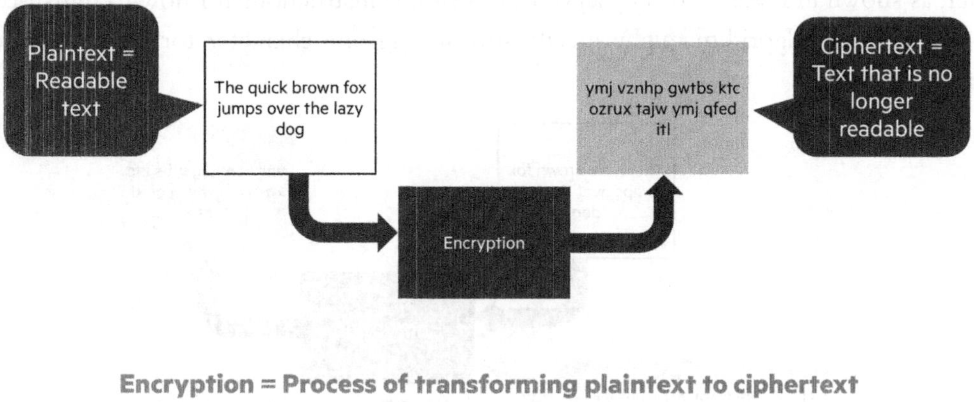

Figure 2-4: Using Encryption for Data Privacy

Encryption is the process of transforming readable data—called plaintext—into unreadable ciphertext, as illustrated in Figure 2-4. The ciphertext can then transit safely across an insecure network. If an unauthorized user eavesdrops on the data, the user will not be able to read the original message.

Decryption

Of course, once the message reaches the intended recipient, that recipient needs to be able to turn the ciphertext back into plaintext to read the message. Decryption reverses encryption, transforming ciphertext back into plaintext, as shown in Figure 2-5.

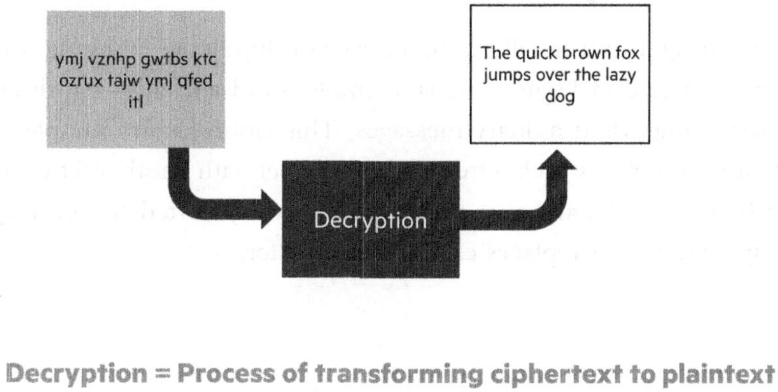

Figure 2-5: Decryption

What Goes into Encryption/Decryption?

How do encryption and decryption transform data? They use an algorithm, sometimes called a cipher, as shown in Figure 2-6. The algorithm is simply instructions for how to transform the data. For example, the algorithm might specify substituting one character for another or transposing certain characters.

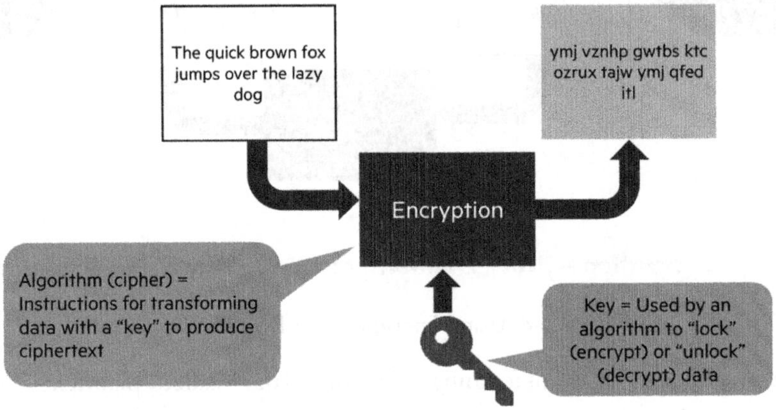

Figure 2-6: What Goes into Encryption/Decryption?

An algorithm works with a key, which is data used by the algorithm to ensure that the same plaintext does not always produce the same ciphertext. The key is critical because hackers could very well know how a particular encryption algorithm works. It is the key that "locks" the ciphertext so that knowledge of the algorithm alone is not enough to allow someone to decrypt the ciphertext. Only someone with the correct decryption key can "unlock" the data and return it to plaintext.

Simple Example of an Algorithm: Caesar Cipher

A simple example of an algorithm can illustrate the relationship between the algorithm and the key. The Caesar cipher, illustrated in Figure 2-7, is a simple—and ancient—algorithm, which Julius Caeser used to encrypt important military messages. This cipher is an example of a substitution algorithm. The algorithm replaces each letter of the alphabet with another letter. It is also called a "shift" algorithm; by writing the alphabet under the alphabet, shifted by several places, you can determine which ciphertext letter replaces each plaintext letter.

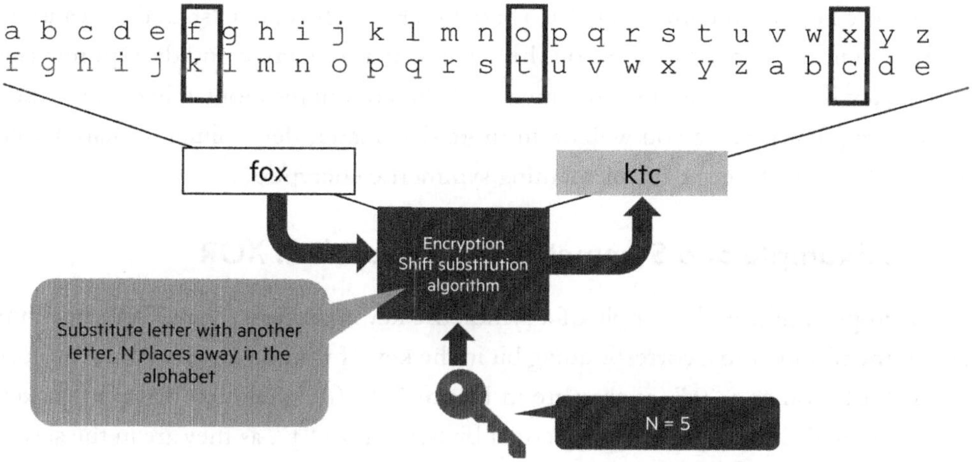

Figure 2-7: Simple Example of an Algorithm: Caesar Cipher

In this example, the algorithm is "substitute each letter with another letter a certain number of places away in the alphabet." The key is the number of places to shift. The ciphertext for "fox" will differ based on the key. When the key is "shift 5," as shown in this example, "fox" becomes "ktc." Decryption simply involves shifting the alphabet in reverse.

Of course, this cipher is quite easily cracked, but it gives you an idea of how encryption works.

Encryption algorithms fall into two broad categories: symmetric and asymmetric.

Symmetric Encryption

With symmetric encryption, the same key encrypts data as decrypts data, as shown in Figure 2-8.

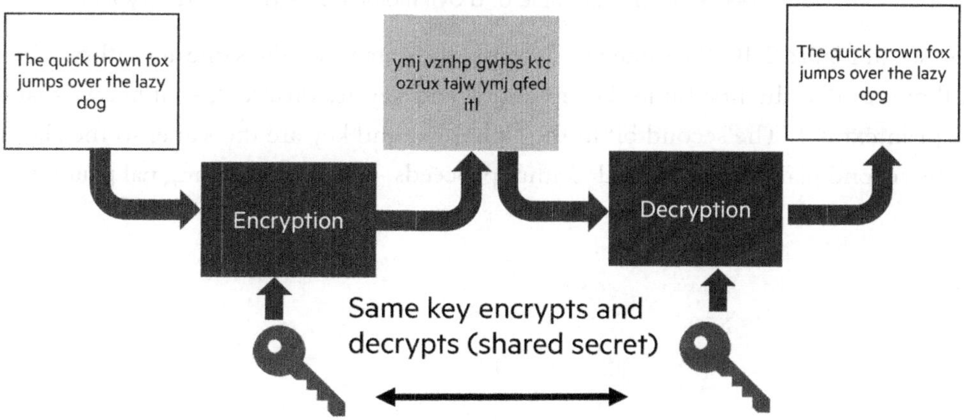

Figure 2-8: Symmetric Encryption

The symmetric key is sometimes called a "shared secret." The key is "shared" because both parties to the communications need to agree on the key in order to encrypt and decrypt messages for each other. The key is "secret" because the privacy of the communications rely on only the authorized parties knowing the key. As you will see in more detail later, determining a shared secret securely forms one of the challenges of implementing symmetric encryption.

A Simple Example of a Symmetric Key Algorithm: XOR

XOR is a simple, but useful example of a symmetric key algorithm. The XOR algorithm compares each bit in the plaintext to a corresponding bit in the key. If the bits are the same, the corresponding ciphertext bit becomes a "0," as they are in Figure 2-9's first plaintext bit and first key bit. If the bits are different the corresponding ciphertext bit becomes a "1", as they are in the second plaintext bit and key bit in Figure 2-9.

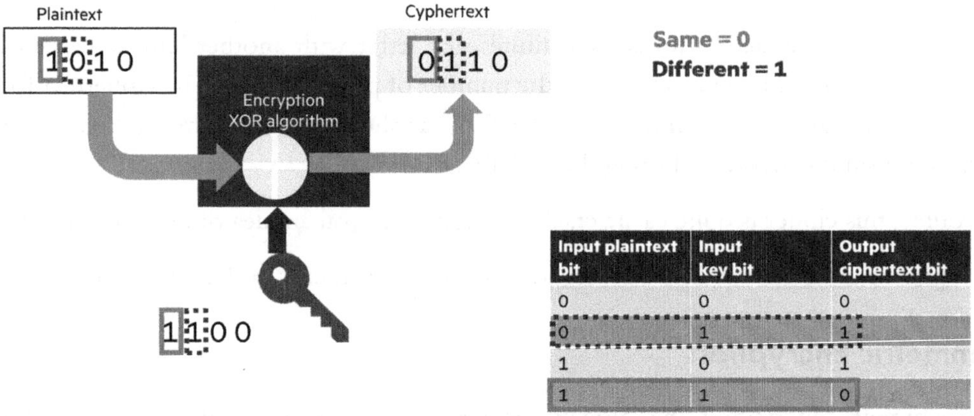

Figure 2-9: A Simple Example of a Symmetric Key Algorithm (Encryption)

As you see in Figure 2-10, the same key decrypts the data using the same algorithm. The decryption algorithm sees that the first bit in the ciphertext and key are different, so it knows that the first bit in the plaintext is 1. The second bit in the ciphertext and key are the same, so the algorithm yields 0 for the second plaintext bit. The algorithm proceeds, producing the original plaintext.

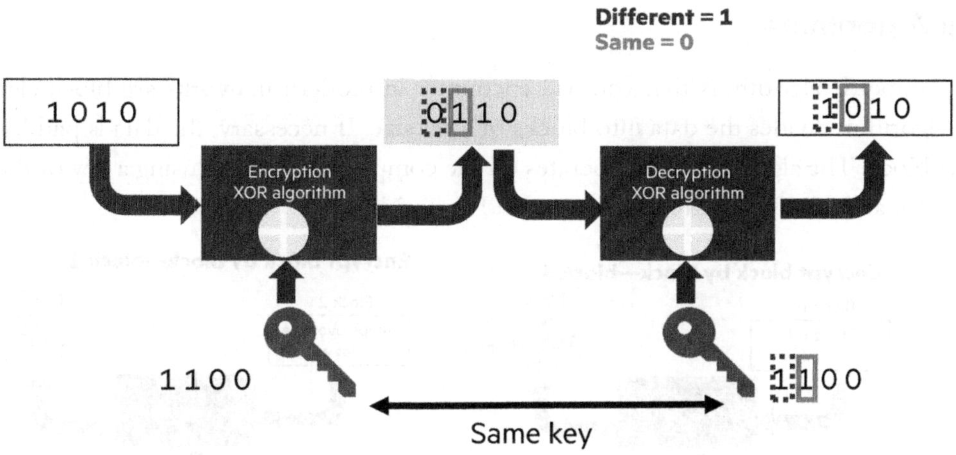

Figure 2-10: A Simple Example of a Symmetric Key Algorithm (Decryption)

Stream Algorithms

XOR is an example of a stream cipher, or algorithm, which operates on the data bit by bit, as shown in Figure 2-11. A stream algorithm requires a key of the same length or longer than the plaintext so that every plaintext bit has a corresponding key bit.

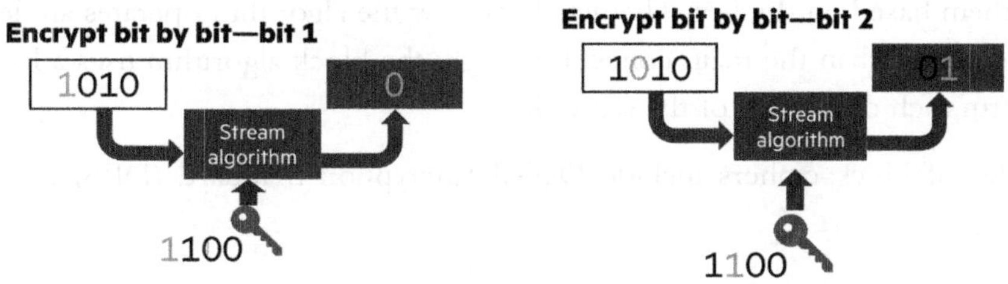

Figure 2-11: Stream Algorithms

Ideally, a stream algorithm would use a random one-time pad (OTP), a shared key that is the same length or longer than the message and never reused. With an OTP, the XOR algorithm, despite its simplicity, is mathematically proven to be uncrackable. However, using OTPs is not typically practical. The overhead of agreeing on such a key in advance for the vast amounts of data that devices need to transmit is prohibitive.

Rivest Cipher 4 (RC4) is a stream algorithm that attempts to simulate OTPs with shorter keys. RC4 expands the shorter key into a pseudorandom keystream of the correct length. This algorithm was the basis of older forms of wireless encryption (Wired Encryption Protocol [WEP] and Temporal Key Integrity Protocol [TKIP]), but does have some known flaws and is not recommended for current use.

Block Algorithms

Most symmetric algorithms that you will encounter in modern networks are block algorithms. A block algorithm divides the data into blocks of a set size. If necessary, the data is padded to fill out the last block. The algorithm then operates on the complete data block using a key of the same size to produce a block of ciphertext, as shown in Figure 2-12.

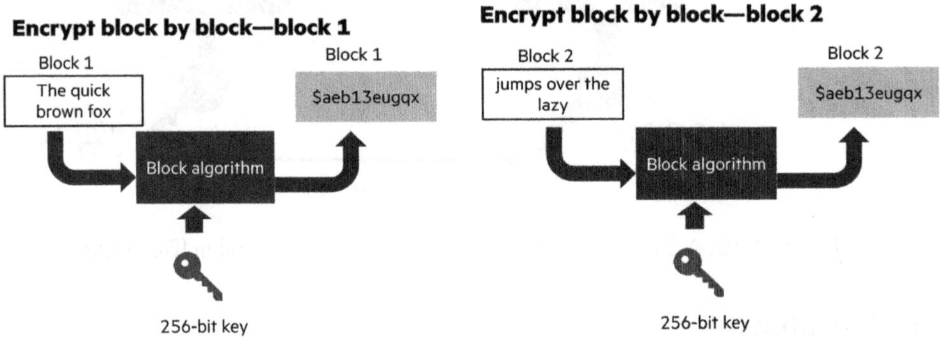

Figure 2-12: Block Algorithms

Exactly how the block algorithm transforms the data depends on the specific algorithm. For example, Advanced Encryption Standard (AES) involves laying the plaintext bytes in a matrix and then shifting them based on the key. The details of how the algorithm operates are less important for you to understand than the main concept, though: the block algorithm uses a key of a certain size to transform each data block of the same size.

Other examples of block ciphers include Digital Encryption Standard (DES) and Triple DES (3DES).

The Risks of Key Reuse with Symmetric Algorithms

A block cipher reuses the same key to encrypt multiple blocks of data in a message. However, reusing keys can cause serious issues. If the same key encrypts the same block of data in the same way, hackers can start to find patterns. Any pattern is an opportunity for a hacker to start cracking the code. As a simplified example, illustrated in Figure 2-13, the hacker could start to guess that a frequently appearing block of ciphertext is a commonly transmitted set of data such as the TCP port for HTTP and the HTTP code for a GET message. Hackers can even send traffic that forces a specific known response to further aid their efforts. With knowledge of the plaintext that corresponds to some of the ciphertext, the hacker is well on the way to cracking the key or deriving more plaintext.

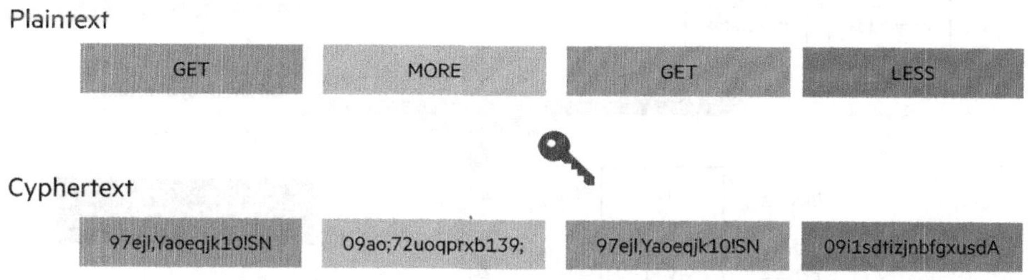

Figure 2-13: Repeated Encryption of Same Block with Same Key

Reusing keys with an XOR-based algorithm also causes issues. Due to how the XOR function works, hackers know that two same-size ciphertext messages XORed together are equivalent to the corresponding plaintext messages XORed together *if* the messages were encrypted with the same key. With any three of these sets of data known, the hackers can derive the fourth, as shown in Figure 2-14. Of course, ciphertext is transmitted freely, so hackers can collect multiple blocks of ciphertext easily. And, as discussed earlier, hackers can guess some plaintext associated with this ciphertext or force the production of ciphertext f*rom known plaintext. Therefore, it is critical to avoid reusing keys to prevent hackers from deriving plaintext trivially.

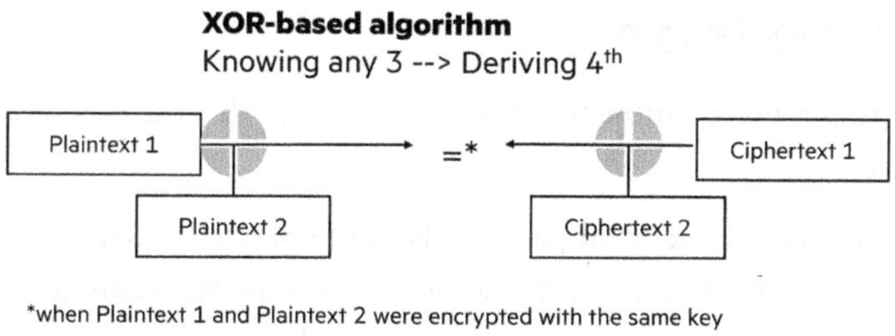

Figure 2-14: Key Reuse with XOR-Based Algorithms, Trivial to Crack

Initialization Vectors (IVs)

To be practical to implement, though, algorithms need to reuse keys. To avoid the same key and set of plaintext producing the same ciphertext, encryption algorithms can use initialization vectors (IVs). An IV is simply a value used to kick off the process in a different way every time that the algorithm encrypts new plaintext. Some algorithms require the IV to be random or, more precisely, pseudorandom as no value is truly random. Sometimes the cipher does not require a random IV, but the IV should still never be repeated. In the latter case, the IV can be called a number used once (nonce).

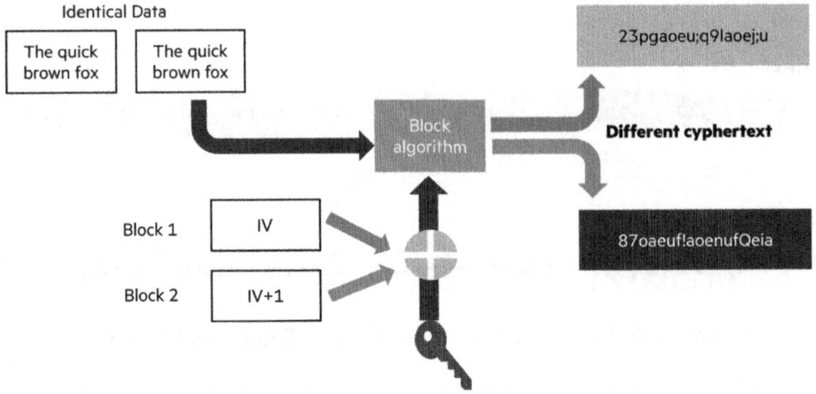

Figure 2-15: Initialization Vectors (IVs)

Exactly how a block algorithm uses an IV depends on something called the mode of operation. A simple example with counter (CTR) mode is shown in Figure 2-15. As you see, the algorithm encrypts an IV with the key and XORs the result with the plaintext to produce the ciphertext. The IV increments for each new block. This ensures that the same plaintext produces different ciphertext, preventing hackers from detecting patterns.

The Importance of Key Length

You should also consider the relationship between key length and the security of data that is encrypted by that key.

You learned about brute force attacks on passwords in the previous chapter. Hackers can similarly run brute force attacks on keys. These attacks do not use any clever techniques to look for patterns or exploit an algorithm's weaknesses. Instead, hackers simply try every combination of bits possible for the key until using that key to decrypt the ciphertext produces legible plaintext.

Because every bit in a binary key can have two possible values, the number of combinations possible for a key equals 2 to the power of the key's length, as shown in Figure 2-16. As you see, a 4-bit key can have 16 possible combinations. Because a hacker could quickly try each of the 16 combinations until one works, a 4-bit key would provide almost no security at all—even if you were using a highly secure encryption algorithm.

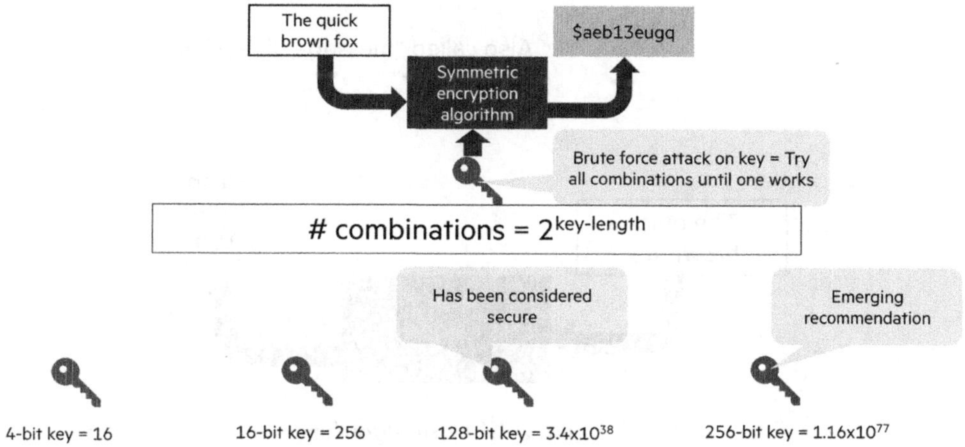

Figure 2-16: The Importance of Key Length

A 128-bit key has 3.4x1038 possible combinations. (For comparison, one trillion is 1x1012.) For many years, 128-bit keys have been considered protected from brute force attack due to the infeasibility of even a supercomputer trying so many combinations. However, with computational power increasing exponentially, security experts are advising the move toward 256-bit keys. Due to the exponential relationship of key length to key strength, a 256-bit key is *not* twice as strong as a 128-bit key, but instead many trillions of times stronger. A 256-bit key *squares* the number of combinations possible, resulting in 1.16x1077.

Hash Algorithms

You now understand how properly implemented symmetric encryption can ensure data privacy. You will next move on to examining the role of hash algorithms in secure communications.

A hash algorithm operates on data of any size to produce a string of fixed length called the hash or message digest, as shown in Figure 2-17. Ideally, the algorithm produces a unique digest for each unique data string, which acts like the data's fingerprint. If the data changes, its "fingerprint" no longer matches, so entities can detect the change.

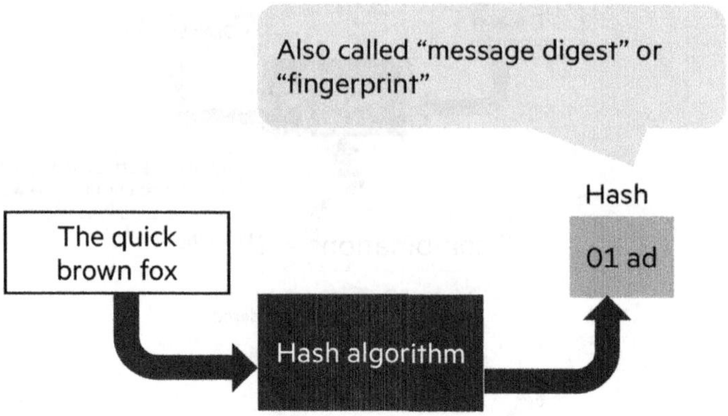

Figure 2-17: Hash Algorithms

In practice, because the hash is typically shorter than the data, some collisions might occur—that is two strings might produce the same hash, as only so many hashes exist. However, it is very important that the algorithm function such that it is computationally infeasible to predict which to messages will have the same hash. In other words, a hacker should not be able to calculate how to change a message without changing its hash as well.

In addition, hash algorithms should be irreversible. That is, no one should be able to determine the original plaintext that produced the hash. For this reason, hash algorithms are sometimes called "one-way" or "trapdoor" functions.

Examples of hash algorithms include Message Digest 5 (MD5), Secure Hash Algorithm (SHA)1, SHA2-256, SHA2-384, and SHA2-512, and SHA3.

Hash-Based Message Authentication Codes (HMAC)

Hashing can help two parties like Bob and Alice to authenticate each other's messages and ensure their integrity. But because anyone can change a message and produce a valid hash for the new message, the hashing algorithm must incorporate a key, or shared secret. A hash created from a message and a key is called a message authentication code (MAC), and only parties that know the key can create valid MACs.

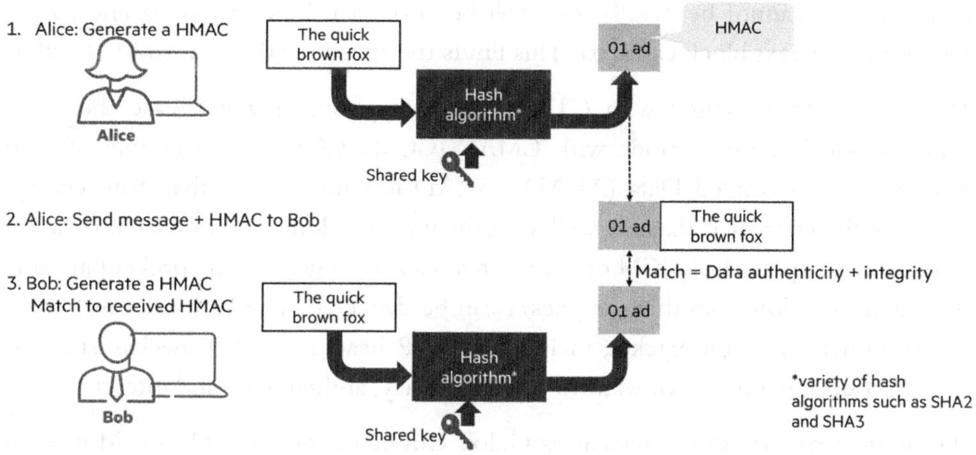

Figure 2-18: Hash-Based Message Authentication Codes (HMAC)

Examine the process, illustrated in Figure 2-18, in a bit more detail.

Alice wants to send a message to Bob. She needs Bob to know that she is the one who sent the message and that the message that he receives is the one that she sent with no changes. She and Bob agree on a key, called an integrity or authentication key, in advance.

The HMAC algorithm takes the shared key and the message as inputs to a hash algorithm and produces the MAC as an output. HMAC can work with a variety of hash algorithms including SHA2 and SHA3. (In more detail, HMAC involves double hashing the message. This makes HMAC resistant to key extension attacks, in which hackers are able to add data to the end of the message without invalidating the hash.)

Alice then appends the MAC to her message and transmits both to Bob.

Bob applies the same functions to the data that he receives using the same key. He should receive the same result. If he does not, he knows that the data has been altered in transit. If he does, he knows that both the message and key are correct—in other words, Alice sent the message, which has not been changed.

Other Ways to Authenticate Messages

Other ways to authenticate messages exist. For example, a block cipher can run in a mode called cipher block chaining (CBC). In this mode, an IV is XORed with the first plaintext block before that block is encrypted. Then each ciphertext block is XORed with the next plaintext block before that block is encrypted. A CBC MAC is the last ciphertext block for the message (all other blocks are discarded). Because data from the complete set of plaintext went into producing this MAC, and the MAC will change if the plaintext changes at all, it functions much like a hash-based MAC.

However, CMAC cannot be parallelized well because each block must be encrypted before the encryption for the next block can start. This limits the speed for data transmitted with CMAC.

CMAC is often implemented with CTR mode encryption, in which case the entire mode of operation is called Counter mode with CMAC (CCM). CCM is an example of Authentication Encryption with Associated Data (AEAD). AEAD is a mechanism that both encrypts data *and* authenticates the encrypted data, as well as authenticates additional associated data that is sent in plaintext. For example, when CCM operates on a packet, it encrypts the packet payload, but leaves the IP header in plaintext so that the packet can be delivered. The IP header is "associated data," and CCM authenticates the packet, including the IP header. AEAD mechanisms are important because, of course, parties often want to ensure privacy, authenticity, and integrity for the data.

Another important AEAD mechanism is Galois Counter Mode (GCM). GCM uses an IV, a key, plaintext that needs to be encrypted and authenticated, and data that needs to be authenticated but not encrypted. It outputs ciphertext, created with CTR mode encryption, and a MAC, created with Galois field multiplication. You do not need to understand the field mathematics. The main point is that the most processor-intensive operations for generating this MAC can be parallelized so that GCM can secure data at higher speeds than CCM.

You could also use the authentication portions for CCM and GCM on their own, in which case they are called CMAC and GMAC, respectively.

How Do Parties Communicate the Symmetric Keys?

As you have looked at the key-based encryption and hashing, you might have wondered how Bob and Alice agree on the keys. Remember these keys are shared secrets. Both Alice and Bob must know them, and they must keep them secret.

They could communicate the keys over a secure communication channel in readiness for using the keys to secure communications over an insecure channel in the future (see the top of Figure 2-19). For example, Bob and Alice could meet in person, agree on random key values, and program these into their devices. Then they could use those keys to encrypt and authenticate future communications.

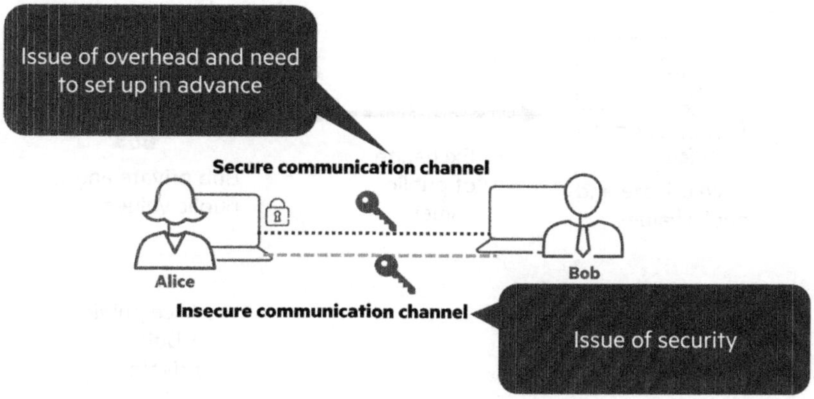

Figure 2-19: How Do Parties Communicate the Symmetric Keys?

However, this approach is not feasible in the real world. It requires every entity to have the ability to securely communicate with every other entity. In the real world, your Web browser might need to communicate securely with multiple new Web servers every day with no other way to reach those servers except over the insecure Internet. In addition, a process that requires so much overhead would lead to reusing keys over a long period; even with IVs, so much reuse is not advisable.

Ideally, Alice and Bob—and any other two entities—should be able to arrive at a shared key over an *insecure* communication channel (see the bottom of Figure 2-19). But how can they do so securely?

Diffie-Hellman (DH) for Secure Key Agreement

Diffie-Hellman exchanges enable Alice and Bob to securely generate a shared key across an insecure network. You will now look at a basic model of how the exchange works, illustrated in Figure 2-20, without getting into the complexities of the math.

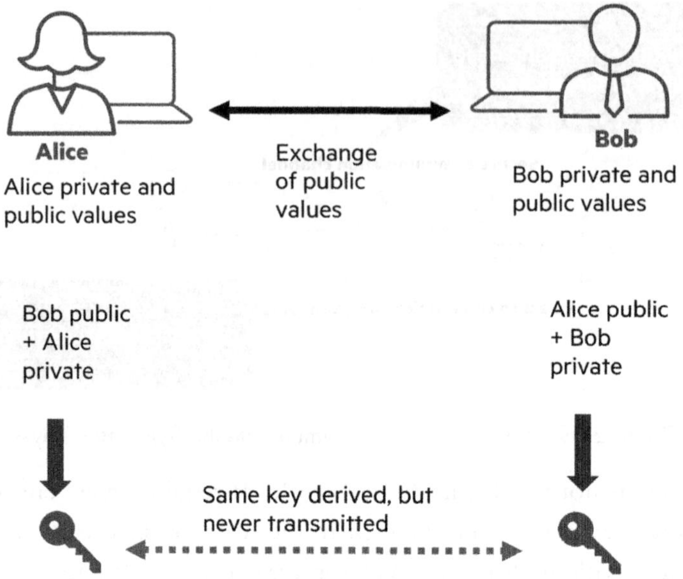

Figure 2-20: Diffie-Hellman (DH) for Secure Key Agreement

Alice and Bob each select a private and public value, which are linked together but in such a way that it is computationally infeasible to derive the private value from the public value. The original Diffie-Hellman used a large prime number and the modulus operation (dividing by the number and finding the remainder) to create the private and public values. The more modern approach involves calculating positions across an elliptic curve (EC). The EC approach has the advantage of providing relatively higher security with the same size (or smaller) values, as compared to prime modulus.

Alice and Bob then exchange their public values but keep their private values secret. They each then perform a mathematical operation on the peer's public value and their own private value. The mathematics of prime modulus or elliptic curve is such that the operation produces the same output using Bob's public value and Alice's private value as it does using Alice's public value and Bob's private value.

Implemented with large enough values, Diffie-Hellman is secure against eavesdropping. You can only generate the shared secret if you have one of the private values, and those are never exchanged. In addition, the private values cannot be derived from the public ones.

Diffie-Hellman Susceptibility to MITM

However, Diffie-Hellman is subject to MITM attacks, as shown in Figure 2-21. This is because there is nothing "special" about the public values that lets Alice and Bob determine their source. Mallory could create her own private and public values. She could then replace Alice's public value with her own and send it to Bob. Similarly, she could replace Bob's public value with her own and send it to Alice. Then Alice and Mallory will generate a shared key, and Bob and Mallory will generate another shared key. Mallory can shuttle communications between Alice and Bob. Because she has the keys, she can read all of the communications and even modify them if she chooses.

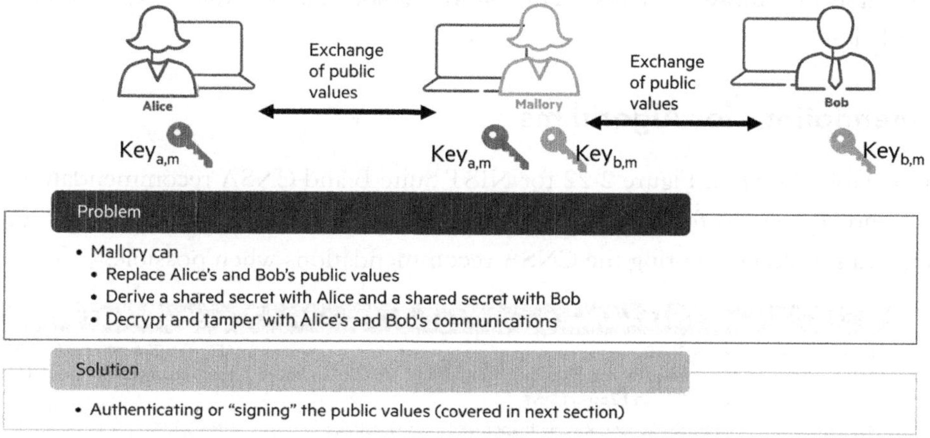

Figure 2-21: Diffie-Hellman Susceptibility to MITM

For this reason, Diffie-Hellman is usually implemented with data authentication or "signing" to enable Alice to prove to Bob that her public value belongs to her and vice versa. Message authentication like HMAC can vouch for data, as you learned, but this would create a "chicken-and-egg" problem. How can Alice authenticate data with a shared key before she and Bob have established a shared key? You will learn how asymmetric encryption can also authenticate data later in this chapter; this is the most common way to authenticate Diffie-Hellman exchanges.

NIST Suite B and CNSA

You have looked at a high level at how symmetric encryption and hash algorithms work. However, you should understand that not all algorithms are equal. The security provided by encryption and key-based hashing depend in large part on the strength of the algorithm, as well as the size of the key used.

To ensure that you are using the correct algorithms and key sizes, you can refer to trusted standards bodies such as the National Institute of Standards and Technologies (NIST). NIST Suite B, which was originally released in 2005 and has had multiple updates, consists of algorithms approved by the US National Security Agency (NSA) for securing various levels of information from non-classified to top secret. Private entities can follow similar guidelines for securing their sensitive data.

In 2018 NIST announced the Commercial National Security Algorithm (CNSA) suite. Quantum computing is promising to bring much more processing power into the reach of users—and hackers. With more processing power at their disposal, hackers will be able to crack algorithms with key lengths that used to be considered secure. The CNSA suite recommends for algorithms and key lengths to address those dangers. This study guide covers the CNSA transition phase recommendations.

Recommendations for Algorithms

Refer to the table shown in Figure 2-22 for NIST Suite B and CNSA recommendations for secure algorithms and key lengths for block ciphers, message authentication, and shared key agreement. At this point, you should be favoring the CNSA recommendations when possible.

	NIST Suite B Approved Ciphers	Specified for CNSA Transition
Block ciphers	AES or 3DES 14 approved modes of operation, including CTR, CCM and GCM; 128-, 192-, and 256-bit keys	AES with 256-bit keys (any of the 14 approved modes of operation)
Message authentication	HMAC, KMAC, or CMAC with approved hash algorithm: • SHA2 (SHA-224, SHA-256, SHA-384, SHA-512, SHA-512/224, and SHA 512/256) • SHA3 (SHA3-224, SHA3-256, SHA3-384, SHA3-512, SHAKE128, and SHAKE256)	HMAC, KMAC, or CMAC with SHA/SHA3-384 or higher
Key agreement	Diffie-Hellman Elliptic Curve Diffie-Hellman (ECDH)	ECDH with Curve-P 384

Avoid when possible:
RC4 or DES for encryption
MD5 for hashing

3DES has known issues

Figure 2-22: Recommendations for Algorithms

For encryption, AES is the preferred algorithm. CNSA recommends 256-bit keys. RC4 is no longer considered secure, nor is DES, a block cipher. 3DES is approved by Suite B, but does have some known issues.

For hashing, you will sometimes see MD5 as an option. But MD5 has known exploits and should be avoided when possible, as hackers can predict how to manipulate data and still make the MAC read as valid. HMAC-MD5 is more secure than MD5; however, it is still recommended that you use HMAC with an approved form of SHA. CNSA recommends SHA/SHA3-384. (The 384 refers to the number of bits in the hash output.)

For key agreement, CNSA recommends Elliptic Curve Diffie-Hellman (ECDH) with at least a 384-bit curve.

Secure Communications: Asymmetric Encryption, Certificates and PKI

Earlier you learned about symmetric encryption, in which the same key encrypts and decrypts data. Next, you will look at asymmetric encryption.

Asymmetric (Public Key) Encryption

Using a similar idea to the public and private values that you examined in the Diffie-Hellman exchange, asymmetric encryption uses two connected values, called a private and public key. Together these keys compose a keypair. Each key in the pair decrypts the data encrypted by the other key. Although the key pair works together, you cannot derive one key simply because you know the other key. The entity that owns the keypair keeps the private key secret but freely distributes the public key. This approach to encryption enables Bob and Alice to encrypt and authenticate data without having to worry about exchanging keys securely.

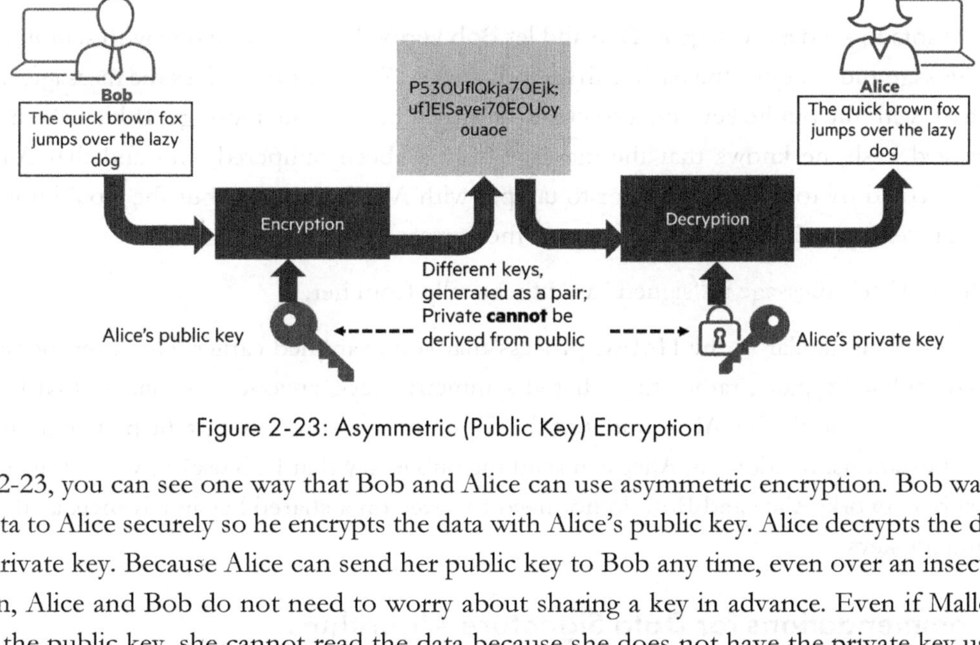

Figure 2-23: Asymmetric (Public Key) Encryption

In Figure 2-23, you can see one way that Bob and Alice can use asymmetric encryption. Bob wants to send data to Alice securely so he encrypts the data with Alice's public key. Alice decrypts the data with her private key. Because Alice can send her public key to Bob any time, even over an insecure connection, Alice and Bob do not need to worry about sharing a key in advance. Even if Mallory intercepts the public key, she cannot read the data because she does not have the private key used to decrypt the data.

Asymmetric Encryption for Data Signing

In Figure 2-24, you see another way that Alice and Bob can use asymmetric encryption. The transmitter can use asymmetric encryption to "sign" data.

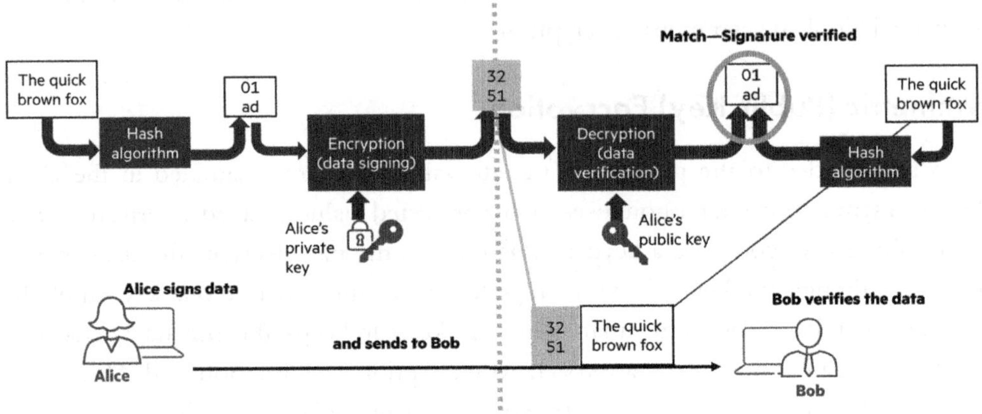

Figure 2-24: Asymmetric Encryption for Data Signing

Alice wants to send a message to Bob and let Bob verify that she is the one who sent it. She hashes the message and encrypts the hash with her private key. When Bob receives the message, he decrypts the hash with the public key and also creates his own hash of the message. If his hash matches the decrypted hash, he knows that the message has not been tampered with and that Alice sent it. Mallory could try to pose as Alice, or to tamper with Alice's message, but she would not be able to make an "encrypted" hash that decrypts to the correct value.

In short, Alice's message is "signed" as authentically from her.

This process is similar to the HMAC process that you examined earlier. However, because it uses private/public keypairs, rather than shared symmetric keys, anyone can validate that the message was sent specifically by Alice and not by Bob or someone else participating in the trusted communications. In addition, Alice can send the public key that Bob uses to verify the data over an insecure network; Alice and Bob do not need to agree on a shared key in advance, as they do with symmetric keys.

Recommendations for Data Signature Algorithms

NIST Suite B and CNSA also dictate secure algorithms to use for data signatures. You can see these recommendations in the table in Figure 2-25.

	NIST Suite B Approved	Specified for CNSA Transition
Digital signature	Digital Signature Algorithm (DSA), Rivest Shamir Adleman (RSA), Elliptic Curve DSA (ECDSA) + approved hash algorithm: • SHA2 (SHA-224, SHA-256, SHA-384, SHA_512, SHA-512/224, and SHA 512/256) • SHA3 (SHA3-224, SHA3-256, SHA3-384, SHA3-512, SHAKE128, and SHAKE256)	RSA with 3072-bit modulus or higher + approved hash algorithm

Figure 2-25: Recommendations for Data Signature Algorithms

Asymmetric Encryption Considerations

As you learned in the previous section, the main dilemma of symmetric encryption is communicating a shared key securely. Diffie-Hellman mostly solves this dilemma, but it still requires data authentication to avoid MITM attacks. But how can Bob and Alice authenticate their Diffie-Hellman public values if they have not yet shared a secret key for doing so? Asymmetric encryption solves this dilemma. Bob and Alice can use a private key to sign data. They can freely distribute the public key for validating the signed data over an insecure network. In fact Bob and Alice can encrypt data for each other too without having to communicate anything over a secure channel. They simply distribute their public keys over an insecure channel. Bob can then encrypt Alice's data with Alice's public key, and Alice can encrypt Bob's data with Bob's public key.

However, asymmetric encryption has some challenges. First, it is quite slow as compared to symmetric encryption. An email application can use public key encryption to encrypt an email with the recipient's public key because users expect a bit of latency with email. However, if a device used asymmetric encryption to encrypt all the data sent over the network, for example, the user would notice latency. This issue makes asymmetric encryption a poor candidate for communicating data in transit in bulk. Instead, as you will see, applications typically use asymmetric encryption to set up trusted communications and then shift to symmetric encryption.

The second challenge of asymmetric encryption probably already occurred to you if you have a suspicious mind. Alice can certainly send her public key to Bob over an insecure network without putting her private key at risk. But how does Bob know that Alice's supposed key really is hers and not one sent by Mallory pretending to be Alice? Bob could communicate with Alice securely in advance and manually install Alice's public key. However, having to do this negates the benefits of asymmetric encryption, as the goal of asymmetric encryption is to enable secure communications over insecure networks without advance communications over a secure channel.

As you will see in the next section of this chapter, digital certificates solve this dilemma.

What is a Digital or X.509 Certificate?

Certificates provide key management: they tie a specific identity to a public key. In other words, they verify to everyone that the public key posing as "Alice's key" really belongs to Alice. And they do this without Bob having to install Alice's public key on his machine in advance, enabling Alice and Bob to set up secure communications on the fly.

Digital certificates follow a standard format based on industry standard X.509, which is why you will sometimes hear certificates called X.509 certificates. You can see some of the key fields within an X.509 certificate in Figure 2-26.

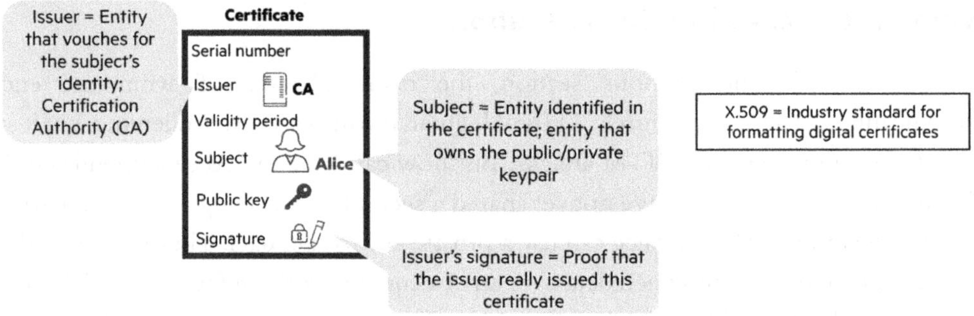

Figure 2-26: What is a Digital or X.509 Certificate?

The subject identifies the entity that owns the certificate and the keypair associated with the certificate.

The public key is the subject's public key.

The issuer identifies the entity that issued this certificate to the subject. The issuer forms a crucial component of the certificate's trustworthiness. Mallory could create a certificate with her own public key and any identity that she wants to claim. But unless the certificate has been issued by a source that you trust, you would know not to trust Mallory's certificate. Just like specific governmental bodies are responsible for issuing documents such as driver's licenses and passports, Certification Authorities (CAs) are responsible for issuing digital certificates. We trust that the government has taken the proper steps to validate the passport holder's identity, so we trust the passport. Similarly, devices and applications trust that the CA has validated the subject's identity, so they trust the certificate issued by the CA.

A certificate also includes the issuer's signature. Much like watermarks help people to validate that the proper governmental body truly created an ID like a driver's license, the certificate signature validates that the issuer claimed by the certificate truly issued the certificate.

The serial number uniquely identifies the certificate among certificates issued by the same CA. The validity period gives the certificate a specified period in which the certificate is valid. The longer that a device can retain a certificate, the less secure the certificate becomes. Perhaps the private key is compromised or the certificate owner is no longer approved for the rights that the certificate provides. Just as you must renew your driver's license or passport periodically, devices must renew their certificates periodically.

You will explore how certificates, signatures, and trust work in detail over the rest of this section.

What Do You Use Certificates For?

You already understand that a certificate binds an identity with a public key (and implicitly with the private key paired with the public key). And you know what you can use private and public keys for: signing data and encrypting data. These are called key usages, and they are actually specified within the certificate. Certificate signing is another example of a key usage. Certificate signing is actually data signing, but for the specialized purpose of issuing certificates. Only CA certificates should have the certificate signing usage.

But you should also consider, from a broader viewpoint, why the device needs to encrypt or sign data and how the device will use the certificate. Much like people might use a driver's license to prove that they can drive a car and also to prove their identity when boarding an airplane, users and devices can use their certificates for multiple purposes. These purposes are defined in extended key usages (EKUs), which can also be included in the certificate, as shown in Figure 2-27. For example, the public key in a certificate with the Secure Email extended key usage can be used to encrypt an email destined to the owner of certificate.

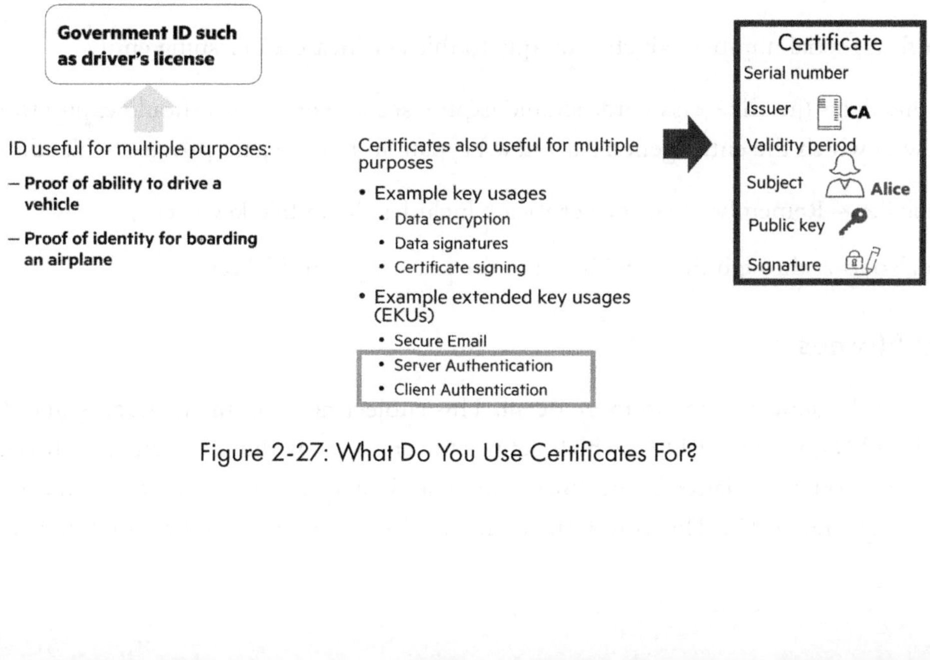

Figure 2-27: What Do You Use Certificates For?

The most relevant EKUs for your purposes are Server Authentication and Client Authentication. To authenticate itself with a certificate, either a server or a client signs data with the private key. By signing the data, the device proves that it owns the private key associated with the certificate and is the source of the signed communications.

Example Certificate

In Figure 2-28, you see what a certificate actually looks like, as displayed on a Windows machine.

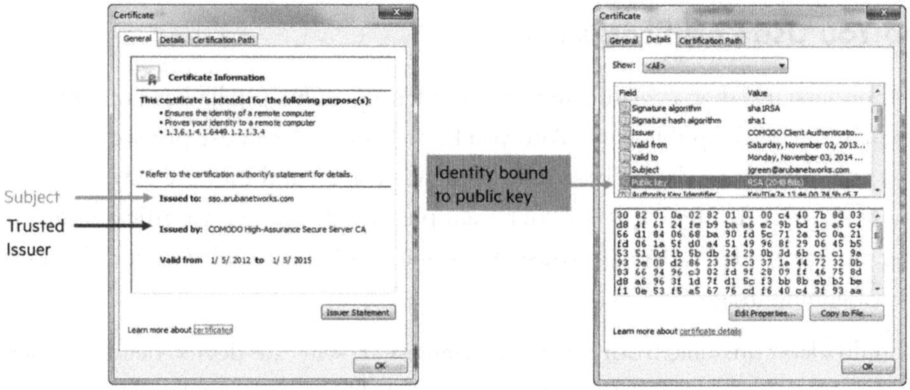

Figure 2-28: Example Certificate

Some of the important fields include:

- Subject name, which identifies the owner of the certificate
- Issuing entity (the name of the CA that signed the certificate)
- Certificate serial number, which is unique to this certificate and issuing entity
- Validity dates (just like passwords should expire, secure certificates should expire too, forcing a renew in which the entity generates a new keypair and once again proves its identity)
- Public key—Remember that the certificate includes the public key itself.

You can also look through these fields to see the key usages and EKUs.

Subject Names

Next look at the subject name in more detail. This subject name is an Abstract Syntax Notation 1 Distinguished Name (ASN.1 DN), which defines the endpoint within a directory such as a Windows domain. The common name is the lowest level and uniquely identifies the subject within the organizational unit (OU). The common name can be a username, a first or last name, a Fully

Qualified Domain Name (FQDN), an IP address—or really any value that the directory uses. The common name exists within an OU, which for a Windows domain is a Domain Group. The organization (O) is typically a domain name. A DN might have two O components such as O=example and O=com. However, Windows domains use DC instead of O for the domain name. The location (L) is typically a city. Above L are state (S) and country (C). The DN might also include email (E).

Example formatting for a DN is:

CN=Alice,OU=Marketing,DC=example,DC=com,L=Santa Clara,S=California,C=US,E=alice@example.com

To identify a certificate holder, an application might look at the certificate's CN or full DN.

The certificate might also include one or more subject alternative names (SANs). A SAN specifies a specific type of ID, such as IP or DNS, and then the value for that ID. SANs are useful because, unlike with CNs, applications can determine exactly what type of ID the SAN is specifying.

CAs and the Public Key Infrastructure (PKI)

Now that you have looked at certificates, you are ready to examine the mechanisms for issuing those certificates.

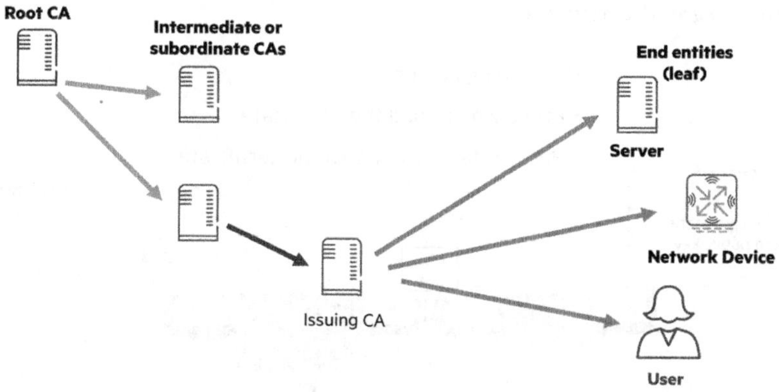

Figure 2-29: CAs and the Public Key Infrastructure (PKI)

Certificates exist within a public key infrastructure (PKI) hierarchy that enables devices to determine which certificates they should trust. Every certificate within the PKI is signed by an entity that vouches for the authenticity of the identity of the certificate holder, as shown in Figure 2-29.

Root CAs are at the top of the hierarchy. They sign their own certificates, called root CA certificates. Root CAs are well-known entities that are widely believed to be trustworthy. For example, DigiCert and Comodo are root CAs. A company can also set up its own internal or private root CA, which is believed to be trustworthy within the company.

As the ultimate source of trust, root CAs must be rigorously secured. The root CA issues just a few certificates itself, issuing and signing certificates for intermediate CAs. This ensures minimal contact between the root CA and other entities. In fact, the root CA is typically kept offline for security purposes. The intermediate CAs, sometimes also called subordinate or signing CAs, are responsible for issuing and signing certificates for end entities such as Web servers, network infrastructure devices, and users. The end entity certificates are sometimes called server certificates, user certificates, or client certificates, depending on the type of entity to which they are assigned. They are also sometimes called "leaf" certificates because they are at the other side of the PKI "tree" from the root.

A PKI could also have multiple tiers of intermediate CAs.

Certificate Signing

Pause a moment to consider what "signing" a certificate actually means. This process, shown in Figure 2-30, is the same that you examined for signing any data with asymmetric encryption. The issuer, such as the intermediate CA, hashes the unsigned end-entity certificate and then encrypts the hash with its own private key. The encrypted hash is the signature that is appended to the certificate, transforming it into a signed certificate.

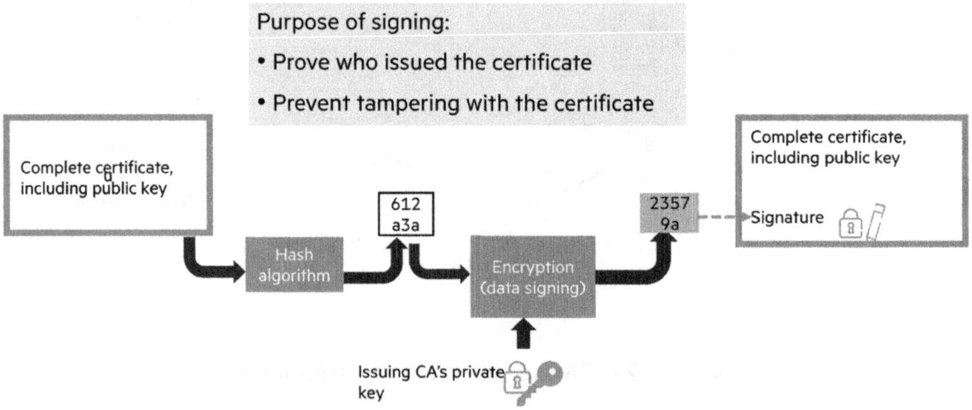

Figure 2-30: Certificate Signing

If a malicious user like Mallory attempted to tamper with her certificate—for example, change her subject name so that she can pose as a trusted website—the alterations would change the hash. Then the certificate hash would no longer match the decrypted signature, and other devices could tell that the certificate was invalid.

Chain of Trust

You can also look at the PKI as a chain of trust, which flows back up through the intermediate CAs to the root CA, as shown in Figure 2-31. Devices and applications can trust the server's certificate because they trust the intermediate CA's certificate. In other words, to validate the server's certificate, applications check the signature in the certificate using the issuing intermediate CA's certificate.

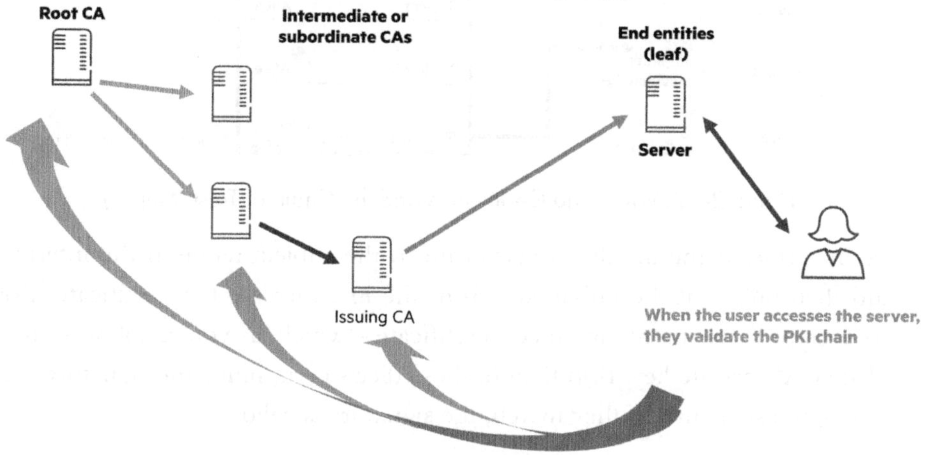

Figure 2-31: Chain of Trust

In turn, devices and application trust the intermediate CA's certificate because they trust the root CA's certificate. Because the root CA certificate is self-signed, applications and devices must accept this certificate as trusted through another mechanism. Typically, they are pre-configured to do so (you will see how later). As you see, this structure solves the practical problem of using asymmetric encryption. Rather than having to learn the public key for every entity with which you might want to communicate over a secure channel in advance, you only have to learn the certificate, and included public key, for a few root CA's.

Validating Certificates and the Chain of Trust

In Figure 2-32, you can examine how this chain of trust works in a bit more detail. Alice is sending her certificate to Bob so that she can prove her identity and distribute her public key. She also includes the certificate of the intermediate CA that issued her certificate. She could optionally include the root CA certificate that issued the intermediate CA's certificate; however, it is best practice to omit it. This is because, as you will see, Bob should have the root CA certificate installed in advance. If he is given the choice of trusting the root CA certificate on the fly, he might make a mistake and trust a root CA that he should not.

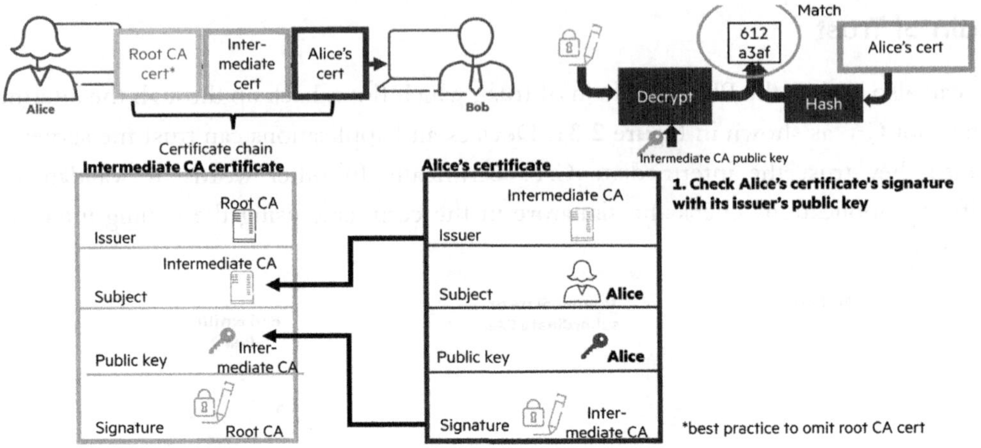

Figure 2-32: Validating Certificates and the Chain of Trust (Step 1)

Bob matches the issuer name in Alice's certificate to the subject name in the intermediate CA's certificate and then pulls out the public key from the intermediate CA certificate. Bob uses this public key to decrypt the signature in Alice's certificate—which, as you recall, was encrypted with the intermediate CA's private key. Bob than hashes Alice's cert, minus the signature, and checks it against the decrypted signature. If they match, the signature is valid.

One of the most common protocols to use certificates is Transport Layer Security (TLS). RFC 5246 for TLS 1.2 requires the end-entity certificate to be sent first and then the intermediate certificate or certificates. The RFC only specifies that the root CA "may" be omitted. However, it is best practice to omit this certificate, and software such as SSL checkers (SSL is an older name for TLS) will trigger alerts if a device includes the root CA in the chain.

How does Bob know that the intermediate CA certificate is valid? He must continue up the chain of trust, checking its signature. In the example in Figure 2-33, the issuer of the intermediate CA certificate is a root CA, so Bob finds the root CA certificate and uses the public key in it to check the signature for the intermediate CA certificate. The root CA certificate was issued by the root CA itself, so the chain stops there. Bob must simply trust this certificate on face value. Typically, this means that the root CA certificate must be installed as trusted in advance.

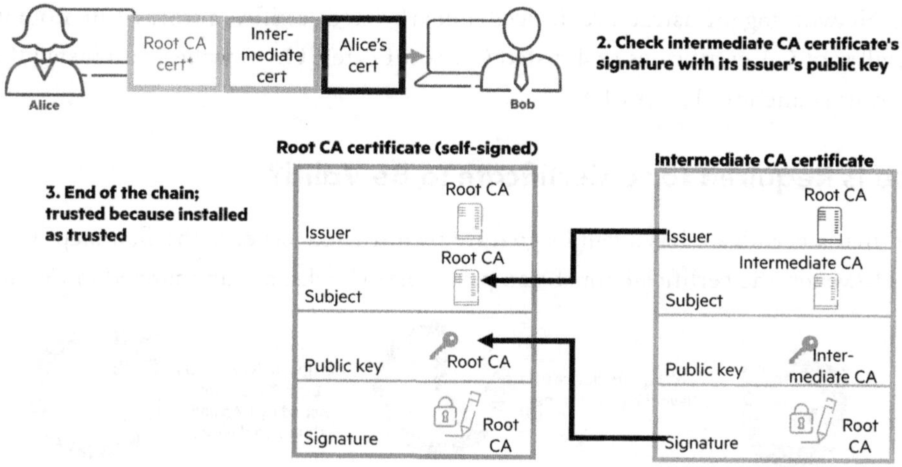

Figure 2-33: Validating Certificates and the Chain of Trust (Steps 2 and 3)

What Does It Mean for a Root CA Certificate to Be Trusted?

In Figure 2-34, you see how a root CA certificate can be installed as trusted on various devices. On Windows machines the certificate is installed in the Trusted Root Certification Authority folder in the certstore. Windows machines come with well-known root CA certificates already installed on them, and admins can choose to install more certificates. Other operating systems have similar certificate stores.

Figure 2-34: What Does It Mean for a Root CA Certificate to Be Trusted?

HPE Aruba Networking infrastructure devices similarly have certificate stores. On mobility devices, you can install root CA certificates as Trusted CA certificates. On switches, you install the root CA certificates in trust anchor (TA) profiles.

What Else Is Required for a Certificate to Be Valid?

Confirming that a certificate has a valid signature by a trusted issuer is the first step in validating a certificate. However, the certificate must also pass other checks, as summarized in Figure 2-35.

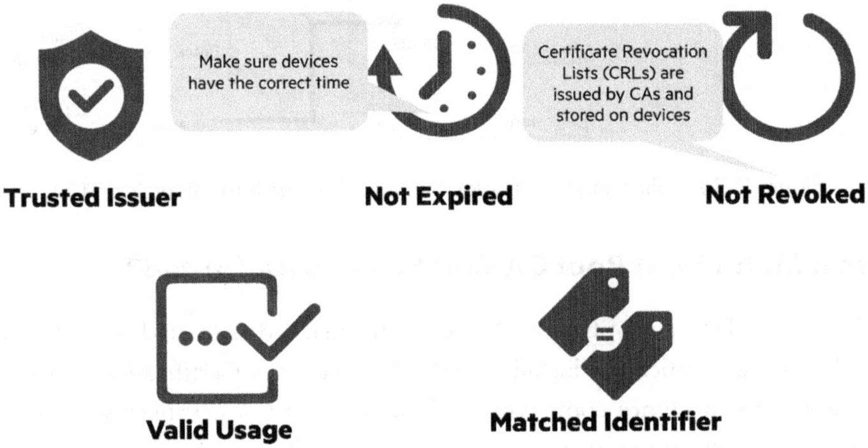

Figure 2-35: What Else Is Required for a Certificate to Be Valid?

To be valid, the certificate must not be expired. In other words, the current date is within the validity period specified by the certificate. This is one reason that it is critical that all devices have correct clocks. Otherwise, the device might start rejecting certificates that are actually valid because the device believes that the current time is outside the certificate's validity period.

In addition, applications and devices check that the certificate's serial number is not on a certificate revocation list (CRL) issued by the CA. CRLs specify any certificates that the CA has decided to revoke because, for example, it determined that the subject obtained the certificate under false pretenses.

The device sending the certificate must also be using the certificate in a way for which that certificate is intended to be used. For example, if the device is using the certificate public key to sign data, the certificate must include the data signature usage.

Finally, devices and applications will attempt to determine if the subject name or SAN matches the identity of the device that sent the certificate. For example, Bob might be communicating with a Web server, which sends Bob's device its certificate. The certificate has an FQDN in its CN and

SAN. Bob's Web browser checks that the FQDN in the certificate matches the FQDN with which he is attempting to communicate.

Certificate Validation in Web Browsers

You need to understand an important additional point about how Web browsers validate the identity in the certificate.

Current versions of browsers such as Chrome and Mozilla Firefox actually check the Web server's FQDN against the SAN, *not* the CN in the main subject field. As shown in Figure 2-36, the browser will present an error if the certificate has only a CN even if that CN matches the server's FQDN and even if the certificate checks out as valid in every other way.

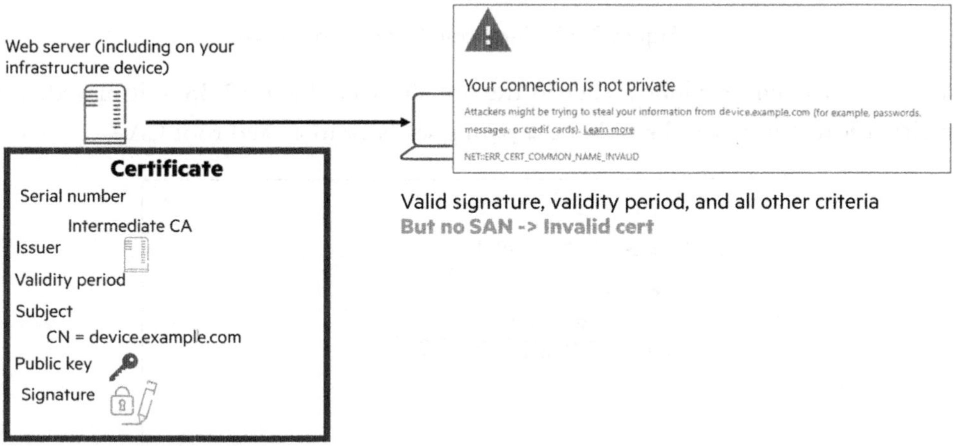

Figure 2-36: Certificate Validation in Web Browsers

You will need to keep this in mind as you create certificates on your devices if you want to use those certificates for HTTPS.

An example of how one Web browser, Chrome, shows that it trusts a Web server's certificate is shown in Figure 2-37. As you see, the browser shows a lock icon. If you want to see more information about the certificate, you can click the icon and then **Certificate**.

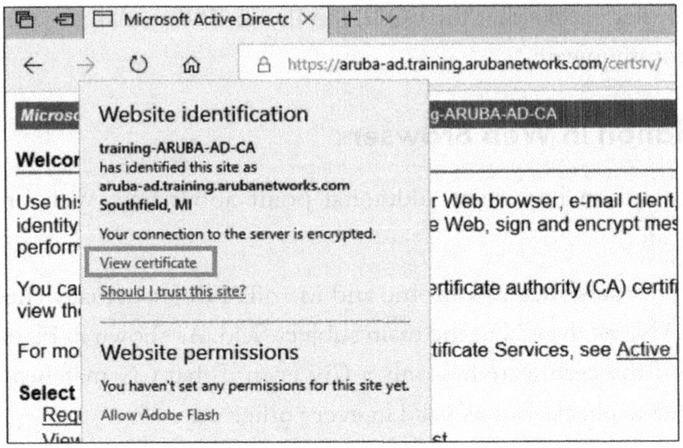

Figure 2-37: Certificate Check in a Browser

On a Windows machine, a window such as the one shown in Figure 2-38 is displayed. You can click the **Certification** Path tab and see that the path goes back to a valid root CA.

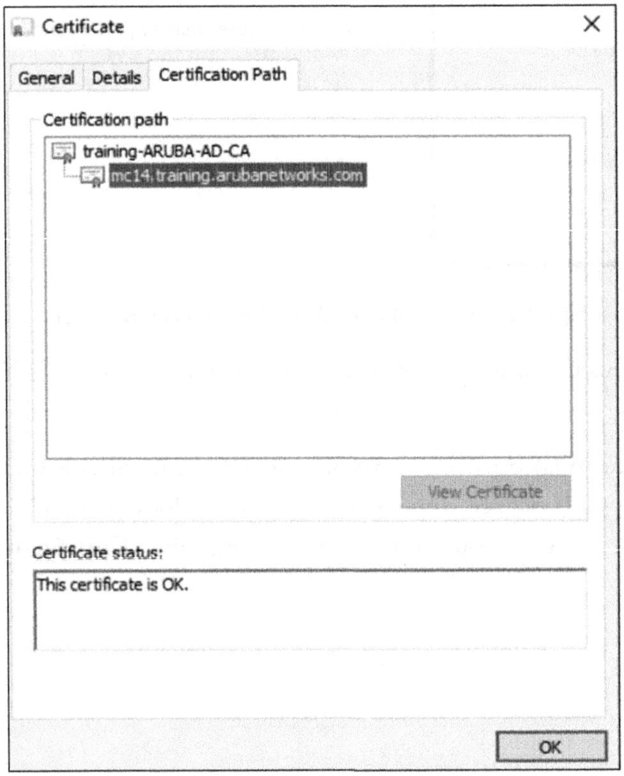

Figure 2-38: Certificate > Certification Path

Public CAs versus Private CAs

You now understand how certificates exist and are validated within a PKI. Next you will consider how devices obtain their certificates. You can have your users and devices obtain certificates from public CAs, your organization's own private CA, or both.

Public CAs are recognized and trusted by virtually all devices. Windows, Apple, Android, and other vendors pre-install these CAs' root certificates in their devices' operating systems. Examples of prominent public CAs include Comodo, DigiCert, and GoDaddy, but other public CAs exist. You would typically install CA-signed certificates on public-facing services so that you know that any devices that interact with those services will trust the certificates automatically. However, you must pay a fee to a public CA to obtain the certificate, which is why organizations are selective in which services use these certificates.

Companies can set up their own private CAs that are specific to their own organization. They often use the Windows Certification Authority (CA) because it is domain aware. However, open-source options such as OpenSSL also exist. A company that does not want to manage its own PKI could also obtain a managed private CA; for example, Globalsign offers such an option.

A private CA can be a cost-effective choice for issuing certificates to many end-users and endpoints. In addition, the company can maintain more control over exactly which users and devices receive certificates, which can establish greater security for internal communications. With a private CA, the company has control over the PKI and can create the intermediate CAs that it requires. In addition, network infrastructure devices and non-public-facing services often use private IP addresses and FQDNs. A public CA cannot issue certificates with a private IP address in the subject name or SAN, based on RFC 1918. Similarly, public CAs do not issue certificates with subject names or SANs that specify private domain names, such as mydevice.example.local. Therefore, a company must use a private CA to obtain certificates for devices and services with private IP addresses or FQDNs.

Self-Signed Certificates

Rather than obtain certificates from CAs, you could theoretically have your devices use self-signed certificates. As shown in Figure 2-39, a self-signed certificate is simply a certificate that a device issues to itself and signs with its own private key. As you learned earlier, root CAs have self-signed certificates. However, within a true PKI, only the root CA should have a self-signed certificate.

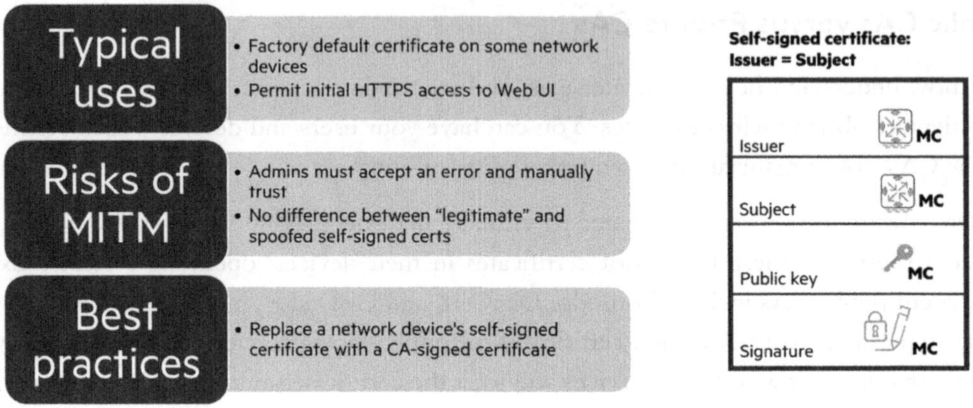

Figure 2-39: Self-Signed Certificates

Some HPE Aruba Networking infrastructure devices come with default self-signed certificates simply so that admins can connect to them using HTTPS for initial configuration. However, because a self-signed certificate is not signed by a trusted CA, the certificate does not truly validate the device's identity. If a network device is using its self-signed certificate, when an admin accesses its Web UI, the admin will need to accept a certificate error in the browser. Getting people used to accepting errors is not good for security because those people might later accept an error when a true threat is occurring. For example, a hacker might try to set up a MITM attack between the admin and the network device, submitting a spoofed, self-signed certificate to the admin's device. The admin has no good way to detect the difference between the "real" self-signed certificate and the spoofed one.

For this reason, HPE Aruba Networking recommends that companies install CA-signed certificates on network infrastructure devices such as Mobility Controllers (MCs) and switches, rather than use self-signed certificates in production.

Online Versus Offline Keypair and Certificate Signing Request (CSR) Generation

Now look at an example of how you can obtain a certificate for a device such as an MC, IAP, or switch. To request the certificate, you must first create a public/private keypair. You must also create a certificate signing request (CSR), which will include the device's subject, public key, and other requests for the certificate.

You have two choices for how you can create the public/private keypair and CSR: online, on the device that will own the certificate, or offline, off of the device that will own the certificate using another application.

When you create the keypair and CSR online, the private key remains on the device—perhaps permanently depending on the device. This can be good for security, but might interfere with certain applications. For example, the company might want to back up the private key so that admins can install the same certificate on new hardware if the original device fails. Or the company might want to deploy the same certificate across multiple devices in a cluster, which requires installing the private key on all of the devices so that the devices can use the certificate.

Some devices require online generation, but on others, you can generate the keypair and CSR offline using another application such as OpenSSL. Your organization might want to create the CSR offline so that security teams can control SANs, key usages, and other fields. Creating the CSR offline also gives you the ability to back up the private key and to deploy the same certificate and private key on multiple devices.

When you create the CSR offline, though, remember that you are creating the private key offline too, which makes it more likely to be compromised. You must take appropriate measures to secure the private key. This could mean encrypting the private key, storing it in a physically secure location, or both.

Creating Certificate Signing Requests (CSRs) and Keypairs

Now look at the process in more detail.

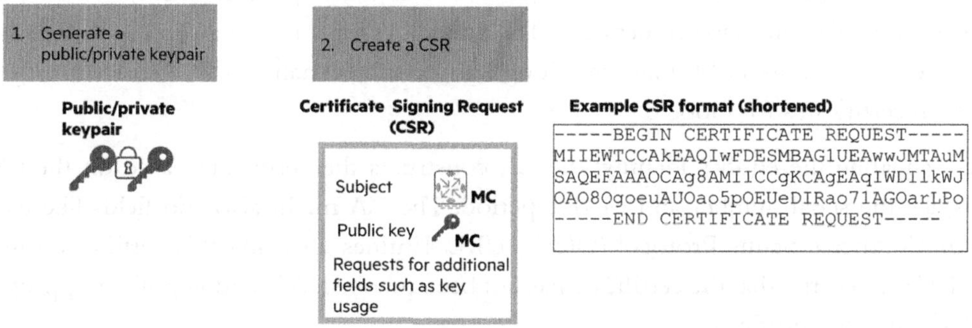

Figure 2-40: Creating Certificate Signing Requests (CSRs) and Keypairs

First, you generate a public/private keypair, which involves creating "entropy." Basically, this means that the device refers to external factors such as clocks or user mouse movements to come up with as random a keypair as possible.

Next you generate the CSR. The CSR includes the subject name and public key. It might also include requests for other fields like the validity period and key usages.

In Figure 2-40, you can see what a CSR, using Base64 format, looks like.

Obtaining Certificates

After generating the keypair and CSR, you submit the CSR to the CA, as shown in Figure 2-41. There are many ways to submit a CSR, depending on the CA type. You might submit the CSR through a website or, for a Windows CA, through the domain.

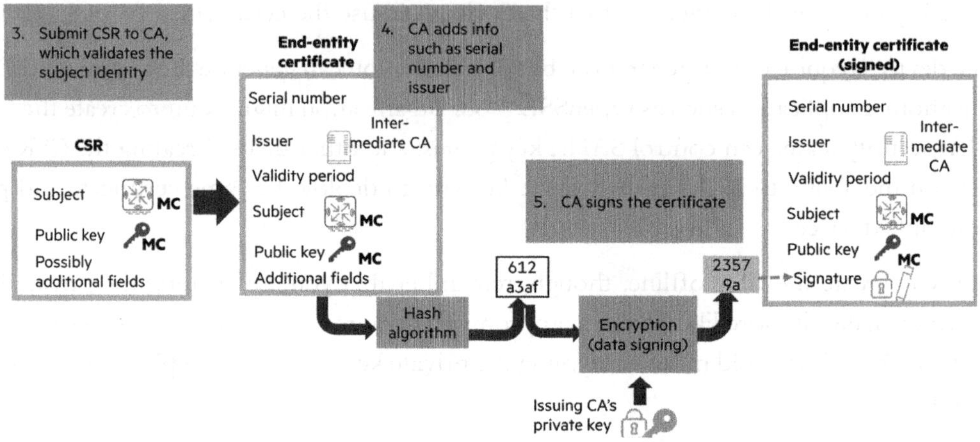

Figure 2-41: Obtaining Certificates

Trusted root CAs are expected to take measures to ensure that the entities requesting a certificate are really who they are. Windows CAs require domain authentication. Each public CA has its own measures for validating subject identities. The public CA must take steps to check, for example, that someone who submits a CSR that uses Google as the subject name or DNS SAN is truly authorized to obtain certificates for Google.

After validating the entity's identity, the CA constructs the certificate based on the CSR, adding fields like the serial number and validity period. The CA might also add fields like usage and an Online Certificate Status Protocol (OCSP) URL. Entities checking this certificate can check the OCSP URL to verify that the certificate has not been revoked. After adding all the appropriate fields, the CA signs the certificate.

The CA then returns the signed certificate to you.

As you see, the private key is never transmitted as part of this process. This key must remain private and secure.

Certificate File Formats

Certificates can be stored in several different types of files. It is important that you understand the file formats so that you can install certificates correctly. The format types fall into two broad categories: binary encoding and Base64 (ASCII) encoding, as shown in Figure 2-42.

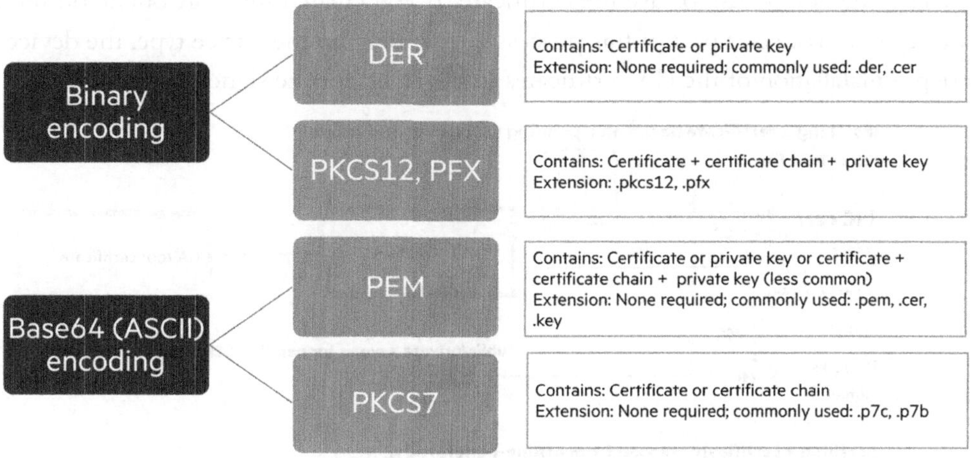

Figure 2-42: Certificate File Formats

Binary encoded files use X.509-defined binary values to encode the certificate. Binary formats include DER. A DER file can include a certificate or a private key. DER files do not require a specific extension, but you will often see them with .der or .cer extensions.

PKCS12 and PFX files also use binary encoding. However, these files bundle the end entity's certificate, all the certificates in the chain back to the root CA, and the private key associated with the end entity's certificate. PKCS12 and PFX files are useful for situations in which someone created the public/private key pair offline, and you need to install both the certificate and the private key on the device that owns the certificate. These files use the .pkcs12 or .pfx extensions. The two types of files are much the same, but PFX is often used on Windows devices.

Base64 encoding uses ASCII characters to encode the certificate. This encoding also marks each component with characters such as -----BEGIN CERTIFICATE----- and ----END CERTIFICATE-----. Perhaps the most common Base64 format is PEM. A PEM file can contain a certificate or a private key. A PEM file can even bundle a certificate, certificate chain, and private key much like a PKCS12 or PFX file; however, this is a less common usage. You will often see extensions such as .pem, .cer, and .key (for key files) for PEM files.

PKCS7 is another Base64 file format. A PKCS7 file contains a certificate or certificate chain and usually uses the .p7c or .p7b extension.

Installing Certificates

In Figure 2-43, you see an example of installing a signed certificate on a device like an MC. You can install a DER or PEM file that includes only the signed certificate. In this case, the device must already have the private key associated with the certificate's public key installed on it. Without the private key, the device cannot use the certificate. If you created the CSR online on the device, the private key was created as part of this process. Depending on the device type, the device might also require pre-installation of the CA certificates in the chain for the signed certificate.

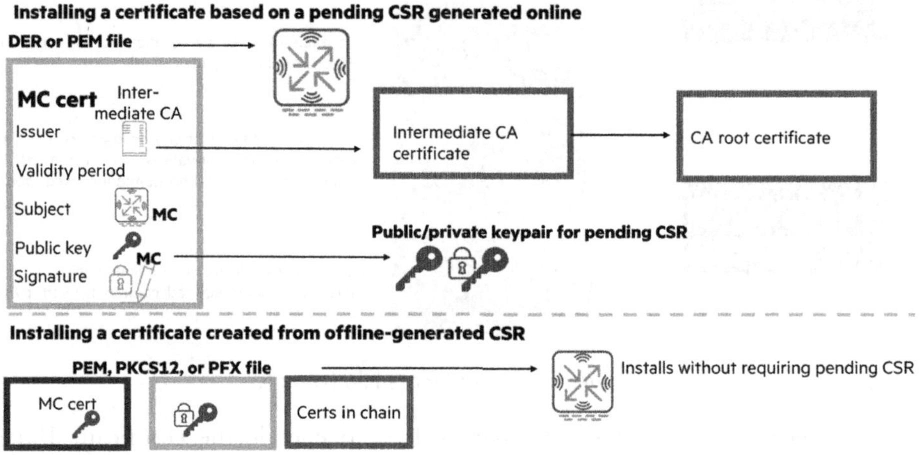

Figure 2-43: Installing Certificates

You can also install a bundle of the device certificate and the associated private key using a PEM, PKCS12, or PFX file. The bundle might further include the CA certificate chain. Because the file has the private key included in it, the signed certificate public key does not need to match with a public/private keypair in a pending CSR. This is the appropriate choice when you or someone else created the CSR offline.

Secure Communications: Transport Layer Security (TLS)

You will next learn about Transport Layer Security (TLS).

Transport Layer Security (TLS)

TLS is used by many upper layer protocols, including HTTPS and RADIUS secure (RadSec), in order to provide a secure communications channel for client-server communications. The TLS channel is often called a secure tunnel. A secure TLS tunnel is simply traffic that is secured by encryption and authentication between two endpoints. (This type of tunnel is different from a tunnel that uses encapsulation to carry possibly unencrypted traffic through a network.)

To create the secure tunnel, the client and server perform a handshake, as shown in Figure 2-44. During the handshake, the server and optionally the client authenticates with a certificate. TLS also uses the certificates and the associated asymmetric encryption to create a shared key. With a shared key in place, the client and server can then switch to symmetric encryption for better speed and performance.

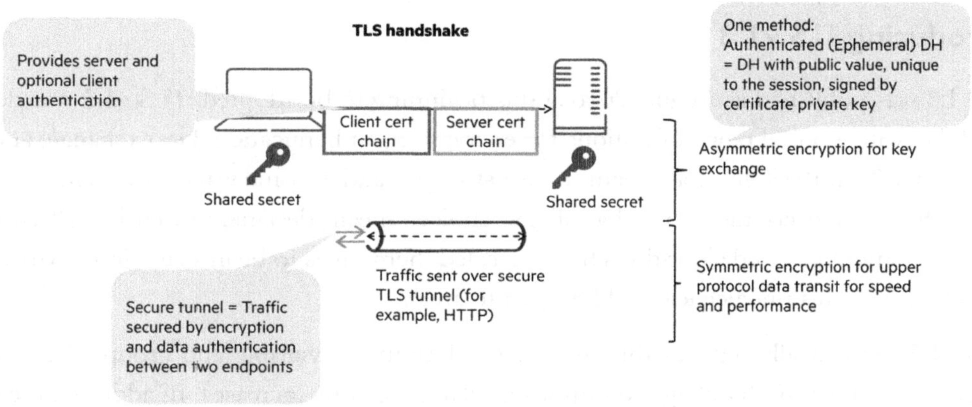

Figure 2-44: Transport Layer Security (TLS)

Consider the process in a bit more detail.

The client and server begin by negotiating the algorithms that they will use to agree on a shared secret, as well as the algorithms that will secure data after the tunnel is created.

The server next sends a certificate chain to authenticate itself to the client. Server authentication protects clients from connecting to rogues—hackers attempting to collect information from the clients or install malware. Client authentication is optional and only performed if the server requests it. Whether the server also enforces client authentication depends on whether the service being protected with TLS is publicly available or restricted to certain users.

Next the client and server generate/exchange the encryption and authentication keys that will be used to secure traffic sent over the TLS tunnel. TLS supports multiple key exchange methods. Often it uses Diffie-Hellman, which you examined earlier, but an "authenticated" version of Diffie-Hellman, which is secure against MITM attacks. The recommended form of authenticated Diffie-Hellman is "ephemeral Diffie-Hellman." With this method, the server selects a Diffie-Hellman public and private value unique for the session. It uses the private key associated with its certificate to sign its Diffie-Hellman public value. This prevents a malicious user from posing as the server and sending its own public value to trick the client into creating a shared key with it. If the server requires client authentication, the client will also sign its Diffie-Hellman public value with its private key. At the end of this process, the client and server have a shared secret.

The client and server then divide this shared secret into encryption and authentication keys. The tunnel is now up. Note that the ability to send correctly encrypted and authenticated data proves that a device generated the same shared key as the other device; this is how a device proves that it really has the private key associated with the certificate that it sent to authenticate. Without that key, it could not have generated the same shared secret.

Introducing TLS v1.3

TLS 1.3 was published in August 2018 and is beginning to be adopted. TLS v1.2 was designed for flexibility with a broad range of options for encryption, authentication, key exchange, and signature algorithms. That flexibility was meant to be a strength, and in some ways it was. However, with too many choices and confused by what all the choices mean, designers often left all choices open. Because some supported algorithms have revealed themselves to be insecure in the years since v1.2 was published, this practice leaves TLS open to attack.

TLS v1.3 dramatically reduces the number of algorithms available and groups them into simple groups of options. With only good options available, security increases. In addition, the handshake is shortened because the client can immediately begin the key exchange rather than start by negotiating algorithms.

Authentication, Authorization and Accounting (AAA)

In this section, you will learn about the authentication, authorization, and accounting (AAA) framework.

Authentication, Authorization, and Accounting (AAA)

You will now move on to examining another key framework for securing networks: Authentication, Authorization, and Accounting (AAA). As you will see, many of the technologies that you already examined, such as certificates, can play a role with AAA, and AAA technologies such as RADIUS can play a role in supporting secure communications.

AAA controls users' access to network services. A service can be many things. In this study guide, it is primarily two things. First it can be access to a management interface such as the Web UI or CLI of a network infrastructure device. This study guide will call this manager AAA. Second, it can be access to a wireless network or access to an Ethernet port. This study guide will call these two types of access user/device AAA.

Controlling access consists of three components:

- **Authentication**—Identify who users (or devices) are and verify that they are who they claim to be
- **Authorization**—Define what services users can use based on their authenticated identity
- **Accounting**—Track what users do

You will look at two AAA protocols: RADIUS and TACACS+.

AAA Roles

The AAA framework includes three devices, shown in Figure 2-45:

- A user (or sometimes device), who is requesting access for a service. This could be a manager for manager AAA or a wireless or wired client for user/device AAA.
- A Network Access Server (NAS), which enforces access to the service. For our purposes, the NAS is an IAP cluster, an MC, or an AOS-CX switch. The NAS is sometimes also called a Network Access Device (NAD) or an authenticator (in context of a protocol called 802.1X, which you will examine in more detail later). The NAS might also be called a RADIUS client (when RADIUS is the AAA protocol), or a TACACS+ client (when TACACS+ is the AAA protocol).
- An AAA server, which verifies users' identities, uses policies to determine for which services users are authorized, and holds accounting information submitted by NASes. The AAA server is often called an authentication server more colloquially. It can also be called more specifically a RADIUS server or TACACS+ server. In this study guide, the AAA server will be HPE Aruba Networking ClearPass Policy Manager (CPPM).

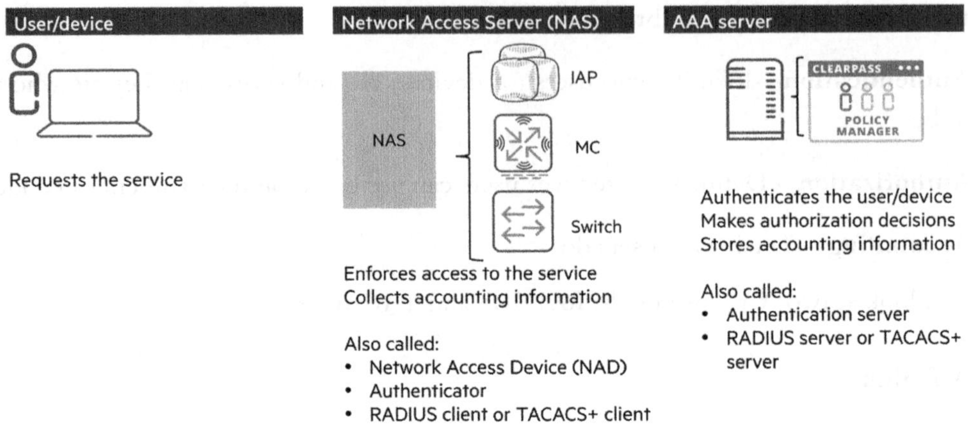

Figure 2-45: AAA Roles

High-Level Overview of RADIUS Authentication/Authorization Process

RADIUS is the most common AAA protocol. It allows users to request many types of services.

Figure 2-46 illustrates the broad strokes of the RADIUS authentication and authorization process. You will look at more detailed processes when you look at using RADIUS to control network access in a later chapter. For now you simply need to understand the broad framework.

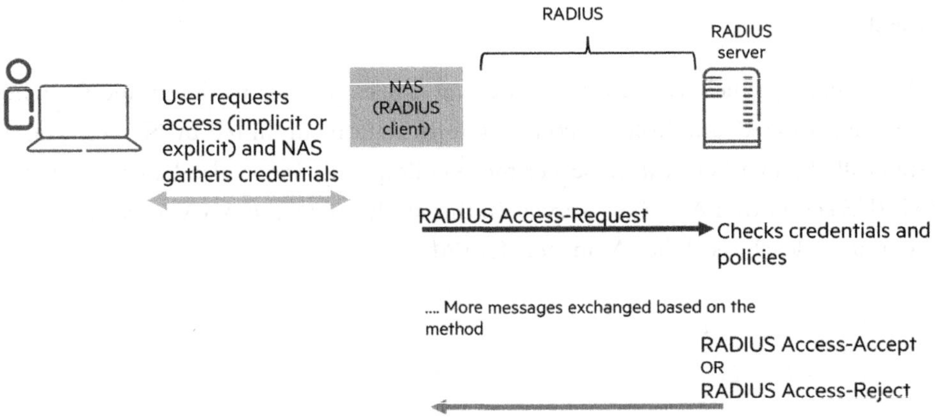

Figure 2-46: High-Level Overview of RADIUS Authentication/Authorization Process

A device attempts to gain access to a service hosted by a NAS. For example, the device attempts to log into the NAS's Web UI. As part of this process, the NAS collects credentials from the user, as dictated by the type of service that the user is accessing. For example, the user submits credentials using login fields on the Web UI login page.

The NAS is configured to use a RADIUS server to authenticate the user on the backend, so it sends a RADIUS Access-Request to the RADIUS server. The request includes information about the user requesting access and the user credentials.

The RADIUS server checks the credentials against its user account database. The server further uses policies to decide if the user is authorized. For example, a user could submit valid credentials, but be requesting a service for which he or she is not authorized. In that case, the RADIUS server would reject the request, just as it would if the credentials were incorrect. Based on the results, the RADIUS server sends an Access-Accept or Access-Reject message to the NAS.

The NAS enforces the decision and typically informs the user of the decision in some way. For example, the NAS lets the user log into the Web UI. Or the NAS prevents the user from logging in and sends a login failure message over HTTP/HTTPS.

RADIUS Authentication Methods

RADIUS can work with several authentication protocols, as shown in Figure 2-47. These protocols include Password Authentication Protocol (PAP), Challenge Handshake Authentication Protocol (CHAP), Microsoft CHAP (MSCHAP), and Extensible Authentication Protocol (EAP).

Auth protocol	Used for	Credentials	Remarks
PAP	AOS-CX switches: – Manager authentication	Username and password	**Lowest security** Can expose passwords to cracking attempts
CHAP	– Some user/device authentication	Username and password	**Lower security** Can expose passwords to more limited cracking attempts Requires access to plaintext passwords on RADIUS server (reversible encryption)
MSCHAP	MCs: – Manager authentication – Some user/device authentication	Username and password	**Low security** Provides mutual authentication Provides better password protection than PAP/CHAP
EAP Multiple methods	User/device authentication (802.1X)	Username and password or certificates (preferred)	**Best security with the proper methods** (EAP-TLS recommended) Protects credentials with TLS

Figure 2-47: RADIUS Authentication Methods

PAP

RADIUS PAP is less secure than the other methods. Many have a misconception that RADIUS PAP transmits the password in plaintext, which is true of PAP on its own, but not of RADIUS PAP. RADIUS PAP does encrypt the password, protecting it from casual sniffing and hacking. However, the method that RADIUS PAP uses to encrypt the password has been discovered to be vulnerable to attack. It can expose both the password and the key used to encrypt the password to hackers.

CHAP

CHAP sends a hash of the password, but this hash is subject to offline dictionary attacks. While RADIUS CHAP is generally preferred to RADIUS PAP, keep in mind that RADIUS CHAP requires the RADIUS server to store the password with reversible encryption, which is not possible for some user databases and requires non-default settings for others.

MSCHAP

MSCHAP is more secure than PAP or CHAP for several reasons. It requires mutual authentication, which means that the RADIUS server must authenticate to the device submitting the credentials. Some MITM attacks involve rogue RADIUS servers that accept all authentication attempts, so mutual authentication is critical to ensure that users are connecting to the trusted network. In addition, MSCHAP does not send the password or even a password hash over the network. Instead it uses a password hash to encrypt a field, which both protects the password and allows the RADIUS server to store it with non-reversible encryption. However, MSCHAP still has known vulnerabilities, principally because its uses MD5 for the hashing mechanism.

Note that, although PAP, CHAP, and MSCHAP, all have known vulnerabilities, they are sometimes the only options available for you. For authenticating managers attempting to log in to the CLI or Web UI, AOS-CX switches support PAP and CHAP. MCs support PAP and MSCHAP. These protocols also provide the backend authentication protocol for when the HPE Aruba Networking network infrastructure authenticates devices with MAC-Auth or authenticate guests through a captive portal. Sometimes the relatively low security is acceptable because the communications run over relatively trusted links. As you will learn, HPE Aruba Networking devices also let you enhance the security when you implement these protocols by using RADIUS Security (RadSec).

EAP

802.1X uses EAP, which supports several methods such as Protected EAP (PEAP) and EAP-TLS. These methods provide mutual authentication and secure encryption of sensitive information. EAP offers the most secure authentication protocols for RADIUS, and 802.1X is the recommended method for authenticating users and devices.

Credentials

The type of authentication protocol determines which type of credentials are supported, as shown in Figure 2-47. Most methods use usernames and passwords, but EAP-TLS authenticates users with certificates, which you learned about earlier.

Supplemental information on RADIUS PAP and CHAP Security

In RADIUS PAP, the NAS creates an MD5 hash of a shared secret, preconfigured on both the NAS and RADIUS server, plus a random 16-octet authenticator value, which is transmitted in plaintext so that the server has it. The NAS then XORs this hash with the password, padded out to the same length of the hash. To launch an attack, hackers can attempt to authenticate and then capture the RADIUS packet with their encrypted password. By XORing the encrypted password with the password that they tried, the hackers can obtain the hash of the shared secret and known authenticator value. They can then run an MD5 dictionary attack to discover the shared secret, at which point they can decrypt all passwords. Even if they are unsuccessful at cracking the shared secret, the hackers can use the valid hash that they discovered to generate valid requests. With these valid requests, the hackers can run a dictionary attack on a user's password.

RADIUS CHAP also uses an MD5 hash to obscure the password—CHAP hashes the password with a random challenge value—so it is vulnerable to MD5 dictionary attacks. However, these attacks expose a single password, not the shared secret that unlocks all keys, so they are more time-consuming and less valuable.

RADIUS Packets and AVPs

RADIUS packets use UDP, which means that the RADIUS client and server do not establish a formal session. They simply send messages to each other and trust that the messages will arrive at the destination. If a client does not receive a response, it resends the request. You can configure timeout and retransmission settings to control how long the client will keep trying for a response.

The well-known UDP port for RADIUS authentication is 1812. (Some legacy implementations use 1645, but 1812 is recommended.)

The packet contents are summarized in Figure 2-48. After the UDP header, the packet includes a code that specifies what type of packet it is, such as Access-Request. The packet also includes an ID, which uniquely identifies the request. All the Access-Request, Access-Challenges, and Access-Accept or Access-Reject messages for the authentication/authorization process share an ID.

Most of the packet consists of attributes, or attribute value pairs (AVPs). An attribute is the name defined for a specific parameter while the value is the setting for that parameter for the current request or response. For example, one attribute is "User-Name;" the value for that attribute might be "Alice." So, the AVP is "User-Name: Alice."

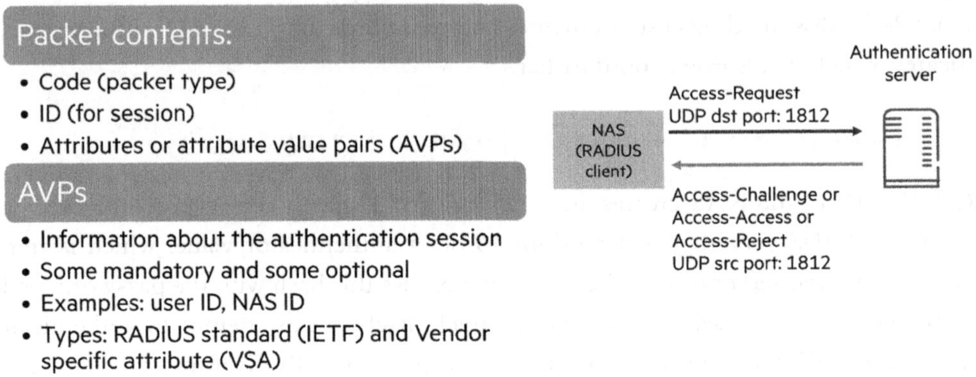

Figure 2-48: RADIUS Packets and AVPs

A few AVPs are mandatory, but RADIUS defines many optional attributes, which the RADIUS clients and RADIUS servers can include as necessary for the situation at hand. The RADIUS standard itself defines many attributes, which are sometimes called RADIUS IETF attributes. Vendors can also define their own attributes, called vendor specific attributes (VSAs).

In practice, the attributes and VSAs are transmitted as numbers, which are defined in a RADIUS Dictionary. For example, instead of including IETF:User-Name, the AVP in the actual packet specifies 0:1 because 0 is the ID for the IETF dictionary and 1 is the value for the User-Name attribute within this dictionary.

AVP Role in Decisions

AVPs permit flexibility for the RADIUS packets. A NAS can choose the relevant AVPs to include for a particular request, and a RADIUS server can do the same with its responses.

As shown in Figure 2-49, Access-Request AVPs give the RADIUS server context for the decisions that it makes. For example, "Service-Type: login" indicates that a user is trying to log into a management interface on the NAS. The Called-Station-ID AVP might include the SSID to which a wireless user connects.

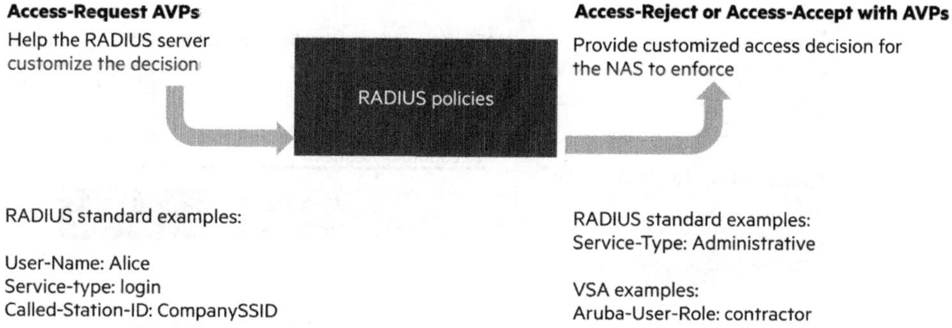

Figure 2-49: AVP Role in Decisions

The RADIUS server can draw on the information provided by the AVPs to make policy-based decisions. For example, a policy might permit a user to access one SSID but not another.

The output of the RADIUS policy is an access decision: either an Access-Reject or Access-Accept. An Access-Accept packet might simply grant or deny access, but it can also include AVPs that customize the services for which the user is authorized. For example, when the RADIUS server authorizes a user to log into a management interface, it could send a Service-Type to say that the user is given administrative (read-write) access.

Vendors can define VSAs to communicate settings that are useful for their solutions. The RADIUS server can then include those VSAs in policies. For example, CPPM can use the Aruba-User-Role VSA to tell MCs, IAP clusters, and AOS-CX switches to which role to assign a connected user. The network infrastructure device can then use that role to adjust how it handles the user's traffic. As you will see, this process helps the network infrastructure to enforce the role-based access controls that are so critical to micro-segmentation and a ZTN.

RADIUS Accounting

RADIUS accounting enables companies to track information about users who have accessed network services, as shown in Figure 2-50.

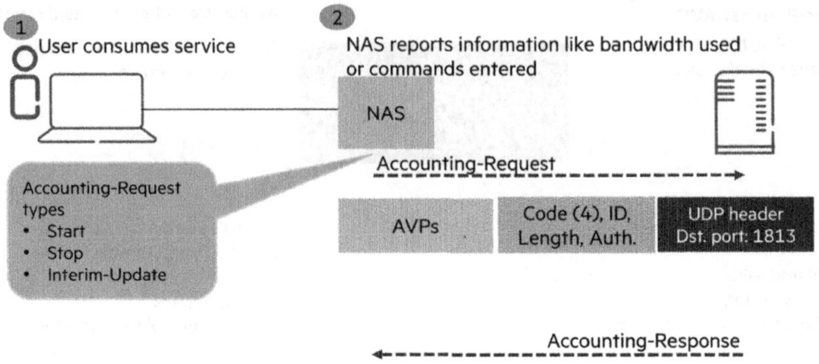

Figure 2-50: RADIUS Accounting

RADIUS accounting takes effect after a user successfully authenticates and is authorized for a service. NAS then creates a session for the user and collects information to send to the RADIUS accounting server. The accounting server is often the same server as the RADIUS authentication server, but can be different.

The NAS sends the accounting message to UDP port 1813, if the RADIUS server is using the standard ports. (Port 1646 is a legacy port for RADIUS accounting, which should not be used). When the RADIUS accounting server receives an Accounting-Request, it must store the information and then return an Accounting-Response as an acknowledgment. (Remember that RADIUS uses UDP, so acknowledgements are not built in.)

A NAS can send Start, Stop, and Interim-Update messages.

A Start message, sent at the start of a session, is optional. This message includes AVPs to describe the user and the session. These AVPs might include the user's identity (User-Name) and location (NAS-IP-Address and NAS-Port).

The NAS can send interim updates at a periodic interval, describing how the user is consuming the services. For example, if the service is network access, RADIUS accounting can track the number of bytes that the user transmits. Or, if the service is management access, accounting can track the commands entered, which can be important for troubleshooting, as well as auditing and compliance purposes. Interim updates are not mandatory, but they can be for network monitoring and control solutions.

Finally, the NAS sends a stop message, summarizing the session when it ends.

RADIUS Security Considerations

It might have occurred to you that a hacker like Mallory could pose as a RADIUS server and send a spoofed RADIUS Access-Accept to grant herself access to services for which she is not authorized.

Similarly, she could try to spoof Access-Request packets to gather as much information as she can from the RADIUS server responses.

RADIUS seeks to prevent such attacks with a shared secret, which is essentially a matching key configured on the NAS and the RADIUS server, as shown in Figure 2-51. (The NAS can have a different shared secret for each RADIUS server with which it communicates, and the RADIUS server can have a different shared secret for each NAS.)

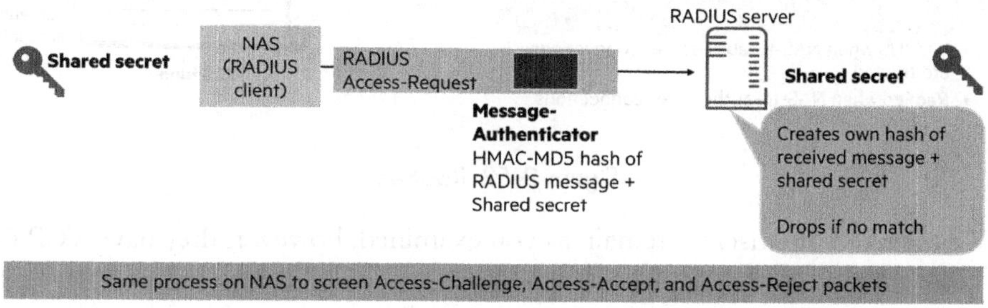

Figure 2-51: RADIUS Security Considerations

One of the RADIUS AVPs that can be included in RADIUS messages is Message-Authenticator. The Message-Authenticator consists of an HMAC-MD5 hash of the entire RADIUS message using the shared secret as a key. When the NAS or RADIUS server receives a RADIUS message, it hashes the packet with the same key and makes sure that its result matches the Message-Authenticator field. If the result does not match, the server or NAS ignores the message.

The Message-Authenticator is required with RADIUS EAP. RADIUS PAP and CHAP use a deprecated method for ensuring data integrity and authenticity, based on the shared secret and MD5 hashes, rather than the more secure HMAC-MD5 hashes. However, all MD5-based hashes have vulnerabilities.

You should also be aware that RADIUS does not encrypt the message, which makes it vulnerable to eavesdropping. (RADIUS PAP does encrypt the password field, but not the rest of the packet.)

In short, RADIUS remains an important protocol for AAA; however, its security mechanism is outdated, so RADIUS should only be deployed on its own on trusted networks.

RadSec

RadSec, formalized in RFC 6614, helps to ensure the security of RADIUS so that you can deploy it over untrusted networks. The RADIUS server and RADIUS client use TLS to authenticate each other with certificates and create a tunnel with strong data encryption and authentication. They then exchange RADIUS messages over that tunnel, as shown in Figure 2-52.

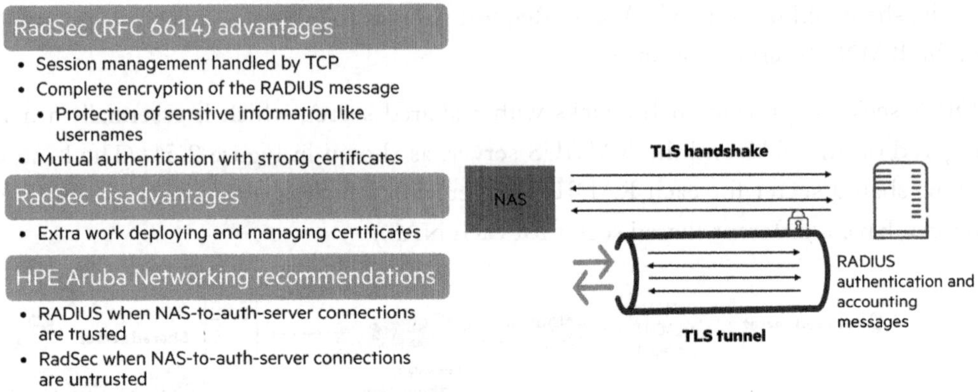

Figure 2-52: RadSec

The RADIUS messages themselves remain as you examined; however, they have TCP headers and use a TCP port, 2083 on HPE Aruba Networking devices, rather than use UDP. RadSec uses a single TLS session for both authentication and accounting.

Because RadSec uses TCP, RADIUS clients no longer need to rely on timeouts and retransmits to determine when a RADIUS server is unavailable. Instead, they can take advantage of the TCP session.

Even more critically RadSec dramatically improves RADIUS security. It encrypts the complete RADIUS message, protecting information such as NAS and client IP addresses, client MAC addresses, client usernames, and other information included in AVPs. In addition, RadSec provides much better guarantee to the NAS that it is communicating with the legitimate RADIUS server and vice versa. A RADIUS shared secret is essentially a password, which means that it is subject to dictionary attacks like passwords, particularly if admins do not set a long and random enough secret. In addition, passwords can be leaked. Certificates provide much stronger authentication. With a proper length for the public/private key pair, the private key is not subject to brute force attacks.

The main disadvantage of RadSec is that you must deploy and manage certificates on all of the network infrastructure devices that act as NASes. This requirement will naturally create more work when you add a network infrastructure device and more opportunities for troubleshooting.

CPPM, MCs, IAPs, and AOS-CX switches support RadSec. HPE Aruba Networking recommends RADIUS when the network connections between the NAS and authentication server are trusted and RadSec when they are not. Different organizations will define what makes network connections trusted or untrusted in different ways, so you should discuss this topic with the organization for which you are working.

TACACS+

TACACS+ is another AAA protocol, the features of which are summarized in Figure 2-53. You will likely use RADIUS more often, so the discussion of TACACS+ will be briefer than that on RADIUS. TACACS+ is used primarily to authenticate users attempting to log into a management interface.

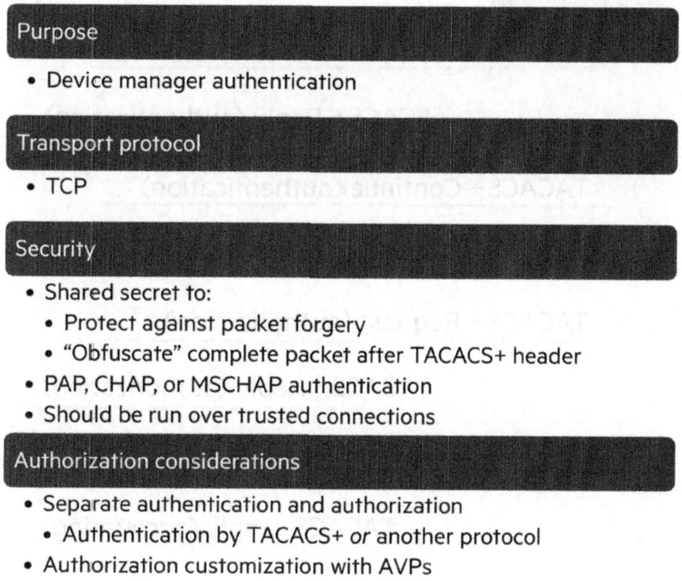

Figure 2-53: TACACS+

TACACS+ is carried over TCP, which handles acknowledgements and retransmissions; therefore, it does not need the timeouts and retransmission settings that RADIUS does. TCP can also detect the difference between a server that is slow to respond and a server that is down.

TACACS+ also provides encryption of the entire message after the TACACS+ header using a shared secret that matches on the NAS and the TACACS+ server. Because the method for encrypting the message does not meet modern standards, the TACACS+ standard refers to this as "obfuscation" rather than encryption.

TACACS+ supports PAP, CHAP, and MSCHAP as authentication methods. CPPM uses TACACS+ PAP. Because TACACS+ does not provide rigorous, modern security, it should be used only over trusted network connections.

TACACS+ separates authentication and authorization more rigidly than RADIUS.

When a NAS needs to verify a manager's identity, it collects the username and password from the management session protocol. It then sends a TACACS+ Start message to the server, as shown in

Figure 2-54. The TACACS+ server checks the credentials and indicates to the NAS whether the user has successfully proved his or her identity.

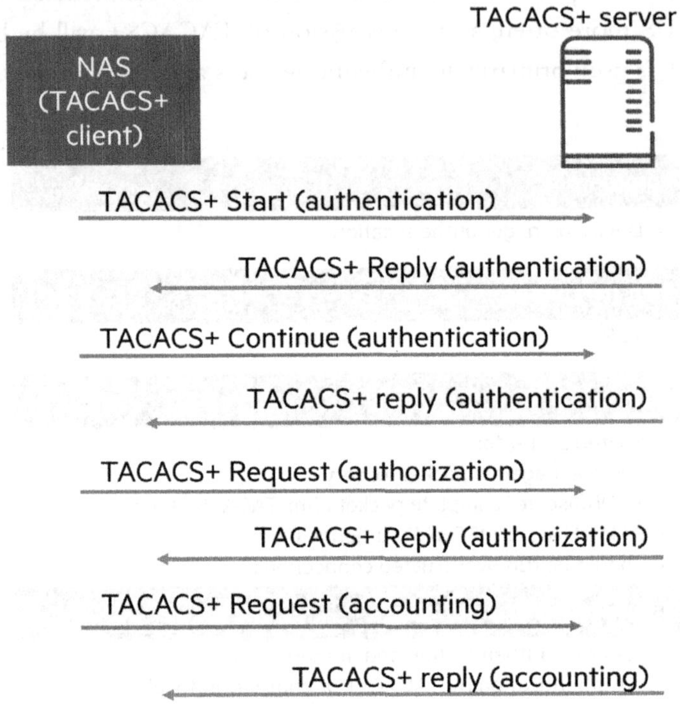

Figure 2-54: Simple Example of TACACS+ Exchange

The NAS sends a separate Request message to authorize a user for specific services. The request includes:

Information about the user's authentication (which may or may not have been performed with TACACS+)

The requested services, indicated by AVPs

The TACACS+ server checks the user information against its policies to determine whether to grant the request. It sends a reply to the NAS with its decision. The authorization status field indicates whether to accept the request. The messages field has messages that are displayed to the user to explain the decision. AVPs can customize the service much like RADIUS AVPs do. For example, the privilege-level AVP controls what level of access the manager has in the management interface (1 to 15 with 15 being full read-write access). A NAS might also be set up to ask the TACACS+ server whether an administrator is authorized to enter particular commands.

Finally, much like RADIUS accounting, TACACS+ accounting enables the NAS to track information about the consumed service.

RADIUS vs TACACS+

While RADIUS and TACACS+ are both AAA protocols, they have some key differences, which are summarized in Figure 2-55.

Attribute	RADIUS	TACACS+
Connection protocol	UDP *TCP for RadSec	TCP
Encryption	None, except password in PAP EAP portion of RADIUS message might be encrypted *Encryption with RadSec	Entire packet (but uses obsolete methods)
Approach to AAA	Combines authentication and authorization	Separates authentication and authorization
Intended use with HPE Aruba Networking products	User AAA Admin AAA	Admin AAA

Figure 2-55: RADIUS vs TACACS+

RADIUS uses UDP for its transport protocol while TACACS+ uses TCP. This means that RADIUS transmissions are less reliable. The RADIUS client does not have an ongoing session with the RADIUS server. The client will need to use RADIUS timeouts and retries to determine when the RADIUS server has become unreachable while TACACS+ can rely on the built-in mechanisms of TCP. However, as you learned, RadSec makes RADIUS more reliable by sending the messages over a TLS tunnel.

RADIUS itself does not encrypt the packet. The exception is that if RADIUS PAP is used, RADIUS encrypts the password field. However, the form of encryption is not secure. Note, however, that when RADIUS EAP is used, the EAP method might provide encryption for EAP messages inside RADIUS packets. In addition, RadSec provides strong encryption for the entire RADIUS message.

TACACS+ encrypts the entire packet. However, the encryption is no longer up to modern standards, so TACACS+ only claims obfuscation and not true encryption.

RADIUS combines authentication and authorization into one exchange. TACACS+ keeps these operations more distinctly separated.

How you use the protocols with HPE Aruba Networking products differs. You can use RADIUS to authenticate users, authorize them to access the network, and run accounting on the connection. You might also use RADIUS to authenticate admins, to authorize them for management access to a device, and to run accounting on the management session. You use TACACS+ only for admin AAA.

Summary

You should now understand the technologies behind secure network communications, as well as some of the major regulations for cyber security. In the next chapter, you will begin with the first step in the HPE Aruba Networking security strategy: establishing a trusted network infrastructure by hardening your devices against attack.

Learning checks

1. To which organizations does PCI DSS 3.2 apply?

    a. Any US federal agency or contractor

    b. Any organization that collects and stores data about EU citizens

    c. Any organization that handles users' personal health data

    d. Any organization that transmits or stores credit cardholder data

2. Match each concept with the correct description.

    Concepts

    a. Stream algorithm

    b. Symmetric encryption

    c. Block algorithm

    d. Initialization vector (IV)

    Descriptions

    __ A value used within the encryption process to help prevent identical key reuse

    __ Encryption algorithms such as AES, which encrypt sets of data that are the same size as the key

    __ The same key encrypts and decrypts data

    __ A form of encryption that encrypts data bit by bit

3. True or false: MD5 is recommended for key-based hashing over SHA2.

   a. True

   b. False

4. Alice hashes her message, encrypts the hash with her private key, and sends the message plus the hash to Bob. What key should Bob use to validate the message?

   a. His own private key

   b. His own public key

   c. Alice's private key

   d. Alice's public key

5. Match each entity with the correct description.

   Concepts

   a. Root CA

   b. Intermediate/subordinate CA

   c. End entity (leaf)

   Descriptions

   __ Entity that issues certificates to end entities

   __ Device that obtains certificates, but does not issue certificates

   __ Entity believed to be trustworthy

6. What are some of the purposes of TLS? (Select two.)

   a. Ensuring that users select good passwords that are resistant to dictionary attack

   b. Establishing encrypted and authenticated communications for upper layer protocols

   c. Submitting CSRs to a CA and obtaining signed certificates

   d. Preventing malicious users from exploiting ARP

   e. Authenticating a server and optionally a client with certificates

7. What level of security does a RADIUS shared secret provide?
   a. It encrypts the RADIUS message but does not ensure the message's integrity and authenticity.
   b. It validates the message's integrity and authenticity but does not provide encryption.
   c. It both validates the message's integrity and authenticity and provides encryption.
   d. It enforces pre-authentication for the NAS to the RADIUS server but does not provide message integrity or privacy.
8. Make notes on what you have learned about certificates. Consider questions such as these:
   - What values do certificates offer for authenticating devices' identity and messages?
   - What does a chain of trust mean?
   - What weak points can exist in certificate deployments?
   - How can you use Windows tools to find a certificate's subject name and SAN?

Note

For answers to all learning checks, refer to Appendix B.

---

3 Hardening AOS-CX Switches

EXAM OBJECTIVES

- ✓ Use best practices for out-of-band management
- ✓ Password-protect switches and use external authentication
- ✓ Disable insecure protocols
- ✓ Enforce physical security

Assumed Knowledge

- Basics of IP networking
- Basics of Ethernet switching
- Basics of 802.11 wireless networks

Overview

This chapter guides you through implementing best practices for hardening your AOS-CX switches. You will learn how to set up out-of-band management and to authenticate managers securely. You will learn how to use secure protocols and how to implement a number of other best practices.

Hardening Overview

You will begin by considering best practicing for hardening network infrastructure devices.

Why Harden Devices?

When switches are not secure, neither is the network that they help to establish. If hackers gain unauthorized management access to a switch, the damage that they can do is nearly limitless.

If hackers simply want to launch a denial of service (DoS) attack, they can delete routes, disable ports, or simply power off the switch. But hackers can take more subtle approaches as well. With access to enough of the network infrastructure, bad actors can reconfigure routes to redirect traffic through their device, which can help them to eavesdrop or even to launch man-in-the-middle (MITM) attacks.

Hackers could also take a number of approaches to eavesdrop on traffic. They could configure mirrors to copy traffic out the port connected to their device. They could clear out switches' MAC forwarding tables, causing the switches to treat all traffic as unknown unicasts and flood all traffic—including out the port -connected to the hacker device.

By collecting IP addresses from the network and engaging in other forms of network reconnaissance, hackers can launch many other kinds of attacks. In short, hardening switches is a necessary measure for security personnel to take to protect their networks against a wide variety of exploits.

Many companies need to comply with local and regional regulations that require specific security measures. By locking down management access and auditing the measures taken, the company can prove that it is securing the network infrastructure properly.

Best Practices Checklist

For the rest of this chapter, you will explore how to complete these best practices:

- Set up out-of-band management network
- Authenticate managers securely
- Enable secure protocols and disable insecure ones
- Ensure physical security and implement other security measures such as Control Plane Policing (CoPP)

For more information, visit the HPE Networking Support site and search for hardening guides.

Out-of-Band Management

You will next learn about establishing out-of-band management.

Network Infrastructure Device Planes

To understand how to secure your switches, you should understand a bit about their software and hardware architecture. Switches divide functions across several planes, as illustrated in Figure 3-1:

- **Management plane**—Each switch has a centralized management plane, which is established in the software running on the switch's main CPU (this CPU is in a modular switch's management module or embedded in a fixed-port switch). The management plane holds the configuration settings. It also hosts management interfaces and interacts with managers.
- **Control plane**—This plane also exists in the software. It runs routing protocols and builds the routing table. It also runs Layer 2 protocols such as Spanning Tree Protocol (STP) and its variants, and it builds MAC forwarding tables. As part of its duties, this plane generates traffic such as routing protocol updates and Bridge Protocol Data Units (BPDUs).

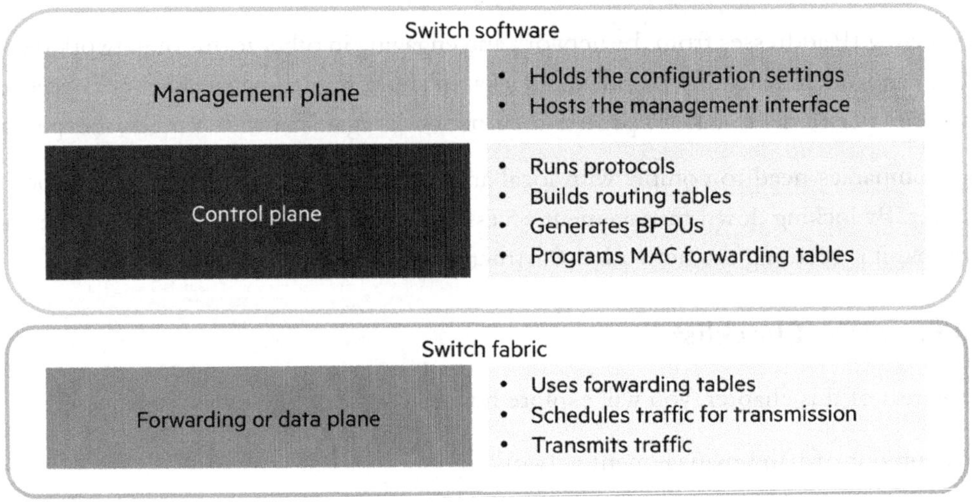

Figure 3-1: Network Infrastructure Device Planes

- **Forwarding plane, data plane, or fabric**—The data plane, also called the switch fabric, exists in the switch hardware. It contains the data channels that receive traffic on one switch port, transmit the traffic through the switch backplane, and send the traffic out another switch port. The fabric includes ASICs, which are purpose-designed circuits for holding hardware forwarding tables. The control plane programs the hardware forwarding tables, which can then make forwarding decisions at the high speeds demanded by 10, 25, 40, and even 100 Gbps connections.

Out-of-Band Management to a Console Port

You can reach a switch's management interfaces in two ways: in-band through connections used for other network traffic or out-of-band through a connection dedicated to management.

Advantages

- Complete separation of management and data planes
- Very difficult to lock yourself out

Disadvantages

- Less flexibility
- Remote access possible, but can be complex

Connection	Protocol	Interface
Console	Serial	CLI (or Menu)

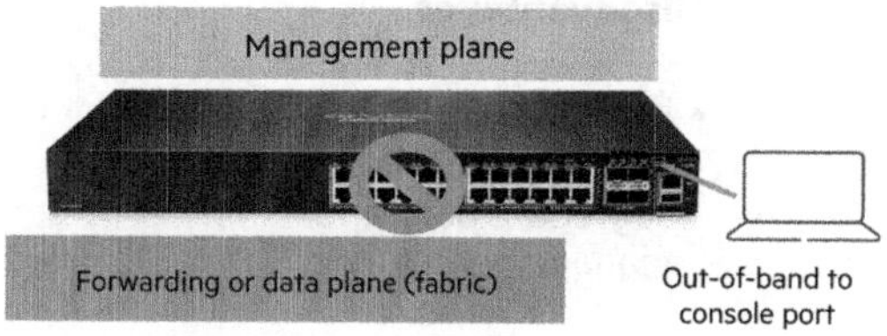

Figure 3-2: Out-of-Band Management to a Console Port

Traditionally, out-of-band management uses a serial connection to a console port on the switch chassis, such as the one shown in Figure 3-2. The console port supports only the CLI and Menu interface, and it connects directly into the switch's management plane. This means that out-of-band management does not require the switch to have an IP address or network connectivity. To open a management session with the switch and access the CLI, you use terminal software such as Tera Term or PuTTY, which are available as freeware, or Microsoft HyperTerminal.

Typically, managers need to have physical access to the switch to reach the console port. If you have properly secured the switches physically, the console port should be secure as well. It is possible to connect console ports to terminal servers, which connect to an Ethernet network. The terminal server accepts incoming Telnet or SSH sessions from managers, translates them to serial, and sends them to the correct console port. However, establishing remote access in this way requires a separate solution which might be unfamiliar to network admins.

In-Band Management

Using in-band management, admins reach the management interface at one of the switch's IP addresses through the same types of ports that are used to carry data for other users on the network (see Figure 3-3). To manage a switch through in-band (networked) access, you must configure the switch with an IP address and subnet mask that is reachable on your network.

Advantages

- No separate infrastructure required

Disadvantages

- Less resistant to traffic issues
- Requires additional settings to segregate data and management traffic

Connection	Protocol	Interface
IP	Telnet	CLI
	SSH	CLI
	HTTPS	Web UI and REST
	SNMP	No interface (accesses MIB objects)

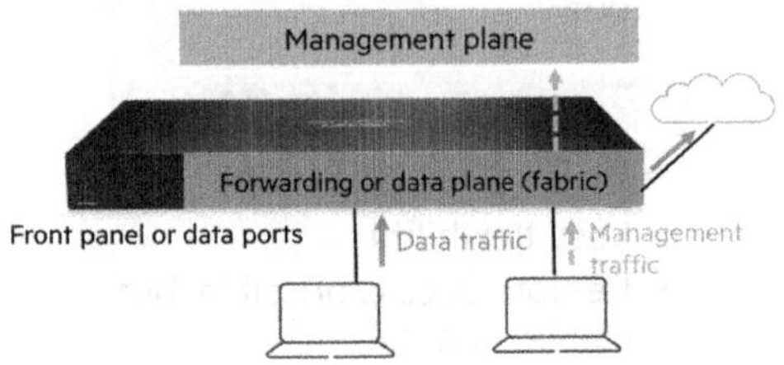

Figure 3-3: In-Band Management

When you access the switch at an IP address, you can access the CLI or Web UI (or the REST interface). To access the Web UI, you simply use a Web browser and HTTPS. For example, if the switch has IP address 10.1.10.10, you will navigate to https://10.1.10.10.

An in-band management session to the CLI requires terminal emulation software, such as PuTTY or Tera Term. This software communicates with the switch, typically by using one of two network protocols. One protocol, Telnet, carries the terminal session in plaintext. The second protocol, Secure Shell (SSH), establishes a secure tunnel for the session using authentication and encryption. Depending on the software version, AOS-CX switches Telnet, SSH, and HTTPS or SSH and HTTPS only.

In-band management is easy to implement and inexpensive. However, management access to the switch can be compromised by issues in the network. In addition, with in-band management, it is technically possible for any user to access a switch's management interfaces. You must take additional steps to segregate data and management traffic and ensure that only the proper users and devices access the management plane.

Ethernet Out-of-Band Management

All AOS-CX switches offer another option for out-of-band management: an Ethernet port called the management (Mgmt) port. Like a console port, this port connects directly to the switch's management plane. But it uses Ethernet, rather than serial, communications, and it has an IP address. This means that you can access the CLI or Web UI at the Mgmt IP address. Use SSH or HTTPS just as you would with in-band access. Because this port uses Ethernet and IP, it provides the convenience that was traditionally associated with in-band access. You can establish an Ethernet management network to connect managers to the switch remotely. The Ethernet management and data networks are completely isolated from each other.

The Ethernet mgmt port combines many advantages of console out-of-band and in-band management. Managers with access to the management network can reach switches' management interfaces securely and remotely from anywhere in the campus.

Sometimes the company will want to extend the management network across the WAN, so that managers can also manage branch networks remotely. For example, a network administrator at the company HQ could access switches at a branch and troubleshoot an issue or perform management tasks like creating a new VLAN. If the company plans to provide external users access into the OOBM network, the company should consider that management stations offer a high value target for hackers and permitting external access to them could expose them to hackers. If you need to do so, you should communicate with the appropriate server teams for best security practices such as strict firewall policies on the station itself and locked down functions.

To achieve the best available security, the OOBM network and the production network should be physically separated. In other words, you should connect the mgmt ports and the management stations to their own switches, other than production core switches. Doing so does require extra effort and equipment. However, management switches can be inexpensive, since 1GbE will provide all the necessary bandwidth for management.

When you use the Ethernet mgmt port, you can configure the switch to reject in-band management access, enhancing security by preventing normal users from ever reaching the management interfaces, without sacrificing the convenience of remote access. You can easily choose which protocols to enable on the OOBM network and which to enable on the data plane, as shown in Figure 3-4. On AOS-CX switches, you tell the switch on which Virtual Routing and Forwarding (VRF) instances particular protocols are enabled. The mgmt VRF corresponds to the mgmt port. Data ports are in the default VRF by default, or admins can create other VRFs and assign ports to them.

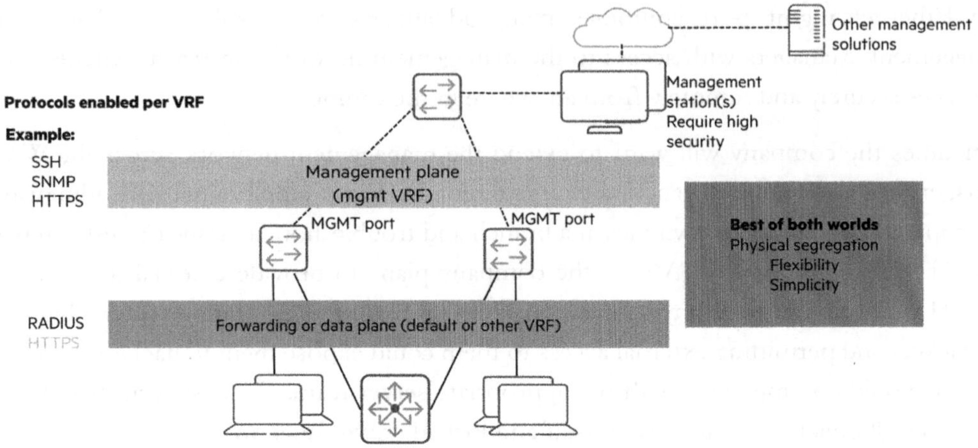

Figure 3-4: Ethernet Out-of-Band Management

This approach makes it easy to specify and determine which protocols the switch accepts where, as opposed to access control lists (ACLs). ACLs are lists of rules that select traffic based on header, and they can be more complex to configure.

Note that enabling a protocol on a VRF applies to traffic using that protocol that is directed to the switch. It has no effect on traffic using that protocol but sent to another destination.

CLI Commands: Management Interface Configuration

The AOS-CX mgmt port is a Layer 3 interface, which means that you associate IP settings directly with it, rather than assign the port to a VLAN and configure IP addresses on the VLAN interface. As shown in the table in Figure 3-5, you can access the port and set an IP address or enable DHCP instead. If you are using DHCP, the OOBM network should have a DHCP server attached.

In configure context

Step	AOS-CX Command
Access the interface	`interface mgmt`

In oobm/management context

Step	AOS-CX command	Notes
Specify static IP address or enable DHCP for the address.	`ip static <IP/prefix length | dhcp>`	Specify a static *or* DHCP address.
Specify a default gateway IP address.	`default-gateway <IP>`	If you are using DHCP, this command might not be necessary.
Specify the static DNS server IP addresses.	`nameserver <IP>`	You can set multiple DNS servers separated by a space.
Enable or disable the interface.	`[no] shutdown`	

Figure 3-5: CLI Commands: Management Interface Configuration

CLI Commands: Locking Down Protocols to the Out-of-Band Management Network

To prevent users from using the data network to access switch management interfaces, and to force them to use the out-of-band management, you must take an additional step. For each protocol, choose whether to enable protocols on the "mgmt" or "default" VRF or both. For example, if you want admins to access the switch CLI only on the mgmt port, enable the SSH server on the mgmt VRF and disable it on the default VRF. See Figure 3-6 for the CLI commands.

In configure context

Step	AOS-CX Command

Lock down management protocols to using the OOBM port or management VRF.	`ssh server vrf mgmt`
	`no ssh server vrf default`
	
	`https-server vrf mgmt`
	`no https-server vrf default`
	
	`snmp-server vrf mgmt`
	`no snmp-server vrf default`

Figure 3-6: CLI Commands: Locking Down Protocols to the Out-of-Band Management Network

Methods for Restricting In-Band Management Access

If the company cannot create a separate OOBM infrastructure, you do have options for locking down in-band access to management protocols on the switch logically.

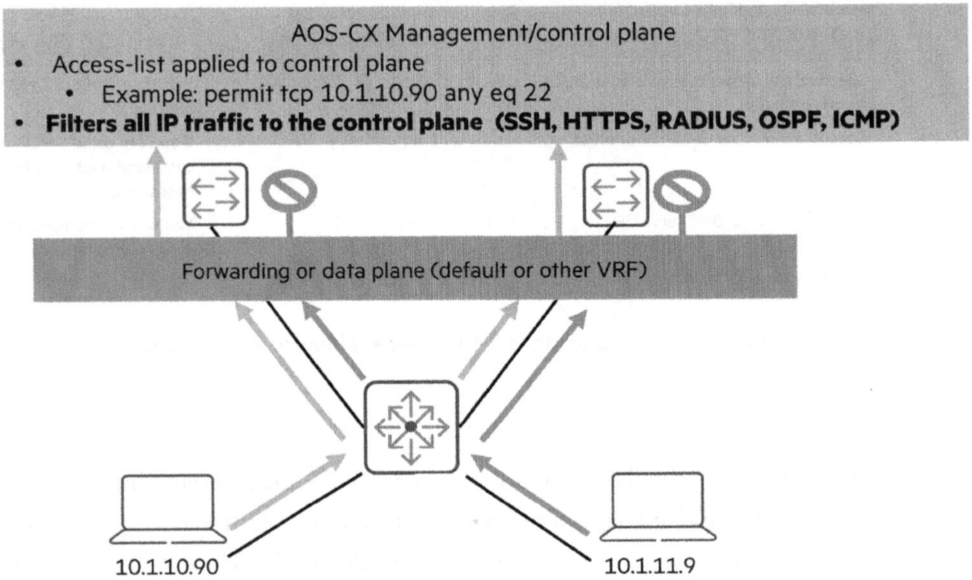

Figure 3-7: Methods for Restricting In-Band Management Access

AOS-CX switches support applying an ACL to the control plane, as illustrated in Figure 3-7. An ACL is a set of rules that examine traffic headers and permit or deny the traffic based on the header. The control plane ACL filters all traffic incoming to the switch itself, rather than just the management protocols such as SSH, Web, and SNMP, and it has an implicit deny all at the end. If you use this option, you must be very careful to determine all the traffic that the switch might need to receive.

Secure Manager Authentication

You now know how to set up a secure connection over which to access switch management interfaces. The next step is configuring the switch to enforce secure authentication of users who reach those interfaces.

Best Practices for Authentication Methods

To prevent unauthorized access, AOS-CX switches require users to authenticate with a username and password. Users can authenticate either through a local account, with unique credentials recognized by the switch itself, or through an external AAA server, either a RADIUS or TACACS+ server, as shown in Figure 3-8.

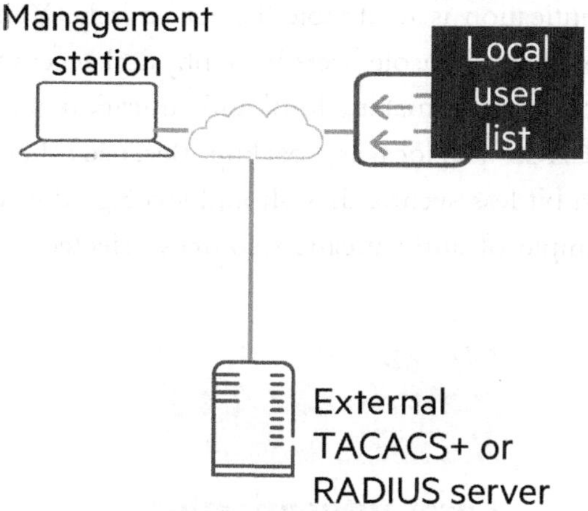

Figure 3-8: Authentication Sources

An external server provides far superior scalability, since one set of credentials can work for many different devices. Using an external server also makes it easier to manage multiple admin accounts. Many enterprise servers have built-in mechanisms for forcing users to create new passwords periodically while local credentials are likely to stay the same for far too long, due to the manual labor involved in renewing them. In addition, if an admin leaves the organization, it is relatively easy to revoke that admin's account on the RADIUS or TACACS+ server. Still, local authentication can be valuable. In the event of a malfunction with a RADIUS or TACACS+ server, local authentication will remain available. Additionally, local authentication requires almost no additional infrastructure.

AOS-CX switches allow customers to specify the authentication type based on the access type—sometimes called AAA channel. The access types define all of the ways that admins can access the switch management interfaces. On AOS-CX switches these are:

- Console
- Telnet
- SSH
- HTTPS

HTTPS encompasses both the Web UI and REST.

You can choose your preferred authentication type for each access type. You can also choose different RADIUS server or TACACS+ server groups for different access types.

The best practice for authentication is to disable Telnet entirely. You should typically use local authentication on the console port. Console users have physical access to the infrastructure, which already provides one layer of security, making local authentication acceptable. Other access types should use RADIUS or TACACS with local as a backup. Because other channels, even if used with mgmt ports, are inherently a bit less secure, they should leverage the scalability and security of an external server. For an example of authentication sources selected for various access types, see Figure 3-9.

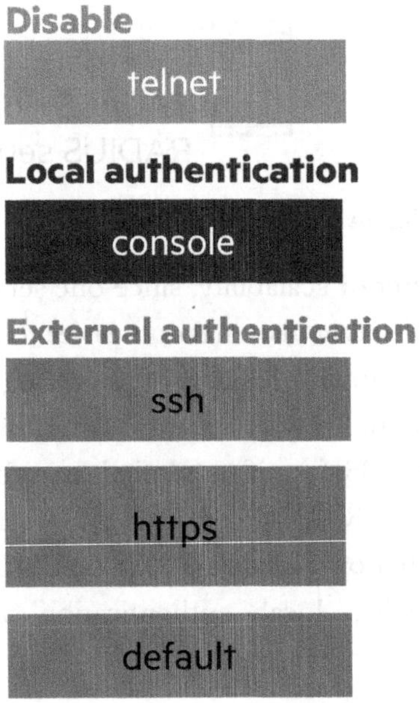

Figure 3-9: Example Authentication Sources Per Access Type

Manager Groups (Roles)

When a manager authenticates, that manager is assigned a role.

AOS-CX switches separate management users into three different default groups, or roles: administrators, operators, and auditors. All types of users may access the switch through the CLI or Web UI (or REST API, as well); however, their privileges differ.

Operators have no configuration privileges. They can only view nonsensitive switch configuration information. In the CLI, operators can only issue show commands, and they can only view settings in the Web UI.

Administrators have full privileges, including full read-write access to the CLI and Web UI. They can configure any settings, including user accounts.

Operators:

- CLI and Web UI: Display-only access

Administrators:

- CLI and Web UI: Full access, including over user accounts

Auditors:

- CLI: Auditor context commands
- Web UI: System > Log page

Figure 3-10: AOS-CX Switch Manager Groups

Finally, auditors have access to CLI commands only in the auditor context (auditor>). Auditors can view event log contents and local accounting log contents. They can also copy command output to a remote server or a local USB drive.

See Figure 3-10 for a summary of the default manager groups.

Admins can create custom manager groups as well, but that configuration is beyond the scope of this study guide.

Enforcing Best Practices for Local Authentication

AOS-CX switches have a default "admin" account, but no default password for it. Instead, the first manager to log in is prompted to set this password. This manager should, of course, establish a secure password using guidelines such as those covered in Chapter 1.

You can also create additional local accounts and assign them to one of the default groups (or roles) introduced earlier.

You can configure password length and complexity requirements to ensure that admins create proper passwords for other accounts. You create a password policy which specifies settings such as minimum length and required characters. You then activate the policy. If, for future passwords, admins try to create a password that does not meet the requirements, the switch OS registers an error and the admin is prompted to create a different password.

CLI Commands: Local Authentication Configuration

For the command for specifying the service export password that encrypts credentials on AOS-CX switches, see the table in Figure 3-11. You also see the commands for setting up password complexity rules and configuring user accounts.

In configuration context

Step	AOS-CX command	Notes
Configure the password used to encrypt credentials in the config.	`service export-password`	Optional; after you enter the command, you will be prompted for a password
Set the minimum manager password length	`password complexity minimum-password-length <1-32> enable`	Set to at least 8 characters (NIST), but preferably much longer
Set the default user passwords	`user admin password [ciphertext | plaintext] [password]`	
Create a new user account and set a password.	`user <username> group <groupname> password [ciphertext | plaintext] <password>`	

Figure 3-11: CLI Commands: Local Authentication Configuration

Understanding How Credentials Are Stored in the Config

All credentials are stored in the configuration with encryption. This applies to manager passwords, as well as credentials such as RADIUS shared secrets. This means that managers cannot see the passwords when they view the configuration.

- Credentials: Stored encrypted in config
- Options:
  - Set encryption key with **service export password**

  Note: Service export password must be set if config applies to multiple switches

Figure 3-12: Understanding How Credentials Are Stored in the Config

The service export password acts as the key for this encryption, as shown in Figure 3-12. If you want to deploy the same config across multiple switches, you should set the service export password in the config. Then all switches will use the same service export password and be able to decrypt the credentials.

Setting up External Authentication for Management Access

To control all forms of management access except the console port, HPE Aruba Networking recommends you use external authentication servers, as shown in Figure 3-13.

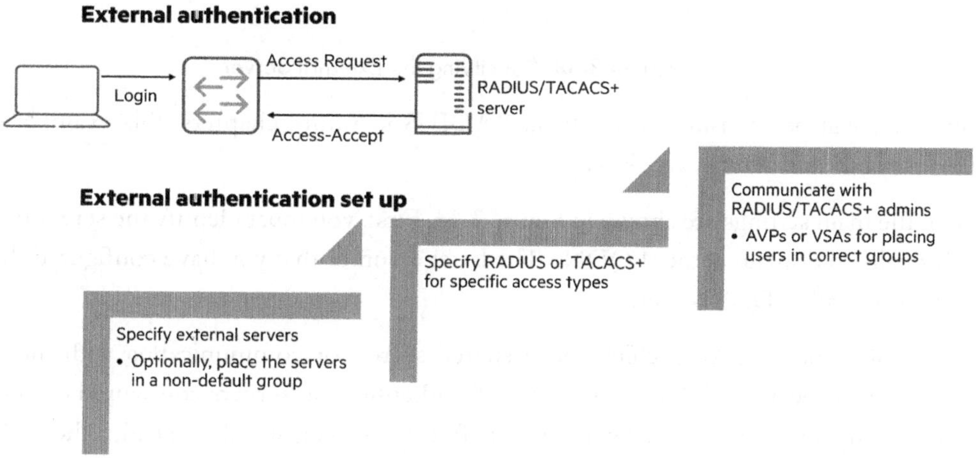

Figure 3-13: Setting up External Authentication for Management Access

The next section will guide you through what you need to know in order to successfully set up external authentication. You will learn how to specify the server. You will next learn how to specify that you want to use that server for particular access types. Finally, you will learn about the Attribute Value Pairs (AVPs) and Vendor Specific Attributes (VSAs) that the server must send to assign the correct role or access level to admins. The authentication server admins will be responsible for configuring these AVPs/VSAs, but you might need to consult with them about the requirements.

Specifying the External Server

To use external authentication, you must specify the authentication server on the AOS-CX switch. You can use RADIUS or TACACS+ to authenticate the managers. The customer might require extra security, based on whether the customer considers the network trusted or untrusted. Extra security options for RADIUS include RadSec or IPsec.

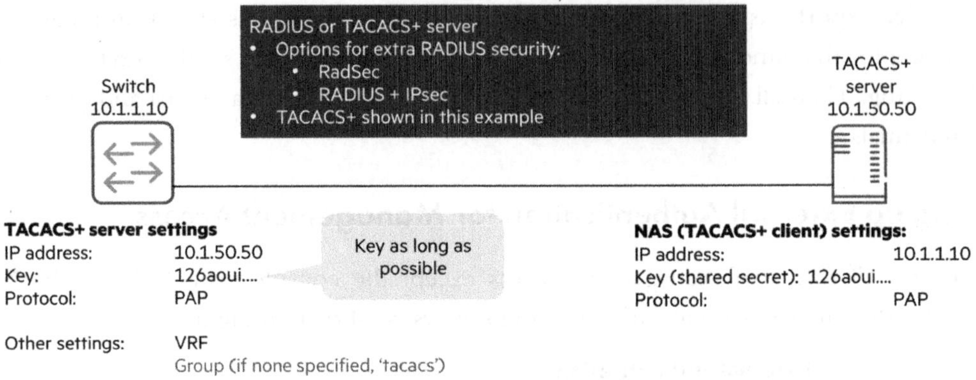

Figure 3-14: Specifying the External Server

Because you will be focusing in depth on RADIUS in future chapters, this example focuses on TACACS+, but the settings are similar.

Some of the main settings are shown in Figure 3-14. First, you must identify the server by IP address or fully qualified domain name (FQDN). The latter assumes that you have configured the switch to communicate with a DNS server.

You must also specify a key, which sets a shared secret for communicating with the TACACS+ server. You can specify a global key used for all authentication servers configured on the switch, or you can specify an individual key when you define the server, which overrides the global key. Set the key as long as possible to enhance security. AOS-CX switches allow keys up to 31 characters in length. You must also specify on which VRF the switch reaches the server.

You can optionally place the TACACS+ server in a manually defined group on AOS-CX switches; if you do not, it is part of the default group, named "tacacs." Similarly, you can place RADIUS servers in a non-default group or leave them in the default "radius" group.

You can also configure the authentication protocol used with RADIUS or TACACS+. The external authentication server must support the same protocol. HPE Aruba Networking ClearPass Policy Manager (CPPM) uses PAP for TACACS+, so you need to match on the switches.

You must verify that the authentication server is similarly set up to communicate with the switch. A TACACS+ server must have the switch IP address configured on it as a TACACS+ client—sometimes called Network Access Server (NAS) or, on CPPM, a network device. The shared secret configured for the switch on the TACACS+ server *must* match the key configured for the TACACS+ server on the switch. Otherwise, the TACACS+ server will not respond to the switch's messages. Requirements are similar for RADIUS servers.

Specifying External Access Methods

After you have specified the RADIUS or TACACS+ servers in the default or appropriate group, you can enable external authentication for specific access types.

Refer to Figure 3-15 for help breaking down the command syntax, which can be complex.

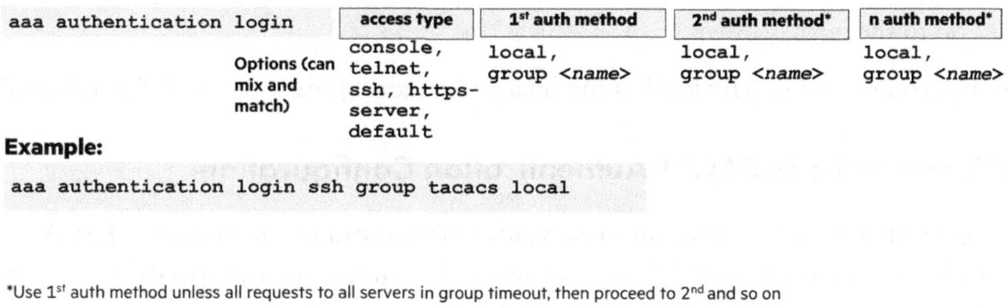

Figure 3-15: Specifying External Access Methods

After specifying **aaa authentication login**, you specify the access type, the primary authentication method, and additional authentication methods.

Access Type

The access types specify how the manager accesses the switch management interface. Options include **console**, **telnet**, **ssh**, **https-server**, and **default**. (But you should disable Telnet.) The default type tells the switch how to authenticate users logging in with any type that does not have its own configuration. For example, if you configure a specific login configuration for SSH, but not for HTTPS, the switch enforces the default configuration when a user accesses the Web UI.

Primary Authentication Method

To enable RADIUS or TACACS+ authentication, specify **group** and the name of the group to which the RADIUS or TACACS+ server or servers are assigned. Specify **local** for authentication to the local admin users.

Additional Authentication Methods

When you configure RADIUS or TACACS+ as the primary authentication method, you must choose a secondary method as well. The switch enforces the secondary method only when it cannot reach any authentication server in the group. For example, you have specified RADIUS server group "managerRadius" as the primary authentication method for SSH. This group has two RADIUS servers. The switch considers the primary method to have failed when its maximum number of retries have timed out for both servers.

At that point, the switch enforces the secondary authentication method (although the switch might still be cycling between the servers in the background).

AOS-CX switches permit more than two authentication methods. If the second one fails, the switch moves on to the next.

Typically, you should specify local as the final authentication method used if all others fail.

CLI Commands: TACACS+ Authentication Configuration

The table in Figure 3-16 shows the commands that you can use to specify a TACACS server and optionally add it to a group on AOS-CX switches. As you see, you can specify the TACACS server's key (shared secret) globally, in which case the key applies to all TACACS servers configured on the switch that do not have an individual key set. The individual key, if set, overrides the global one.

In configure context

Step	AOS-CX command	Notes
Set a global key	tacacs-server key [plaintext | cipher] <*secret*>	Optional; use a global key if multiple TACACS servers use the same key
Specify one or more TACACS servers	tacacs-server host <*FQDN* | *IP*> [vrf <*VRF*>]	Remember to configure the desired primary server first Setting the key is optional if you set the global key (overrides global key) Set the VRF to the VRF used to reach server
Create a TACACS server group and assign a server to the group	aaa group server tacacs <*group name*> server <*FQDN* | *IP*>	Only required if not using the global group
Configure TACACS with local fallback for the recommended access forms.	aaa authentication login <ssh | https-server> group <*group name*> local	Remember to set local for the secondary (or tertiary) method.

Figure 3-16: CLI Commands: TACACS+ Authentication Configuration

You also see how to configure TACACS with local fallback as the authentication method for SSH and web access.

Assigning Manager Roles on an AOS-CX Switch

You will now look at external authentication works in a bit more detail. When a manager attempts to log in, the AOS-CX switch sends an Access-Request to the RADIUS or TACACS+ server. If the manager succeeds authentication, the server sends an Access-Accept. For the user to receive access, that Access-Accept must also include a VSA or other attribute that indicates the user's access level.

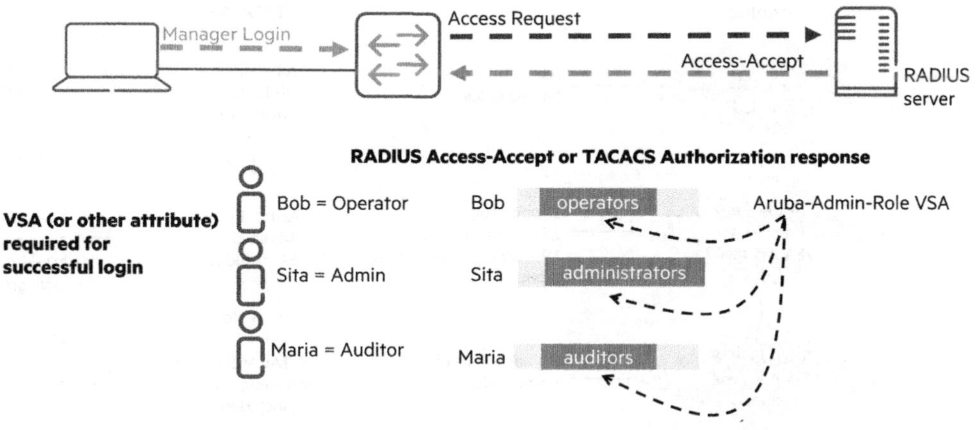

Figure 3-17: Assigning Manager Roles on an AOS-CX Switch

Preferably, you should have the server send an Aruba-Admin-Role VSA. The value for the VSA is the name of one of the manager groups that you looked at with local authentication. For example, if the Aruba-Admin-Role VSA is "operators," the user is assigned to the operators group and receives those privileges. If the Aruba-Admin-Role VSA is "administrators," the user is assigned to the administrators group and receives full privileges, and so on. (See Figure 3-17.)

AOS-CX switches support the Aruba-Admin-Role VSA for both RADIUS and TACACS+ messages.

Order of Priority for AVPs/VSAs on AOS-CX

The Aruba-Admin-Role VSA is recommended because it is simple to use and flexible. However, AOS-CX switches support other VSAs as well. If multiple VSAs or AVPs are included, the Aruba-Admin-Role VSA takes precedence. If that VSA is not present the Aruba-Priv-Admin-User VSA takes precedence. This VSA assigns roles based on numbers from 1-15 and 19. 1 assigns the user to the operators group, 15 to the administrators group, 19 to the auditors group, and 2-14 to any manually configured local groups that have those IDs.

For RADIUS, if neither Aruba VSA is present, but a RADIUS standard Service-Type AVP is, that AVP takes effect. The Service-Type AVP values have these effects:

- 7 = Access permitted at operator level (NAS-Prompt-User)
- 6 = Access permitted at administrator level (Administrative-User)
- Null or other value = Access denied

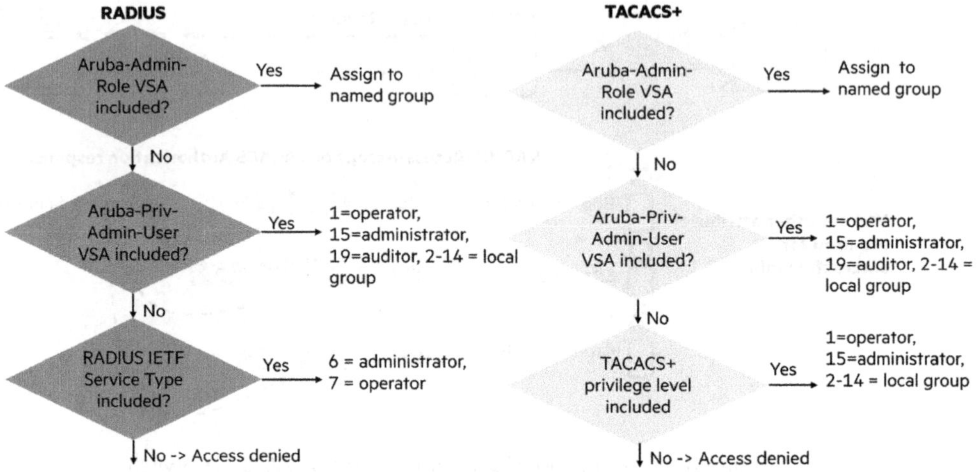

Figure 3-18: Order of Priority for AVPs/VSAs on AOS-CX

Similarly, the TACACS+ privilege level takes effect for TACACS+ if neither of those VSAs are sent. The privilege levels have these meanings:

- 1 = Access permitted at operator level
- 15 = Access permitted at administrator level
- 2-14 = A level that can be associated with a local group

See Figure 3-18 for a summary of these rules.

In Figure 3-19, you can see a manager authentication record, as shown within a CPPM tool called Access Tracker. This TACACS+ record shows up when a network device like a switch sends a TACACS+ request to authenticate a management user.

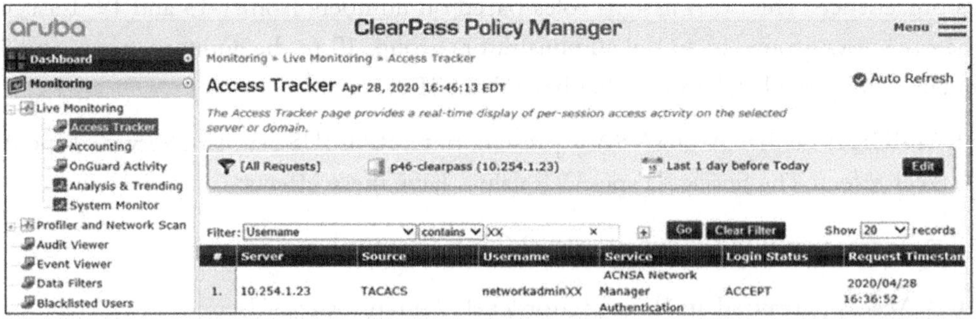

Figure 3-19: CPPM Access Tracker

If you click the record, you can open it and see that CPPM has granted the request in this example, as shown in Figure 3-20.

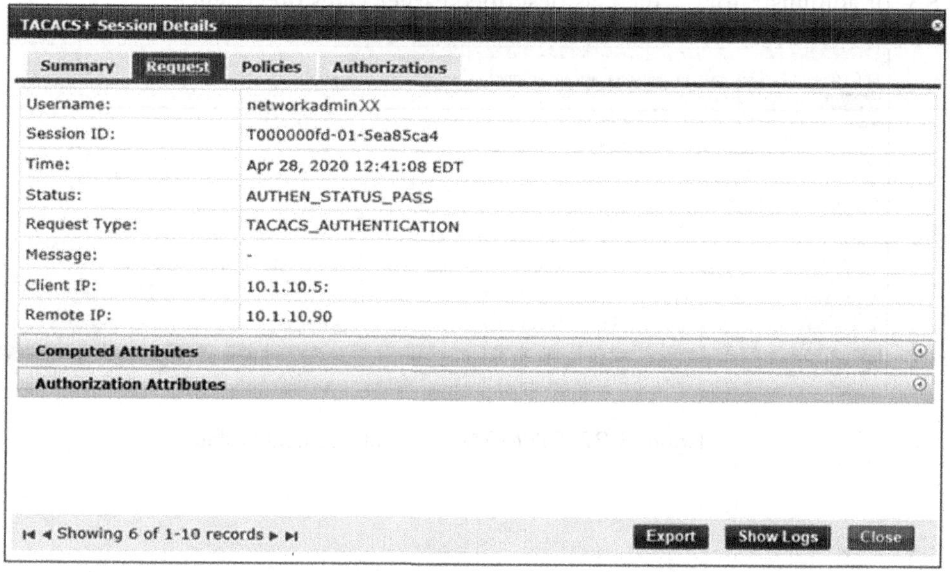

Figure 3-20: CPPM Access Tracker TACACS Record

As shown in Figure 3-21, the Policies tab indicates that CPPM assigned this user to a specific profile, called "assign switch tacacs admin role" in this example.

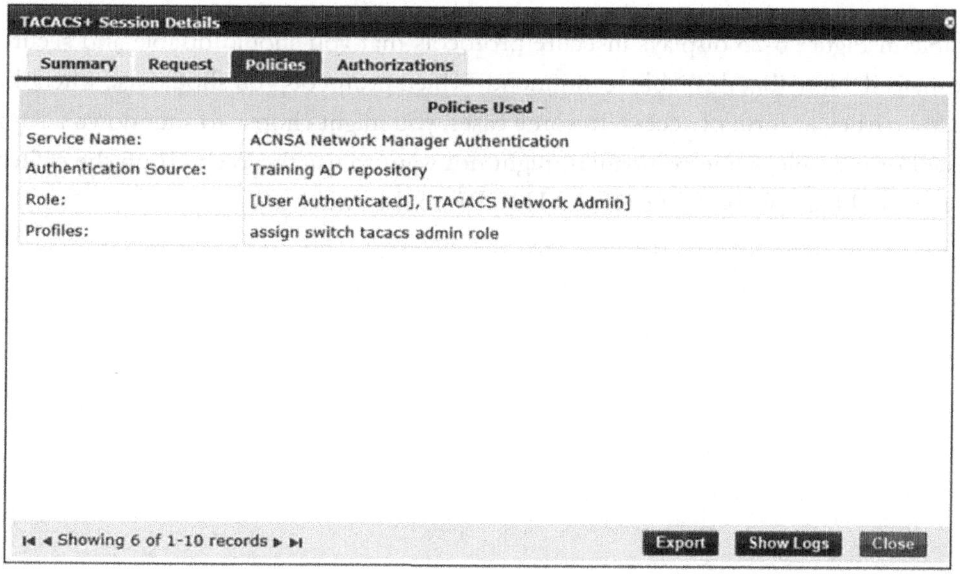

Figure 3-21: CPPM Access Tracker TACACS Record (Policies Tab)

The settings associated with this profile are shown in Figure 3-22. As you see, the privilege level is set to 15, which corresponds to "manager" level access. The profile also assigns an Aruba-Admin-Role VSA, of administrators, which, as described earlier takes precedence.

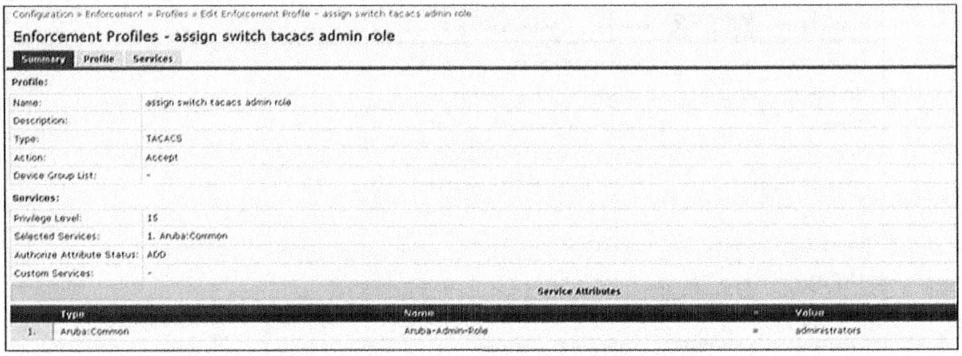

Figure 3-22: CPPM TACACS Enforcement Profile

Secure Protocols

You have established a secure out-of-band-management network. You have set up authentication. You must also ensure that managers can only use secure protocols.

Use Secure Protocols

The table in Figure 3-23 displays insecure protocols that you should disable and secure protocols to use instead. Note that this table recommends which protocols you should use when the company requires a particular form of access. In some cases, you might choose to shut down a form of access entirely. For example, some companies might not want to use the Web UI on the switches. In this case, you would disable both HTTP and HTTPS on the switches.

Access type	Protocol	Security	Enabled by default on AOS-CX
CLI	Telnet	Insecure	
	SSH	Secure	X (mgmt + default VRF)
File transfer through serial	TFTP	Insecure	
	SFTP	Secure	X (mgmt + default VRF)
Web	HTTP	Insecure	
	HTTPS	Secure	X (mgmt + default VRF; self-signed cert)
SNMP	SNMPv1/v2c	Insecure	
	SNMPv3	Secure	

Figure 3-23: Use Secure Protocols

Telnet vs SSH

Both Telnet and SSH provide access to the switch CLI over TCP/IP. And both support username- and password-based authentication of clients. However, while SSH encrypts communications, Telnet does not, meaning that data delivered over Telnet can be stolen much more easily than data delivered over SSH. In addition, SSH supports mutual authentication—that is, authentication of both the server (in this case, the switch) and the client. Because the client authenticates the switch, a manager such as Alice can know that she is truly connecting to the switch and *not* to hacker Mallory who might try to insert herself between the management station and the switch. In this way, mutual authentication protects against MITM attacks. SSH also provides options for public-key-based authentication rather than password-based authentication.

Given the substantial security advantages of SSH, it is best practice to leave SSH enabled and disable Telnet entirely. On AOS-CX switches, SSH is enabled by default on the "mgmt" and "default" VRF; if you are using the Mgmt port to establish an OOBM network, you can disable SSH on the "default" VRF.

AOS-CX switches further secure SSH by periodically renegotiating new keys for the session after an hour or 1 GB of data is transmitted, whichever comes first. You can adjust these values (see your switches' configuration guide).

Earlier versions of AOS-CX software (before 10.08.1021) do not support Telnet. For later software versions, for security, you should make sure that no one enables this feature.

TFTP versus SFTP

TFTP enables clients to receive files from TFTP servers or send files to those servers. However, TFTP sends files in plaintext and does not provide authentication. AOS-CX switches support SFTP as a secure alternative. SFTP transfers files over a secure SSH connection, which provides authentication and encryption. SFTP is always enabled when admins are using SSH.

HTTP versus HTTPS

HTTP transmits data in plaintext. HTTPS uses Transport Layer Security (TLS) to establish a secure tunnel for transmitting data. The switch acts as an HTTPS server, which uses a certificate to authenticate to the client and set up a secure tunnel. The manager submits credentials over that tunnel, and the rest of the session is encrypted as well.

AOS-CX switches do not support HTTP, only HTTPS. HTTPS is enabled by default on the default and mgmt VRFs using the self-signed certificate. You should replace the self-signed certificate with a CA-signed one, though.

SNMPv1/v2c versus SNMPv3

SNMPv1/v2c transmit community strings, originally meant to act as "passwords," in cleartext, so they are insecure. SNMPv3 provides data authentication and privacy with secure algorithms and keys.

AOS-CX switches have the public community configured on them by default, but SNMP is not enabled on any VRFs by default. If you need to use SNMP, you should configure SNMPv3 only for security.

Set Up SSH Securely

You will now learn how to set up secure SSH management for your switches. A brief review of SSH will help.

When the management client sends a request to open an SSH session with the switch, the switch responds and starts setting up a secure tunnel, as shown in Figure 3-24. The switch and client must agree on algorithms to use to secure the tunnel. The switch also authenticates itself to the client with its SSH public key. In other words, the switch sends the public key and it also signs data with that key, proving that it has the associated private key. As long as the client accepts the switch's authentication, the devices establish a secure tunnel. The management client then authenticates across that tunnel. The client can authenticate with a username and password or with an SSH private/public keypair like the switch. This study guide focuses on username and password authentication.

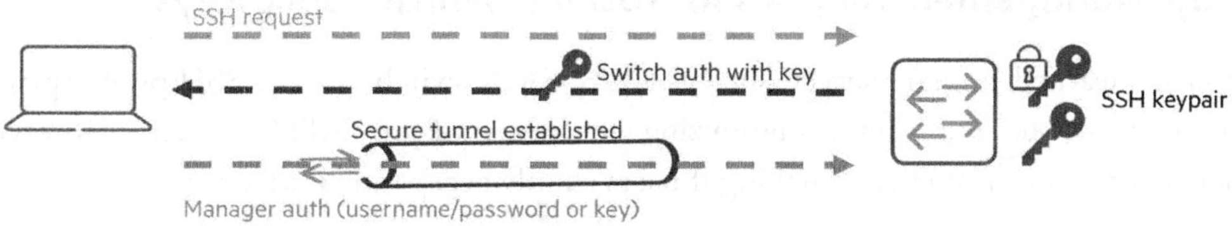

Figure 3-24: Set Up SSH Securely

AOS-CX switches support SSH out of the box, but you need to take steps to ensure that you are using SSH in a secure way.

First, you need to configure the desired authentication type as discussed earlier, whether you use external authentication, as recommended, or local accounts.

Next, you should disable Telnet and make sure SSH is enabled (it is by default in the software versions covered in this study guide). If you are using the mgmt VRF on an AOS-CX switch, remember to disable SSH on the default VRF.

You should also check the SSH algorithms and disable non-certified ones.

Then you should generate a new SSH keypair.

Finally, you need to set up management clients to trust the switch's public key.

CLI Commands: SSH Configuration

Use the table in Figure 3-25 as a reference for configuring SSH on AOS-CX switches.

In configure context

Step	AOS-CX Command	Notes
Disable Telnet and TFTP.	`no telnet server vrf <name>`	Telnet might be supported on AOS-CX depending on the software version
Generate a new SSH keypair.	`ssh host-key {ecdsa [ecdsa-sha2-nistp256 | ecdsa-sha2-nistp384 | ecdsa-sha2-nistp521] | ed25519 | rsa [bits {2048 | 4096}]`	
Enable SSH.	`ssh server vrf <VRF>`	Enabled by default; only required if SSH has been disabled earlier or enabled on the wrong VRF.

Figure 3-25: Set Up SSH Securely

When you generate a keypair, remember that the more bits, the more secure.

Set Up Management Stations to Trust the Switch Public Keys

As you just learned, an SSH server, such as an AOS-CX switch, uses its SSH public/private keypair to authenticate to the client, minimizing the risks of an MITM attack. However, mutual authentication does not mean anything if users simply accept any public key.

SSH public keys are not traditionally distributed in the certificates, which, as you learned in Chapter 2, help devices verify whether a public key belongs to the entity to which they think that key belongs. Instead, you typically need to set up the trust manually. (In fact, some HPE Aruba Networking devices *do* support certificates for SSH, but this is an advanced topic, not covered in this training.)

For traditional SSH keys, you have a few options, which are summarized in Figure 3-26.

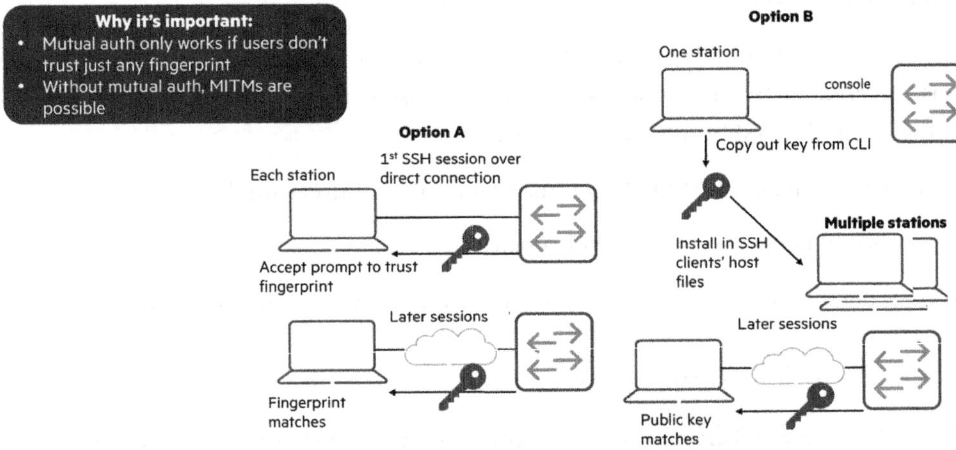

Figure 3-26: Set Up Management Stations to Trust the Switch Public Keys

A: Connect Directly and Accept Fingerprint

The first time that you use SSH to connect a management station to a switch, you can connect the device directly to the switch Mgmt port so that you are sure the connection is secure. When you open the SSH session, the client will show you some hexadecimal characters, tell you that this is a new "fingerprint," and ask if you want to accept the fingerprint. The fingerprint is a hash of the switch's public key, which the client receives as part of the session establishment. (As a shorter version of the public key, the hash is easier for you to identify as correct than a full public key.) In this case, you know that the fingerprint—and associated public key—truly belong to the switch because you are directly connected to the switch. You simply choose to trust the fingerprint, and the connection continues.

Your management client stores the fingerprint. If later you attempt to connect to the switch with SSH, and the device on the other end sends a new public key, the management client will detect that the fingerprint does not match the key and alert you. This will give you a heads up that you are not connecting to the device that you think you are and a possible MITM attack is underway—or that someone has updated the switch's keypair, in which case you need to verify that this action was authorized.

Of course, hackers could try to use the switch's public key, which is freely distributed, and which would make the fingerprint match. But remember that the switch signs data with its private key during the session establishment, so hackers cannot successfully establish a session with a public key that they do not own.

The advantage of this method of establishing trust is that it is simple. However, for true security, you must connect every new management station directly to every switch, which can be tedious or infeasible in some circumstances.

B: Copy out the Public Key and Install as Trusted on Management Clients

To address these issues, you might take a different approach. You can save multiple switches' public keys in a file and then add those keys as trusted keys on all of your management stations.

To obtain a public key securely, connect one management station directly to the switch CLI on the console or Mgmt port. (In the latter case, you can accept the fingerprint on this single station.) View the public key (AOS-CX command: **show ssh host-key**). Copy the text into a file.

You can then import this key as trusted into your management clients. The precise instructions depend on the SSH client that you are using. For example, Linux stores trusted SSH server public keys in a "known_hosts" folder to which only root users have access. Tera Term stores the keys in a file referenced in the SSH settings. Refer to your client's documentation for more details.

While it does not matter if an unauthorized party obtains the public keys themselves, you do not want anyone to be able to tamper with the file and insert an unauthorized key. Take appropriate security measures to secure the file such as storing it on machines and locations that only managers can access.

You will now explore an example of installing a public key through a file on Tera Term:

1. You can put the key in a text file, preceded by the IP address of the switch that owns the key, as shown in Figure 3-27.

Figure 3-27: Host Key File

2. Load the SSH key file in Tera Term.

    a. In Tera Term, select **Setup > SSH** (See Figure 3-28).

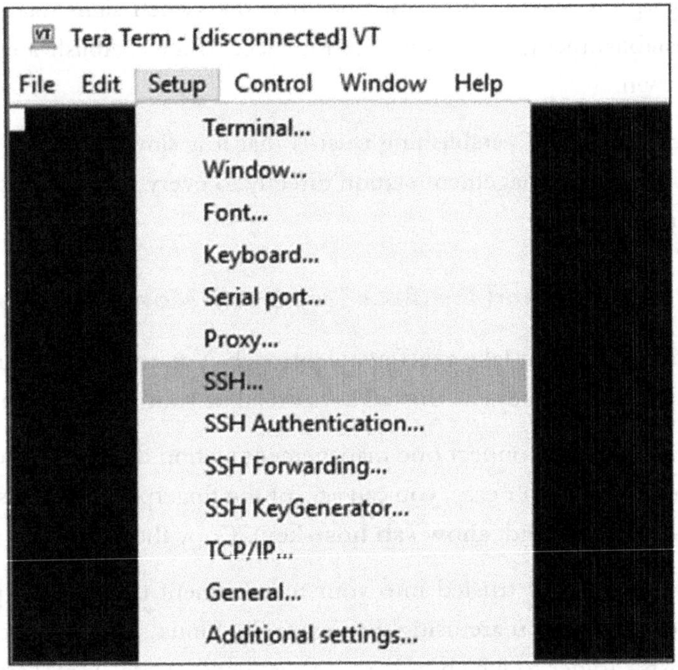

Figure 3-28: Tera Term Setup

    b. Click **Read-only files** in the SSH Known Hosts section and browse to your file (See Figure 3-29).

    c. Click **OK.**

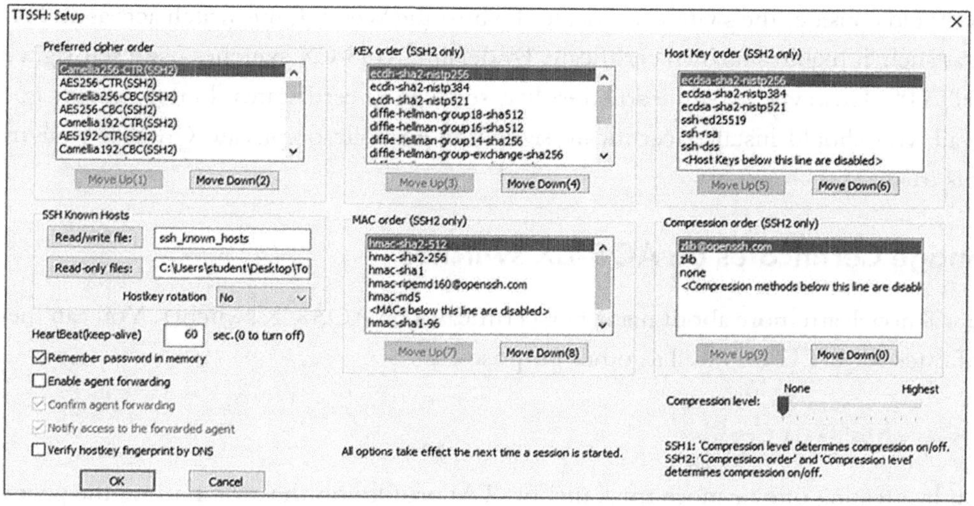

Figure 3-29: Tera Term SSH settings

3. Now when you use Tera Term on this station to access the switch, Tera Term will already trust the public key.

Set Up Secure Access to the Web UI

For the last form of management access, you will learn how to set up secure access to the Web UI. (See Figure 3-30.)

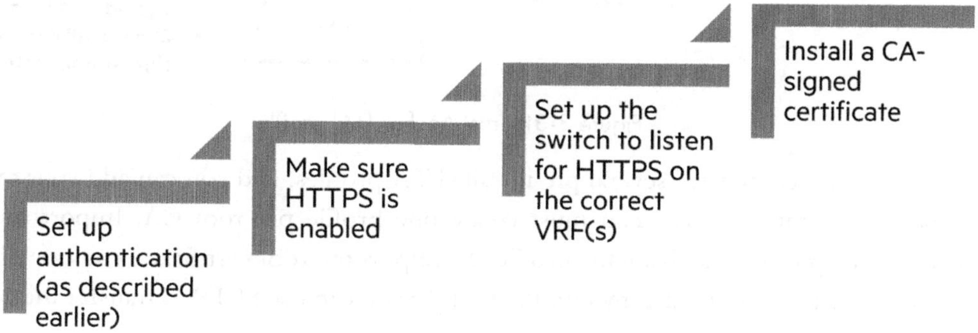

Figure 3-30: Set Up Secure Access to the Web UI

First, you must set up authentication as you learned how to do earlier.

You then verify that the HTTPS server is enabled (as it is by default on AOS-CX switches). You also choose the VRF or VRFs on which to enable the HTTPS server; if managers are accessing the switch on the Mgmt port, use the mgmt VRF and disable HTTPS on the default VRF.

You should consider the switch certificate. To host the Web UI, the switch acts as an HTTPS server and, as such, it requires its own certificate. By default, AOS-CX switches use a self-signed certificate for HTTPS. However, as you learned earlier, self-signed certificates do not provide good security. Instead, you should install a certificate signed by a public or private CA that your management clients trust.

Manage Certificates on AOS-CX switches

You will now learn more about managing certificates on AOS-CX switches. You can then use those certificates for HTTPS as well as other purposes.

Trust Anchor (TA) profile

Begin by creating one or more trust anchor (TA) profiles on the switch using the process outlined in Figure 3-31.

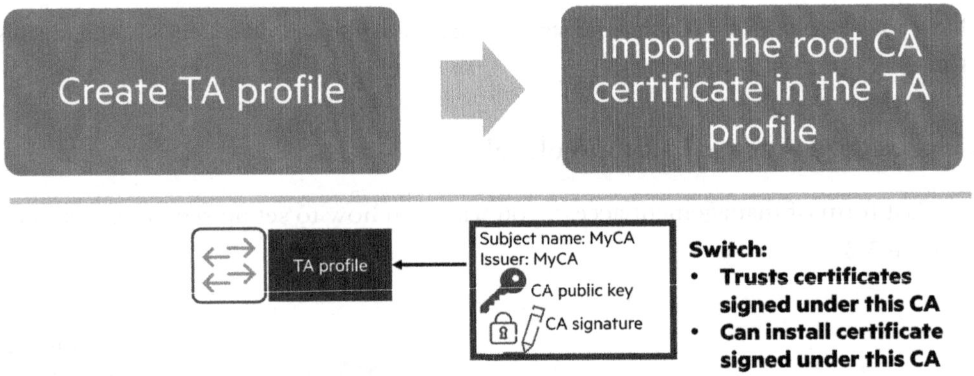

Figure 3-31: Trust Anchor (TA) profile

AOS-CX switches come with several pre-installed TA profiles, and you can add your own for the CAs that your company uses. You must create one profile per root CA. Import the root CA certificate in the profile to activate the profile. To import the root certificate, you can either copy a Base64 formatted certificate directly into the CLI, or you can use SFTP to transfer the certificate.

Once the CA certificate is installed, the switch can trust certificates signed by that CA or intermediate CAs under it. The switch can also install its own certificate signed by that CA or an intermediate CA under it.

Install a Provided Certificate

You have several options for installing a certificate on an AOS-CX switch. First, you will learn how to install a certificate that someone else has created and provided to you. See Figure 3-32 for an outline of the process.

To work for HTTPS, the certificate should have a DNS SAN that matches the switch hostname. (Alternatively, it could have an IP SAN that matches the switch IP address, but many companies prefer to omit IP addresses from certificates.) The certificate file must use PEM format. In addition to the signed certificate, the PEM file must include the encrypted private key, created when the admins created the certificate signing request (CSR).

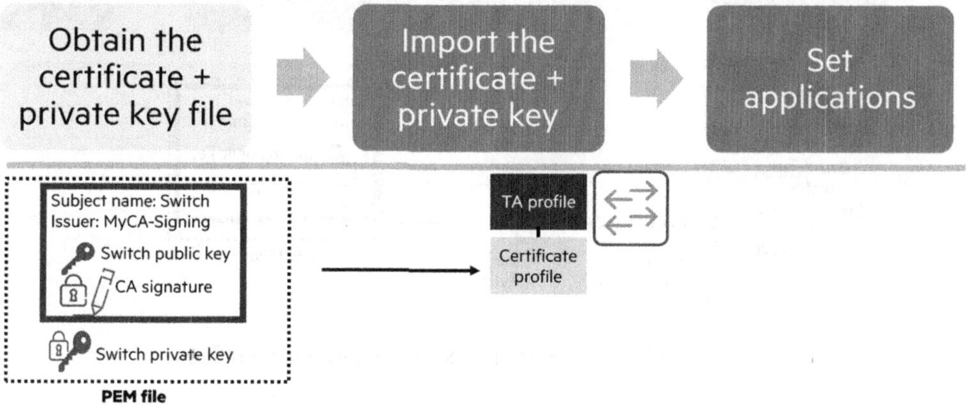

Figure 3-32: Install a Provided Certificate

To import the provided PEM file to the switch, begin by creating a certificate profile on the switch. Then use the **import** command to import the certificate, specifying the TA profile for the CA that signed the certificate. Again, you can import the certificate by copying the text of the PEM file directly into the CLI. Or you can use SFTP to transfer the PEM file.

You will be prompted to specify a passphrase for the private key. You should get that passphrase from the admins who provide you the certificate file.

After you install the certificate, you must specify the desired applications for the certificate. For example, if you want the switch to use the certificate for its HTTPS server, you enter this command: **crypto pki application https-server certificate <*name*>**.

Create a CSR and Import the Certificate

Rather than create the CSR offline, you can create it directly on the switch, following the process outlined in Figure 3-33.

To create the CSR on the switch, begin, as before, by creating a certificate profile. Then in that profile specify the subject CN. You can optionally specify the subject OU, city, state, and other values. Note, however, in this case, that you cannot set SANs. If your use case requires SANs, obtain the certificate from a CSR created offline, as described earlier.

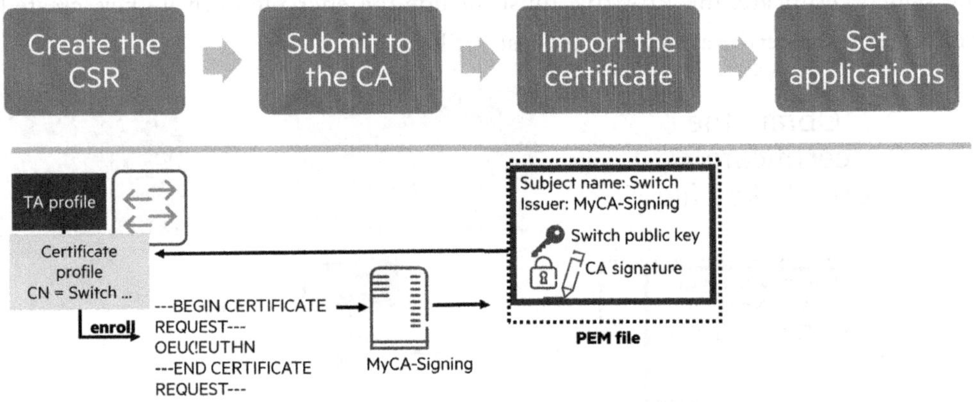

Figure 3-33: Create a CSR and Import the Certificate

Then indicate that you want to enroll from the terminal, which generates a certificate signing request (CSR) on the switch. The CSR specifies the subject name for the certificate, and its creation also triggers the switch to create a public/private keypair for the certificate. When you generate a CSR in the CLI, the terminal shows a BASE 64 formatted request, which you can copy into a file and submit to your CA.

When the CA returns the signed certificate, you import it to the switch under the certificate profile as you previously learned how to do. However, in this case, the PEM file only includes the signed certificate, not the private key which is held on the switch itself. The switch matches the public key with the stored private key and install the certificate. Successful installation also requires that the CA that signed the certificate can be traced back to a TA profile on the switch.

After you import the certificate, you can set applications for it, as discussed previously.

CLI Commands: Certificate Management

The table in Figure 3-34 shows the CLI commands that you can use to generate a CSR and install a certificate.

In configure context

Step	AOS-CX Command	
Create a TA profile	crypto pki ta-profile <*name*>	
Install a TA certificate.	crypto pki ta-profile <*name*> ta-certificate import [terminal | *URL*]	
Create a certificate profile.	crypto pki certificate <*name*>	
Create a CSR	crypto pki certificate <*name*> subject common-name <*cn*> [*other fields*] key-type {ecdsa curve-size <*size*> | rsa key-size <*size*> enroll terminal	
Install a switch certificate	crypto pki certificate <*name*> import <terminal | *URL*> ta-profile <*TA name*>	
Set applications.	crypto pki application <*app name*> certificate <*cert name*>	Use the help to see a list of application names

Figure 3-34: Manage Certificates on AOS-CX switches

Enrollment over Secure Transport (EST)

You can also use Enrollment over Secure Transport (EST) to manage certificates on AOS-CX switches more automatically (as of version 10.06.0001).

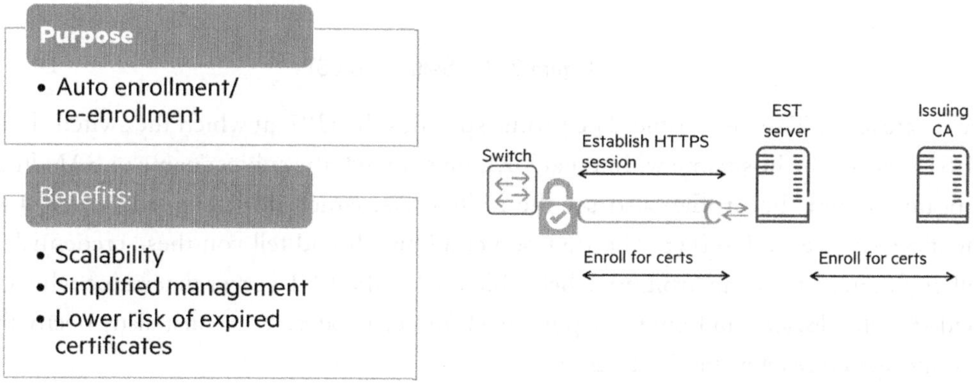

Figure 3-35: Enrollment over Secure Transport (EST)

When you use this option, the switch creates an HTTPS session with an EST server, as shown in Figure 3-35. This server helps the switch to enroll or re-enroll for certificates. This option is preferred for wider scalability because it simplifies the deployment of certificates. Automatic re-enrollment before a certificate expires also decreases the risks that admins will forget to renew a device's certificate, which would cause functions that use that certificate to fail. Using EST does require the deployment of an EST server, which can communicate with the issuing CA. Explaining how to set up the EST infrastructure is beyond the scope of this study guide. (ClearPass Onboard can act as an EST for endpoints.)

Setting up EST

To set up your AOS-CX switch to use EST, follow the process outlined in Figure 3-36.

Begin by installing the CA certificate that signed the EST server's HTTPS certificate. Also ensure that the AOS-CX switch and the EST have synchronized time by setting up authenticated NTP to the same server (covered later in this chapter). Finally, if you plan to specify the EST server by FQDN, specify a DNS server on the switch.

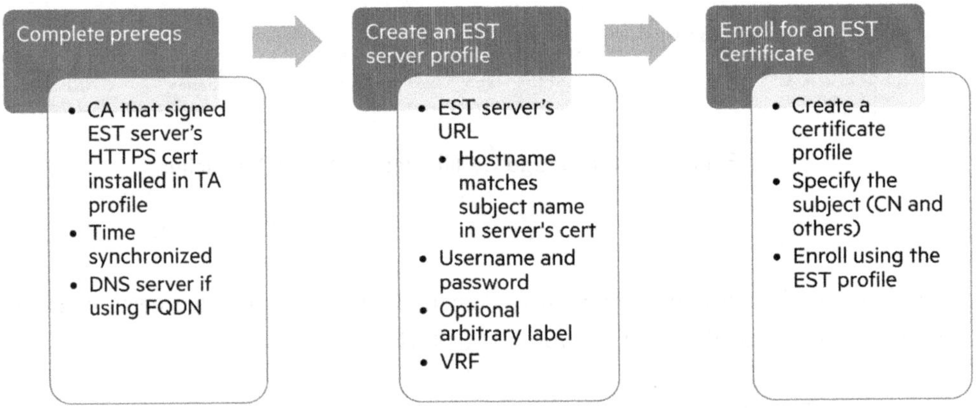

Figure 3-36: Setting up EST

Then create an EST server profile. The profile specifies the URL at which the switch should contact the EST server. The hostname within that URL must match the subject name or SAN in the server's certificate. Within the profile, also set the switch username and password, which it will use to authenticate to the EST server. The EST server admin should tell you these credentials, as well as whether you need to set an arbitrary label, which helps the EST server determine which services to provide for the device and how to help it enroll for certificates. You must also specify the VRF on which the switch reaches the EST server.

You then configure a certificate profile just as you learned to do before. Make sure to specify the subject name settings that you desire for the switch's certificate. Then use the **enroll** command with the **est-profile** option, specifying your EST profile.

Note also that AOS-CX switches support NetEdit for automating many processes; you can learn about NetEdit in other HPE Aruba Networking training.

Configure SNMP securely

Simple Network Management Protocol (SNMP) is an application layer protocol that helps network administrators manage and monitor their network. SNMP-managed networks have several key components, as shown in Figure 3-37. Each managed device, such as a switch, has an "agent," a software module that enables management functions. The managing device is called the Network Management Station (NMS) or SNMP server.

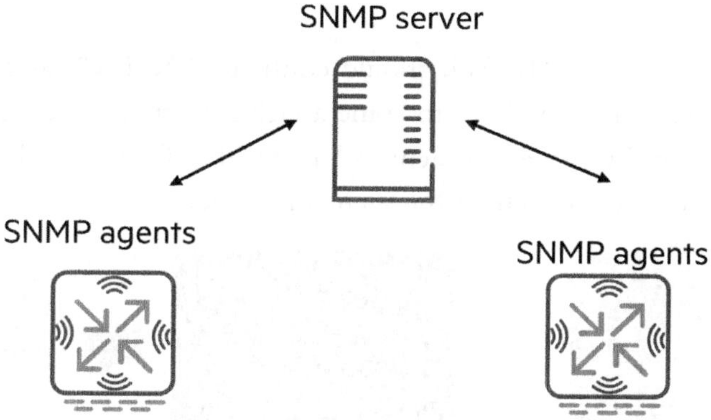

Figure 3-37: SNMP Components

In this section of the chapter, you will learn to set up SNMP securely.

Disable SNMPv1/v2

While SNMP is a useful tool, its earlier versions suffer from poor security. The first two versions send critical data in clear-text, creating a serious vulnerability for a packet sniffing exploit. Community strings were meant to act as authentication. The agent only accepts write requests with the correct read-write community or read requests with the correct read-only or read-write community. However, because community strings are sent in cleartext, they do not provide true authentication. If hackers can spoof read and write requests, they can collect information about the network and even make configuration changes to it.

For that reason, best practice is to disable SNMPv1 and v2c.

On AOS-CX switches, SNMP is not enabled on any VRFs by default so it is implicitly disabled. If you plan to use SNMPv3, you should enable the SNMP-server on the desired VRF or VRFs. You should then specify that you want the switch to listen to SNMPv3 only. Refer to Figure 3-38 for a summary of SNMPv1/v2c vulnerabilities and recommendations.

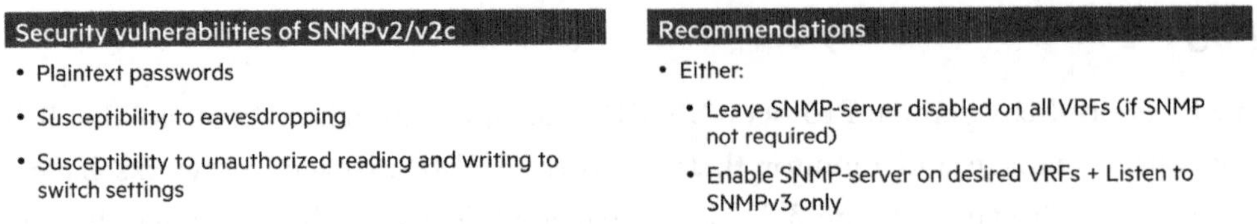

Figure 3-38: SNMPv1/v2c Vulnerabilities and Recommendations

Set Up SNMPv3

If the customer wants to use SNMP, you should configure SNMPv3, as summarized in Figure 3-39. While SNMP v3 has susceptibilities to some attacks, it offers secure encryption to prevent eavesdropping. It also offers truer authentication in the form of shared encryption and authentication keys, which must match on the agent and server.

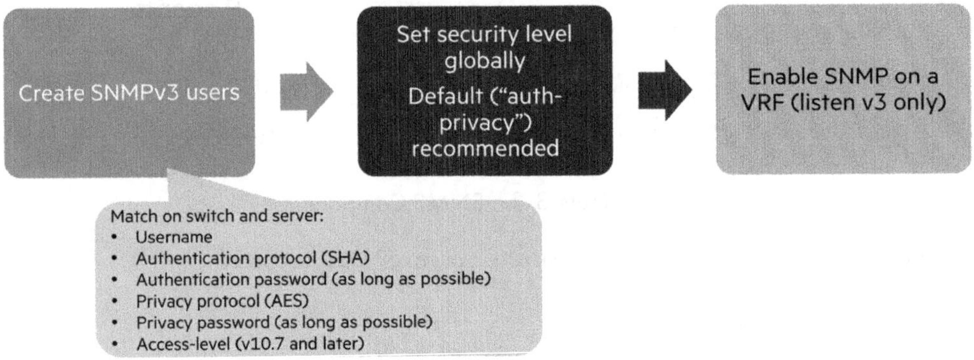

Figure 3-39: Set Up SNMPv3

Setting up SNMPv3 can involve many pieces. For this study guide, you will focus on the simplest setup.

You must create one or more SNMPv3 users, specifying an authentication algorithm and password and a privacy algorithm and password. SHA is recommended as the best authentication algorithm available for SNMPv3 and AES as the best encryption algorithm available for SNMPv3. Do *not* use DES or 3DES for the encryption algorithm.

Note that authentication refers to *data* authentication, not user authentication; an SNMPv3 user is authenticated by knowing the same authentication and privacy keys as those configured on the switch. The authentication and privacy settings are optional but recommended. Without them the SNMP messages are not secured. The SNMP server must be configured with a matching SNMPv3 user with the same username, authentication algorithm, authentication password, privacy algorithm, and privacy password.

In version 10.4, SNMPv3 users were granted access to the full switch MIB except certain excluded access control MIBs. Version 10.7 and later added the option to set each user's access-level to read-only or read-write.

You can set the security level globally to either "auth," which requires SNMPv3 users to use data authentication or "auth-privacy," which requires SNMPv3 users to use data authentication and encryption. The "auth-privacy" option is enabled by default and recommended.

To finish enabling SNMPv3 on AOS-CX switches, enable SNMP on the desired VRF or VRFs. Then, as discussed earlier, set SNMP to listen to SNMPv3 only.

CLI Commands: SNMPv3 Configuration

Use the commands shown in Figure 3-40 to disable SNMPv1/v2c and enable SNMPv3.

In configure context

Step	AOS-CX Command	Notes
Disable SNMPv1/v2c.	`snmp-server snmpv3-only`	
Create SNMPv3 users.	`snmpv3 user <username> [auth <md5 | sha> auth-pass <ciphertext | plaintext> <auth key>] [priv <des | aes> priv-pass <ciphertext | plaintext> <priv key>] [access-level <ro | rw>]`	Prefer SHA and AES; make sure to match the username, algorithms, and keys on the SNMP server.
Delete SNMPv3 users.	`no snmpv3 user <username>`	
Enable SNMP on a VRF.	`snmp-server vrf <name>`	

Figure 3-40: CLI Commands: SNMPv3 Configuration

For AOS-CX, SNMP is disabled on all VRF by default. If you want to use SNMPv3, delete the default "public" community so that the switch does not have any SNMPv1/v2 communities. Then create SNMPv3 users. Finally, enable SNMP on the correct VRF, "mgmt" or "default," depending on whether the switch reaches the SNMP server on the Mgmt port or a data plane interface.

Physical Security and Other Measures

This chapter concludes with a discussion of physical security and a few additional security enhancements.

Ensure Physical Security

Physical access to a network infrastructure device provides unauthorized personnel more opportunities to disrupt the network and gain unauthorized access. If nothing else, a hacker can remove power from key devices, causing a devastating DoS attack. The company should have a security policy that dictates appropriate measures for physical security; in fact, many regulations require such policies.

Recommendations for Physical Security

Typically, all network infrastructure devices should reside in locked closets. The more secure the lock, the more secure the devices. Biometric devices such as retina scanners are more secure because they actually detect an authorized manager as opposed to anyone who has obtained the key—whether legitimately or illegitimately. The extra security does come at extra cost and increased management, though; the scanner must be configured to admit new managers. Biometrics are more commonly used in data centers than in campus closets.

You should also consider disabling unused switch ports to prevent users from connecting unauthorized devices to them. Some customers, however, will prefer that you leave ports enabled but configure authentication on them such that the port automatically blocks access or grants customized access based on the user or device. (You will learn more about authentication later in this study guide.)

Recommendations for Physically Insecure Devices

In the rare case that you cannot physically secure a switch, you should make doubly certain to secure the switch as recommended throughout this chapter. For example, you might want to use stronger measures for authenticating users connecting through the console.

You might also want to secure a physically insecure switch's uplink to prevent a hacker from disconnecting the switch and replacing it with the hacker's own device. A hacker who can connect their device in a network infrastructure VLAN can gain unauthorized access to valuable resources. MACsec, which sets up authentication and encryption between infrastructure devices, is one option for securing the switch uplink, but this feature is outside the scope of this training.

Also note that some AOS-CX switches have a Reset button. This button's behavior depends on the front-panel-security factory-reset setting. By default, factory-reset is disabled. In that case, pressing the reset button, holding it for five seconds or less, and releasing the button causes a reboot. Pressing and holding the reset button for five to 20 seconds causes a hard reset. Pressing and holding the reset button longer than that has no effect. Depending on the switch software version (check your switch's Fundamentals Guide), you might be able to enable front-panel-security factory-reset. In that case, pressing and holding the reset button between 20 and 25 seconds causes the switch to reset and revert to factory defaults. HPE Aruba Networking recommends leaving this setting disabled particularly when you cannot ensure physical security.

Enable Control Plane Policing (CoPP)

As you learned earlier, a switch's control plane runs protocols such as spanning tree and routing protocols. It is vital to the health of a network that these protocols run properly. The switch must be able to process routing protocol updates, for example, so that it can learn the correct ways to route traffic. However, hackers can exploit these protocols to implement DoS attacks. For example, a hacker could flood a switch with so many routing protocol updates that the switch's CPU is tied up, preventing it from functioning properly and even compromising managers' ability to access the switch management plane and deal with the issue. Such floods can also occur through accident or carelessness, such as the introduction of a network loop that multiplies protocol packets.

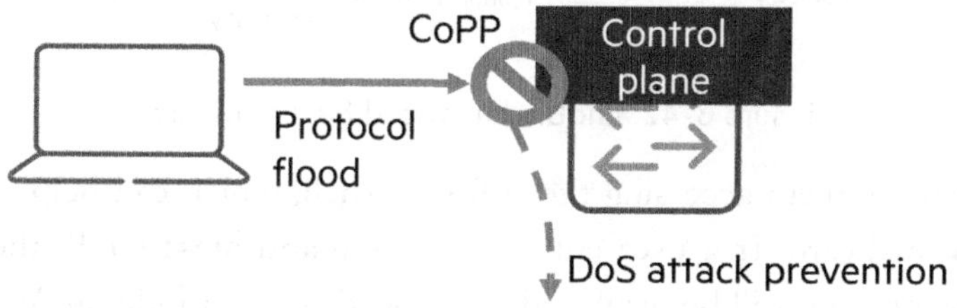

Figure 3-41: Enable Control Plane Policing (CoPP)

Control Plane Policing (CoPP) helps to protect the switch's control plane from such issues. CoPP monitors traffic incoming into the control plane based on class. In this context, a class is a kind of traffic such as a spanning tree Bridge Protocol Data Unit (BPDU) or routing protocol packet. The CoPP policies specify a rate limit for that class in packets per second, as illustrated in Figure 3-41. The rate limit allows the switch to receive sufficient packets for a protocol that is functioning normally (in fact, more than a sufficient number). However, the switch will drop excessive packets to protect its CPU for other important functions. On AOS-CX switches, a CoPP policy can further define a burst size, which permits the traffic class a short burst beyond the set rate limit. The AOS-CX switch policy also defines a drop priority. If the switch becomes overwhelmed, it will drop the protocols with a lower priority first, allowing the switch to maintain the traffic for the most important protocols.

All AOS-CX switches, as of the release of this study guide, support CoPP, which is enabled by default on the data plane (front panel) ports. The switch comes with a factory default CoPP policy, which applies the recommended rate limits, burst settings, and drop priority for each class. You can create up to 32 additional custom CoPP policies, though only one can be active at a time. However, again, it is recommended that you leave the default policy in effect in most situations.

Please note that CoPP is different from the control-plane ACL. A control-plane ACL determines which traffic is allowed to arrive on the control plane. CoPP rate-limits allowed traffic.

Enable Enhanced Secure Mode

Enhanced Secure Mode implements a number of operating differences in software feature support, how commands are executed, and the way parameters are displayed.

To enable this mode on AOS-CX switches, you must boot the switch to the ServiceOS mode and then enable enhanced secure mode, as shown in Figure 3-42.

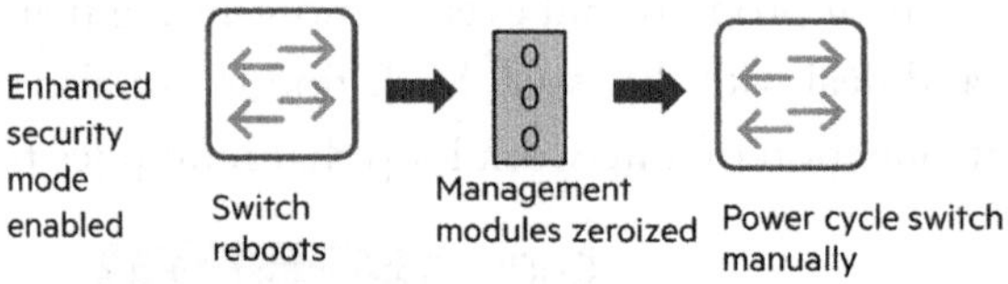

Figure 3-42: Enable Enhanced Secure Mode

This mode prevents users from accessing the software shell, which can help to prevent malicious users from installing malware. If a user enters the command **start-shell**, the command will be rejected and an error message will be displayed. Enhanced secure mode on AOS-CX switches also locks down options in the ServiceOS to prevent hackers from infiltrating the switch software.

CLI Commands: CoPP and Enhanced Secure Mode

The table in Figure 3-43 shows the commands for using CoPP and enhanced secure mode.

Step	AOS-CX Command	Notes
Enable default CoPP policies	`apply copp-policy <default | name>`	**Default policy is enabled on AOS-CX by default**
View CoPP information.	`show copp <configuration | default configuration | <policy_name> configuration | statistics>`	**Viewing the configuration shows which policy is applied. Viewing a policy's configuration shows the settings for that policy. Viewing statistics shows how many packets for each class are passed and dropped.**
Enable Enhanced Secure Mode	`boot system serviceos` `secure-mode enhanced`	**You must boot the ServiceOS (an enable mode command) and then enable enhanced mode from there.** **Important: Enabling enhanced mode resets the configuration.**

Figure 3-43: CLI Commands: CoPP and Enhanced Secure Mode

For CoPP, remember that the default policy is already applied by default. You only need to manually apply a policy if have created a non-default policy. A CoPP policy must always be applied. If you remove the current CoPP policy, the switch applies the default policy. The table also shows the commands with which you view which CoPP policy is applied, a particular CoPP policy's configuration, and a particular CoPP policy's statistics.

The table also shows the commands for booting the switch to its ServiceOS and then enabling enhanced secure mode. It is important to remember that enabling enhanced mode resets the configuration.

Authenticated Network Time Protocol (NTP)

It is critical that network infrastructure devices have the correct time. Without a correct time, issues such as these can occur:

The switch is subject to replay attacks, in which hackers capture valid traffic and resend it. Such attacks can in turn help hackers launch MITM attacks or gain unauthorized access.

The switch cannot correctly validate certificates or use its own certificates.

The switch cannot successfully communicate with a RADIUS server, due to timestamp discrepancies.

The switch logs are inaccurate and difficult to use—and inadmissible for forensics. Imagine trying to troubleshoot an issue across multiple network infrastructure devices when you cannot accurately tell when a particular event occurs. When you are trying to respond to security events, inaccurate clocks become even more troubling. You cannot use network device logs with inaccurate timestamps as part of a forensic record or audit trail.

While you can set the clock on a switch manually, HPE Aruba Networking recommends that you use a Network Time Protocol (NTP) server from which switches and other network devices takes their clock. To use NTP authentication, you must deploy a private server. Ideally, the NTP server should receive its time from a satellite, which would mean that the server or servers would need to be connected to a supported GPS device and wired to a window to receive the time from the satellite.

With a private NTP server, you can—and should—use NTP authentication. NTP authentication uses a shared integrity key that matches on the clients, such as switches, and the NTP server. The server uses this key to sign its messages, which ensures that a hacker cannot spoof the server's IP address and set an incorrect time on the switch. View Figure 3-44 for a summary of the differences between unauthenticated and authenticated NTP.

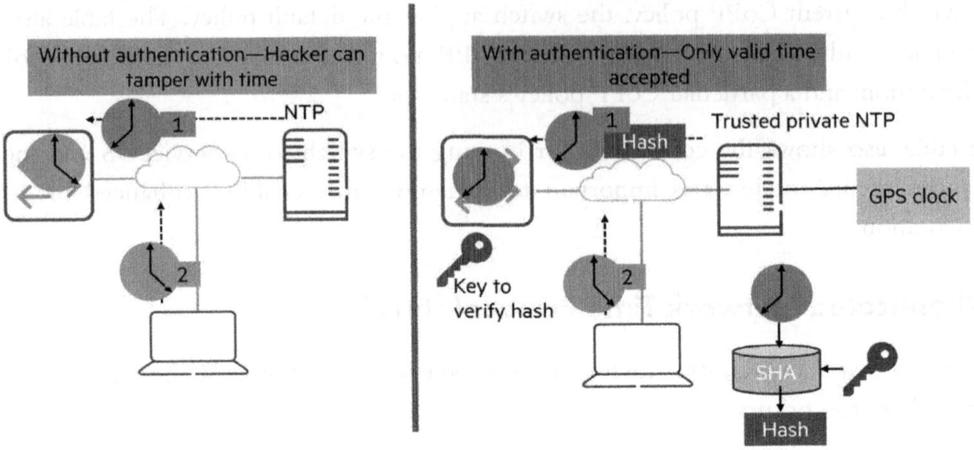

Figure 3-44: Authenticated Network Time Protocol (NTP)

AOS-CX switches supports MD5 or SHA-1 based authentication for NTP; SHA-1 is preferred over MD5.

CLI Commands: Authenticated NTP

In Figure 3-45, you see the commands for setting up NTP with authentication.

In configure context

Step	AOS-CX Command	Notes

Enable NTP authentication and set the NTP authentication key.	`ntp authentication`
	
	`ntp authentication-key id <ID> sha1 [ciphertext] <key-value> trusted`	

Specify the NTP server and the associated key ID.	`ntp server <IP> key-id <ID> [iburst]`	iburst mode enables faster synchronization.
Specify the VRF for NTP.	`ntp vrf <VRF>`	
Enable NTP.	`ntp enable`	

Figure 3-45: CLI Commands: Authenticated NTP

Take Additional Hardening Actions

You will now look at a few more settings for hardening switches.

Idle Session Timeout

Configure the idle session timeout to a nonzero value. A timeout of zero means that the session will never timeout. If a manager forgets to log out of a session manually, an unauthorized user who gains access to the management device could hijack the session. In addition, a switch can only handle a certain number of management sessions (the number depends on the switch). If sessions never time out, all of the sessions could get tied up, preventing managers from accessing the switch.

In AOS-CX, the session timeout default value is 30 minutes. It can be set between 0 (disabled) and 43,200 minutes (30 days). This feature does not apply to the Web UI.

Banners

Additionally, you should set a banner that explains that access is only for authorized users. In some regions it is easier to prosecute unauthorized users who gain or attempt to gain management access when the company can prove that the user should have known that access was not permitted. You can set such as banner with the message of the day (MOTD) banner, which displays before users attempt to log in.

You can also configure an exec banner, which displays after a successful login. See Figure 3-46 for an example.

```
The device you have accessed belongs to example.com. Only
authorized users are allowed to access this device
Switch login: admin
Password:
There was 1 failed login attempt since the last successful
login
Last login: 2024-08-19 14:52:59 from the console
User "admin" has logged in 18 times in the past 30 days
```

Figure 3-46: AOS-CX Exec Banner

Note that, when you successfully login, AOS-CX switches also always inform you about the previous successful and unsuccessful login attempts. This behavior can help to alert you to issues. For example, if you are the only authorized manager, you would want to know if someone tried to access the switch, even unsuccessfully.

Lockouts

You can set lockouts that will lock a user out from further management access attempts after too many authentication failures. As you learned previously, lockouts can be a valuable tool in preventing dictionary attacks against passwords. However, hackers can also exploit lockouts. If a hacker can guess a manager name and enter an incorrect password multiple times, the hacker can lock the legitimate manager out of the account. For this reason, you should use non-default manager usernames. You should also be thoughtful in setting the lockout time. A lockout time on the order of minutes can make dictionary attacks infeasible, but still allow legitimate admins to accidentally lock themselves out once and still do their jobs.

CLI Commands: Timeouts and Banners

Refer to Figure 3-47 for the commands for setting timeouts and banners.

In configure context

Step	AOS-CX Command	Notes
Set the idle timeout period for console, SSH, and Telnet	cli-session timeout <0-43200>	Set in seconds 0 disables the timeout, which is not recommended in the real world.
Set a banner to explain authorization	banner <exec | motd> *<delimiting character>*	The delimiting character is any character of your choice (which the banner will not include). After entering it, you enter text for the banner, including multiple lines. You then enter the delimiting character again to end. You do not need to configure anything for AOS-CX switches to show information about previous logins. They show this information by default after a successful login.

Figure 3-47: CLI Commands: Timeouts and Banners

Understand the Trusted Platform Module (TPM)

AOS-CX switches typically include a FIPS 140-2 compliant Trusted Platform Module (TPM). (Check whether your model does in the device's datasheet.) The TPM is a dedicated hardware component within the switch for storing sensitive data such as the switch's private keys for certificates. AOS-CX switches also come with a TPM certificate already installed on them. This certificate is signed by a CA operated by HPE Aruba Networking.

The switch's TPM also provides secure boot, as illustrated in Figure 3-48.

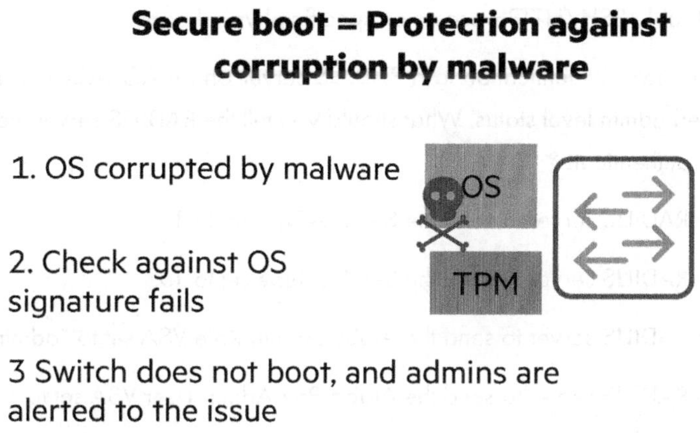

Figure 3-48: Understand the Trusted Platform Module (TPM)

If a hacker can succeed in installing malware in the switch OS, the hacker can obtain a powerful and hard-to-detect backdoor into the network. Secure boot prevents this from happening. The TPM stores a hash of each OS component signed by a secure key. Whenever the switch boots, the bootloader checks each OS component against this signature. If the OS has been corrupted, the component does not match the signature. The switch will not boot, and admins are alerted to the issue.

Summary

Congratulations! You can harden AOS-CX switches against attack. You can segregate management from data traffic to create a secure management environment. You can authenticate your managers securely and ensure that they use secure protocols to manage the switches. You can enforce physical security for switches and protect their control plane.

Learning checks

1. You have set up an Ethernet out-of-band management network and connected AOS-CX switch Mgmt ports to it. What additional step should you take to ensure that users in the production network cannot access the management interfaces (Web UI or CLI)?
   a. Apply a control plane ACL on the mgmt VRF.
   b. Enable management protocols on the mgmt VRF and disable them on the default VRF.
   c. Create a list of authorized management IP addresses.

d. Enable and disable SSH/HTTPS access on specific physical ports.

2. You are configuring SSH authentication to a RADIUS server on an AOS-CX switch. You want Bob and Alice to be assigned admin level status. What should you tell the RADIUS server manager about the AVP to send when they authenticate?

   a. Configure the RADIUS server to send the Service-Type set to 1.

   b. Configure the RADIUS server to send the Service-Type set to 15.

   c. Configure the RADIUS server to send the Aruba-Admin-Role VSA set to "administrators."

   d. Configure the RADIUS server to send the Aruba-Priv-Admin-User VSA set to 1.

3. Which of the following is a list of the secure protocols that HPE Aruba Networking recommends using?

   a. Telnet, TFTP, HTTP, SNMPv1/v2c

   b. SSH, SFTP, HTTP, SNMPv3

   c. SSH, SFTP, HTTPS, SNMPv3

   d. Telnet, SFTP, HTTPS, SNMPv1/v2c

4. You are using SSH to access your switches. What is one recommendation for configuring the management station to authenticate the switch?

   a. View the switch's public SSH key in the CLI and copy the key into the management station SSH client's known hosts file.

   b. Configure the switch's username and password in the known hosts file for the management station's SSH client.

   c. Configure the management station to communicate with a RADIUS server to check the switch's credentials.

   d. Configure the management station to communicate with a TACACS server to check the switch's credentials.

Note

For answers to all learning checks, refer to Appendix B.

---

4 Hardening AOS Devices

EXAM OBJECTIVES

✓ Understand why hardening is important

✓ Implement HPE Aruba Networking hardening best practices, such as disabling insecure protocols, implementing password protection and using external authentication, enforcing physical security, using Control Plane Security (CPsec)

Assumed Knowledge

- Basics of IP addressing
- Basics of Dynamic Host Configuration Protocol (DHCP)
- Basics of 802.11 wireless networks

Overview

It is crucial that you create a trusted mobility environment. This chapter briefly covers the options you can provide for FIPS-compliant hardware and software. You will then learn in detail how to lock down administrative access to your HPE Aruba Networking mobility devices, lock down services on those devices, and use CPsec to secure communications between APs and MCs.

Introduction

This chapter begins with a discussion of best practices for hardening the wireless infrastructure.

Why Harden the Wireless Infrastructure Devices

You have learned how failing to secure management access to switches creates an untrusted environment and leads to insecure communications. The same holds true for the wireless infrastructure. If hackers can gain management access to an MC, they can cause all sorts of havoc.

Among other issues, they can disrupt communications and cause a denial of service (DoS) attack. Or they change firewall policies to grant themselves greater access to the network than they should have.

It is also critical to harden the communications between APs and MCs. If hackers can read these communications, they can conduct a great deal of reconnaissance on the network. And if hackers can inject their own, unauthorized control communications, they could even disrupt the network. For example, they could launch a DoS attack by ordering the APs to shut down.

In short, it is imperative for companies to protect MCs' and APs' management planes and the control traffic between them. (You will learn about securing user access and users' data traffic in another chapter.)

Note that throughout this chapter, "AOS device" will refer to any device running HPE Aruba Networking Wireless Operating System (AOS) software such as an HPE Aruba Networking MC or an HPE Aruba Networking Instant AP (IAP).

Best Practices Checklist

To successfully harden wireless devices, you should follow the best practices checklist shown in Figure 4-1, which you will explore in-depth in the rest of this chapter.

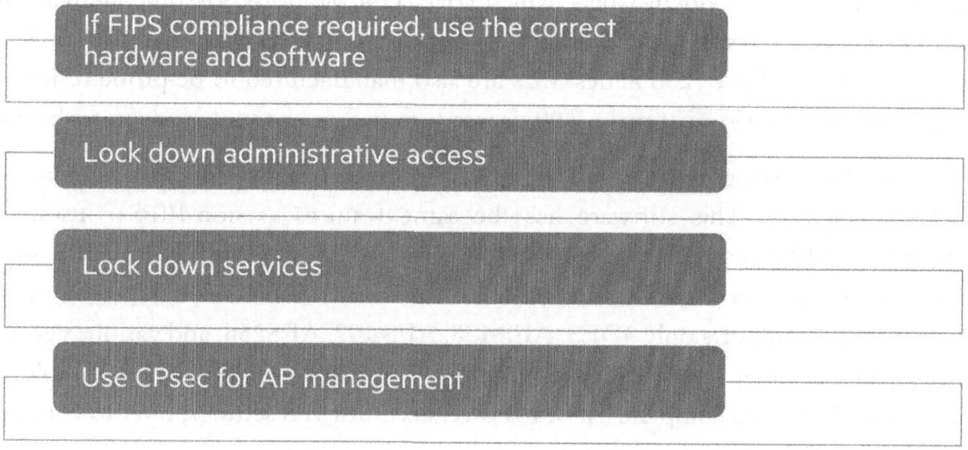

Figure 4-1: Best Practices Checklist

As you learned earlier, Federal Information Processing Standards (FIPS) are a set of US regulations that dictate measures for securing data with which non-military federal agencies and contractors must comply. If your company falls in this area, make sure that you are using FIPS-compliant HPE Aruba Networking hardware and software.

The rest of the best practices apply to all companies. You should lock down administrative access by disabling local authentication and configuring external authentication through a Remote Authentication Dial-In User Service (RADIUS) or Terminal Access Controller Access Control System Plus (TACACS+) server, such as HPE Aruba Networking ClearPass Policy Manager (CPPM). You should also lock down services to protect the HPE Aruba Networking mobility devices' management and control plane. Furthermore, you should use CPsec for AP management, as is enabled by default.

FIPS Compliance

You will begin by briefly examining ways to ensure FIPS compliance.

FIPS/TAA Hardware and Software

If your company needs to comply with FIPS, HPE Aruba Networking makes it easy for you to verify that the mobility infrastructure meets the regulatory requirements. First make sure to select FIPS/TAA models for all devices. FIPS prohibits the use of hardware manufactured in China out of concerns that the hardware might include backdoors included by the Chinese government. HPE Aruba Networking FIPS/TAA hardware is manufactured in the US or Singapore. This hardware also comes with tamper-evident labels, which cover console ports. If someone has attempted to access the device's console port, admins can easily see and know that the device could be compromised. FIPS/TAA 7200 Series MCs are also manufactured to be probe resistant; anything inserted inside the chassis must make a 90 degree turn before encountering internal components.

HPE Aruba Networking also provides separate software builds for devices that need to be FIPS/TAA-compliant. This software has the same features as non-FIPS software. However, support for non-FIPS-compliant algorithms is removed. For example, the FIPS software only supports OpenSSL version 2.0.2 (rather than 1.0.1 too) and TLS version 1.2 (rather than versions 1.0 and 1.1 too). It supports only 3DES, AES 128, AES 192, AES256, and various versions of SHA (no RC4, MD5, or other non-certified algorithms). The FIPS software further disables Telnet and only allows SSHv2. For a complete list of FIPS versus non-FIPS settings, refer to your devices' *User Guides*.

Lock Down Administrative Access

You will now look at how to lock down administrative access.

Best Practices Checklist

Locking down administrative access includes protecting the console port to prevent unauthorized management access. It might also involve configuring access policies to control management access to the MC. You should additionally set up external authentication and authorization of managers. Finally, to ensure the use of external authentication, you will need to effectively disable local authentication. These administrative access best practices are summarized in Figure 4-2.

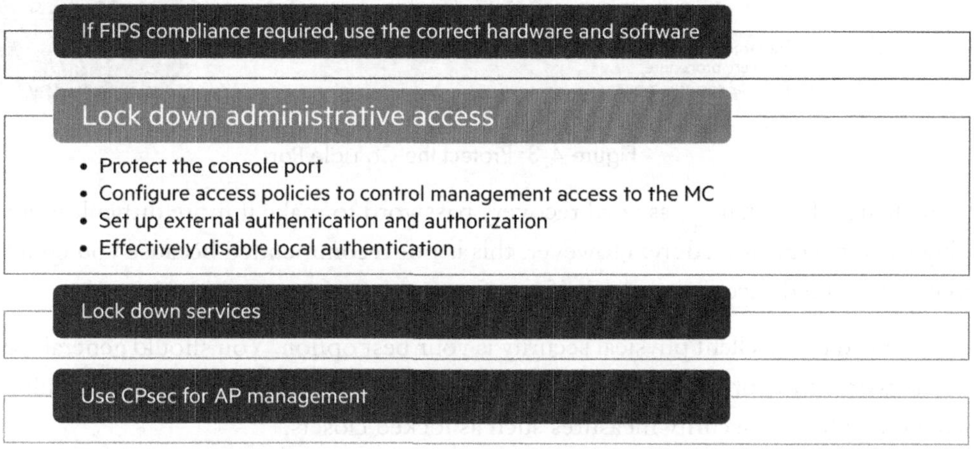

Figure 4-2: Best Practices Checklist

Protect the Console Port

The serial console port on an MC or AP is considered a privileged interface. (See Figure 4-3 for example locations of these ports.) An attacker with physical access to an MC's console port can easily compromise the MC using the password recovery procedure, which is widely known and is given out by HPE Aruba Networking Technical Support frequently when customers lose access to administrative passwords. The password recovery procedure can be used to change the password for the "admin" account – it does not reboot the MC or change any other configuration. However, once hackers have changed the admin password, they can do whatever they want on the MC.

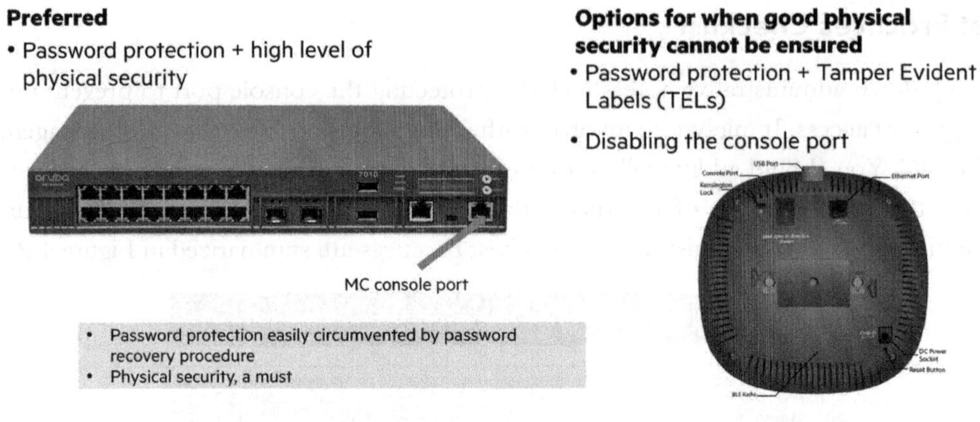

Figure 4-3: Protect the Console Port

You can change the default password recovery password to make it more difficult to attack an MC by using the recovery procedure. However, this is not recommended because you could easily lock yourself out of the device permanently if you forget the recovery password.

In short, providing excellent physical security is your best option. You should generally deploy MCs at the network core or sometimes aggregation layer, where the company should already be implementing physical security measures such as locked closets.

Physically securing APs might be more difficult because APs must be deployed where they can provide a good signal. Often APs are installed high up and relatively out of reach. If you cannot provide good physical security for an infrastructure device, HPE Aruba Networking recommends applying tamper-evident labels (TELs) across the console port. TELs can be obtained from a wide variety of commercial sources and are provided with FIPS/TAA models. To get the most benefit from TELs, routinely inspect the device to see if the label has been removed.

Alternatively, you might disable the console port on the physically insecure device through the software (an option beginning with AOS 8.0). This will prevent the password recovery procedure from being available. However, this command only disables the serial port while AOS is running. If the device is rebooted, the boot ROM may still be accessed through the console port, which can be used to boot alternative configuration files, boot alternative software images, or to erase data from flash memory.

HPE Aruba Networking Mobility Device Control Plane vs Datapath

You have secured the HPE Aruba Networking device's console port, but most management traffic arrives on the device's IP address. You need to consider how to protect the device from unauthorized types of access. In this chapter, you will now focus on strategies for protecting the control plane.

To understand what protecting the control plane means, you need to understand what the control plane is, as opposed to the datapath. (You will examine ways to secure the datapath in other chapters.) The control plane receives any traffic that is destined to the infrastructure device's own IP address. This can be management traffic such as HTTPS traffic to an MC's Web UI or Secure Shell (SSH) traffic to an MC's command line interface (CLI). It can also be traffic that the device sends and receives as part of its functions as a network infrastructure device. Just a few examples of control plane traffic are RADIUS communications between an MC and a RADIUS server, communications between APs and MCs, and communications between MCs and a Mobility Conductor (MCR). In Figure 4-4, the solid arrows indicate traffic destined to the MC control plane.

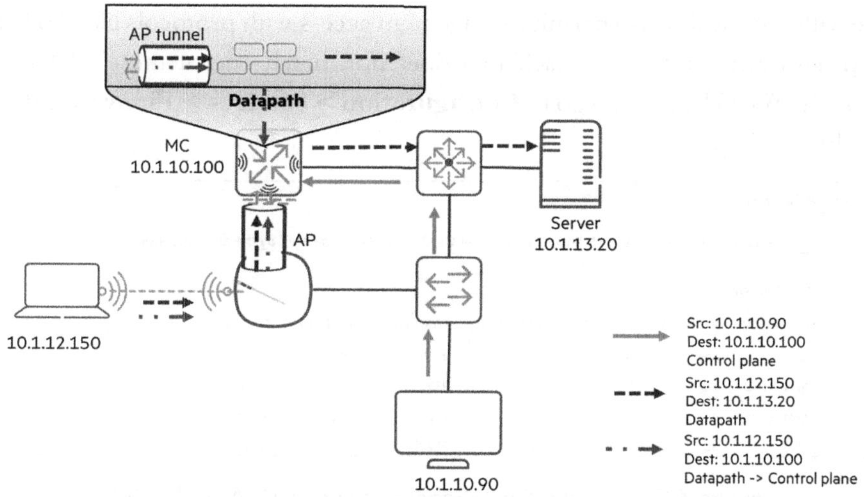

Figure 4-4: HPE Aruba Networking Mobility Device Control Plane vs Datapath

Datapath traffic, on the other hand, is traffic from clients on "untrusted ports" such as wireless local area networks (WLANs). The HPE Aruba Networking device is responsible for controlling this traffic and then forwarding towards its final destination. Typically, the clients are wireless clients. IAPs receive their traffic directly and bridge it themselves. MCs receive the wireless client traffic from tunnels established with APs. They can also receive wired client traffic from tunnels established with switches and forward that on the datapath. In Figure 4-4, the dashed arrows indicate datapath traffic.

Client traffic in the datapath must pass through the AOS firewall, which implements role-based policies which you will learn about in later chapters.

Most often clients are sending traffic to devices on the other side of the firewall, such as a server in the LAN or in the Internet. However, datapath traffic can pass to the control plane, if the client sent the traffic to the MC's own IP address. The dotted line arrow in Figure 4-4 provides an example of such traffic.

Adjust the Control Plane ACL (Firewall ACL Allowlist)

You can also implement rules called access control lists (ACLs) on control plane traffic. These rules filter packets arriving on the MC control plane, based on fields in the packets' headers.

MCs come with a default control plane ACL, called the firewall ACL allowlist in the Web UI. The rules in this ACL permit traffic to specific protocols and TCP or UDP ports that the device needs to function. For example, they permit the TCP and UDP ports that the controller uses to communicate with APs. Rules also permit management access with protocols like SSH and HTTPS. The MC drops all traffic directed to itself that does not match one of these ACLs. To view the default rules on the Web UI, you can go to **Configuration > Services > Firewall** and expand **ACL allowlist** (see Figure 4-5).

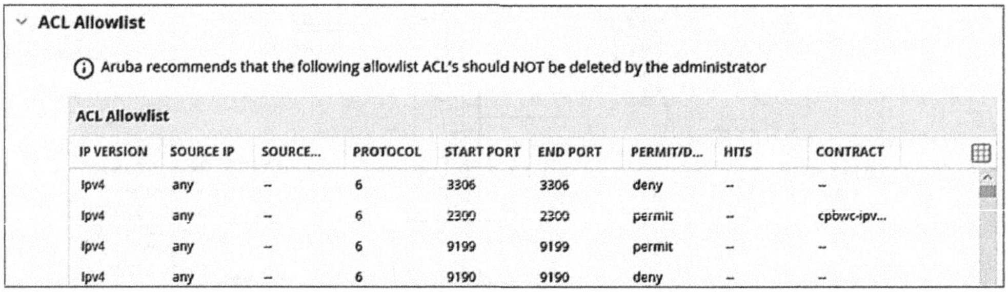

Figure 4-5: Configuration > Services > Firewall > ACL allowlist

Because the MC needs to support these traffic types to function, you should not delete default rules unless you have a strong understanding of the rules' purpose and know that the device does not require that type of traffic for the customer's solution.

By default, the ACLs permit traffic from all sources. To enhance security, you *can* create new rules that specify specific source addresses for specific protocols. For example, you can establish a management network on 10.1.5.0/24 and then permit TCP 22 (SSH) only from 10.1.5.0/24 (see Figure 4-6). After you have tested that the new, source-specific, rules are functioning, you can delete the corresponding default rules.

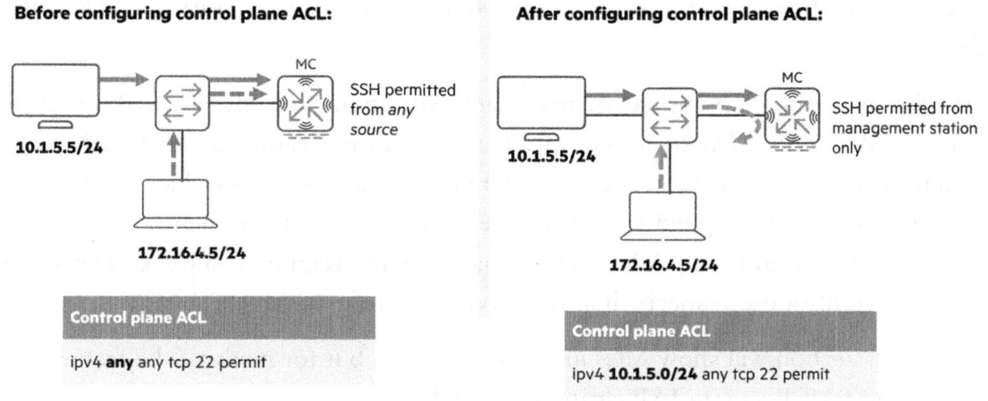

Figure 4-6: Adjust the Control Plane ACL (Firewall ACL Allowlist)

Secure Manager Access to the Web UI

You will next examine best practices for securing manager access to the HPE Aruba Networking mobility devices' Web UI. This discussion focuses on securing access to the Web UI for a standalone MC or MCR solution, but many similar concepts apply to IAPs as well (refer to the IAP documentation for details).

First, you should understand that it is best practice to authenticate managers to an external server rather than an internal list of users. An external authentication server like CPPM can use a variety of authentication sources, including Active Directory and other Lightweight Directory Access Protocol (LDAP) servers. Such solutions have built-in methods for forcing users to change their passwords periodically and for revoking the account of a manager who no longer works at the company. External authentication provides greater centralization and scalability. When you use CPPM as the authentication server, you also gain CPPM's abilities to make nuanced decisions about which role to assign to a manager; CPPM can take a variety of information into account, beyond identity, such as location and time.

To enable external authentication for management users, you must set up a RADIUS or TACACS server group on the MC, which specifies one or more servers. Then you must configure admin authentication to use that server group.

When you use RADIUS to authenticate managers to CPPM, you should generally use Microsoft Challenge Handshake Protocol (MSCHAP). When you use TACACS to authenticate managers to CPPM, you must use Password Authentication Protocol (PAP). As you recall from earlier, these protocols do not provide rigorous security on their own. Therefore, you need to ensure that the path between the network infrastructure device and the authentication server is trusted. If this is not the case, you should add protection. For example, you can use RADIUS with RADIUS Security (RadSec).

In addition to setting up secure manager authentication, you should make sure to install a Certification Authority (CA)-signed certificate. As you learned earlier, a self-signed certificate makes it so admins must accept errors when they connect to the Web UI—a bad habit and against best security practices. A self-signed certificate does not provide true authentication of the device or protection from man-in-the-middle (MITM) attacks. In this section, you will also learn how to install a CA-signed certificate and specify it as the server certificate for the Web UI.

Note that this section will show MCs in the illustrations, but for many of the features, a similar set up applies to an MCR's or an IAP cluster's Web UI.

Create the Authentication Group

Refer to Figure 4-7 to see how to create an authentication group in the Web UI. Go to the **Configuration > Authentication > Auth Servers** page. Click + in the Server Groups list to add the group. Name the group and click **Submit**.

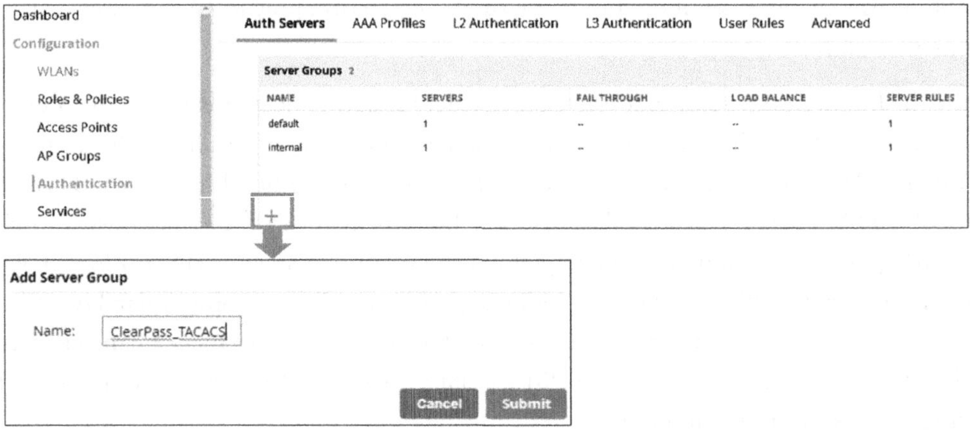

Figure 4-7: Secure Manager Access to the Web UI

Create the Authentication Server

The process for creating an authentication server is illustrated in Figure 4-8.

You can then select the new group, scroll down to the servers list, and click +.

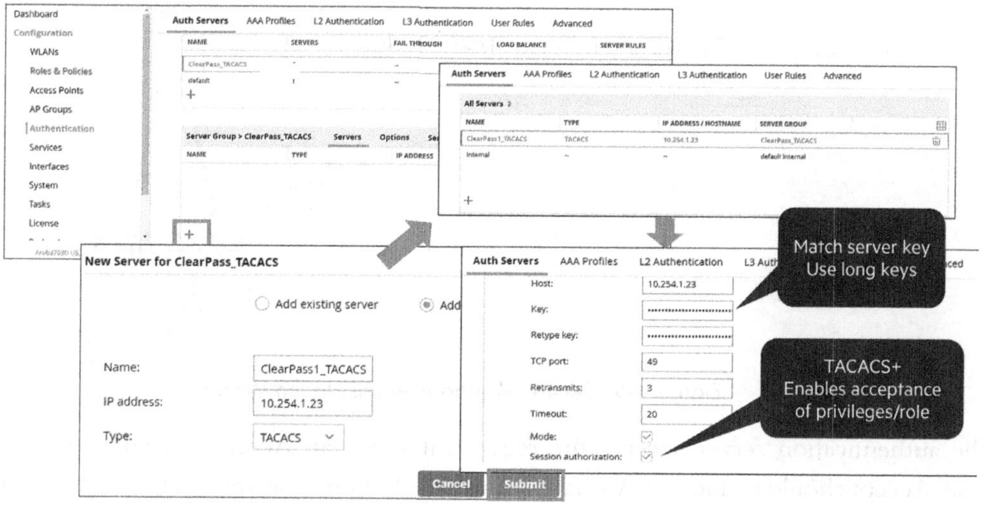

Figure 4-8: Secure Manager Access to the Web UI

You can choose to add an existing server to the group or create a new server and add it. In this example, you are adding a new server. Specify the name and IP address or fully qualified domain name (FQDN). You can select TACACS or RADIUS for the type. In this example, you are setting TACACS. Click **Submit**.

After you create the server within the group, you must select it. Scroll down to its detailed settings and set the key to match the shared secret on the authentication server. It is best practice to set a very long (at least 25 characters) pseudorandom string of characters, which you coordinate with the server admins.

The Mode check box means that the server is enabled; keep it selected. For a TACACS+ server, make sure to also select the **Session authorization** check box, which enables the device to accept manager role assignments from this server.

How the External Authentication Server Assigns Managers to Roles

When a manager attempts to log in to the mobility device's management interface, the device sends an Access-Request to the authentication server. The example in Figure 4-9 shows CPPM as that server.

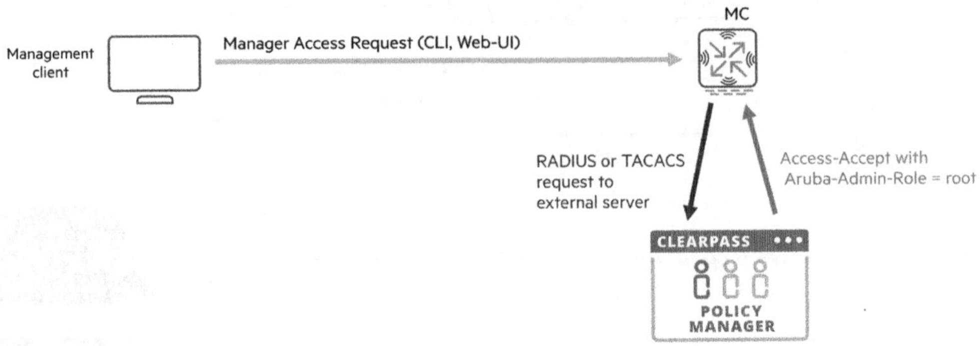

Figure 4-9: Secure Manager Access to the Web UI

If the authentication server accepts the request, it sends an Access-Accept to the device. This Access-Accept should include an Aruba-Admin-Role Vendor Specific Attribute (VSA). This VSA's value is the name of the role for the manager. If this VSA is not included, or if it includes an invalid role name, the device gives the authenticated manager the default role set in the Admin Authentication settings. To prevent elevated access in case the server is configured incorrectly, set the default role to a low privilege level such as "guest-provisioning."

You will now see an example of CPPM sending an Aruba-Admin-Role VSA in response to a TACACS+ authentication request. In Figure 4-10, you can see the record for the TACACS+ request in CPPM Access Tracker.

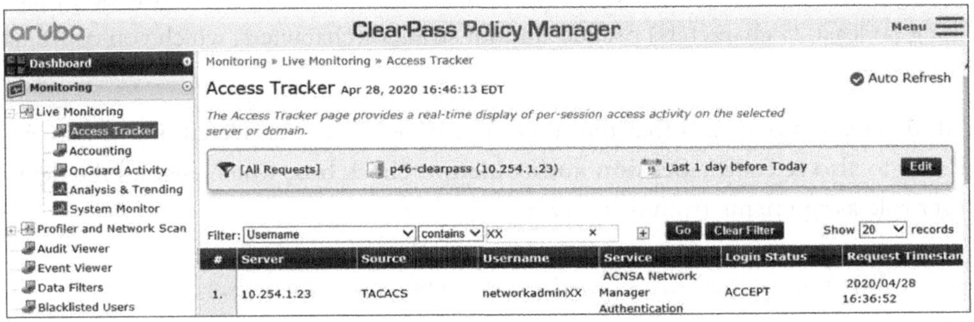

Figure 4-10: CPPM Access Tracker

The output in Figure 4-11 indicates that the user passed authentication.

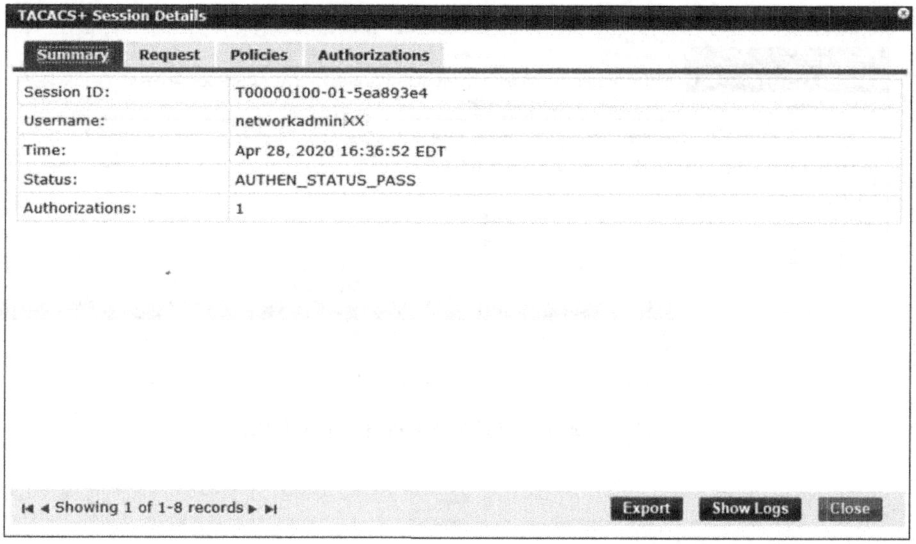

Figure 4-11: CPPM Access Tracker TACACS+ Record (Summary)

In Figure 4-12, you can see that CPPM assigned the manager to the "ArubaOS Wireless – TACACS Root Access profile."

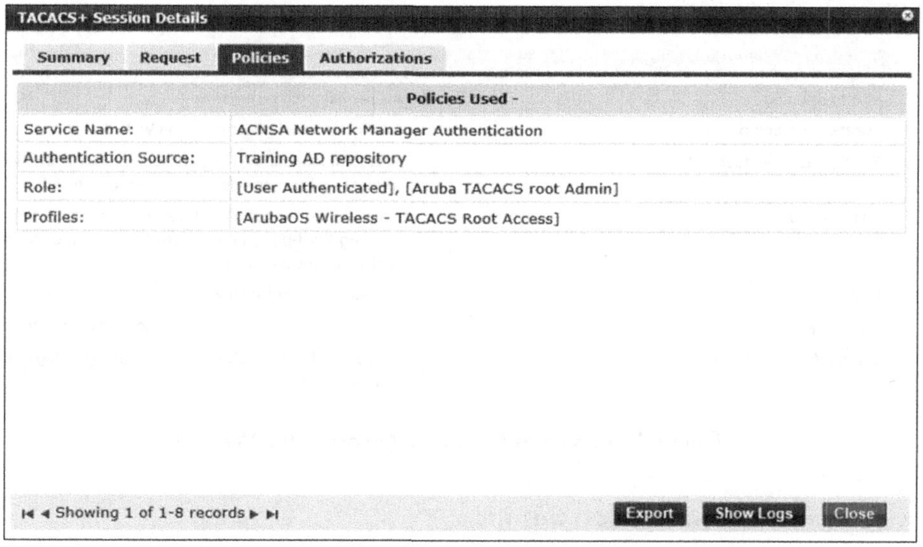

Figure 4-12: CPPM Access Tracker TACACS+ Record (Policies)

In Figure 4-13, you see that this profile tells CPPM to send the Aruba-Admin-Role VSA set to "root."

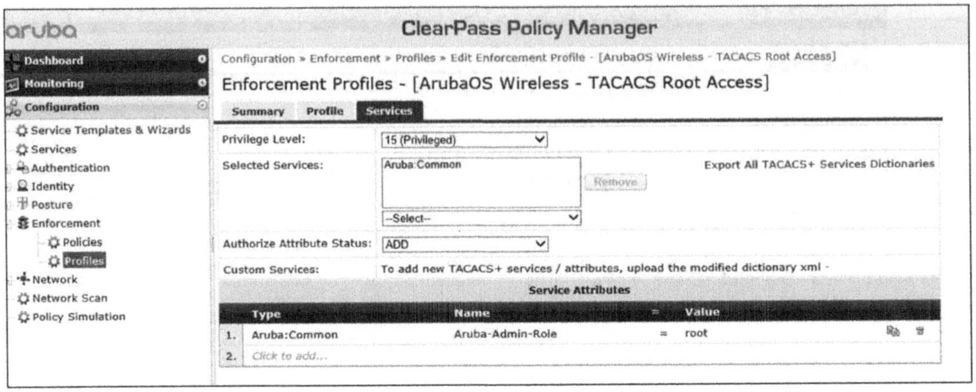

Figure 4-13: TACACS+ Profile in CPPM

Management Roles

The administrator's role determines the management and monitoring rights that the admin receives. The predefined management roles for HPE Aruba Networking mobility devices are: ap-provisioning, guest-provisioning, location-api-mgmt., nbapi-mgmt, root, read-only, and standard. Refer to the table in Figure 4-14 for more information about each management role's specific permissions.

Management Role	Permissions
ap-provisioning	AP provisioning commands
guest-provisioning	Configuration of guest users via WebUI
location-api-management	Access to location API information, intended for allowing third-party location appliances to collect the info
nbapi-mgmt	Access to northbound API (NBAPI) functions, intended for allowing third-party applications to collect monitoring info such as application health
root	All management functions
read-only	Access to CLI show commands and WebUI monitoring pages
standard	All management functions except making changes to management users

Figure 4-14: Secure Manager Access to the Web UI

Configure Admin Authentication Settings

In Figure 4-15, you see where you enable admin authentication, specify the server group that you created earlier, and set the default role. You can use MSCHAPv2 if you are using RADIUS for the authentication protocol. However, if you have set up TACACS+ to CPPM, you should leave this check box clear and use PAP.

Configuration > System > Admin

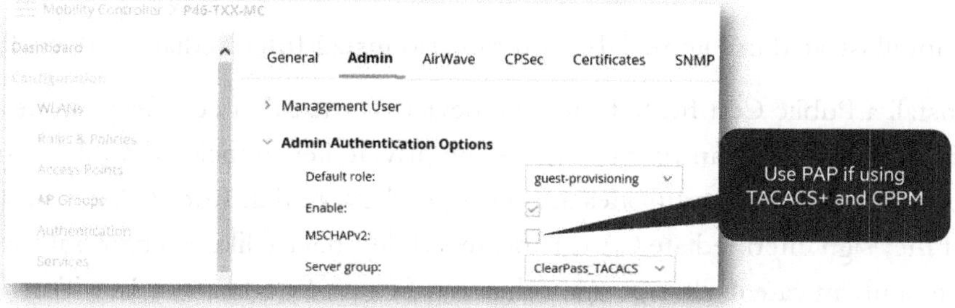

Figure 4-15: Secure Manager Access to the Web UI

MC Certificate Management

To prevent administrators from seeing certificate errors when they log into the Web UI, you must install a CA-signed certificate on the network infrastructure device. Before you learn how to do so, you should understand a bit about managing certificates on HPE Aruba Networking Mobility device like an MC.

You can install several types of certificates on MCs; you can see some of the most common types in Figure 4-16.

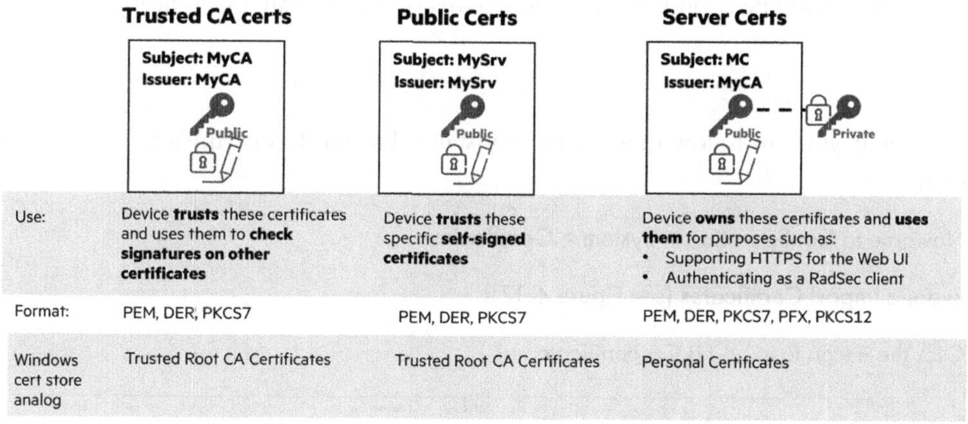

Figure 4-16: MC Certificate Management

You can install Trusted CA certificates, which function much like the certificates installed in the Trusted Root CA Certificates stores on a Windows client. The MC uses the public key in a Trusted CA certificate to check the signature for other certificates. For example, if the MC is using RadSec, it needs to verify the certificate for the RadSec server. If the RadSec server sends a certificate signed by MyCA, the MC uses the public key in the MyCA certificate, which must be installed as a Trusted CA cert, to verify the certificate. If the signature is valid, the MC trusts the signature.

Although not illustrated in Figure 4-16, you can also install Intermediate CA certificates.

You can install a Public Certificate to tell the device to trust that certificate alone. In other words, the MC trusts a peer that can prove it owns the private key associated with the public key in the certificate. Typically public certificates are self-signed. Note that Root CA certificates are also self-signed, but they sign intermediate CA certificates while a device like a server only uses its self-signed certificate to authenticate itself, not sign other certificates. In other words, a Trusted CA certificate lets the device trust many certificates. A public certificate only lets the device trust that one certificate.

You can also install Server Certificates on the MC. The device owns these certificates. That is, it has the public *and* private key for them. The Server Certs are analogous to the Personal Certificates store on Windows. An MC can use its Server Certificates for a wide variety of purposes. For example, it can use a Server Certificate to authenticate as a server and establish a TLS tunnel to provide HTTPS access to the Web UI. It can also use a Server Certificate to authenticate as a RadSec client and for many other purposes not covered in this study guide.

MCs support additional certificate types that are outside the scope of this training. For example, OCSP Responder and OCSP Signer certificates enable the device to use Online Certificate Status Protocol (OCSP) to check on the revocation status of a certificate dynamically.

Installing a Trusted CA Certificate through the AOS UI

In more detail, you can follow these steps to install a Trusted CA certificate on an MC through the Web UI:

1. Navigate to **Configuration > System > Certificates**.
2. Expand **Import Certificates** (see Figure 4-17).
3. Click the + icon to import a CA certificate.

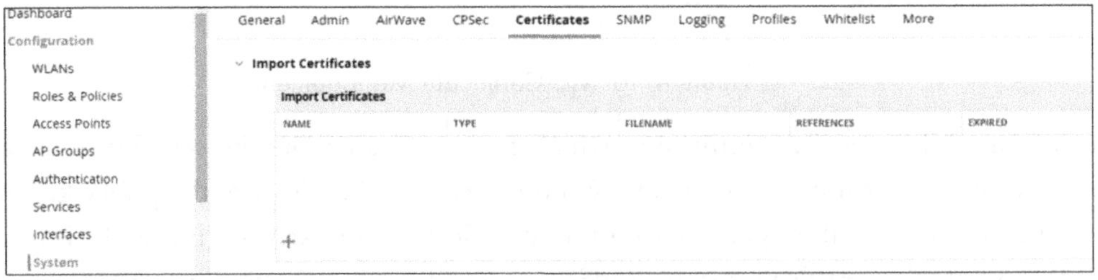

Figure 4-17: Configuration > System > Certificates > Import Certificates

4. Scroll down to the **New Certificate** section (see Figure 4-18).
5. For **Certificate Name**, type an intuitive name.
6. Click **Browse** and find the CA certificate file.
7. Set the **Certificate format**.
8. Set **Certificate type** to **TrustedCA**.
9. Click **Submit**.

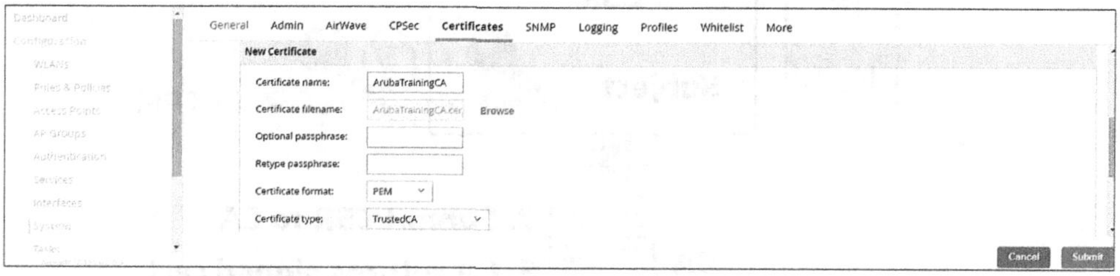

Figure 4-18: Configuration > System > Certificates > New Certificate (TrustedCA)

10. Verify that the certificate is installed. (See Figure 4-19 for an example.)
11. Apply the pending changes in the Web UI.

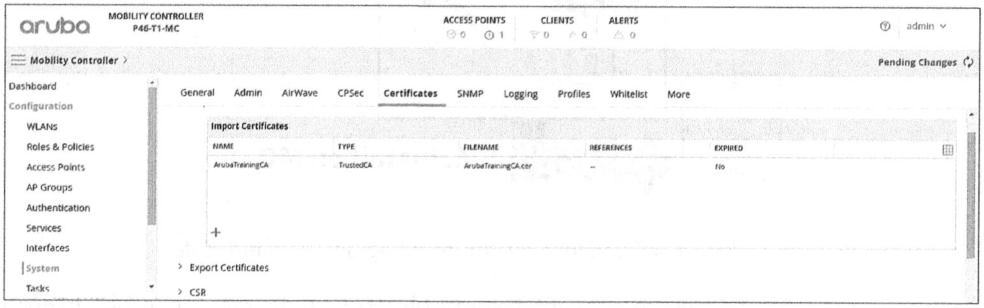

Figure 4-19: TrustedCA Certificate Installed

Methods for Obtaining Server Certificates on an MC

To review, you have several choices for installing server certificates on an MC.

Option 1: Online

You can create a certificate signing request (CSR) on the device, as shown in Figure 4-20.

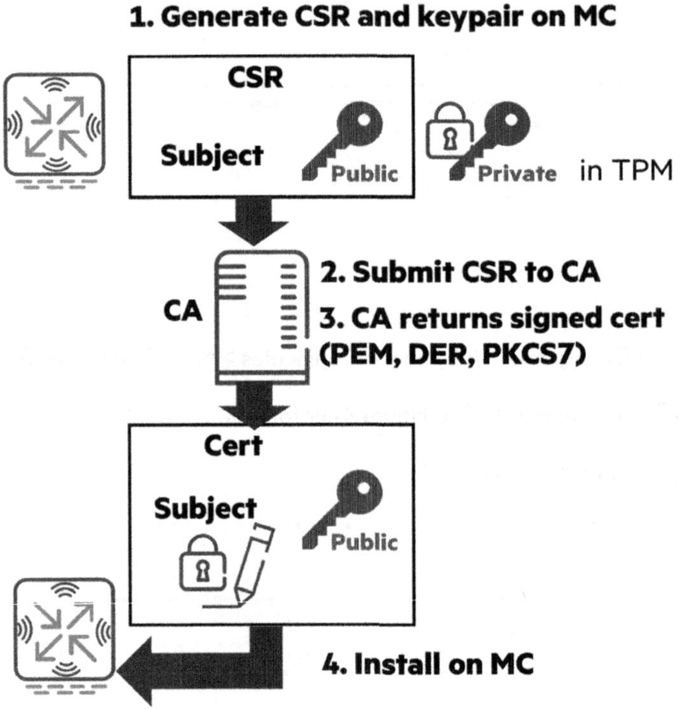

Figure 4-20: Option 1 for Installing an MC Certificate

Begin by creating the CSR, which also creates a public/private keypair. The MC keeps the private key safely in its Trusted Platform Module (TPM). Copy the CSR output into a file; the CSR includes only the public key. You can then submit the CSR file to a CA and obtain a signed certificate. You then import the signed certificate as a PEM, DER, or PKCS7 file, which the device matches to the private key in the pending CSR.

Option 2: Offline (Generally Preferred)

Alternatively, you or your company's security team can create the CSR and public/private keypair offline, as shown in Figure 4-21. This is the generally preferred method. It allows you or the security team to create a CSR with subject alternative names (SANs) and to request precise key extensions for the certificate. It also lets you archive the private key. With access to the private key, you can deploy the same certificate across multiple devices. For example, when you implement captive portal, you typically need a public CA-signed certificate so that any device, whether part of your organization or not, will trust that certificate. Using the same certificate across multiple devices in a cluster can simplify and save costs.

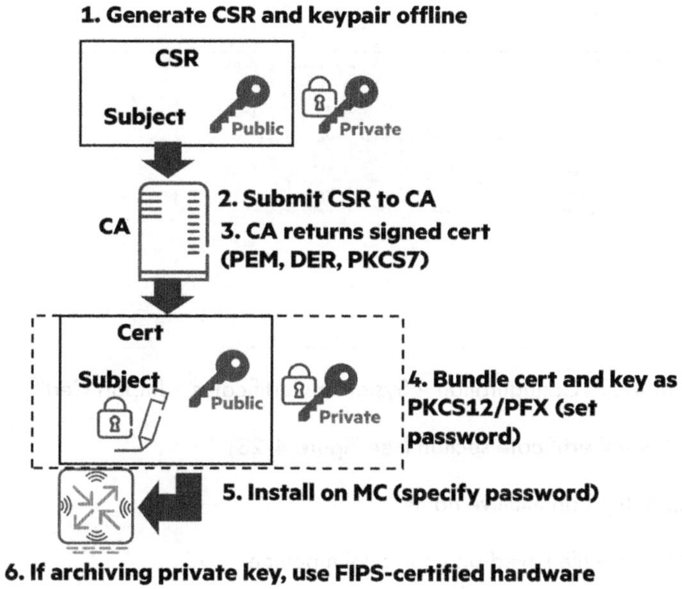

Figure 4-21: Option 2 for Installing an MC Certificate

After creating the offline CSR, you or the security team submit the CSR (*not* the private key) to the CA. When you receive the certificate back, you bundle the certificate with the private key in PKCS12 or PFX format. You then install the file on the MC.

Because creating the CSR offline makes the private key available outside the network infrastructure device, you must take care to protect the private key, as well as any file that includes the private key, with a strong password. When you install the PKCS12 or PFX file on the MC, you will be prompted to enter the password associated with that file. After you install the file on the MC or MCs, you should either destroy the private key and PKCS12 or PFX file or archive them on physically secure, FIPS-certified hardware with strong encryption protections.

In more detail, these are the steps for installing a certificate/private key created offline through the AOS Web UI:

1. Navigate to the **Configuration > System > Certificates** window.
2. Scroll to the **Import Certificates** section (see Figure 4-22).
3. Click +.

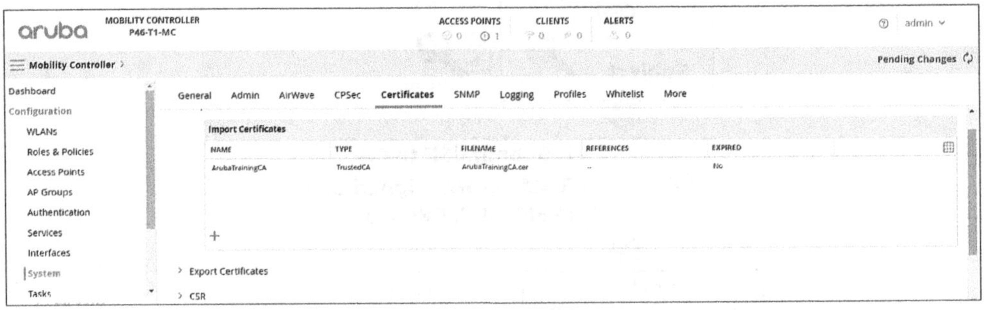

Figure 4-22: Configuration > System > Certificates > Import Certificates

4. Scroll down to the **New Certificate** section (see Figure 4-23).
5. For **Certificate name**, type an intuitive name.
6. Click **Browse** and find the file saved on your local machine.
7. Enter the password that you set when you created the PFX file in the **Optional passphrase** and **Retype passphrase** fields.
8. Set **Certificate format** to **PFX**.
9. Set **Certificate type** to **ServerCert**.

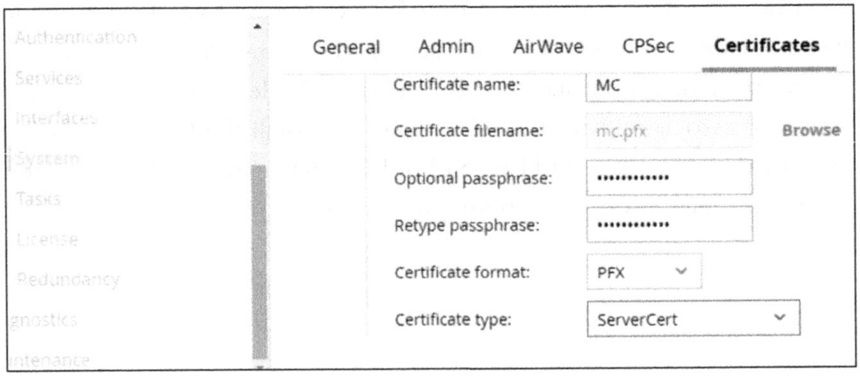

Figure 4-23: Configuration > System > Certificates > New Certificate (ServerCert)

10. Click **Submit**.

11. Verify that the certificate is installed. (See Figure 4-24 for an example.)

12. Deploy the pending change in the Web UI.

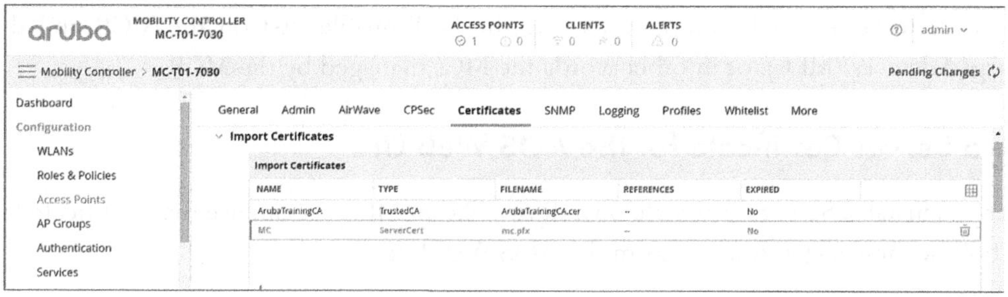

Figure 4-24: ServerCert Installed

Using EST on MCs

MCs also support Enrollment over Secure Transport (EST) to automate certificate enrollment and re-enrollment, as shown in Figure 4-25. To review, this option is preferred for wider scalability, fewer ongoing management demands, and lower risks that a certificate will expire and disrupt functions that rely on that certificate.

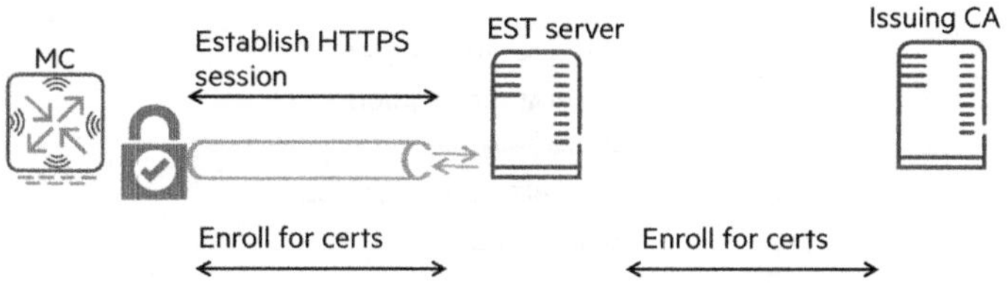

Figure 4-25: MC Certificate Management

To set up your MC to use EST, begin by installing the CA certificate that signed the EST server's HTTPS certificate. Also ensure that the MC and the EST have synchronized time by setting up authenticated Network Time Protocol (NTP) to the same server (covered later in this chapter).

Then create an EST template, which includes information such as the EST server's IP address or FQDN (which must match the subject name or SAN in the server's certificate), an organizational unit (OU) for the MC, a challenge password, and optional arbitrary labels, arbitrary enrollment labels, and arbitrary re-enrollment labels. These labels help the EST server determine which services to provide for the device and how to help it enroll for certificates. Your EST server admin should tell you the proper labels, if required.

After you create the EST profile, you can enable it on the **Configuration > System > Certificates** page. The MC will then establish an HTTPS session with the EST server and automatically enroll for a server certificate. The EST server will manage communicating with the issuing CA at the backend.

If you are using an MCR solution, activate the EST profile first on the MCR and then on the managed devices (MDs), or in other words the MCs managed by the MCR.

Use a Server Certificate for the AOS Web UI

After you install a Server Certificate on a device like an MC, you can use that certificate for the Web UI. Follow these steps to do so from the AOS Web UI:

1. Navigate to **Configuration > System > Admin.**
2. Expand **Admin Authentication Options.**
3. Scroll to **WebUI Authentication** (see Figure 4-26).
4. From the **Server certificate** drop-down menu, select the name of the certificate that you installed.
5. Click **Submit.**

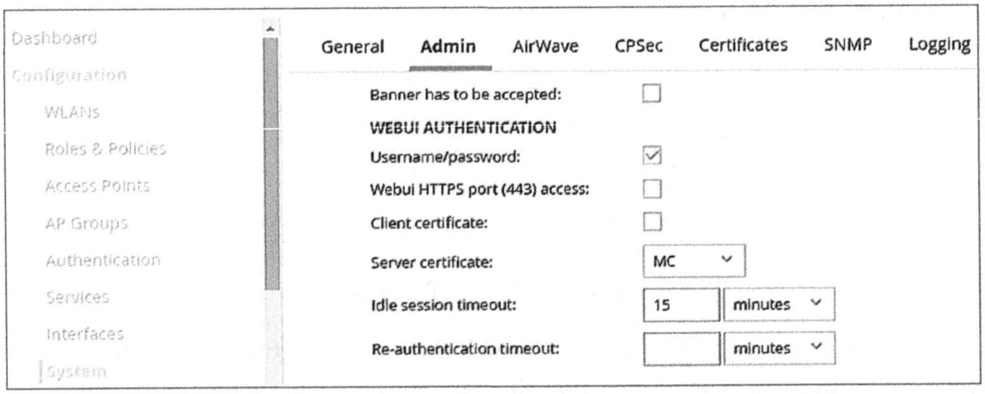

Figure 4-26: Configuration > System > Admin > Admin Authentication Options

6. Deploying the pending changes. After that point, the Web UI will start using the new certificate to authenticate itself for HTTPS.

Use Certificate-based Authentication for the Web UI

Once you have a certificate on the MC, you can choose to use it for functions like the Web UI.

Step 1: Set up certificates

For additional security, you can implement certificate-based authentication for managers who log into the Web UI. The managers must connect from a device with a valid management user certificate. The manager can then access the Web UI without submitting additional credentials.

To set up this form of authentication, you should deploy user certificates for managers on approved management clients. These certificates should identify the admins in the subject name and in a principal name SAN. You must then upload the root CA certificate for the manager certificates as a Trusted CA cert on the device. Make sure that the MC has a Server Cert which it uses for the Web UI; also verify that management clients trust the root CA for this certificate.

Step 2: Configure Web UI Settings on the MC

Next, you must configure the MC to use client certificates. You can do so from the **System > Admin > Admin Authentication** options window (see Figure 4-27). You can keep password and certificate authentication enabled at the same time, which could be a good idea while you check that certificate authentication is working. After you have certificate authentication working, you can disable password authentication for the Web UI, while leaving the local account enabled for console access as a backup.

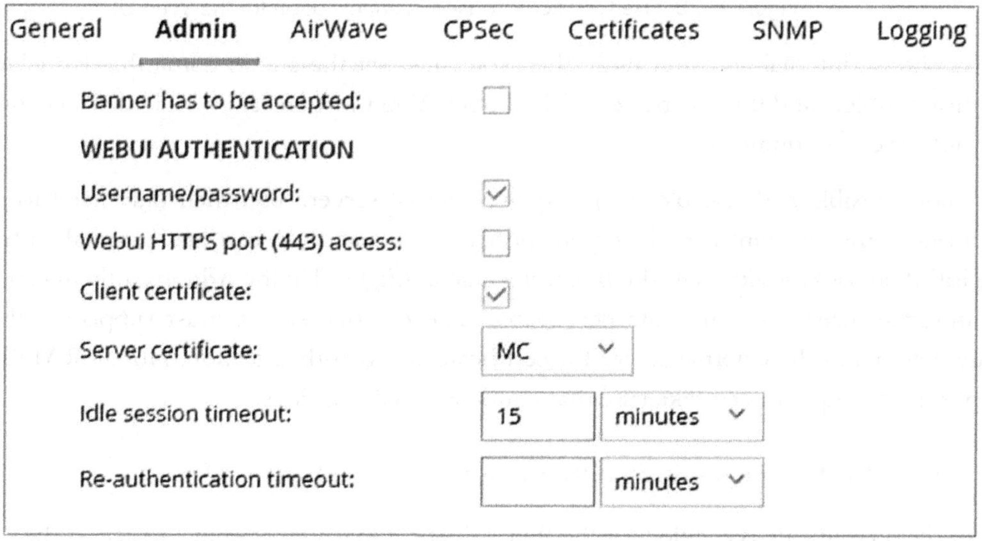

Figure 4-27: Use Certificate-based Authentication for the Web UI

Step 3: Set Up Certificate Admin Accounts

For the last step, create certificate admin accounts by adding a user from the page shown in Figure 4-28. Select the Trusted CA cert, which the MC will use to verify the manager client certificates. Then choose either internal or external authentication.

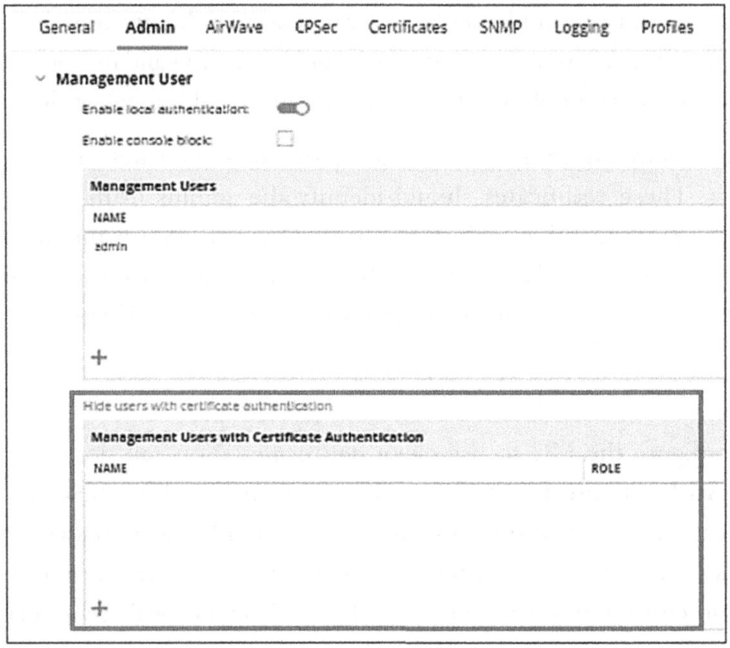

Figure 4-28: Use Certificate-based Authentication for the Web UI

If you choose internal authentication, then you must set the admin name that is included in the certificate subject and the certificate serial number. You must create an individual certificate admin account for every admin user.

It is more flexible and scalable to specify an external server. You then only need to create one certificate admin account for which you specify the Trusted CA cert and external authentication. The MC then uses the external admin server group configured in the Admin Authentication options to authorize managers with valid certificates. The external server must support authorize-only requests to act as the external server for certificate-based authentication. Not all RADIUS servers support these requests, so make sure that yours does. CPPM does.

Process of Certificate-based Authentication for Web UI

When client certificate is enabled for the Web UI, and a manager attempts to access the UI, the MC (or MCR) requests the client certificate during the TLS handshake for establishing the HTTPS session. The client sends the certificate. Depending on how the certificate is stored on the client, the user might need to enter a PIN, which would add an additional layer of security. The MC uses the installed Trusted CA certificate to verify the signature of the client certificate. If the certificate is valid, the MC sends the principal name in the certificate to the RADIUS server in an authorize-only request. The RADIUS server checks whether the name is authorized and, if so, sends an Access-Accept with the admin's role specified in the Aruba-Admin-Role VSA.

Refer to Figure 4-29 for an illustration of this process.

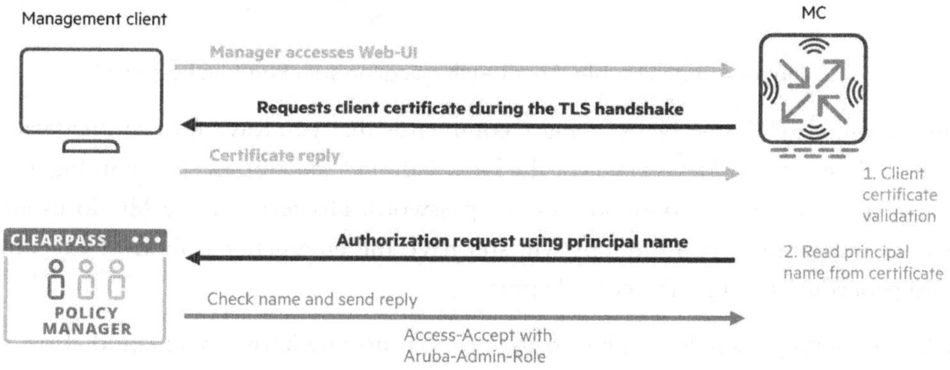

Figure 4-29: Use Certificate-based Authentication for the Web UI

As with password-based authentication, if the Access-Accept does not include a valid role in this VSA, the MC applies the default admin role, which should be set to a low privilege level.

"Disable" Local Authentication with Random Password

You can completely disable local authentication for MCs and MCRs. However, in the event of an external authentication server becoming unavailable, nobody would be able to access the device's management interface. You could also use redundant authentication servers to address this problem; however, communication issues on the MC or in the network could still put admins at risk of being locked out of the MC.

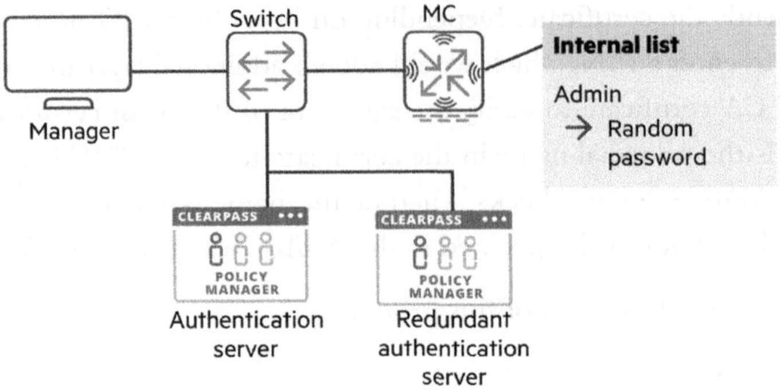

Figure 4-30: "Disable" Local Authentication with Random Password

For this reason, HPE Aruba Networking recommends that you leave local authentication enabled and set a randomly created password for the local "admin" account, as shown in Figure 4-30. Then under normal conditions, nobody knows the password. However, if the MC loses access to the external authentication server, admins will still have the emergency option of executing the lost password procedure through the console port.

Or, instead of setting a single, random password that no one knows, you could also set a random password for each device. You would then lock those unique passwords in a safe, secured with a known password.

Lock Down Services

You will now learn how to lock down services.

Best Practices Checklist

If your network requires SNMP, use only SNMPv3. You also need to set up authenticated Network Time Protocol (NTP) to ensure that the devices' clocks and timestamps are synchronized. You should further use Control Path Defense policies to protect the MC's control plane. You should also enable a few best practice global firewall settings. These best practices are summarized in Figure 4-31.

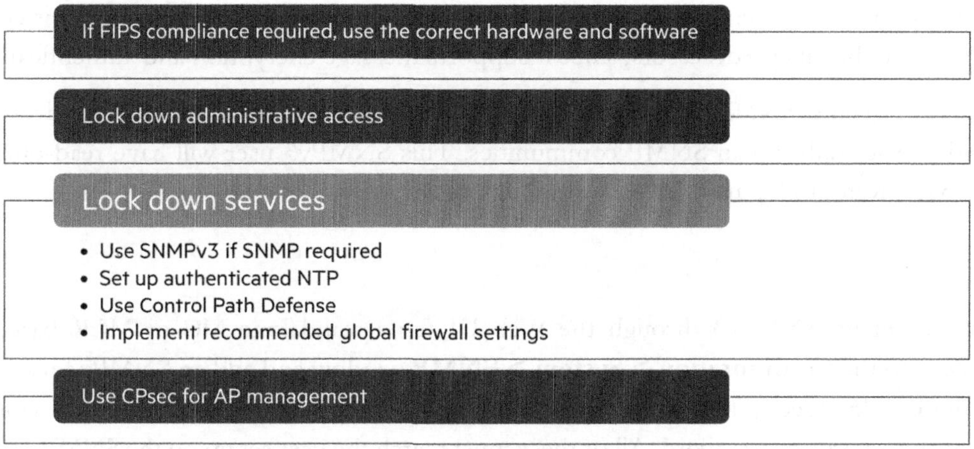

Figure 4-31: Best Practices Checklist

SNMPv3 on HPE Aruba Networking Mobility Devices

HPE Aruba Networking mobility devices, including MCRs, MCs, and IAPs, support SNMP for monitoring purposes only. That is, they provide read access to SNMP servers, as shown in Figure 4-32. The devices can also send SNMP traps to trap receivers.

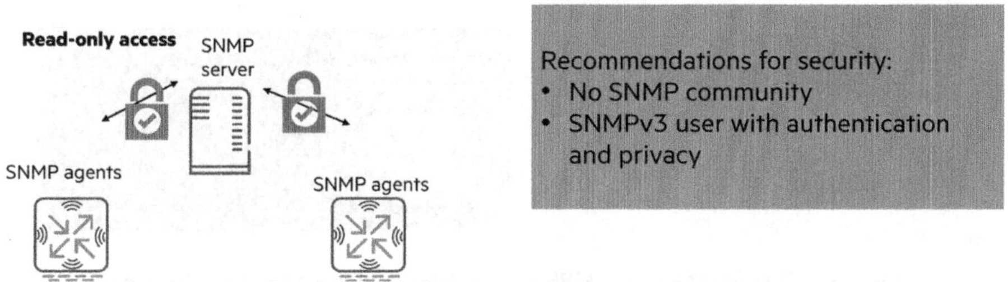

Figure 4-32: SNMPv3 on HPE Aruba Networking Mobility Devices

Recall what you learned in Chapter 3 about setting up secure SNMP access. SNMP is a useful tool, which allows network management systems to poll devices for information such as port configuration, status, and interface counters. SNMP Message Information Bases (MIBs) define object IDs (OIDs) for the various types of information that the SNMP server collects from the device. In essence, they enable the SNMP server and SNMP agent to decode each other's communications. You can obtain the AOS MIBs for your specific software version at the HPE Networking Support site.

SNMPv1 and SNMPv2c suffer from poor security. Critical data is sent in clear text and would-be passwords are sent in plaintext, making them nearly useless as passwords. Furthermore, it is fairly easy to tamper with or copy messages. SNMPv3 is not perfectly secure either, but it is considerably more secure than its predecessors, since it supports message encryption and authentication.

To set up secure SNMP monitoring on HPE Aruba Networking mobility devices, create an SNMPv3 user rather than SNMP communities. This SNMPv3 user will have read-only access to the device. Refer to Figure 4-32 for a summary of these recommendations.

Using SNMPv3

You can set up SNMPv3 through the Web UI for a standalone MC or MCR based solution. Navigate to the **Configuration > System > SNMP** window and add an SNMPv3 user. As shown in Figure 4-33, specify the username, authentication protocol, authentication password, privacy protocol, and privacy password. All of these must match the user set up on the SNMP server exactly (protocol type as well as password).

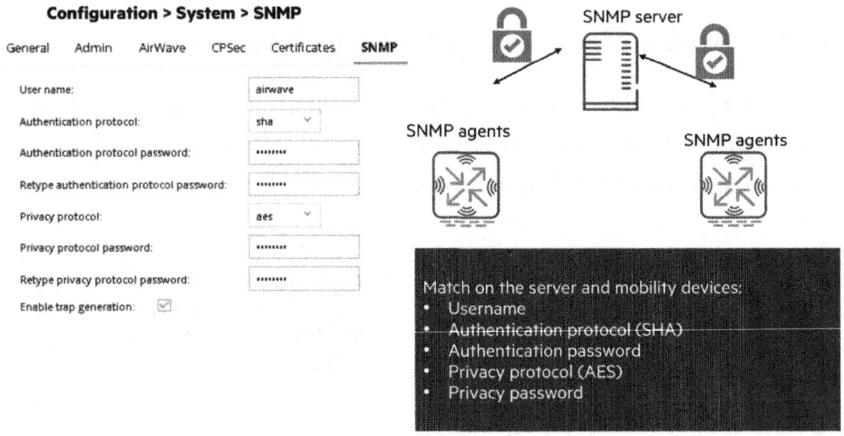

Figure 4-33: SNMPv3 on HPE Aruba Networking Mobility Devices

If you choose to enable trap generation, also specify the IP address of the SNMP server as a trap receiver.

Setting up Authenticated NTP

As you also learned in Chapter 3, network infrastructure devices must all share the same, correct time, or else issues can arise with certificates and authentication servers. Devices can become susceptible to replay attacks. And logs are only useful for troubleshooting and forensics if they have accurate timestamps.

To ensure an accurate clock, you should connect network devices to a Network Time Protocol server and then set up authenticated NTP. By using NTP, your network's devices will all use a central time source. By authenticating NTP, you will ensure that an attacker cannot intercept NTP communication and supply a device with a false clock, as shown in Figure 4-34.

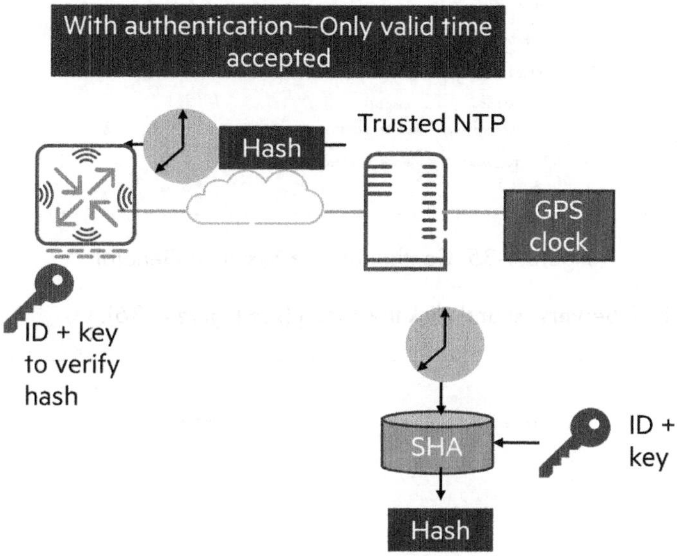

Figure 4-34: Setting up Authenticated NTP

NTP authentication uses a symmetric key that is configured on both the HPE Aruba Networking mobility device and the NTP server.

To enable NTP in the MC or MCR Web UI, you must be at the managed device level. Then follow these steps.

1. Navigate to **Configuration > System > General** (see Figure 4-35).
2. Scroll down and expand **Clock**. (If you do not see Clock, make sure that you are at the individual controller level.)
3. Make sure that **Set clock** is set to **Using NTP**.
4. Set the **Time zone**.

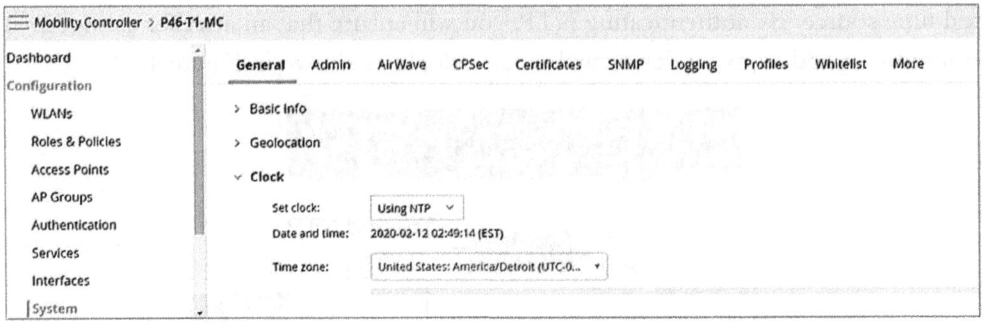

Figure 4-35: Configuration > System > General

5. Scroll down to the NTP Servers list and click the + icon (see Figure 4-36).

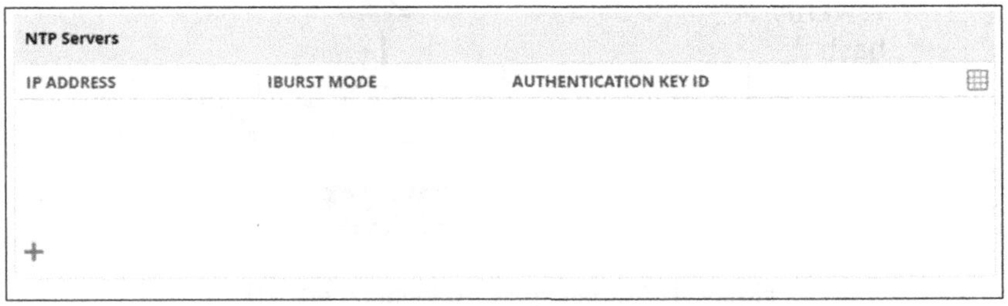

Figure 4-36: Configuration > System > General (NTP Servers)

6. Scroll down and specify the settings for the NTP server such as those shown in Figure 4-37.

7. Use the slider to enable **Use NTP authentication**.

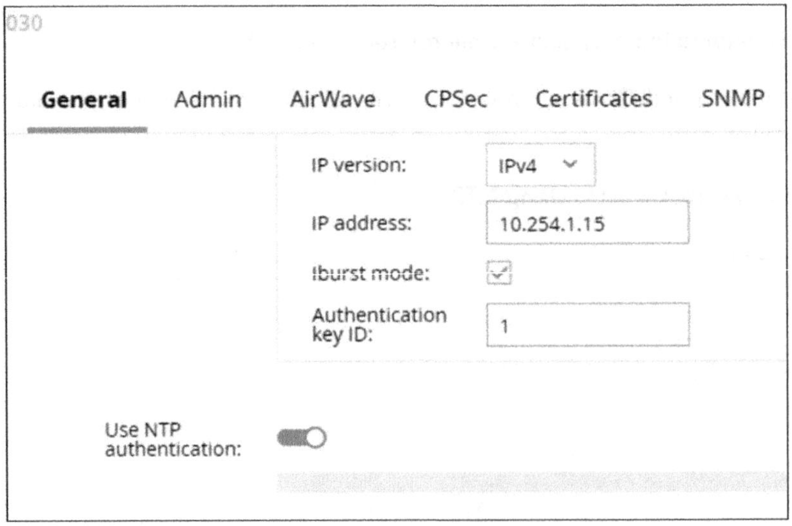

Figure 4-37: Add NTP Server

8. Scroll down to the NTP Authentication Keys list and click the + icon (see Figure 4-38).

Figure 4-38: NTP Authentication Keys

9. Scroll down and configure the key. Make sure to match the authentication key ID, algorithm, and key value specified on the server. (See Figure 4-39.)

10. Select the **Trusted** check box.

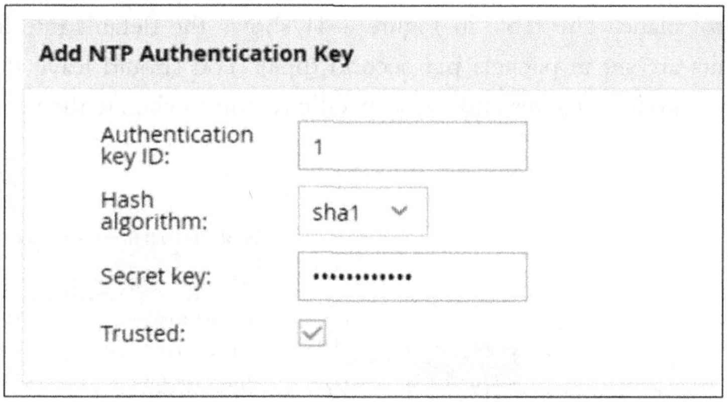

Figure 4-39: Add NTP Authentication Key

11. Click **Submit**.

12. You will see a warning about rebooting the MC.

13. Deploy the pending changes.

14. Reboot the MC as recommended.

    a. Navigate to **Maintenance > Software Management > Reboot** (see Figure 4-40).

    b. Make sure that the **Save configuration before reboot** check box is selected.

    c. Click **Reboot**.

    d. Confirm the reboot.

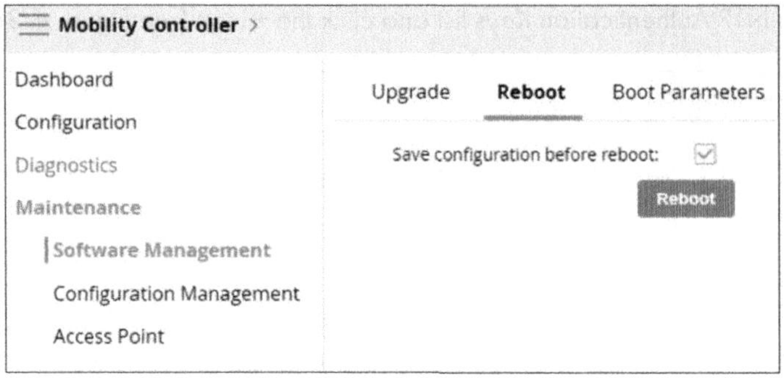

Figure 4-40: Reboot an MC

Control Plane Defense

Hackers might attempt to overwhelm the MC control plane, which would cause a DoS attack by preventing the MC from forwarding traffic. By default, MCs mitigate these attacks by rate limiting traffic to the control plane. The table in Figure 4-41 shows the default rate limits in AOS 8.6 software; these limits are set in packets per second (pps). You should leave the default settings unless your security team investigates and has a specific reason to change them.

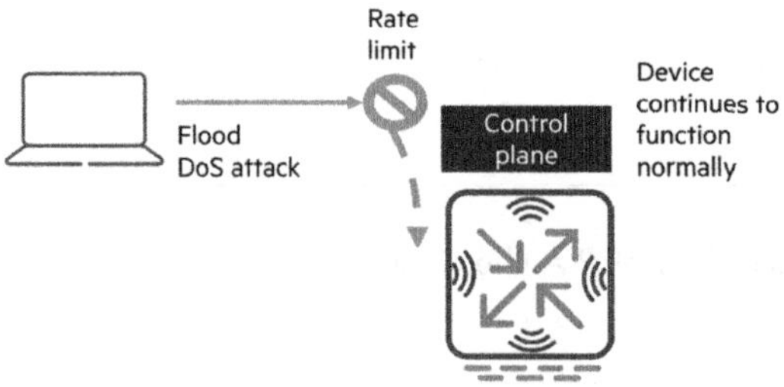

Leave defaults unless reason to change

Total rate limit to control plane

Parameter	Default value in 8.6 (pps)
Untrusted unicast	9765
Untrusted multicast	3906
Trusted unicast	65535
Trusted multicast	3906
Route traffic	976
Session mirror traffic	976
Auth process traffic	976
VRRP traffic	512
ARP traffic	3906
L2/Other	1953

Figure 4-41: Control Plane Defense

MCs can further monitor traffic per-client and police potential DoS attacks. The table in Figure 4-42 shows the type of traffic that the MC ca Monitor. For example, an ARP attack is a flood of ARP messages. A CP attack is a flood of any packets to the device control plane. Limits are set in packets per 30 seconds. Recommended values for attacks that you want to monitor are indicated in the table. If the MC detects that a client violates these limits, it logs the violation. For ARP and CP attacks, the MC can also mitigate the attack by taking an action such as blacklisting the client.

Monitor attacks per untrusted user

Parameter	Default value in 8.6 (per 30 seconds)	Recommended value (per 30 seconds)
Monitor ping attack	None	120
Monitor TCP Syn attack	None	960
Monitor IP session attack	None	960
Monitor/police ARP attack	None	960
Monitor/police gratuitous ARP attack	50	50
Monitor/police CP attack rate	None	3000

Figure 4-42: Control Plane Defense (Attacks per Untrusted User)

Implement Global Firewall Settings

Certain firewall settings can be applied globally, across all traffic on a system. HPE Aruba Networking recommends enabling three of these features to harden the network.

Deny inter user traffic

When you enable the option for denying inter-user traffic, wireless users cannot engage in any form of peer-to-peer communication, because any traffic from one user destined for another is dropped. (See Figure 4-43.) This feature reduces the chance of malicious authorized users—or hackers who have gained authorized access—from attacking other users. Remember the wireless ARP poisoning attack that you saw in Chapter 1, which involved a malicious authorized wireless user. If the malicious wireless user cannot send traffic to other wireless users, the attack is thwarted.

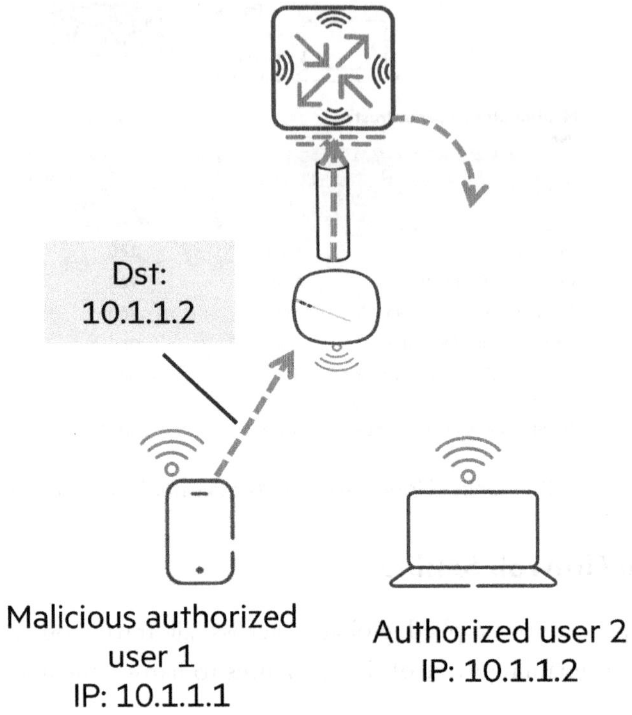

Figure 4-43: Deny Inter User Traffic

Denying inter user can also help to prevent malware from spreading.

In short, you should typically enable this feature for a hardened network. Occasionally a company will have a use case that requires inter-client communications. In this case, you should leave this feature disabled globally and enable it per-SSID instead. Create an SSID only for the clients that require inter-client communications and leave the deny-inter-user traffic policy disabled on that SSID. Then deny inter-user-traffic on all other SSIDs.

An alternative policy blocks only non-IP traffic between users. This ban on non-IP, peer-to-peer communications includes ARP traffic, so you should enable proxy ARP on VLANs if you want to use this feature to block non-IP traffic but enable IP communications.

Prohibit IP spoofing

You should also implement "prohibit-IP-spoofing." This feature is enabled by default. IP spoofing is when an attacker statically configures their device with an IP address that belongs to another device on the network. This is particularly useful for MITM attacks, as well as other exploits. HPE Aruba Networking mobility devices maintain a user table with information about all clients connected on untrusted interfaces, such as wireless connections. This information includes IP address and MAC address, which the device can find by snooping the DHCP communications that assign the IP address to the client. If the mobility device detects that a second client is using an IP address that is associated with another client, the device drops the second client's traffic, as shown in Figure 4-44.

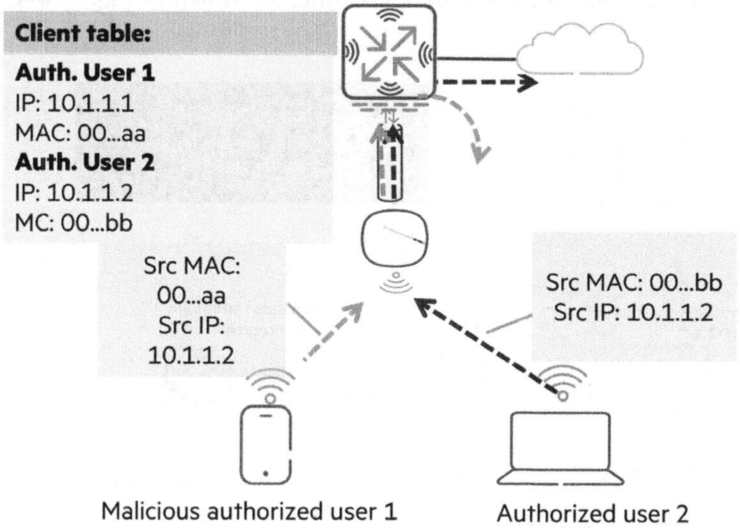

Figure 4-44: Prohibit IP Spoofing

Prohibit ARP spoofing

You should also prohibit ARP spoofing. Recall that hackers can respond to ARP messages for IP addresses that they do not own to "poison" other devices' ARP tables and link their MAC address to the IP address of another device. Prohibiting ARP spoofing is disabled by default. Enabling it prevents any user on an untrusted interface from sending ARP responses or gratuitous messages for any IP/MAC address combination other than its own.

Enforce DHCP

Along with denying inter-user traffic and prohibiting IP spoofing on the firewall, you should enforce DHCP negotiation in the AAA profiles associated with your WLANs.

Typically, wireless clients should be using the IP settings assigned to them by a DHCP server. If a device tries to use an IP address of its own choosing, this could be a sign that the user is attempting to spoof an IP address for malicious purposes.

Enforcing DHCP requires a wireless client to complete a DHCP negotiation before the HPE Aruba Networking mobility device allows the client to send or receive traffic. The device drops any traffic for clients that try to use an IP address other than one they were assigned during their DHCP negotiation, as shown in Figure 4-45. This feature also prevents clients from sending any traffic using a statically assigned IP address, rather than a DHCP-assigned one, as shown in Figure 4-45 under label 3.

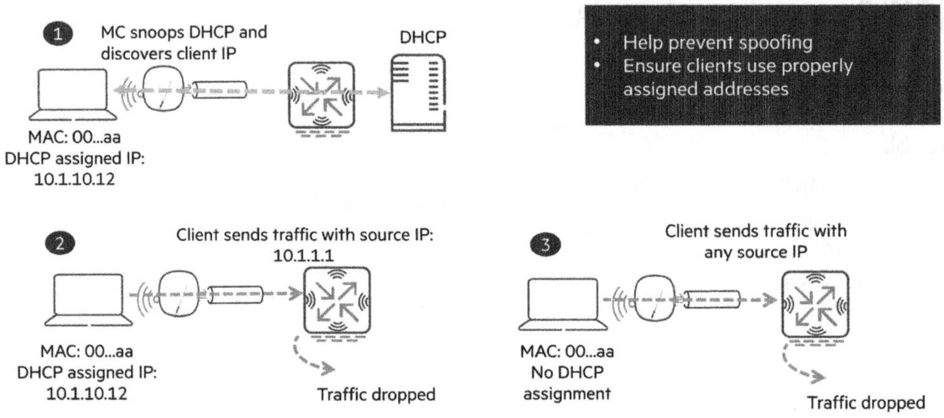

Figure 4-45: Enforce DHCP

Secure Mobility Device-to-Device Communications

For the final section in this chapter, you will learn how to use Control Plane Security (CPsec) to protect control plane communications between APs and MCs.

Why CPsec

A campus AP (CAP) establishes multiple Generic Routing Encapsulation (GRE) tunnels to send user traffic to the MC and to receive heartbeats from its MC. A CAP also has a PAPI channel with the MC. (PAPI stands for Process Application Programming Interface, and it uses UDP port 8211.) All control information passes over the PAPI channel. An unprotected control channel opens many risks. Hackers can eavesdrop on the messages. They can perform network reconnaissance and gather information about your network's settings. If the control channel does not have data integrity protections, hackers can insert their own control messages, shutting down APs or creating open networks. Perhaps less concerning, but still against company policies, users can also insert their own rogue APs and use them to extend access where it is not permitted.

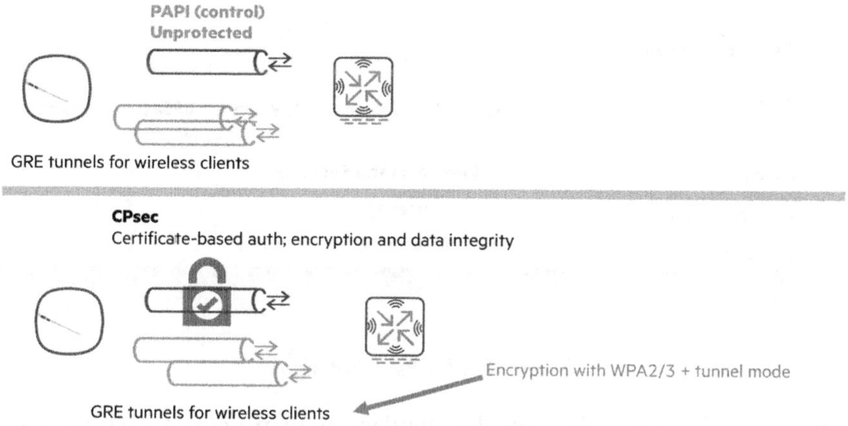

Figure 4-46: Why CPsec

CPsec protects control traffic between APs and MCs with secure IPsec tunnels, as shown in Figure 4-46. Both the AP and MC enforce certificate-based authentication of the peer before permitting the tunnel to come up. Using Encapsulating Security Payload (ESP), the IPsec tunnel also encrypts PAPI data and preserves its integrity as it crosses the tunnel. IPsec with ESP is widely recognized as a secure protocol, and CPsec offers superior security for the MC-to-AP control plane.

Note that CPsec does not affect the tunnels that carry user traffic. However, data in these tunnels can be protected with the SSID's encryption. To ensure that it is, use Wi-Fi Protected Access (WPA) 3 (preferably) or WPA2 for the SSID's security method. Also use tunnel mode for the SSID. Tunnel mode tells APs to send all traffic to the MC so that the MC can apply security measures like firewall policies and deep packet inspection (DPI). It further specifies that the APs send the traffic with 802.11 encryption intact, which results in end-to-end encryption of client traffic. Because the AP never decrypts traffic, you do not need to consider it as a point of potential vulnerability for eavesdropping. You will learn more about these security measure in a later chapter.

CPsec Settings

CPsec is enabled by default, and you should retain this setting. When you configure CPsec through an MCR or on individual MCs, you have the option to enable automatic certificate provisioning or not. For the best security, you should retain the default setting, which is disabling automatic provisioning (see Figure 4-47). Disabling auto cert provisioning does not mean that you have to install certificates on every AP manually. Instead, it means that you must add each AP that you want to make part of your mobility solution to the AP allowlist manually. You can create this allowlist by manually authorizing APs as MCs detect them or by uploading a allowlist to the MCR or standalone MC.

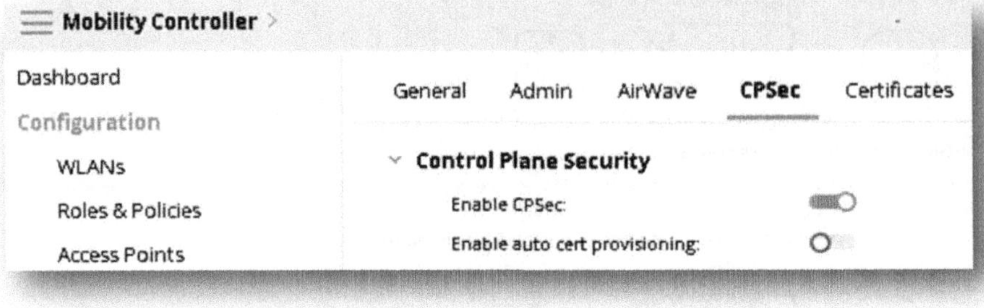

Figure 4-47: CPsec Settings

When you deploy an AP, it goes through the standard boot process to receive its image, and it establishes a PAPI channel, as shown in Figure 4-48. The MC informs the AP that CPsec is required. If auto cert provisioning is disabled, the MC also checks the AP's serial number against its allowlist. If the AP is on the allowlist and has no certificate, the MC then accepts a public key from the AP and issues the AP a certificate with this key. The AP reboots. The AP and MC then establish an IPsec tunnel using this certificate for authentication. All subsequent control traffic is sent through this secure tunnel.

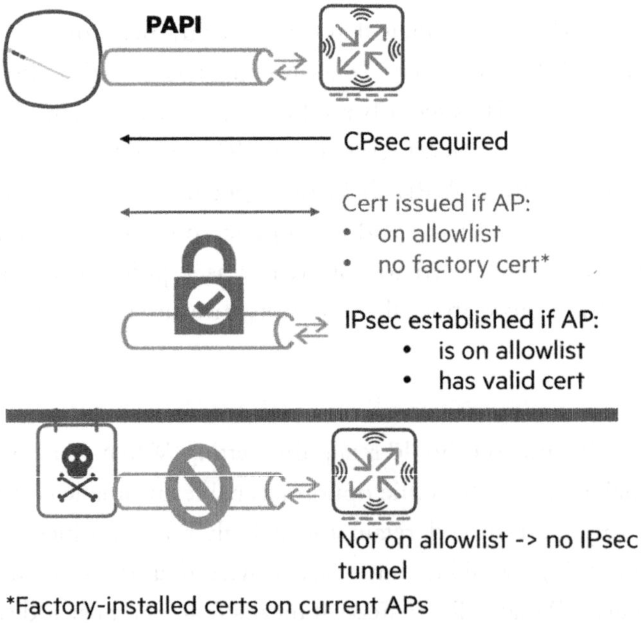

Figure 4-48: CPsec Process

But current HPE Aruba Networking APs have factory-installed certificates in a TPM. These certificates are signed by an HPE Aruba Networking CA certificate. The HPE Aruba Networking

CA root certificate is also factory installed as a Trusted CA certificate on MCs. The subject name within the AP's certificate matches the AP's serial number, validating that the AP is sending its legitimate serial number. When an MC tells the AP that CPsec is required, the AP can immediately skip to the step of establishing an IPsec tunnel, using its factory-installed certificate to authenticate. The MC only accepts valid certificates that have subject names that match the serial numbers listed in the allowlist. If an AP not in the allowlist establishes a PAPI control channel with the MC, the MC does not accept it as a managed device.

Certificate-based authentication is much more secure than password-based authentication, establishing a secure groundwork for your network.

The description that you just read applies when auto cert provisioning is disabled. If you enable auto cert provisioning, you do not need to maintain a allowlist; instead MCs issue certificates to and accept certificates from any HPE Aruba Networking APs that contact them. This feature increases the chance of rogue APs joining the solution. But some companies might choose to use the auto method for faster setup during the initial solution deployment. In this case, you would turn on auto cert provisioning when the customer begins to install APs and then turn it off after the solution deployment.

Secure End-to-End Communications

You should also understand that HPE Aruba Networking automatically provides security for MCR-to-managed device (MD) communications in a solution that uses an MCR. (MDs include all devices managed by MCR such as MCs.) When you deploy an MC as an MD and point it toward an MCR, you set an IPsec key. The MC and MCR automatically establish an IPsec tunnel to secure all of their communications, as shown in Figure 4-49.

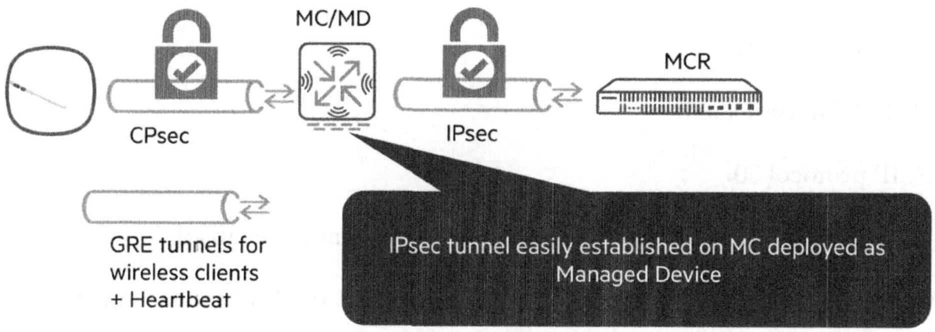

Figure 4-49: Secure End-to-End Communications

In this way HPE Aruba Networking secures communications across the infrastructure end-to-end from APs to MCs to MCR.

Ports for Communications

Sometimes the AP-to-MC or MC-to-MCR communications must pass through an external, third-party firewall. A list of ports to open on that firewall are provided in the sections below for your reference:

Communication Between APs and MC

- CPsec (UDP port 4500)
- PAPI (UDP port 8211); All APs running as Air Monitors (AMs), which are security devices that monitor L2 wireless communications, require a permanent PAPI connection to the MC.
- DNS (UDP port 53); An AP can use DNS to discover an MC
- FTP (TCP port 21)
- TFTP (UDP port 69). All CAPs; If there is no local image on the AP or if the image needs to be upgraded (for example, a new AP), the AP will use TFTP to retrieve the initial image. For remote APs, upgrade the image only by FTP and not TFTP.
- SYSLOG (UDP port 514)
- GRE (protocol 47)

Communications between MCs (MDs) and MCR

- PAPI between MCR and a managed device is encapsulated in IPsec
- IKE (UDP port 500)
- NAT-T (UDP port 4500)
- ESP (IP protocol 50)
- IP-IP (protocol 94) and UDP port 443 if Layer-3 mobility is enabled
- GRE (protocol 47) if tunneling guest traffic over GRE to DMZ managed device.
- TCP port 1723 for PPTP tunnel-based VPN
- TCP port 2323 (required for backdoor debugging)

Management access between an admin's station and a managed device

- HTTPS (TCP ports 443 and 4343)
- SSH (TCP port 22)

HPE Aruba Networking Mobility TPMs

HPE Aruba Networking mobility devices, including MCs and APs, have Trusted Platform Modules (TPMs).

TPMs store a signed "fingerprint" for the OS to enable secure boot. The bootloader checks the OS against the fingerprint in the TPM. If the OS is compromised, the bootloader will detect the problem. Because compromised malware could enable hackers to silently collect decrypted user traffic, as well as reconfigure the network based on their desires, secure boot forms a critical component of the HPE Aruba Networking trusted network infrastructure.

A TPM also stores the private keys associated with the device's certificates. As you recall, you cannot export private keys created when you create a CSR online on a device such as an MC; that is because the key is locked within the TPM. HPE Aruba Networking APs come with secure, factory-installed certificates, each signed by an HPE Aruba Networking CA and associated with a private key stored in the AP's TPMs. The factory-installed certificates enable new APs to support CPsec out-of-the-box without needing MCs to install certificates on them.

As a final note, when the HPE Aruba Networking mobility device uses a FIPS version of the software, you can zeroize the TPM if the device needs to be decommissioned. Many companies have security policies that require complete wiping of data on decommissioned hardware to prevent, for example, someone from taking the hardware and compromising the private key. Because zeroizing the TPM wipes out data that the device needs to function, you should *only* do this when you are decommissioning the device.

Obtain More Information

You can find more information at this link: https://networkingsupport.hpe.com/

From the main menu select **Software and documents**. Search for "hardening guide" and download the correct version for your software.

Summary

Congratulations! You can now implement best practices to keep your HPE Aruba Networking mobility infrastructure secure.

Learning checks

1. True or false: If the console port on an MC or AP is unprotected, attackers can easily gain access using the password recovery procedure.
   a. True
   b. False
2. True or false: HPE Aruba Networking recommends entirely disabling local authentication
   a. True
   b. False
3. Which of these are reasons to avoid using SNMPv1/v2c? (Select all that apply.)
   a. Passwords are sent in plaintext
   b. Critical data is sent in clear text
   c. Hackers can easily send packets with valid community strings
   d. The protocols place a strain on compute resources
   e. The protocols are difficult to configure
4. What security is provided by implementing CPsec between APs and MCs?
   a. CPsec encrypts and authenticates the PAPI control channel and in the GRE tunnels that carry user traffic.
   b. CPsec authenticates the PAPI control channel but does not encrypt it.
   c. CPsec encrypts and authenticates the PAPI control channel.
   d. CPsec encrypts and authenticates the GRE tunnels that carry user traffic and authenticate the PAPI control channel.

5. Consider these questions related to securing access to an MC's control plane. Take notes on your thoughts.
   - Through what protocols can users and management solution access the MC's data and configuration settings?
   - What are the risks of unauthorized users accessing the MC through particular protocols?
   - What strategies would you take toward locking down access to the correct users and devices?

Note

For answers to all learning checks, refer to Appendix B.

---

5 Secure LAN Protocols

EXAM OBJECTIVES

✓ Implement spanning tree protection features

✓ Use DHCP snooping and ARP protection to secure the network

✓ Enforce authentication for routing protocols

Assumed Knowledge

- Basics of IP addressing
- Basics of Ethernet switching
- Basics of Spanning Tree Protocol (STP), Rapid STP (RSTP), and Multiple STP (MSTP)
- Basics of Dynamic Host Configuration Protocol (DHCP)
- Basics of Addressing Resolution Protocol (ARP)

However, if you need more information on STP, RSTP, and MSTP, see "Appendix C: Supplemental Information."

Overview

Many standard protocols that are used to support fundamental network connectivity were designed for a time when the internal network was considered trusted. But in a zero trust world, that is no longer the case. You need to protect your Ethernet LAN by protecting critical protocols.

Hackers can try to attack spanning tree and routing protocols to interfere with how traffic is forwarded. They can also attack DHCP or exploit ARP to implement man-in-the-middle (MITM) attacks. They might attack routing protocols to redirect traffic or cause Denial of Service (DoS) attacks.

This module teaches you how you can set up your AOS-CX switches to prevent these attacks.

Implement Spanning Tree Protections

In this section, you will review spanning tree protocols and learn about best practices for securing them.

Spanning Tree Normal Operation

Many Ethernet networks use spanning tree protocols such as Rapid Spanning Tree Protocol (RSTP) or Multiple Spanning Tree Protocol (MSTP) to manage redundant links. This discussion will refer to both protocols as "spanning tree." (An appendix provides more details about the differences between the protocols.)

An Ethernet broadcast domain cannot have loops. Loops prevent switches from learning the correct forwarding path for particular MAC addresses. They also cause broadcasts, multicasts, and unknown multicasts to loop around the network, multiply, and loop again. This is called a broadcast storm and can take up so much bandwidth that it rapidly brings down a network.

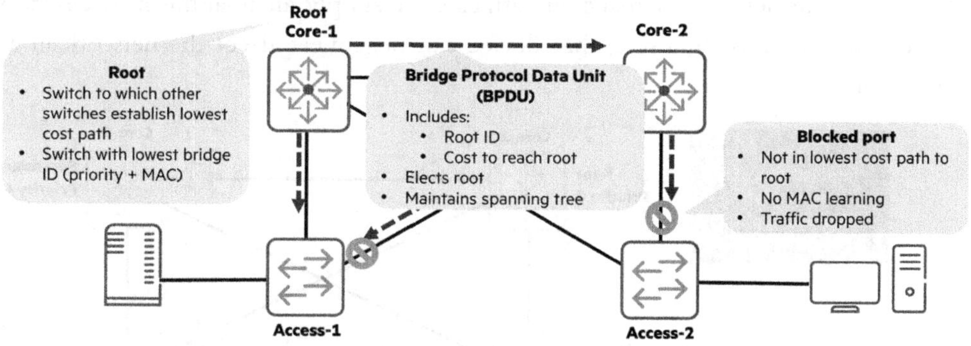

Figure 5-1: Spanning Tree Normal Operation

Spanning tree establishes a single shortest path between each switch and a root switch. (See Figure 5-1.) The root switch is the switch with the lowest bridge ID, which consists of a configurable priority value plus the switch's MAC address. In other words, the switch with lowest configured priority value—0 is the lowest option—becomes root. MAC address acts as a tie breaker only if multiple switches have the same lowest priority value.

Spanning tree uses Bridge Protocol Data Units (BPDUs) to maintain the spanning tree. Among other fields, a BPDU includes the bridge ID of the root and the cost of reaching root from the port that transmitted the BPDU. When a switch first starts running spanning tree or first connects to a network, it believes that it is root. It transmits BPDUs with its own bridge ID in the root ID field. When the switch receives a "superior BPDU," which has a root ID with a lower value, the switch accepts that root and begins putting that root ID in its BPDUs instead. In this way, all switches soon agree on a root. As you see in Figure 5-1, BPDUs propagate from the root and down through the spanning tree.

A spanning tree switch blocks any upstream ports that are not part of the switches' lowest cost path to the root, eliminating loops. Blocked ports do not learn MAC addresses, and they do not accept traffic. For example, if Core-2 in Figure 5-1 floods a broadcast to Access-1 and Access-2, the access switches drop the broadcast, which they will also receive from Core-1. The blocked links remain available for use if a change occurs in the topology. For example, if the link between Access-1 and Core-1 fails, the blocked link between Access-1 and Core-2 will transition to forwarding.

Spanning Tree Vulnerability: Rogue Device Becomes Root Bridge

By design, spanning tree allows any network device to join the spanning tree. While this design's openness helps to eliminate all loops, it leaves choices about which links are active and which are blocked vulnerable to manipulation from unauthorized devices. Because BPDUs have no authentication, it is easy for rogue devices to send legitimate BPDUs and join the spanning tree. Whether the rogue device is a malicious attacker or simply an unauthorized device such as an unmanaged switch added by users, that device can negatively affect the network in a significant number of ways.

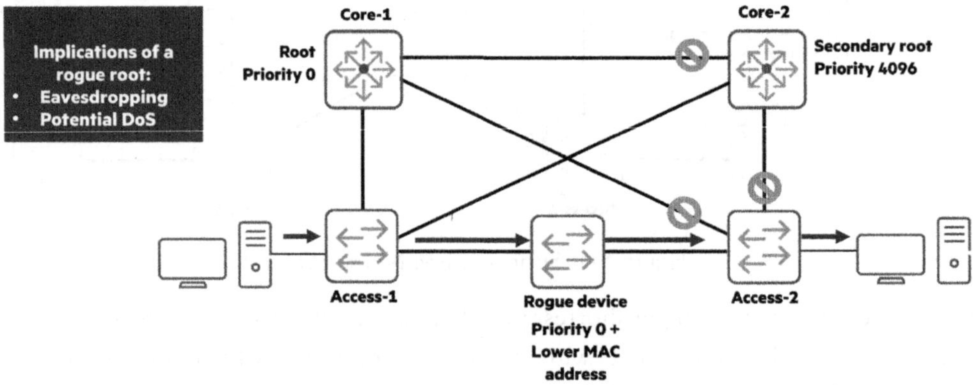

Figure 5-2: Spanning Tree Vulnerability: Rogue Device Becomes Root Bridge

The unauthorized device can cause the wrong links to be active and even become the tree's root.

In Figure 5-2, you see what might happen if a rogue switch joins the spanning tree. The rogue switch has a priority of 0, and it happens to have a lower MAC address than the desired root. Therefore, Access-1 and Access-2 accepts its BPDUs as superior BPDUs.

Based on the superior BPDU, the access switches change the port connected to the rogue device to its root port, and they advertise the superior BPDU to Core-1 and Core-2. Core-1 and Core-2 reconverge their topology based on the rogue device being the root. They block the high-speed link between them. Now all traffic passes through the rogue device, which can let the rogue device eavesdrop on traffic. Because the links to the rogue device might be lower speed, a partial Denial of Service (DoS) attack could also occur.

BPDU Guard

AOS-CX switches offer BPDU guard to prevent this situation from occurring in a network that uses MSTP or RSTP. (This feature might also be called BPDU protection.) This feature operates at the port level and is best implemented on edge ports, which should not be receiving BPDUs.

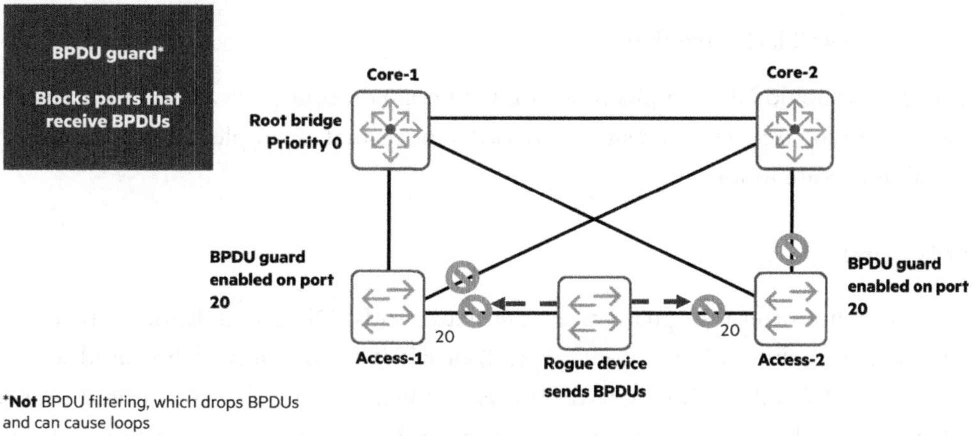

Figure 5-3: BPDU Guard

BPDU guard monitors a port for incoming BPDUs. If the port receives a BPDU, the switch shuts down the port, protecting the network from the rogue device, as shown in Figure 5-3. You can configure the number of seconds that the port remains disabled (range of 0 to 65535). Configuring a value of zero (0) specifies an infinite time out period. In other words, the port remains disabled until manually re-enabled.

The switch also logs an event when BPDU guard blocks the port.

AOS-CX switches also support running spanning tree in Rapid Per-VLAN Spanning Tree Plus (RPVST+) mode, which is intended for interoperability with Cisco switches. The switches support PVST protection, a similar feature to BPDU guard, when they run in this mode.

BPDU guard versus BPDU filtering

Do not confuse BPDU guard with BPDU filtering.

BPDU filtering effectively disables spanning tree on a port. It drops all incoming BPDUs and also prevents the transmission of BPDUs on a port. The port also always remains in the forwarding state. Therefore, BPDU filtering, implemented on an edge port, would prevent a rogue switch from taking over as root by dropping the rogue switch's BPDUs. However, because BPDU filtering forces the port into a forwarding state, it could allow loops to occur.

Edge ports that are vulnerable to rogue switches are often also vulnerable to situations such as these, which could create loops:

- Users connect an unauthorized switch that does not support spanning tree. The users loop ports together on that switch.
- Users loop wall jacks together.

Therefore, you should NOT implement BPDU filtering on edge ports. BPDU filtering is best suited to disable spanning tree on *trusted* switch-to-switch ports that are implementing routing and that you know will not create loops.

Root Guard

AOS-CX switches offer root guard as an alternative to BPDU guard. Root guard works when the switch operates in either MSTP mode (typical) or RPVST+ mode (which is used for compatibility in Cisco networks). Like BPDU guard, this feature helps ensure that the desired root remains the root. However, rather than blocking the port when the port receives *any* BPDUs, root guard only blocks the port when the port receives superior BPDUs that could change the topology (see Figure 5-4).

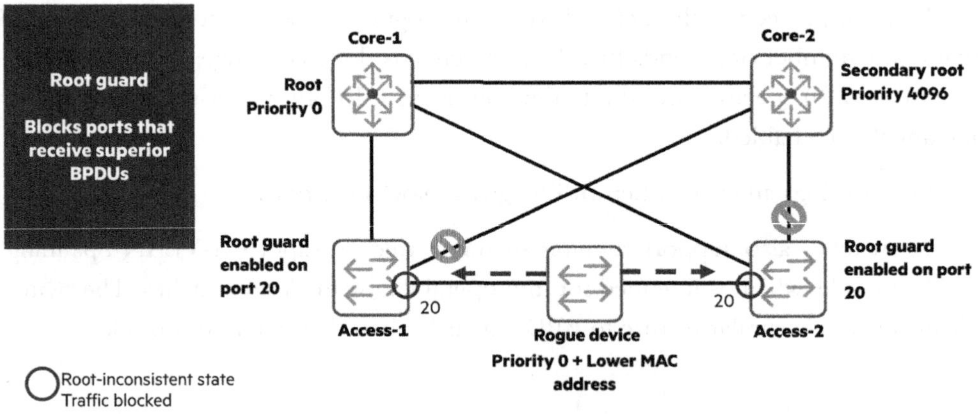

Figure 5-4: Root Guard

If the switch simply ignored the superior BPDUs and remained in a forwarding state, it could allow a loop to form through the rogue root device. Therefore, root guard also changes the port's state to the root-inconsistent STP state if the port receives a superior BPDU. The port will not forward any traffic while it is in this state. Root guard will change the port state from root-inconsistent to its previous state when the port stops receiving superior BPDUs.

Root guard can be useful on switches that might need to connect to downstream switches, but you want to prevent issues if those downstream switches are misconfigured.

You configure root guard on ports that should never receive BPDUs with information about the root—in other words, all ports except those that connect to the root and secondary root (directly or indirectly). For example, in Figure 5-4, you would enable root guard on all of the Access-1 switch ports except those that connect to the Core-1 and Core-2 switches. Likewise, you would enable root guard on all of the Access-2 switch ports except those that connect to the Core-1 and Core-2 switches.

CoPP on Spanning Tree

An MSTP BPDU storm can results from a loop in a "dumb" switch that someone connects to an access switch without authorization (see Figure 5-5). The loop causes BPDUs, as well as other traffic, to multiply. When a BPDU storm bombards a switch on which spanning tree is enabled, the switch's CPU usage rises because the switch must process all of the BPDUs. This can cause the switch to stop performing other functions correctly. The broadcast storm becomes even more damaging to the network and also more difficult to fix, as managers cannot access the switches' management interfaces and fix the problem. AOS-CX switches can mitigate this problem by dropping excessive BPDUs, as shown in Figure 5-5. The loop is still present, but managers can at least access switches' management interfaces and take action, such as shutting down the ports with the loop.

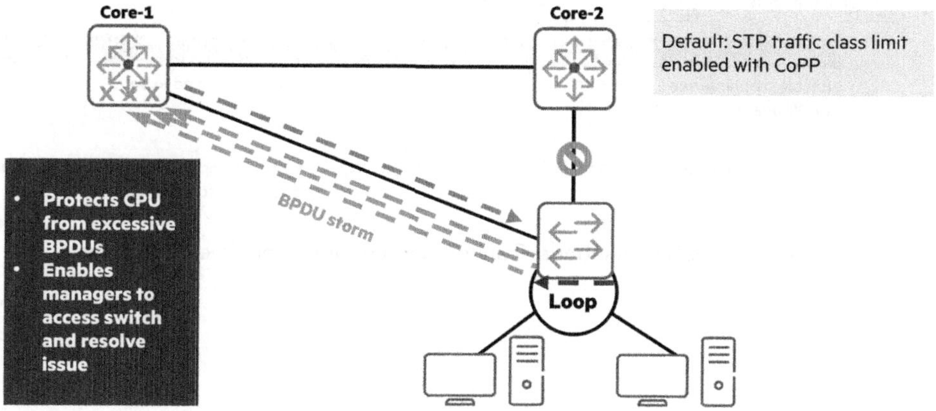

Figure 5-5: CoPP on Spanning Tree

You learned about control plane policing (CoPP) earlier and how it can rate limit specific classes of traffic to the control plane. AOS-CX switches have a default CoPP policy that limits the number of STP packets per second (in 10.04 the default limit is 1500 pps with a burst of 1500 packets). You should keep this policy in place.

Summary of Spanning Tree Protections

This table in Figure 5-6 reviews the features that you have examined in this section.

	Recommended feature	Alternative feature
Edge ports	BPDU guard Prevents rogue switches from connecting	Root guard Prevents rogue or misconfigured switches from becoming root
Global	CoPP rate limit for **stp**	—

Figure 5-6: Summary of Spanning Tree Protections

You should typically enable BPDU guard on edge ports to prevent rogue switches that are implementing spanning tree from connecting to the network. Root guard is an alternative technology that prevents rogue or misconfigured switches from becoming root. If a switch sends BPDUs that are not superior BPUDs, however, the switch is allowed to connect.

Globally the switch should implement CoPP rate limiting on BPDUs, as well as other traffic classes, as you learned in a previous module.

CLI Commands: Configuration of Spanning Tree Protections

Refer to the table in Figure 5-7 to find the commands for enabling spanning tree protections.

In configure context

Step	AOS-CX Command	Notes
Enable spanning-tree BPDU guard	`spanning-tree bpdu-guard`	Implement on edge ports
Configure root guard	`spanning-tree root-guard`	Optional; Implement on edge ports as an alternative to BPDU guard

Figure 5-7: CLI Commands: Configuration of Spanning Tree Protections

Use DHCP Snooping and ARP Inspection

In this section, you will learn how to protect DHCP and ARP.

DHCP Review

Clients use Dynamic Host Configuration Protocol (DHCP) to obtain valid IP settings for the network to which they are connected. An example of this process is shown in Figure 5-8.

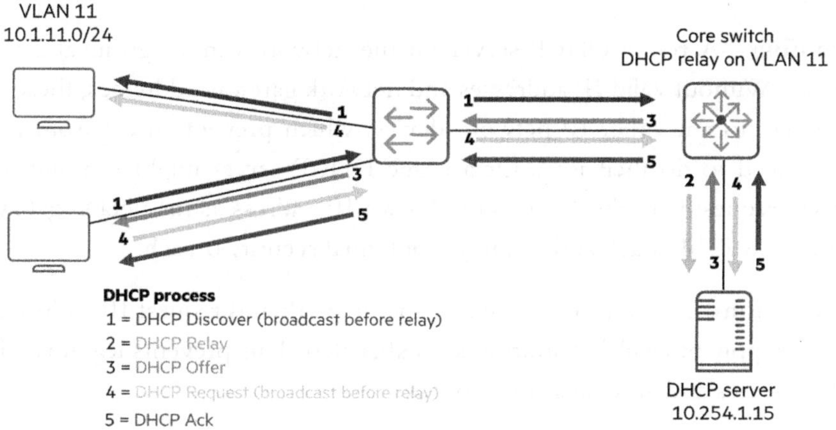

Figure 5-8: DHCP Review

When a client first connects to the network, it sends a DHCP Discover as a broadcast. As you see, the access switch floods this switch to the core and also to other clients in the VLAN. In this network, as is often the case, the DHCP server is not deployed in the client VLAN. So, the core switch acts as a DHCP relay. It forwards the DHCP Discover as a unicast to the DHCP server in a different VLAN.

The DHCP server returns a DHCP Offer with settings such as an IP address that will belong to the client, the IP address of the client's default gateway on the network, and a DNS server IP address. The Offer also includes a lease time, specifying how long the client can use these settings before renewing them. The server directs the offer to the relay, which then directs the offer to the client, typically as a unicast to the client's MAC address.

The client then broadcasts a DHCP Request, in which it formally requests the settings sent by the server. The core switch relays this request as before. The server leases the IP address to the client so that no other device can use it and responds with a DHCP Ack, which is again relayed to the client.

DHCP Attacks

Because DHCP was designed to work in the trusted internal network, it does not provide authentication or access controls. A DHCP server has no way of verifying that the client requesting an address is a legitimate client on the network, and the DHCP client has no way of knowing if the DHCP server that offers it an address is a legitimate server. As a result, DHCP is vulnerable to attacks from both rogue clients and servers.

You need to protect your network from two types of common DHCP attacks:

- **Address spoofing**—A rogue DHCP server on the network can assign invalid addresses to network devices. Without valid IP addresses and network gateway addresses, these devices are unable to connect to any other IP network device, which prevents users from reaching the resources they need to do their jobs. Or a rogue DHCP server might respond to a client's DHCP discover request with the rogue server's own IP address as the default gateway, setting up the client for a MITM attack and creating a potential security breach.
- **Address exhaustion**—An attacker can access the network and request IP addresses until the DHCP server's supply of available addresses is exhausted. This prevents legitimate hosts from receiving IP addresses and accessing the network.

Using DHCP Snooping to Protect Against DHCP Attacks

You will now learn how to use DHCP snooping to protect against these attacks.

DHCP Snooping Introduction

When possible, you should implement support for DHCP snooping to protect your network against address spoofing and address exhaustion attacks. Because, on its own, the straightforward DHCP process offers no security, your switch takes the role of a security guard, overseeing DHCP exchanges and ensuring that endpoints act as they should.

AOS-CX switches support DHCPv4 and DHCPv6 snooping. You should configure both v4 and v6 snooping to secure all DHCP communications. The example configuration in Figure 5-9 focuses on v4 for simplicity.

You should typically implement DHCP at the *access layer*.

Access AOS-CX config

```
vlan 11
    dhcpv4-snooping
    dhcpv4-snooping max-
    bindings 4

interface lag 1
   dhcpv4-snooping trust

dhcpv4-snooping
```

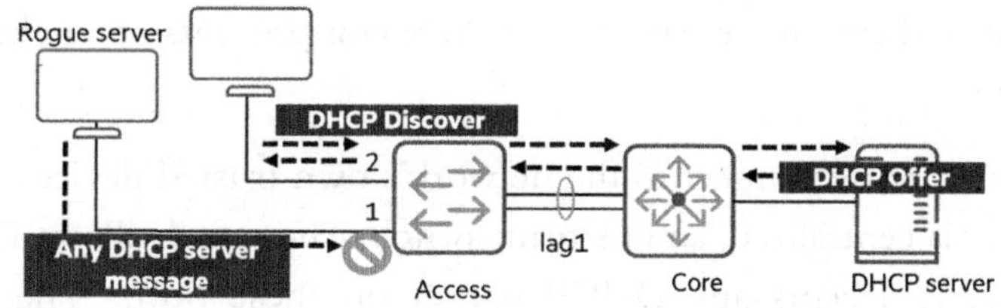

Ports	Status	Port behavior
1-2	Untrusted (default status)	Accepts valid DHCP client packets Drops DHCP server packets Does not forward DHCP client packets from other ports
Lag1	Trusted	Accepts and forwards any DHCP packet

Figure 5-9: Using DHCP Snooping to Protect Against DHCP Attacks

How DHCP Snooping Works

With DHCP snooping, the switch distinguishes between trusted and untrusted ports. When you enable DHCP snooping on a VLAN, all interfaces in that VLAN become untrusted by default. You must configure trusted ports manually.

The switch filters DHCP packets on untrusted ports and determines whether or not they are allowed.

For example, only DHCP servers should send Offer and Acknowledgement packets, so if the switch detects these types of packets on an untrusted port, it immediately discards them. On the untrusted ports, the switch only accepts DHCP client packets:

- Discover
- Request
- Decline
- Inform
- Release

The switch also verifies information in the DHCP clients' Discover and Request packets against the packet headers before allowing the packets onto the network. For example, the switch drops packets in which the source MAC address does not match the DHCP client MAC address—a sign of spoofing.

In addition, although the switch accepts DHCP client packets on untrusted ports, it does not forward them on these ports. This behavior prevents other clients from hearing each other's DHCP traffic.

Trusted ports connect to the network's own trusted devices, such as your DHCP server or, if the server is centralized, an upstream switch. The switch allows DHCP packets to flow freely on these ports. It accepts any DHCP packet on these ports. The switch also forwards DHCP client broadcasts on these ports.

DHCP Snooping Configuration

DHCP snooping operates on an individual VLAN level. Generally, you enable this feature on endpoint VLANs, especially untrusted ones such as guest VLANs, on access switches. To enable DHCP snooping, you must enable the feature both globally and on the desired VLANs. You then set the trusted port.

You can additionally set the max-binding setting for DHCP snooping. This limits the number of IP addresses the client can request, preventing address exhaustion attacks. In this example, you have set the limit to 1, but you can set it higher if, for example, a computer and VoIP phone connect on the same port.

In the example shown in Figure 5-9, the access switch, which is implementing DHCP snooping, forwards the DHCP traffic to the core switch, which relays the traffic. However, in some networks, the access switch might be routing traffic and acting as the relay. In this case, the port (or link aggregation) that forwards DHCP Discover and Request packets and receives DHCP server packets does *not* support the VLAN on which DHCP snooping is enabled. However, you *must* still make that port a trusted port for DHCP snooping.

Authorized Servers

You can increase the level of security that DHCP snooping provides by configuring authorized servers. When you configure authorized server addresses, the switch must receive packets from DHCP servers on a trusted port *and* the packets must have a source address that is on the authorized server list. If the server address is not on the list, even if the port is trusted, the switch considers the packet invalid and drops it, as shown in Figure 5-10.

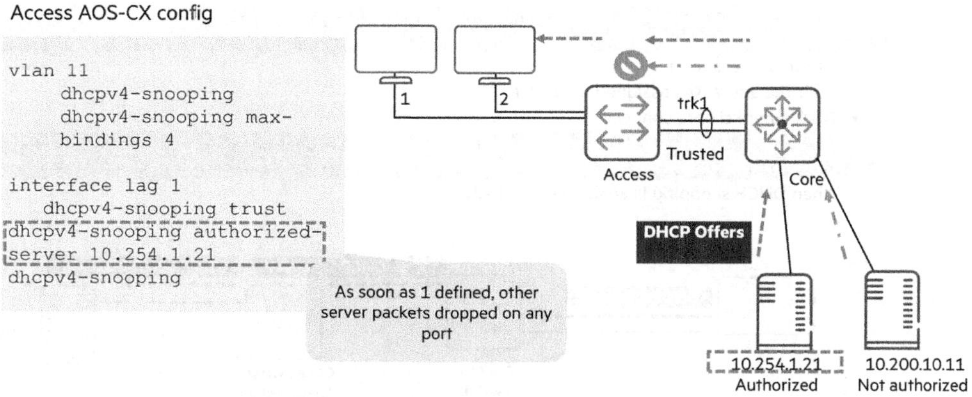

Figure 5-10: Authorized Servers

Configuring authorized servers is optional. If you do not configure them, your switch will consider all DHCP server packets that arrive on trusted ports to be valid. You can configure multiple authorized DHCP servers.

Inserting Option 82 with DHCP Snooping

You've been introduced to how DHCP snooping blocks DHCP attacks by filtering DHCP packets on untrusted ports. DHCP snooping can also facilitate the functions of DHCP itself.

DHCP's option 82 helps your DHCP server to apply specialized policies when assigning endpoints IP addresses and other configuration information. For example, you often want certain ranges of IP addresses to be associated with certain areas. And a service provider's DHCP server might limit a certain switch port to a set number of IP addresses, ensuring that a subscriber network does not consume too many addresses.

To a DHCP server, however, all incoming DHCP packets look alike. With option 82, your access switch acts as the DHCP server's eyes, adding the information that the DHCP server needs to select the correct policy. This information includes the switch's IP or MAC address (remote ID) and the physical switch port on which the request arrived (circuit ID).

A limitation of option 82 is that switches can traditionally insert option 82 information only when they are acting as relays for DHCP requests. This limitation comes into effect in two scenarios:

- A core switch acts as the DHCP relay, and does not have all of the information that the access switch has, such as the client's specific port ID. (This scenario is illustrated in Figure 5-11.)
- The DHCP client is on the same subnet as the DHCP server, so your switch does not need to act as a relay.

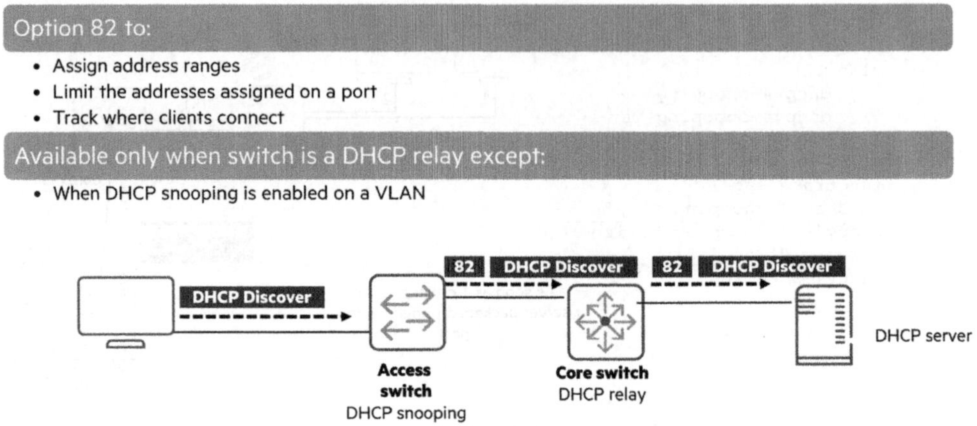

Figure 5-11: Inserting Option 82 with DHCP Snooping

With DHCP snooping enabled for a VLAN, however, your switch pays attention to all DHCP packets on untrusted ports, so the switch can always insert option 82 whether or not it is the relay, as shown in Figure 5-11.

Filtering Option 82 from Connected Devices

Option 82 is a valuable tool, but one that can be hijacked by endpoints in an untrusted network. For example, a hacker can create an option 82 to manipulate the DHCP server into sending the wrong configuration.

Switches can snoop for option 82 inside of DHCP requests from untrusted ports. When it detects this option, it implements the untrusted policy (see Figure 5-12). This policy performs one of three actions:

- The switch can drop the request entirely (drop).
- The switch can forward the request as it is, keeping the option (keep).
- The switch can forward the request but replace the unauthorized option 82 with its own option (replace).

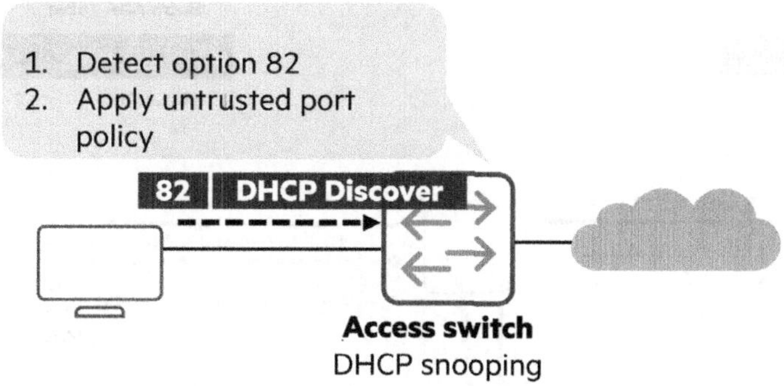

Figure 5-12: Filtering Option 82 from Connected Devices

The untrusted policy is global for all VLANs and interfaces. However, it applies only to untrusted ports. The switch accepts DHCP packets with any options on trusted ports.

By default, DHCP snooping uses the drop option. It detects and drops any DHCP request that includes option 82 and is received on an untrusted port. You should preserve this behavior whenever your switch connects directly to untrusted clients, and an improper option 82 indicates a potential attack.

If you want to take a slightly more permissive approach, controlling option 82 on the switch, but still allowing clients to receive DHCP settings if they include option 82, choose the replace option.

You should use the keep option when your solution requires devices on untrusted ports to set option 82. This is a less common scenario, but some companies might have specific devices that need to use DHCP option 82.

Review ARP Snooping, Poisoning and DoS attacks

To review, ARP helps devices map each IP address on a subnet to the MAC address of the device that owns that IP address. Most IP devices update their tables every time they receive an ARP packet even if they did not request the information. This makes ARP vulnerable to attacks such as ARP poisoning, ARP snooping, and denial of service (DoS).

ARP poisoning occurs when an unauthorized device sends an illegitimate, falsified ARP response, and other devices use the response to change their ARP tables.

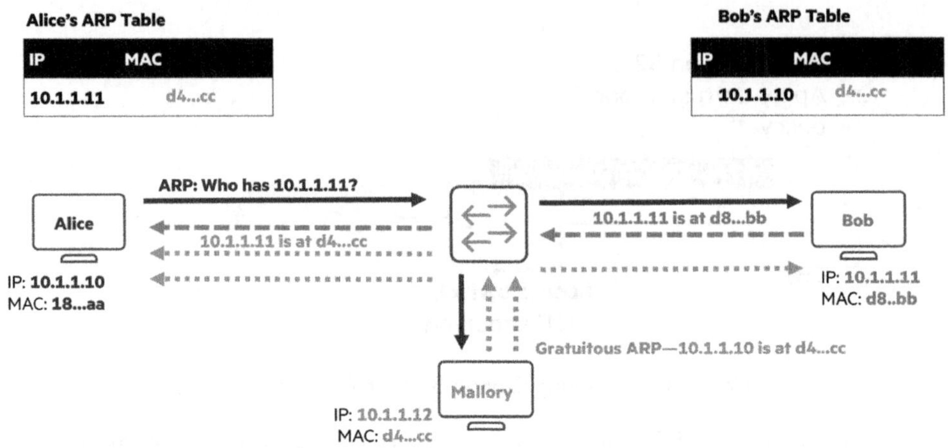

Figure 5-13: Review ARP Snooping, Poisoning and DoS attacks

In the example shown in Figure 5-13:

1. Alice's device broadcasts an ARP request for Bob's IP address.
2. Mallory sends an ARP response that maps her device's MAC address to Bob's IP address. Bob might also send a response, but Mallory can send multiple responses to override his.
3. Mallory also sends a gratuitous ARP packet posing as Alice. This poisons Bob's ARP table and also makes Bob direct responses to her device, not Alice's.

When Alice's and Bob's devices update their ARP tables with the spoofed entries, their ARP tables are considered "poisoned." Because Bob's IP address is matched with Mallory's MAC address, all IP traffic that Alice wants to send to Bob goes to Mallory instead. And the same occurs when Bob wants to send traffic to Alice.

Mallory is now a "man-in-the-middle." She can capture information such as usernames and passwords, email messages, and other confidential company information.

ARP poisoning can also take the form of unsolicited ARP responses and can lead to DoS attacks. For example, Mallory can poison the ARP table for all devices on the subnet by associating the network gateway's IP address with the MAC address of a station. Because the station does not have access to outside networks, outgoing traffic is prevented from leaving the network. The station may also be overwhelmed by the outgoing traffic.

ARP Inspection

AOS-CX switches can use ARP inspection to protect your network against ARP poisoning attacks. You enable the feature per VLAN.

When you set this up in the real world, you should make sure that you have planned and followed the guidelines covered throughout this chapter before you actually enable ARP inspection.

Like the DHCP snooping feature, the ARP inspection feature allows you to designate trusted and untrusted ports. By default, all ports in an ARP inspection VLAN are untrusted.

IP-to-MAC binding table

IP	MAC
10.1.10.10	18...aa
10.1.10.11	d8...bb
10.1.10.12	d4...cc

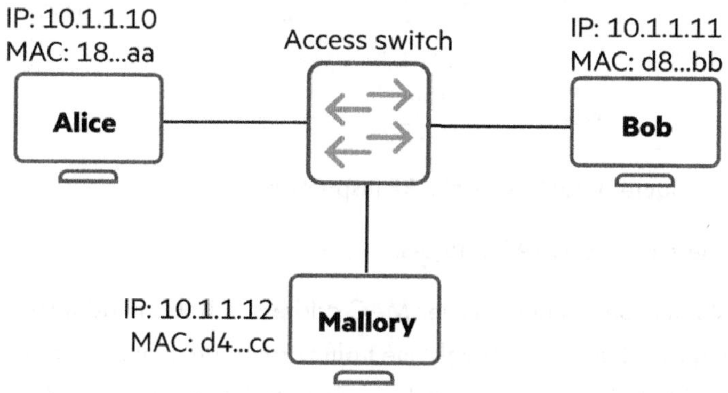

Figure 5-14: ARP Inspection

As shown in Figure 5-14, the switch checks ARP requests and responses on all untrusted ports; it:

- Examines all ARP requests and responses received on that port
- Verifies that each of these packets has a valid IP-to-MAC address binding
- If the binding is valid, the switch updates its local ARP cache or forwards the packet to the appropriate destination. If the binding is invalid, the switch simply drops the packet, preventing other devices from receiving it and being tricked by the faulty information.

Results of ARP Inspection

You can see how ARP inspection protects the network in Figure 5-15.

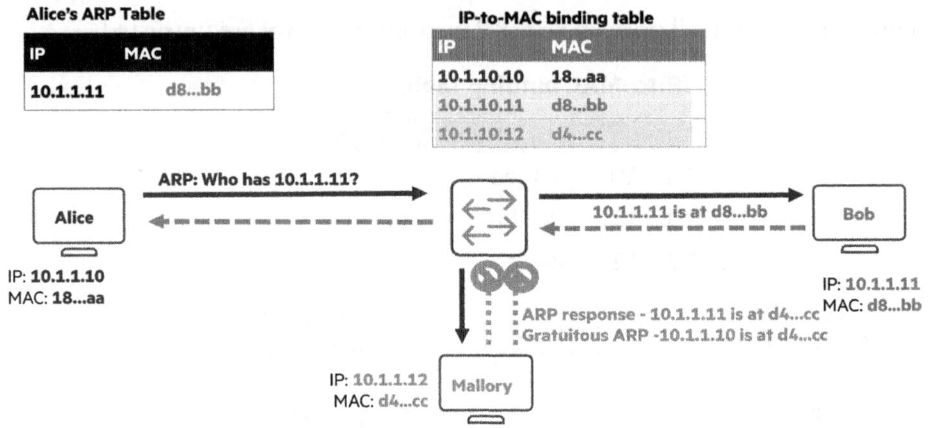

Figure 5-15: Results of ARP Inspection

1. Alice broadcasts an ARP request for Bob's MAC address.
2. Mallory responds with an ARP response that maps her MAC address to Bob's IP address. Mallory also sends a gratuitous ARP posing as Alice. Now, though, the traffic arrives on an untrusted port in a VLAN on which ARP inspection is enabled. Therefore, the switch checks both ARP messages and sees that the IP-to-MAC address binding is incorrect. The switch drops the packets.
3. Bob sends his legitimate ARP response. This response also arrives on an untrusted port in an ARP inspection VLAN. However, its IP-to-MAC address binding is correct, so the switch forwards it as normal.

Now Alice and Bob can exchange traffic.

DHCP Snooping and ARP Inspection

In most VLANs, devices receive their IP addresses using DHCP and these addresses can change. The switch needs a way discover the correct MAC to IP bindings automatically and dynamically.

DHCP snooping fulfills this function, as shown in Figure 5-16.

DHCP snooping builds and maintains a DHCP snooping-binding table, which tracks the MAC address, IP address, lease time, VLAN number, and interface information that corresponds to each DHCP lease. Using this table, the switch can learn which IP addresses should legitimately send traffic on which ports and which MAC address can legitimately use an IP address. Therefore, you must enable DHCP snooping as part of configuring ARP inspection.

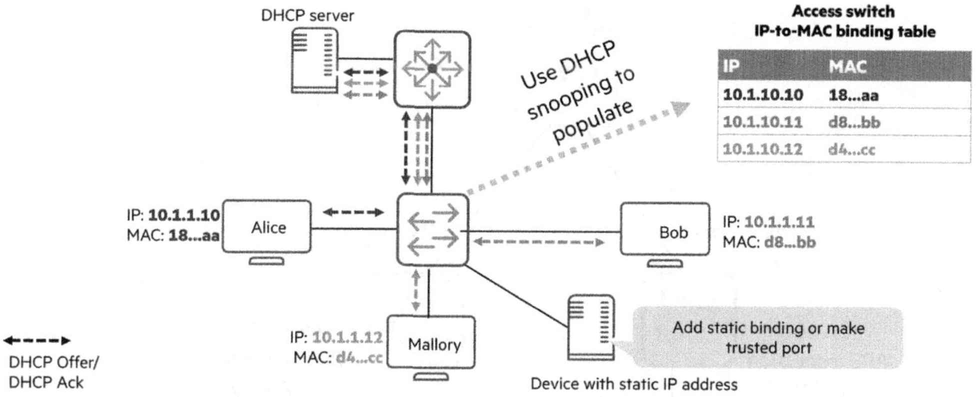

Figure 5-16: DHCP Snooping and ARP Inspection

If you are not using DHCP to assign IP addresses in the protected VLAN, you can configure static IP-to-MAC address bindings, and the switch will use this information to verify ARP packets. However, most networks, do, of course, use DHCP.

Even when you are using DHCP and DHCP snooping, you may need to add a couple static IP-to-MAC address bindings to the DHCP snooping table. The static bindings allow the switch to verify IP-to-MAC bindings for devices that have been assigned static IP addresses. You must also specify the interface for the static binding.

In an active network, you should generally wait at least a week after enabling DHCP snooping to enable ARP. This ensures that the table is properly built up.

For increased security and protection against any form of IP spoofing, you can implement dynamic IP lockdown (**ip source-lockdown**). This feature goes further than ARP inspection and enforces the correct source IP address and MAC address for all IP traffic on a port. Like ARP inspection, dynamic IP lockdown requires DHCP snooping to populate the IP-to-MAC binding table. For more details on implementing dynamic IP lockdown, refer to your switch's Access Security Guide.

Guidelines for Using ARP Inspection

Generally, you should configure ARP inspection as close to the network edge as possible, as shown in Figure 5-17. If all of your access layer switches implement ARP inspection, you usually do *not* configure it on aggregation and core switches, which are presumably protected physically and connect only to trusted devices. Your environment might have different needs, though.

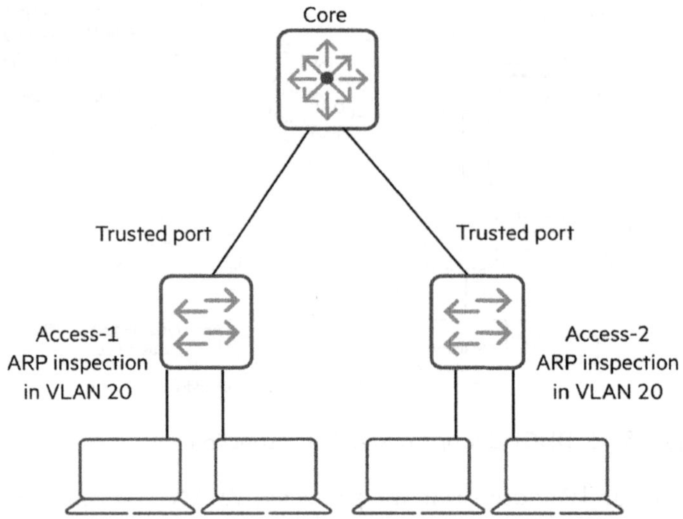

Figure 5-17: Guidelines for Using ARP Inspection

On switches that enforce ARP inspection, make upstream ports, connected to the aggregation layer or core, trusted ports. The switch will not check the ARP requests and responses that it receives on the trusted ports. When switches across the access layer implement DHCP snooping and ARP inspection, each switch only learns the DHCP bindings for clients connected on its own untrusted ports. Therefore, the switch is only capable of monitoring the ARP traffic for hosts connected on these ports. It must trust the ARP traffic arriving from other parts of the network, or it will find that the traffic has no binding and interfere with legitimate communications. In any case, allowing switches to trust traffic that arrives from other switches eliminates double filtering of traffic and reduces the burden on each switch.

If your switch connects to a less intelligent downstream switch that does not implement DHCP snooping and ARP inspection, you might decide to leave the port connected to that switch as untrusted. You must make sure, though, that all DHCP requests will flow through your switch so that it can construct the proper binding table. If you choose to make the port trusted, it is better to place the less intelligent switch in its own less-trusted VLAN to prevent untrusted ARP packets from passing to the protected network.

CLI Commands: Configuration of DHCP Snooping and ARP Inspection

Refer to the table in Figure 5-18 for commands for setting up DHCP snooping and ARP inspection.

In configure context

Step	AOS-CX Command	Notes
Enable DHCP snooping globally	dhcpv4-snooping	
Enable DHCP snooping on a VLAN	vlan <*ID*> dhcpv4-snooping	**DHCP snooping must be enabled both globally and on the desired VLAN(s)**
Configure a trusted port for DHCP snooping	interface <*ID*> dhcpv4-snooping trust	**All ports in DHCP snooping VLANs are untrusted by default**
Configure authorized DHCP servers	dhcpv4-snooping authorized-server <*IP_address*>	**Required only if you want to filter DHCP server messages based on IP address, as well as trusted port**
Choose how the switch treats DHCP requests with option 82	dhcpv4-snooping option 82 untrusted-policy <drop | keep | replace>	
Enable ARP inspection on VLANs	vlan <ID> arp inspection	**DHCP snooping is a prerequisite for ARP inspection**
Create static IP to MAC binding	ipv4 source-binding <*VLAN_ID*> <*IP_address*> <*MAC_address*> <*port_ID*>	**Optional; required only for devices with static IP addresses that are on untrusted ports**
Configure a trusted port for ARP inspection	interface <*ID*> arp inspection trust	**All ports in ARP inspection VLANs are untrusted by default**

Figure 5-18: CLI Commands: Configuration of DHCP Snooping and ARP Inspection

Authenticate Routing Protocols

In this section, you will learn how to configure network infrastructure devices to authenticate routing protocols.

Dynamic Routing Protocols

The network infrastructure uses routing protocols to determine where to route traffic. Routing switches that run the routing protocol exchange protocol packets to discover neighboring routers and communicate routing information. Using this information, the routing switches build their route tables. The routing switches then use those route tables to determine how to route traffic to its destination.

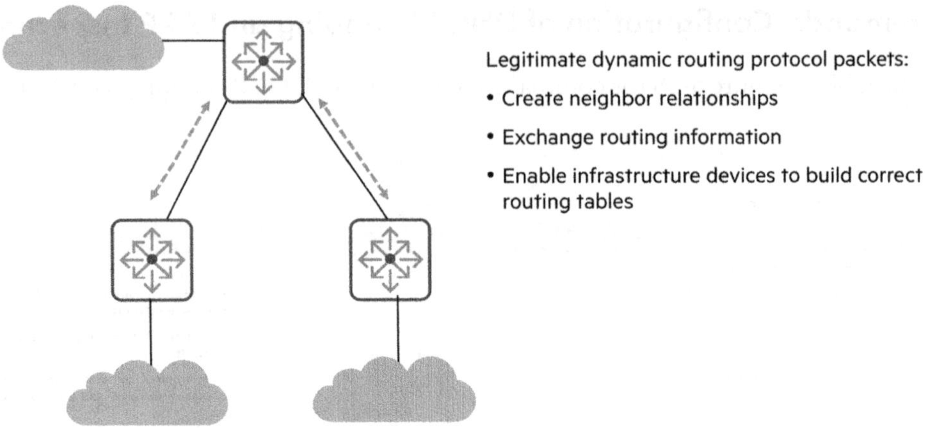

Figure 5-19: Dynamic Routing Protocols

As summarized in Figure 5-19, routing protocol packets are very powerful. They provide the basis for routing switches' and routers' forwarding decisions. If they have the correct information, all is well. But the wrong information can cause widespread issues.

Risks of Routing Protocol Manipulation

If hackers insert their device into the routing protocol, they can manipulate network infrastructure devices into installing incorrect routes, as shown in Figure 5-20.

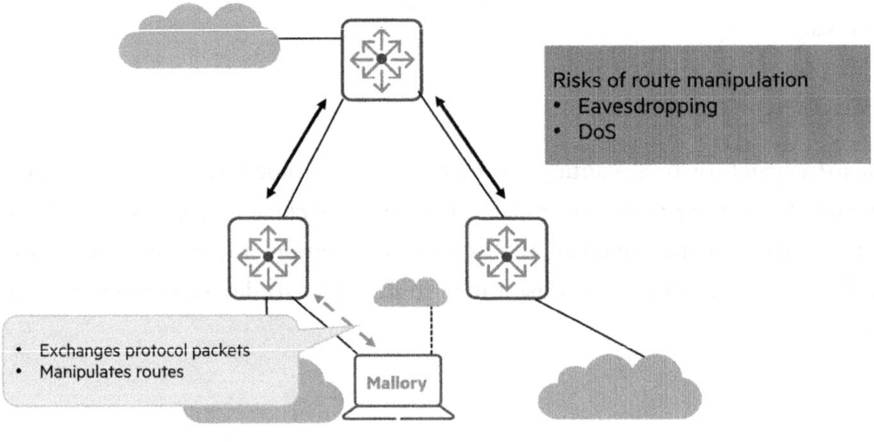

Figure 5-20: Risks of Routing Protocol Manipulation

For example, a hacker might cause the network infrastructure to route traffic through the hacker device rather than on the correct path. The hacker can then collect and eavesdrop on traffic. If the hacker device has another connection through which it can continue routing traffic, the hacker can eavesdrop silently—although the network might slow down depending on how much traffic the hacker is redirecting. If the hacker device is not able to continue routing the traffic correctly, a DoS attack occurs. Endpoints can no longer reach devices in other networks.

Many networks use Open Shortest Path First (OSPF), which uses broadcasts to discover neighbors, making it particularly vulnerable to attack.

Strategies for Securing Routing Protocols

As summarized in Figure 5-21, you have a few options for enhancing the security of the routing protocols.

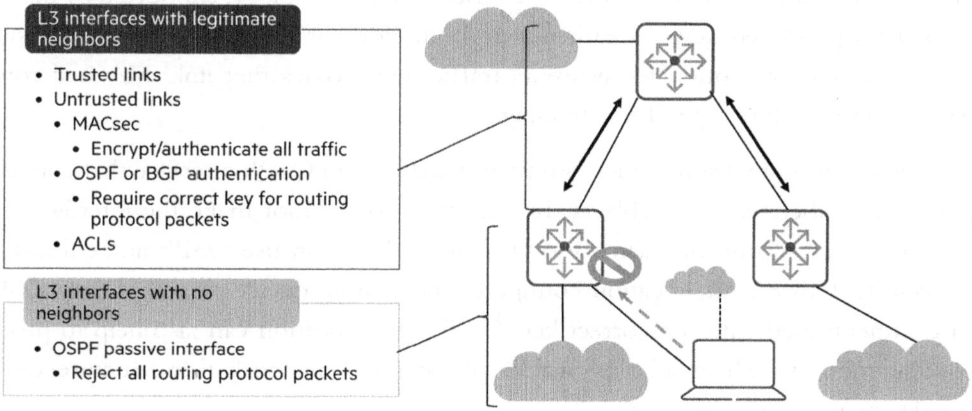

Figure 5-21: Strategies for Securing Routing Protocols

As you choose the appropriate option, you must consider several factors:

- Does the Layer 3 interface that you securing connect to legitimate routing neighbors? Or does it only connect to potentially untrusted clients? In the former case, you must ensure that the switch accepts routing protocol packets from the *correct* neighbors while in the latter, you might configure the switch to reject all packets.
- How trusted are the links for the Layer 3 interface? The less trusted the link, the more security required.

The sections below detail these considerations in the context of securing particular routing protocols.

OSPF

Begin by considering OSPF, which is the most common routing protocol for internal networks. First you should look at each Layer 3 interface that runs OSPF. Often the routing topology is designed such that some Layer 3 interfaces are used to support endpoints, but not to connect to other routers (stub networks) while others are used to connect to other routers or routing switches (transit networks).

You can configure the Layer 3 interfaces that are stub networks as "passive interfaces." The subnet associated with a passive interface is advertised in the protocol, but the OSPF routing device does not accept OSPF packets on this interface and does not form neighbor relationships on it. As you see, this protects the routing switch from accepting OSPF packets from the hacker device.

Now look at transit networks. Often routing designs dedicate one transit network for each link between OSPF routing devices. You need to assess that link. Is it trusted? If not, you should enhance its security. One option is MAC security (MACsec). This technology, supported by many AOS-CX switches, uses preshared keys to authenticate both sides of the link, which then establish data authentication and encryption to secure *all* traffic that crosses that link. MACsec configuration, however, is beyond the scope of this training.

You should be aware that sometimes a routing switch has an OSPF interface that connects to both endpoints and other OSPF neighbors. In that case, you cannot make the interface passive, and MACsec can only secure the switch-to-switch links. You can use OSPF authentication for best effort security. OSPF authentication configures the routing device to accept only OSPF packets that are authenticated with the correct key. OSPF authentication can also help to protect against misconfigurations in which admins accidentally connect network infrastructure devices to an unintended peer.

You might also want to add ACLs to enhance security, permitting routing updates only from legitimate routing devices on the network. This module introduced you to some basic ACL concepts, but more advanced ACLs are beyond its scope.

BGP

If your network uses Border Gateway Protocol (BGP), you need to make similar plans to secure the protocol. However, BGP generally runs on more trusted links. First, it does not generally run on interfaces with endpoints. Second, BGP requires admins to manually configure neighbors' IP addresses. You need to decide if the links between BGP routing devices are trusted and, if not, whether MACsec and/or BGP authentication could be good options.

OSPF Authentication

When an OSPF interface implements authentication, it includes authentication data in all OSPF packets that it transmits. It also checks this data in all OSPF packets that it receives and rejects any packets that fail the check. In other words, only OSPF routing devices configured with the same authentication keys can establish neighbor relationships.

A packet's authentication field specifies the type (as the AuthID), so receiving routing devices know how they are expected to authenticate the packet. They only accept packets that are using the same authentication type configured on their interface. You can choose from several authentication options:

- **Null**—The default authentication type, in which case OSPF interfaces do not enforce authentication. The OSPF packets have an authentication field, but the field does not include any data. An OSPF interface implementing null authentication processes all valid OSPF packets.
- **Simple**—The OSPF interface places the authentication key in every packet's OSPF header. Other OSPF routers, also implementing this type, examine this field. They only accept OSPF packets for which the field matches their own key. Because packets include the key in plaintext, anyone who can collect an OSPF packet transmitted on the network can learn the key. You should use this form of authentication only if you primarily aim to eliminate misconfigurations and routing devices inadvertently connecting on the wrong networks. You should *not* use it to secure your OSPF system from unauthorized routers and hackers.
- **Cryptographic**—The OSPF interface hashes the OSPF packet data and the key. It then places the hash at the end of the packet. The authentication field includes the key ID, a length for the hash, and a cryptographic sequence number; however, it does not include the key itself, protecting it from eavesdroppers. A router receiving the OSPF packet hashes the packet with its secret key, identified by the key ID. If the hash matches, the router knows that the transmitting router has the correct secret key for this key ID, so it accepts the packet. The cryptographic sequence number provides a degree of protection against replay attacks. An OSPF interface only accepts packets with a sequence number equal to or higher than the number for the most recent accepted OSPF packet from this neighbor. The sequence number might be a counter, but is usually based on the system clock. In either case, the number is always increasing. If a hacker obtains a legitimate OSPF packet and attempts to retransmit it later, the packet's cryptographic sequence number is no longer valid, and the other routers on the network reject it. A hacker could replay a packet in the window before the legitimate neighbor sends a packet with a higher sequence, but this window is quite small. Cryptographic authentication provides authenticity, but not privacy: the contents of the packet are not encrypted. AOS-CX switches support MD5 or (in 10.07 and later) HMAC-SHA for the cryptographic hashing algorithm. Because MD5 is deprecated, you should use HMAC-SHA.

An example setup for cryptographic authentication on AOS-CX switches is shown in Figure 5-22. You set the key ID and key string (secret) directly on the Layer 3 interface. The key ID and key string must match on all routing switches on that interface, as shown in Figure 5-22. Note that the passwords shown in the example are for illustrative purposes *only*. You should select secure passwords.

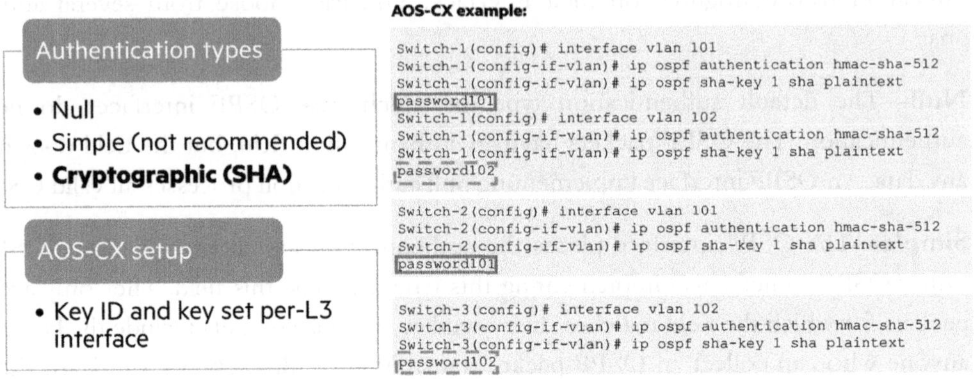

Figure 5-22: OSPF Authentication

BGP Authentication

AOS-CX switches support several options for authenticating BGP sessions.

BGP itself supports authentication. To use BGP authentication, you simply add a password when you define the neighbor. The BGP neighbors use this password to hash the TCP payload of packets that carry the BGP messages, appending the hash to the packet. If the neighbors' passwords do not match, the TCP session for the BGP connection fails to establish. However, this form of authentication uses the MD5 algorithm, so it is less secure.

When possible, you should use TCP Authentication Option (TCP-AO), which is supported in AOS-CX v10.11 and higher. As shown in Figure 5-23, this option leverages TCP authentication and is mutually exclusive with BGP authentication (a password defined on the neighbor). To use TCP AO, set up a keychain on the switch globally. Add at least one key to this keychain. Set the cryptographic algorithm to an HMAC-SHA option, such as **hmac-sha-512**. Set the key to a secure password. Then when you set up BGP neighbor, use the **ao** option and reference the keychain that you configured. *Disable* the **ao accept-ao-mismatch** option to force the neighbor to include the AO option for a connection to establish. Make sure that the BGP neighbor is set up to support AO and to use a keychain with the same ID, algorithm, and key string.

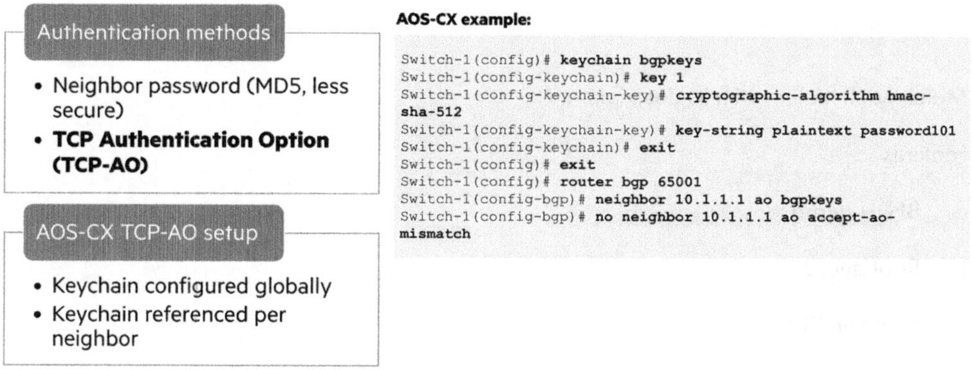

Figure 5-23: BGP Authentication

CLI Commands: Configuration of Routing Protocol Authentication

Use the commands shown in Figure 5-24 as a reference for configuring authentication on the routing protocols used in your network.

In configure context

Step	AOS-CX Command	Notes

Enable OSPF authentication on a layer 3 interface.	`interface <L3_interface_ID>`
	
	`ip ospf authentication [hmac-sha-256 | hmac-sha-384 | hmac-sha-512]`
	
	`ip ospf sha-key <ID> sha <ciphertext | plaintext> <password>`	

Enable BGP with TCP AO on a neighbor.	`keychain <name>`
	`key <ID>`
	`cryptographic-algorithm [hmac-sha-256 | hmac-sha-384 | hmac-sha-512]`
	`key-string <ciphertext | plaintext> <password>`
	
	`router bgp <AS>`
	`neighbor <IP> ao <keychain>`
	`no neighbor <IP> ao accept-ao-mismatch`
		Set the remote AS for the neighbor and then .

Figure 5-24: CLI Commands: Configuration of Routing Protocol Authentication

Summary

Congratulations! You can now protect basic network infrastructure protocols from attack.

Learning checks

1. Match each feature with its description.

   Features

   a. BPDU guard

   b. Root guard

   c. CoPP on STP

   Descriptions

   __ Prevents strain on CPU

   __ Prevents an unauthorized device from becoming root

   __ Prevents rogue spanning tree devices from connecting

2. True or false: DHCP includes native protections to enable clients to authenticate the DHCP server as legitimate.

   a. True

   b. False

3. How does an AOS-CX switch learn dynamic MAC-to-IP bindings for implementing ARP protection?

   a. MAC authentication

   b. SNMP

   c. DHCP snooping

   d. OSPF

4. What basic protections can you implement to prevent unauthorized devices from altering your networks' routing tables? (Select two.)

   a. ARP snooping

   b. OSPF or BGP authentication

   c. OSPF passive interface

   d. ARP protection

   e. BPDU guard

Note

For answers to all learning checks, refer to Appendix B.

6 Network Authentication and Encryption Technologies

EXAM OBJECTIVES

✓ Describe network authentication options

✓ Describe wireless security options, in particular how WPA3 enhances security for wireless networks

Assumed Knowledge

- Basics of IP addressing
- Basics of Ethernet switching
- Basics of 802.11 wireless networks
- Basics of encryption (from Chapter 2 of this book)

Overview

The chapter begins by presenting common forms of network authentication, including MAC-Auth and captive portal, but with a deep dive into 802.1X.

Then, turning to wireless security specifically, you review the differences between open WLANs, WPA/WPA2/WPA3-Personal WLANs, and WPA/WPA2/WPA3-Enterprise WLANs. You will also learn how WPA3-Personal and WPA3-Enterprise resolve security issues in WPA2-Personal and WPA3-Enterprise.

Types of Network Authentication

This section introduces you to various types of network authentication.

Why Network Authentication

You have learned about HPE Aruba Networking's overall strategy for delivering security in zero trust environments. As summarized in Figure 6-1, a key component of this security strategy involves network authentication. By identifying every user and device at the internal perimeter, a company gains visibility into their environment. Such visibility lays the foundation for control. First, the company can take action to control access to the internal network, ensuring that only authorized users and devices can connect. Next, the company can enhance this control by assigning the appropriate access controls to each client, based on knowledge about who is using the client, as well as other factors such as device type and location.

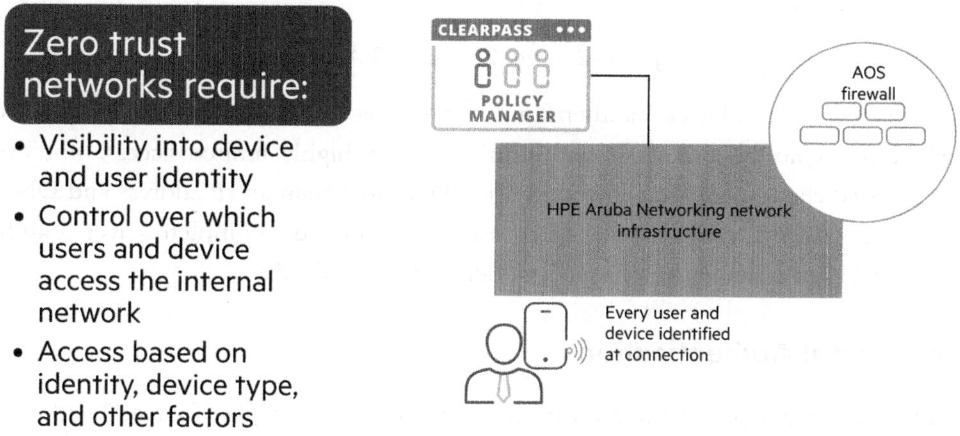

Figure 6-1: Why Network Authentication

This chapter focuses on the first steps—the industry technologies that authenticate clients and grant them access to the network. Later chapters will explain how an HPE Aruba Networking network extends companies' control through role-based policies applied to authenticated clients.

Layer 2 vs Layer 3 Authentication

You can divide network authentication methods into two broad categories: Layer 2 and Layer 3.

As shown in Figure 6-2, Layer 2 authentication methods authenticate users and devices before they receive an active Layer 2 connection. If the client fails authentication, the client is disconnected at Layer 2, preventing the client from sending or receiving any traffic at all. MAC authentication (MAC-Auth) and 802.1X are examples of Layer 2 authentication methods.

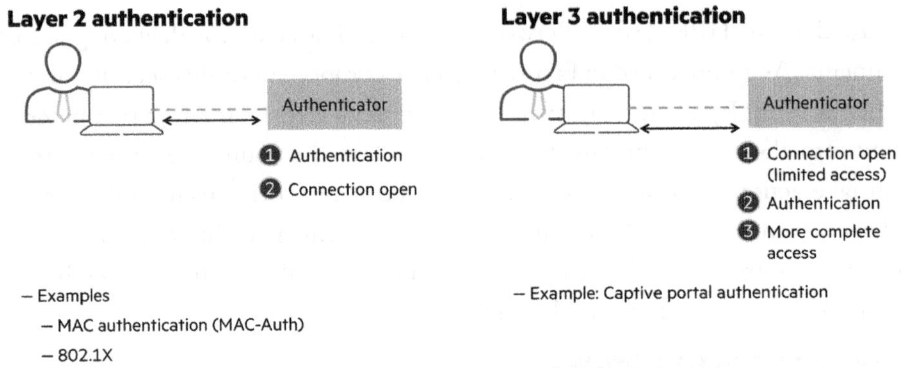

Figure 6-2: Layer 2 vs Layer 3 Authentication

As shown in Figure 6-2, Layer 3 authentication takes place after the client receives an active Layer 2 connection. Typically, the client will initially receive highly limited access. For example, the unauthenticated client might be allowed to use DHCP to obtain an IP address and DNS to resolve hostnames. To receive a higher level of access, the client must complete Layer 3 authentication. Captive portal is an example of a Layer 3 authentication method.

Captive Portal Authentication

Look briefly at captive portal authentication, as the primary example of Layer 3 authentication.

You are probably quite familiar with captive portal authentication, which is often implemented for guest wireless networks in locations such as airports, hotels, and coffee shops. Captive portal authentication provides an easy way for users to log in to a network without special requirements on the part of their devices, nor expertise on the part of the users. A user simply connects to an open network. The user then opens a Web browser, which is directed to a login page, in which the user can enter credentials or sometimes simply accept an agreement to log in. The user's client then receives more complete access to the network.

The process by which the network implements captive portal can have many nuances based on how the company wants to implement the feature. You can get a basic idea of the process from Figure 6-3.

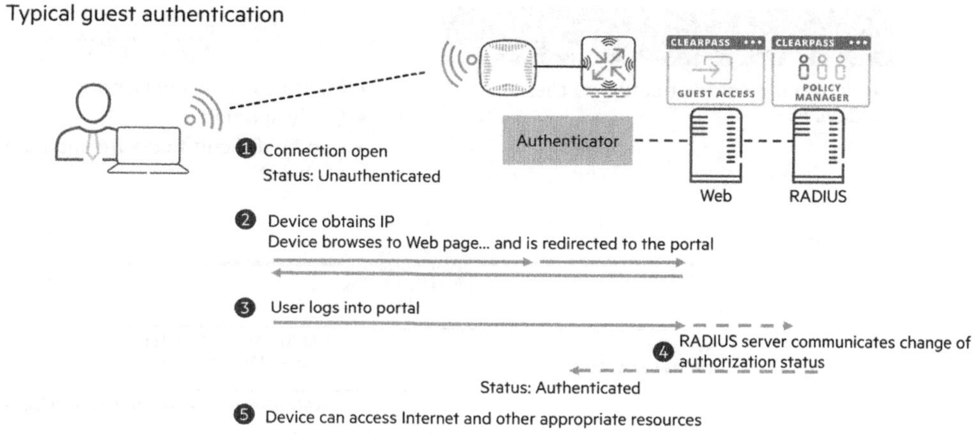

Figure 6-3: Captive Portal Authentication

In this example, the authenticator is an HPE Aruba Networking Mobility Controller (MC), which controls the access point (AP) to which a client connects. (HPE Aruba Networking Instant APs [IAPs] and AOS-CX switches also support implementing captive portal to an external server). MCs support an internal captive portal, but can also integrate with HPE Aruba Networking ClearPass, as shown here.

When a new client connects to the captive portal WLAN, the MC grants the client limited access. The client obtains an IP address and attempts to browse to a web page. The communications flow through the AOS firewall, which redirects the client to a login page on ClearPass Guest. The user logs in, and ClearPass Guest authenticates the user against ClearPass Policy Manager (CPPM). CPPM informs the MC of the user's new authentication status, and the MC assigns the user a new role based on that status. Now the authenticated user can browse to the Internet or use other network resources, as dictated by the user role.

If you want to learn more about captive portal authentication, you can attend other HPE Aruba Networking training.

MAC-Auth

MAC Authentication (MAC-Auth) is one example of a Layer 2 authentication method. A network infrastructure device that implements MAC-Auth checks the source MAC address in a client's incoming traffic. Sometimes the authenticator has a local list of authorized MAC addresses. Sometimes, as shown in Figure 6-4, the authenticator submits the MAC address to an authentication server, which uses the MAC address to determine whether to grant the client access. For example, the server might have a allow list of MAC addresses that belong to the company; it only grants access to MAC addresses on that list.

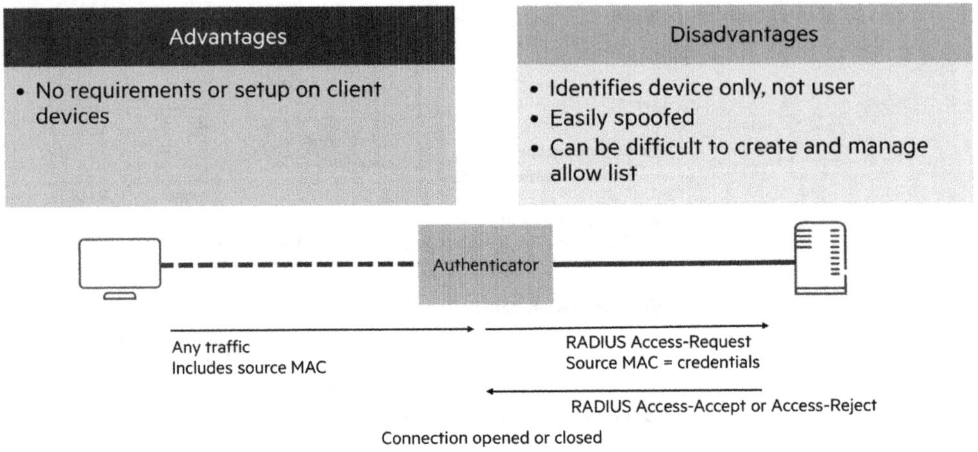

Figure 6-4: MAC-Auth

Because wireless clients always include their MAC address in 802.11 or Ethernet frames, the client does not need to be configured to submit credentials. Therefore, this authentication method provides an option for devices that do not have a user interface, sometimes called "headless devices," or support for 802.1X.

However, MAC-Auth does not provide rigorous security. First it identifies the device only, not the user on the device. Worse, it is often trivial for a hacker to discover MAC addresses used in your system, whether by eavesdropping or accessing a legitimate device. The hacker can then easily spoof an authorized MAC address. For example, the Windows OS lets you change the MAC address in the NIC properties. Finally, it can be difficult to create and manage a allow list of the MAC addresses of all the company's authorized devices. Apple and Android devices are starting to randomize their MAC addresses when they connect to different SSIDs, making it even harder to create a definitive allow list.

Rather than a security measure, MAC-Auth is perhaps most useful for simplifying the provisioning for devices such as printers, Internet of Things (IoT) devices, IP phones, and security cameras.

802.1X Overview

In this study guide, you will focus most on another Layer 2 authentication option: 802.1X, which is recommended for secure authentication in enterprises.

802.1X, an Institute of Electrical and Electronics Engineers (IEEE) standard for Network Access Control (NAC), enables the network infrastructure to authenticate and authorize endpoints for network access. 802.1X requires users to authenticate as soon as connection is established. As shown in Figure 6-5, in a wired network, 802.1X takes effect as soon as the Ethernet connection comes up. In a wireless network, it takes effect after the initial 802.11 association.

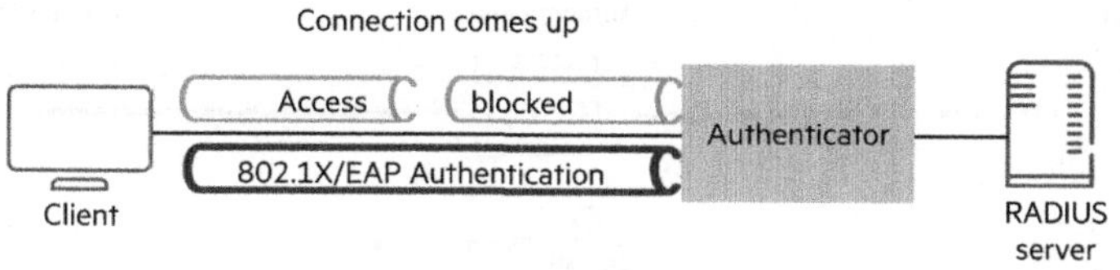

Figure 6-5: 802.1X Overview

To restrict the unauthenticated client's access, 802.1X divides the port into two logical ports. The controlled port permits all types of traffic. Its state depends on authentication status. Both sides of the connection should block the controlled port until the peer has successfully authenticated. After successful authentication, all types of traffic may flow through the controlled port from the point of view of 802.1X. (However, the network infrastructure can further restrict access with firewall policies and ACLs.) The uncontrolled port is always open, but it permits only Extensible Authentication Protocol (EAP) packets. It does not carry *any* other form of traffic, even DHCP messages and DNS messages. 802.1X allows an endpoint to authenticate *before* it has acquired an IP address.

802.1X Roles

The 802.1X process includes three parties, as shown in Figure 6-6. To understand these parties, you will consider an analogy in which a person is attempting to get into an exclusive, very important person (VIP) event in a secure building.

Supplicant

The supplicant is the 802.1X component that runs on the endpoint attempting to connect to the network. In our analogy, Bob, the person attempting to get into the event, is the supplicant. He is the one who wants to gain access to something.

In the real world, 802.1X supplicants are software that run on client devices. For example, Windows and MAC operating systems have native 802.1X supplicants, as do smartphone operating systems such as Android. Admins can also install third-party supplicant software on endpoints.

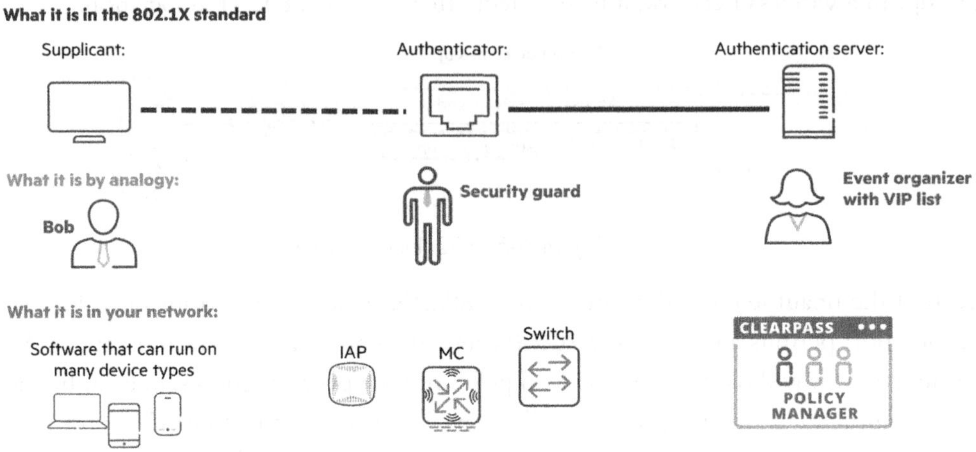

Figure 6-6: 802.1X Roles

Authenticator

The authenticator controls network access. It enforces decisions made by the authentication server, either permitting or denying access. In our analogy, the authenticator is the security guard at the building in which the event is being held. The security guard is responsible for guarding the building and only allowing authorized people in. Similarly, in 802.1X the authenticator stands at the internal network perimeter and either grants or denies access to the network. The authenticator could be a switch, an AP in an IAP cluster, or an MC.

Authentication Server

The authentication server directs the authenticator and ultimately takes responsibility for access decisions. You can think of the authentication server as the event organizer. She has the VIP list and tells the security guard who to let in.

In 802.1X, the authentication server is a RADIUS server. To make its access decisions, the RADIUS server checks users' credentials, but also considers other factors. Is the user connecting at the right time or location? An authentication server may, for example, deny users with legitimate credentials if they are trying to connect outside of regular work hours. The decisions all depend on the authentication server's policies.

For HPE Aruba Networking networks, it is recommended that you use CPPM, which includes a powerful RADIUS server among many other features.

802.1X Process: High-level Analogy

You will now look at the 802.1X process, summarized in Figure 6-7. To help you understand the process, return to Bob, who is trying to get into the VIP event.

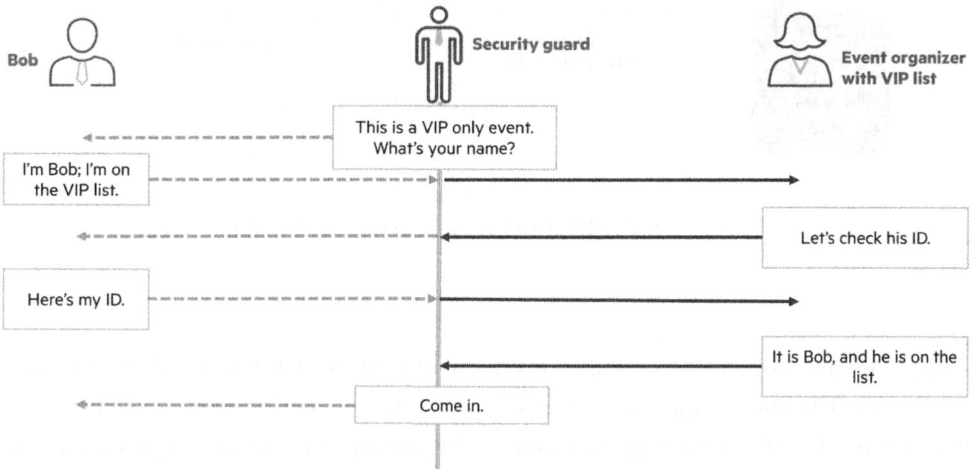

Figure 6-7: 802.1X Process: High-level Analogy

Bob comes to the building and is met by the security guard. The guard informs him that the event is closed to the public and asks for Bob's name. Bob gives him his name and says that he is on the VIP list.

The security guard has an audio/visual communication channel open with the event organizer, who is inside the building. He's going to act as a courier for everything that Bob says, passing it on to the event organizer and similarly, passing on the event organizer's requests to Bob. The guard informs the event organizer that a person claiming to be Bob is outside. The event organizer tells the guard to collect his ID. The guard passes on the message and has Bob hold his ID up to the camera for the event organizer to check. Checking the ID and her list, she sees that the person at the door is indeed Bob, who is on the VIP list. She tells the security guard to let him in, which the guard does.

802.1X Process: Communication Protocols

In a moment, you will see how this process translates to 802.1X. To understand the process, you need to understand the two communication protocols used in 802.1X. (See Figure 6-8.)

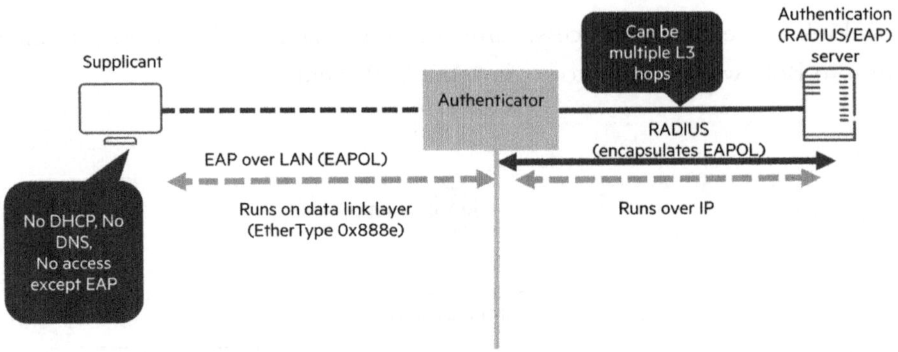

Figure 6-8: 802.1X Process: Communication Protocols

EAPOL

The supplicant and authenticator communicate with EAP over LAN (EAPOL) messages. EAPOL operates at the data link layer and uses EtherType 0x888e. Note that, despite the name, even wireless connections use EAPOL messages. In practice, the messages are often simply called EAP messages.

802.1X almost entirely shuts down the Layer 2 link for unauthenticated clients. The client cannot even send DHCP or DNS messages into the network. The authenticator accepts *nothing* except EAP. This means that usually the client has no IP address. It can only communicate on the local link. The supplicant is either directly connected to the authenticator, or, if the authenticator is an MC, an AP tunnels the supplicant's messages to the authenticator.

RADIUS

The authenticator reaches the RADIUS server over an IP network, and multiple Layer 3 hops can exist between the devices.

The authenticator acts as a relay between the supplicant and the RADIUS server much like the security guard in the previous example. The authenticator encapsulates EAP messages from the client with RADIUS Access-Requests, which it sends to the RADIUS server. Similarly, the RADIUS sever encapsulates EAP messages for the client in RADIUS-Challenge messages, which it sends to the authenticator. The authenticator decapsulates the EAP messages and passes them on to the client.

In this way, the supplicant and authentication server act as the two endpoints in an EAP exchange with the authenticator transporting messages between them.

Note that, in RADIUS terms, the authenticator is the Network Access Server (NAS), Network Access Device (NAD), or RADIUS client. However, the rest of this discussion will use "authenticator" to refer to the device for consistency.

802.1X Process: EAP Framework

You are now ready to understand the 802.1X process for authenticating clients.

EAP, the protocol 802.1X uses for authentication, defines a flexible framework into which you can fit authentication methods that meet the needs of your environment and security policies. Understanding the EAP process can help you to pinpoint and troubleshoot issues.

This process is illustrated in Figure 6-9.

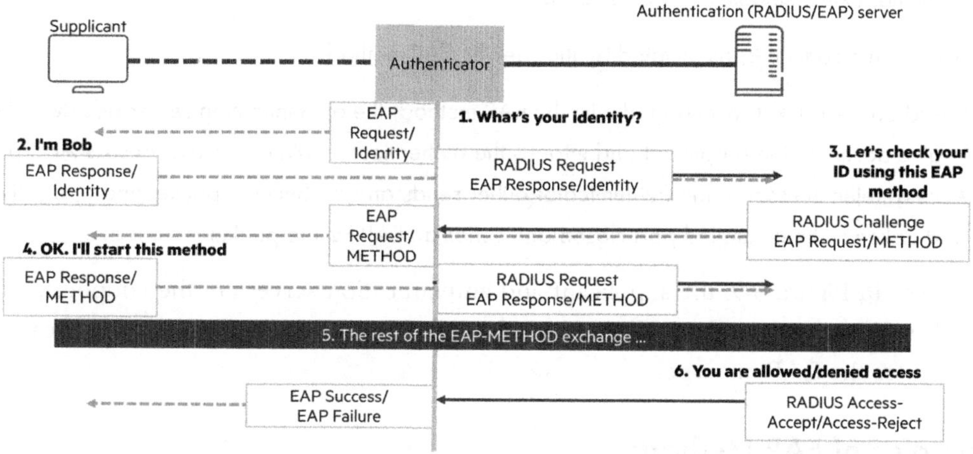

Figure 6-9: 802.1X Process: EAP Framework

The process begins at the point at which the supplicant and authenticator establish a connection. They have both blocked the controlled port so that only EAP traffic can flow. Regardless of the particular EAP method, EAP follows these general steps:

1. The authenticator issues a challenge in the form of an EAP Request/Identity packet. This frame asks the client to identify itself but not to send any other information. (Alternatively, the supplicant can initiate authentication by sending an EAP-Start packet.)

2. The supplicant responds to the challenge with an EAP Response/Identity packet, which typically includes a username. The authenticator forwards the EAP Response/Identity packet, in RADIUS, to the authentication server. Some EAP methods have supplicants submit a false username at this point. In this case, users give their true username only after a secure, encrypted session has been established. This protects the username from hackers. As you learned earlier, exposing the username to hackers can leak

valuable information and also let the hackers lock a user out of the network by submitting incorrect credentials for the username.

3. The server sends an EAP Request, encapsulated in a RADIUS Access-Challenge. This request initiates an EAP method for exchanging credentials. The details of this exchange, such as the particular messages and kind of credentials, depend on the EAP method. Two common EAP methods, which you will examine in more detail in a moment, are EAP-TLS and PEAP.

4. If the supplicant supports the EAP method proposed in the server message, the supplicant continues the process, using that EAP method to verify its identity. If the supplicant does not support the proposed EAP method, it sends an EAP NAK. As you see, it is the supplicant and authentication server that negotiate the EAP method and that must support the same method. The authenticator does not need to support any particular EAP methods; it is only a courier.

5. The process continues, as dictated by the specific EAP method.

6. Based on the packets exchanged with the EAP method, the authentication server decides whether or not to give access to the supplicant and informs the authenticator with a RADIUS Access-Accept or RADIUS Access-Reject message. The authenticator either sends an EAP Success packet and opens the controlled port or keeps the controlled port closed and sends an EAP Failure packet.

As you see in Figure 6-9, the supplicant and authentication server are the endpoints of the EAP-TLS, PEAP, or other EAP method exchange. The authenticator simply encapsulates and decapsulates the EAP messages.

Examples of EAP Methods

Many specific EAP methods exist within the flexible framework. Some common ones are listed in Figure 6-10. One important point is that each of these methods provides mutual authentication—both the server and supplicant authenticate. It is important for the supplicant to authenticate the server to avoid man in the middle (MITM) attacks in which the supplicant connects to a hacker device rather than the legitimate network. EAP-TLS is the preferred method and the one that you should implement for enterprise networks.

EAP method	Credentials	Comments
EAP-TLS	Server = Certificate Supplicant = Certificate	• Recommended option • Supported in many supplicants, including Windows native • Uses TLS handshake to authenticate server and supplicant
Protected EAP (PEAP)	Server = Certificate Supplicant = Username and password (typical)	• Uses TLS handshake to create secure tunnel (outer method) • Uses a less secure authentication method such as MSCHAPv2 inside the tunnel (inner method)
EAP-Tunneled TLS (EAP-TTLS)	Server = Certificate Supplicant = Username and password (typical)	• Similar to PEAP in concept
EAP-Tunneled EAP (EAP-TEAP)	Server = Certificate Supplicant = Username and password or certificate	• Also uses TLS to create a secure tunnel for carrying an inner method • Enables "chaining" of machine and user authentication with tokens

Figure 6-10: Examples of EAP Methods

EAP-Tunneled Layer Security (TLS)

Considered one of the most secure EAP methods, EAP-TLS uses a TLS 1.2 handshake to authenticate the server and supplicant with digital certificates. The Windows native client supports EAP-TLS, as do many other supplicants. ClearPass also supports EAP-TLS. While deploying certificates to clients can be challenging, methods exist for easing and automating this deployment. You will look at some of these methods later.

Protected EAP (PEAP) and EAP-Tunneled TLS (TTLS)

PEAP was developed to provide some of the security of EAP-TLS without forcing stations to have digital certificates. It is the option that the Windows native client uses when the user logs in with a username and password.

Tunneled Transport Layer Security (TTLS) is a similar, but less commonly used, method. (TTLS was developed by Funk Software and Certicom; PEAP was developed by Microsoft, Cisco Systems, and RSA Security.) ClearPass supports both PEAP and EAP-TTLS.

TTLS and PEAP function in similar ways. Both methods involve a two-phase authentication process. In the first phase, the outer method creates a secure tunnel using TLS. The authentication server then initiates a second authentication method, called the inner method, inside the secure TLS tunnel. PEAP and EAP-TTLS can use several different options for the inner method. (In fact, they can even use EAP-TLS as the inner method; however, doing so does not typically make sense because the goal of PEAP and TTLS is to protect a less secure inner method.) EAP-Microsoft

CHAPv2 (MSCHAPv2) is the most common option. EAP-MSCHAPv2 is not secure enough to run over an untrusted network but is acceptable when protected by the TLS tunnel.

EAP-Tunneled EAP (EAP-TEAP)

Like PEAP and EAP-TTLS, EAP-TEAP uses TLS as an outer method to create a secure tunnel. It then uses an inner method to authenticate the client. Depending on the inner method implemented, the client can authenticate with a username/password or a certificate. EAP-TEAP also supports "chaining" of multiple authentications, which can be useful when a company wants to ensure that users connect on authorized machines. The client can authenticate with machine credentials first, proving the device is authorized. The client obtains a token during this authentication. The client can then authenticate with user credentials, including the token from the machine authentication, proving that both the user and device are authorized. EAP-TEAP is supported in Windows 10 and higher and ClearPass 6.9 and higher.

EAP-TLS Process

You will now look at the EAP-TLS exchange in more detail, as understanding it can help you to troubleshoot.

1. As shown in Figure 6-11, the authenticator sends an EAP Request/Identity packet to the supplicant.
2. The supplicant sends an EAP Response/Identity, which the authenticator encapsulates in a RADIUS Access-Request and passes on to the RADIUS authentication server.
3. The RADIUS server replies with an EAP Request/TLS, encapsulated in a RADIUS Access-Challenge. This is the beginning of the TLS process.
4. The supplicant responds with a "ClientHello" record in an EAP Response/TLS, which is encapsulated by the authenticator in another RADIUS Access-Request.
5. The RADIUS server answers with several TLS records in an EAP Request/TLS, encapsulated in another RADIUS Access-Challenge. These messages included a "ServerHello," the server certificate, a request for the supplicant's certificate, and a "HelloDone."
6. The supplicant validates the server's certificate. It checks that the signature is valid and created by a trusted Certification Authority (CA). It also checks the subject name within the certificate and ensures that it matches a server configured as trusted.

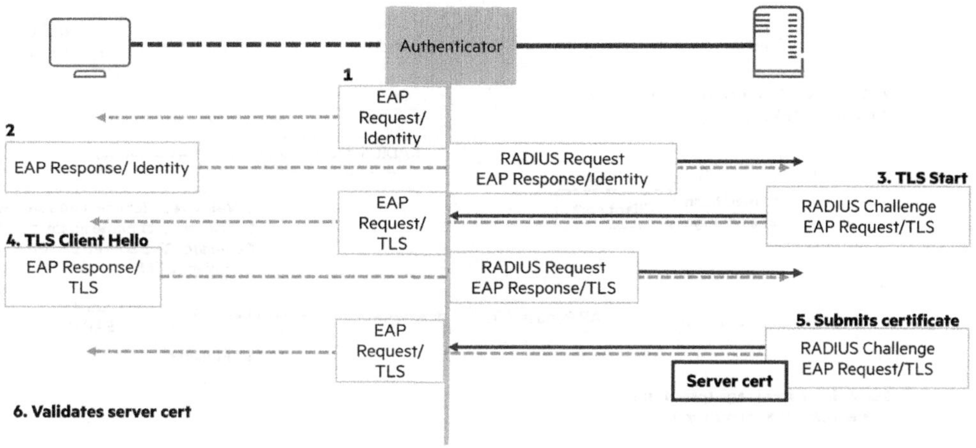

Figure 6-11: EAP-TLS Process (Steps 1-6)

7. As shown in Figure 6-12, if the server's certificate is valid, the supplicant responds with several TLS records in EAP Response/TLS messages. These records include the supplicant's certificate, as well as a verification that the supplicant really owns the certificate. Remember, Mallory could intercept Bob's certificate and submit it as her own. The supplicant submitting Bob's certificate needs to prove that it has the associated private key. The supplicant does this by hashing the exchange to this point and signing that hash with the private key.

8. The server validates the supplicant's certificate. It then uses the public key in the certificate to decrypt the signed hash and check that against its own hash of the conversation. In this way, the server to verifies that the supplicant with which it is communicating really owns the certificate. The server also checks the subject name. It uses this name and its policies to determine if the user is authorized for network access at this time and location.

9. If the server permits access, it completes the TLS handshake. As you learned in a previous chapter, TLS enables the client and server to generate the same shared secret with protection against eavesdropping and MITM attacks; EAP-TLS is no different. The server secures the final message of the handshake using the encryption and integrity keys derived from this shared secret. By doing so, the server gives implicit proof that it owns the private key associated with its certificate. TLS is designed such that, otherwise, the server would not be able to generate the shared secret.

10. The supplicant validates the server's Finished message.

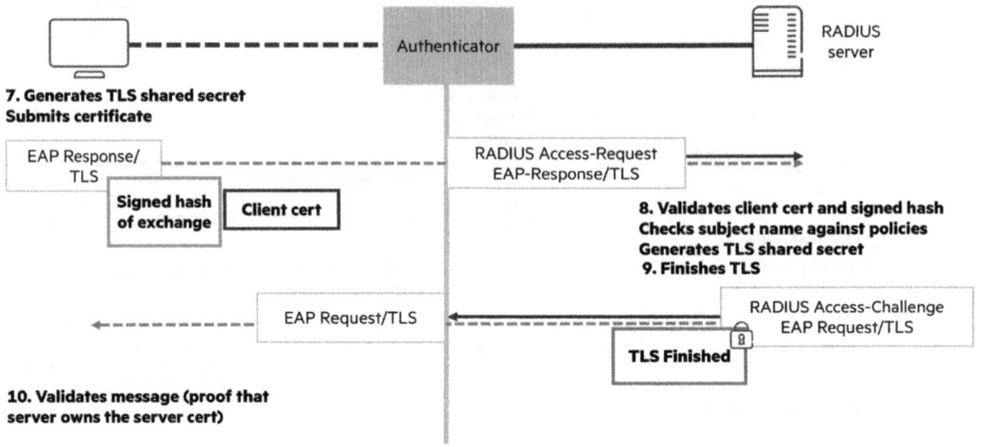

Figure 6-12: EAP-TLS Process (Steps 7-10)

11. As shown in Figure 6-13, if the validation succeeds, the supplicant sends a final EAP Response/TLS message, which the authenticator encapsulates in a RADIUS Access-Request.
12. If authentication is successful, the RADIUS authentication server sends a RADIUS Accept-Accept frame.
13. The authenticator sends an EAP Success to the supplicant. The authenticator and supplicant then open the connection completely.

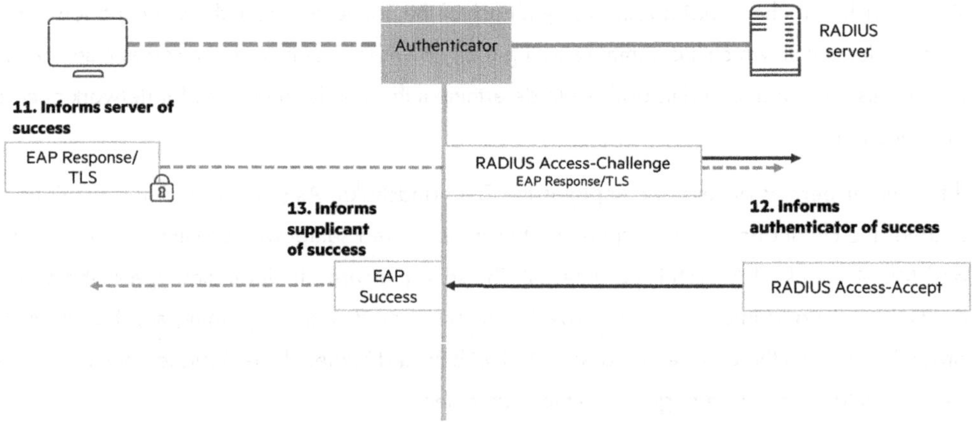

Figure 6-13: EAP-TLS Process (Steps 11-13)

Why EAP-TLS vs PEAP

Now that you have looked at EAP-TLS more closely, consider why HPE Aruba Networking recommends it over PEAP. PEAP's main advantage is that it is simpler to deploy. However, EAP-

TLS has clear advantages in security. EAP-TLS is both widely supported and is the most secure option for EAP authentication.

EAP-TLS provides mutual authentication with Public Key Infrastructure (PKI) digital certificates on both sides. Mutual authentication ensures that clients only connect to trusted network infrastructure.

Digital certificates provide better security than passwords. They use strong asymmetric encryption to authenticate entities and are validated by trusted CAs. Although you can set up requirements to force users to create "stronger," and therefore more secure, passwords, you cannot prevent users from telling people their password or writing it on a paper displayed in plain sight. Generally, it is harder for users to leak certificates inappropriately; systems administrators might, for example, implement security measures to prevent users from exporting a certificate's private key.

Hackers often run dictionary attacks on passwords, phish for passwords, or use other social engineering techniques to get users to leak their passwords. Although no credentials can be totally secure, digital certificates are more resistant to many of those common attacks. Someone who steals a smart card might be able to gain access to the certificates within, but the private keys are usually protected by an additional layer of security such as a password or PIN, which the user must enter before the supplicant can use the certificate. The user certificates installed on an end-user device such as a Windows, Mac, or Linux device are associated with the user account. The certificates are only available for that user when that user has successfully logged in. In addition, if someone steals a user device that contains certificates, the user can immediately report the theft, allowing the private CA to revoke the compromised certificate.

Finally, PEAP is more susceptible to attacks in which improperly configured clients expose their credentials to hackers. As shown in Figure 6-14, in this attack the hacker sets up a rogue AP, which could be a soft AP, that lures the client to connect to it using a legitimate SSID. The AP also has a RADIUS server. The client in this example is improperly configured, so it trusts *all* RADIUS servers. It completes the outer PEAP exchange and starts the MSCHAPv2 exchange. MSCHAPv2 can be cracked, so the hacker uses the information submitted by the client in this exchange to crack the user's password. While failing to configure an EAP-TLS supplicant to trust only certain RADIUS servers can also expose the client to MITM attacks, at least it does not compromise the client certificate, since certificates are meant to be distributed to anyone, trusted or untrusted.

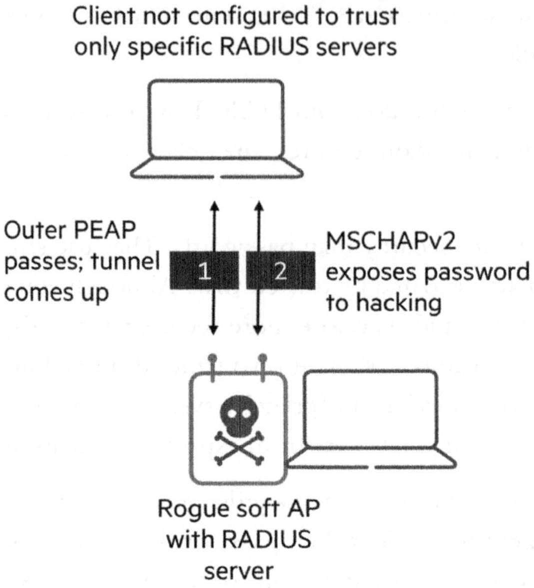

Figure 6-14: Why EAP-TLS vs PEAP

EAP-TLS Requirements

Understanding that EAP-TLS is the best option, consider the requirements for implementing it.

First, you should be aware that both the software supplicant and the authentication server must support EAP-TLS—not all supplicants support all EAP methods. But the native supplicants in many clients, including Windows and MAC clients, do support EAP-TLS.

Both the supplicant and the RADIUS server must have a valid CA-signed certificate installed on them. For enterprise deployments, you will typically have a private CA issue these certificates so that you can better control which devices can connect.

The supplicant and RADIUS server must both also trust the CA certificate that signed the peer's certificate. The supplicant should also be configured with specific trusted RADIUS server subject names. The supplicant will check the subject name in the certificate submitted by a RADIUS server, and only continue the authentication process if the server's certificate has the correct subject name and is signed by a trusted CA. This is an important step that admins sometimes neglect to consider. Sometimes admins configure supplicants to ignore the certificate, to accept any valid certificate regardless of the subject name, or to prompt the user to authorize new certificates. These actions can allow clients to connect to rogue RADIUS servers, which then launch MITM attacks.

The RADIUS server also needs to trust the root CA certificate for supplicant certificates. It further requires policies that authorize users for access based on the subject names in their certificates.

Note that the authenticator does not need to support any particular requirements, or even the EAP-TLS method; it simply shuttles the messages between the supplicant and authentication server.

EAP-TLS Configuration on Windows Client

The steps below explain how to configure EAP-TLS on a NIC on a Windows client for your reference. The steps might differ slightly depending on the OS version, but should be similar to the ones provided below. In a Windows domain, GPOs or other mechanisms will often dictate these settings for all domain computers, so you do not have to configure them on clients on by one.

1. Open the Network Connections window (**Control Panel > Network and Internet > Network Connections**).
2. Right-click the NIC that you want to configure and select Properties. (See Figure 6-15.)

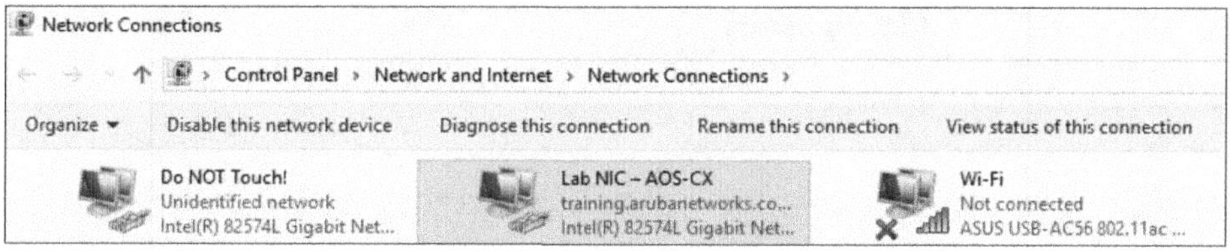

Figure 6-15: EAP-TLS Configuration on Windows Client

3. Click the **Authentication** tab. (See Figure 6-16.)
4. Select the **Enable IEEE 802.1X authentication** check box.
5. To select EAP-TLS, choose **Microsoft: Smart Card or other certificate**.
6. Click **Settings**.

Figure 6-16: EAP-TLS Configuration on Windows Client

7. The Smart Card or other Certificate Properties window is displayed. (See Figure 6-17).

8. Choose **Use a certificate on this computer**. (Alternatively, if your users have smart cards with certificates on them, select **Use my smart card**.)

9. It is important to configure the supplicant to authenticate the RADIUS server to prevent MITM attacks. To do so, follow these steps:

    a. Select the **Verify the server's identity by validating the certificate** check box.

    b. Select the **Connect to these servers** checkbox and specify the subject name in the RADIUS server's RADIUS/EAP certificate. In the example in Figure 6-17, the RADIUS server's name is clearpass.training.arubanetworks.com.

    c. Choose the specific root CA that signed the RADIUS server's certificate.

d. Make sure that the **Don't prompt user to authorize new servers or trusted certificate authorities** check box is selected. Selecting this check box can protect users who might click past errors and warnings.

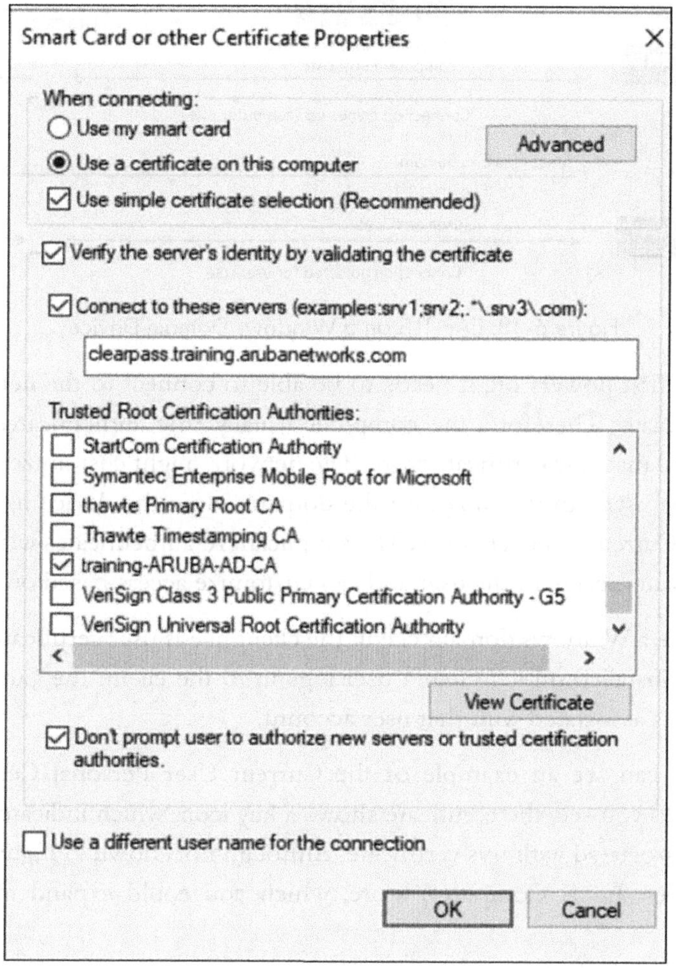

Figure 6-17: EAP-TLS Configuration on Windows Client

10. Click **OK** and **OK** again until you have closed the window.

EAP-TLS on a Windows Domain Device

You will now explore a simple example of a device that might need to authenticate to the network with EAP-TLS: a user's computer, which is joined to a Windows domain. (See Figure 6-18.) The Windows device has two certificate stores, one stores certificates for the local machine and the other stores certificates specific to a user account on that machine.

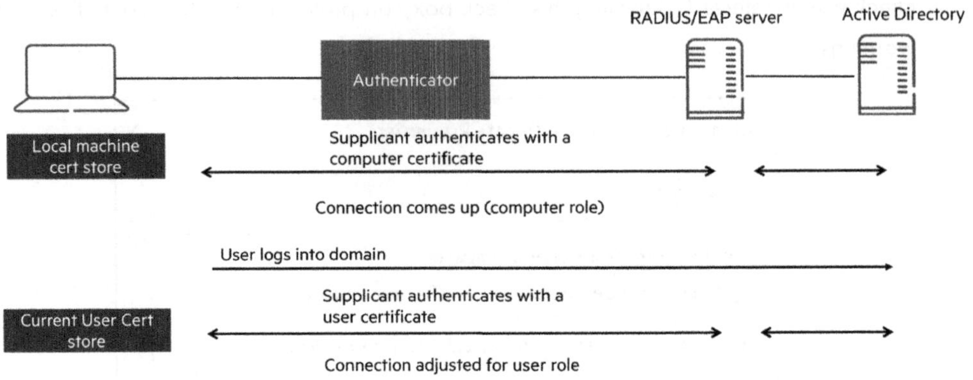

Figure 6-18: EAP-TLS on a Windows Domain Device

When the computer first powers on, it needs to be able to connect to the network so that a user can log into the domain. Therefore, the computer usually first authenticates using a computer certificate in the Local machine certificate store. The network might customize the connection with a computer role. The user can then log into the domain. Now the device has access to the user certificate inside the Current User cert store. The supplicant re-authenticates with the user certificate. The network now grants access to the user and can customize access based on the user role.

If multiple users share a Windows domain client, the client needs user certificates for each user that might use the client installed on it. When a user logs in to the client, the Current User store will include the certificates associated with that user account.

In Figure 6-19, you can see an example of the Current User Personal Certificates store on a Windows machine. As you see, the certificate shows a key icon, which indicates that the computer has the private key associated with this certificate. Although not shown in Figure 6-19, the machine also has a Certificates (Local Computer) store, which you could expand to see the computer certificates.

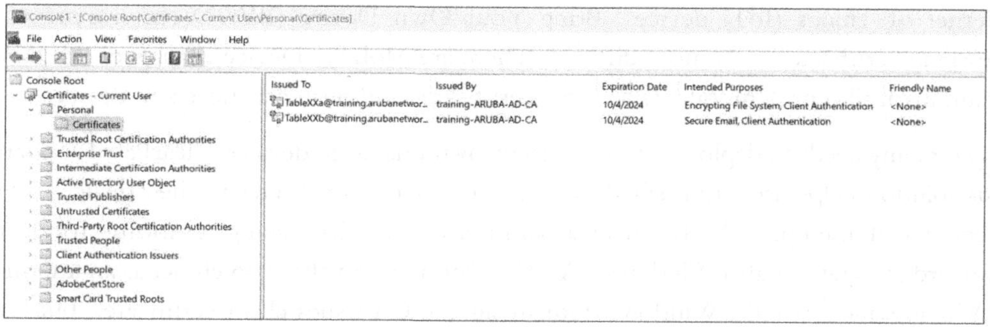

Figure 6-19: Certificates Store on a Windows Machine

Some Options for Deploying Certificates on User Devices

As you have learned, ensuring that devices have the correct machine and user certificates can be challenging. Maintaining a large number of certificates requires specialized software and trained IT staff. In addition, in environments that have multiple operating systems, installing certificates on all clients can be difficult. There are many options for deploying and managing certificates. You see just a few of those options outlined in Figure 6-20.

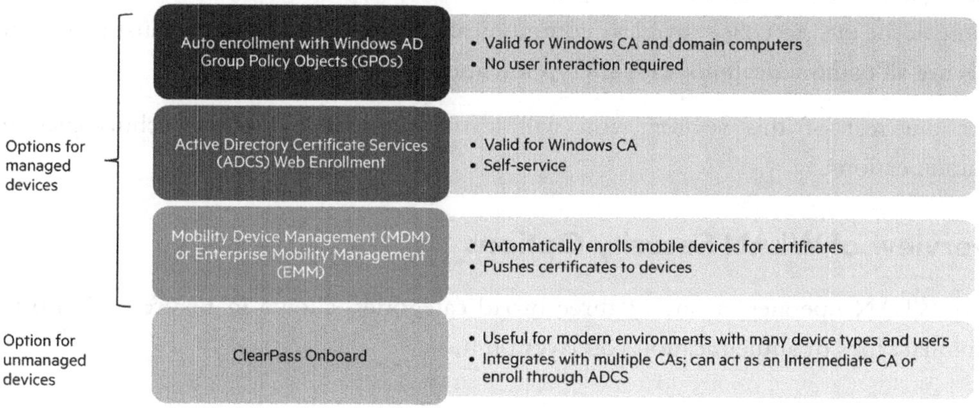

Figure 6-20: Some Options for Deploying Certificates on User Devices

Typically, customers set up a private CA to issue the certificates required for EAP-TLS. Microsoft Windows CA is a common option for Windows domains that use Active Directory (AD). AD administrators can set up group policy objects (GPOs) to automatically push down certificates to devices joined to the domain. The Windows CA also supports Active Directory Certificate Services (ADCS) Web Enrollment, which enables users to obtain certificates through a self-service portal.

However, the Windows CA options are geared for a world in which the only devices are computers joined to the domain. A modern network has far more types of devices, including smartphones and

Internet of Things (IoT) devices. Bring Your Own Device (BYOD) environments can make managing certificates even more difficult. There are Mobility Device Managers (MDMs) that help to automatically enroll mobile devices for certificates and push certificates to devices.

If a company needs to deploy certificates to many unmanaged devices, ClearPass Onboard can help. This solution helps users to easily download the correct certificates to their devices. Onboard can integrate with multiple CAs. It can act as an Intermediate CA, issuing certificates for devices. When Onboard integrates with a Windows CA PKI, Onboard can alternatively act as an intermediary with ADCS, in which case the Windows domain signing CA issues client certificates, but the Onboard client makes the process of obtaining the certificate transparent to the user.

WLAN Security

You will now learn about WLAN security technologies.

Need for Secure Wireless Communications

Let's briefly review what you learned about secure communications in Chapter 2. Communicating parties want to know that their communications are private and authentic. Because it is trivial for anyone in the area to collect wireless communications, wireless networks must take extra measures to secure all communications with encryption and authentication.

Over the rest of this section, you will learn about the wireless technologies that secure communications.

Overview of WLAN Security Options

Every WLAN operates in one of three broad categories, shown in Figure 6-21, no matter what additional types of authentication you layer on it.

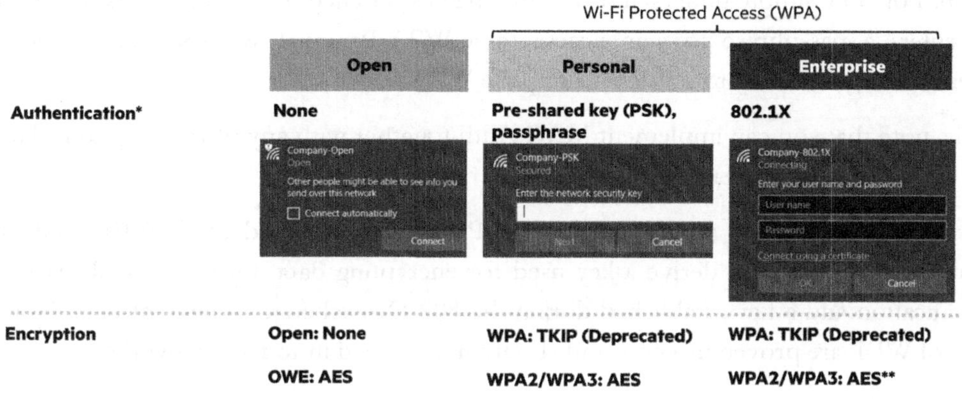

*Option to add MAC-Auth
**Different AES key lengths and modes of operation in WPA2 and WPA3
WEP = Deprecated legacy option not covered in this course

Figure 6-21: Overview of WLAN Security Options

802.11 wireless LANs (WLANs) can be open, in which case users do not have to submit authentication credentials to be allowed to connect to the network. When the client attempts to associate with the WLAN, the AP simply permits the association. Traditionally, open networks do not provide encryption. However, Opportunistic Wireless Encryption (OWE) now provides Advanced Encryption Standard (AES) encryption for open networks. You will learn more about OWE later in this section.

Alternatively, the WLAN is secured with Wi-Fi Protected Access (WPA). WPA comes in two flavors. WPA-Personal requires users to enter a passphrase, which is a shared secret for all users who are authorized to connect to the WLAN. The passphrase is sometimes called a pre-shared key (PSK) because it is used as the foundation for encrypting and authenticating traffic in the WLAN.

WPA-Enterprise, the recommended option for enterprises, uses 802.1X. With 802.1X, a client authenticates to a backend authentication server using individual credentials, which could be a username and password or a digital certificate.

Both WPA-Personal and WPA-Enterprise provide encryption as well as authentication. WPA comes in three versions, and the version determines the supported encryption algorithm. WPA originally used Temporal Key Integrity Protocol (TKIP), which was based on RC4 encryption. WPA and TKIP are now deprecated. WPA2 and WPA3 both support AES, although, as you will learn, WPA3 can impose different requirements for the AES key length and mode of operation. Note that later versions can also have backward compatibility with earlier versions. For example, a WPA2 WLAN could permit clients to use TKIP or AES; however, for security, it *should* only allow AES.

The WPA versions have additional differences, which you will explore in detail throughout this section. For now simply understand that the user experience for all three versions is similar. The user enters a passphrase to gain access to a WPA-Personal WLAN, but submits individual credentials to gain access to a WPA-Enterprise WLAN.

Finally, note that you can implement MAC-Auth together with any of these options. However, the main focus for this chapter is the base security options.

Wired Equivalent Privacy (WEP) was a pre-WPA security standard. Static WEP used a passphrase to authenticate users and derive a key used for encrypting data. Dynamic WEP used 802.1X for authentication and a key established during the 802.1X exchange to encrypt data. However, both forms of WEP are proven insecure and should not be used in today's networks.

Overview of 802.11 Connection Steps

Before you look at each security option in more detail, it will be useful for you to review the steps in establishing an 802.11 connection. These steps are illustrated in Figure 6-22.

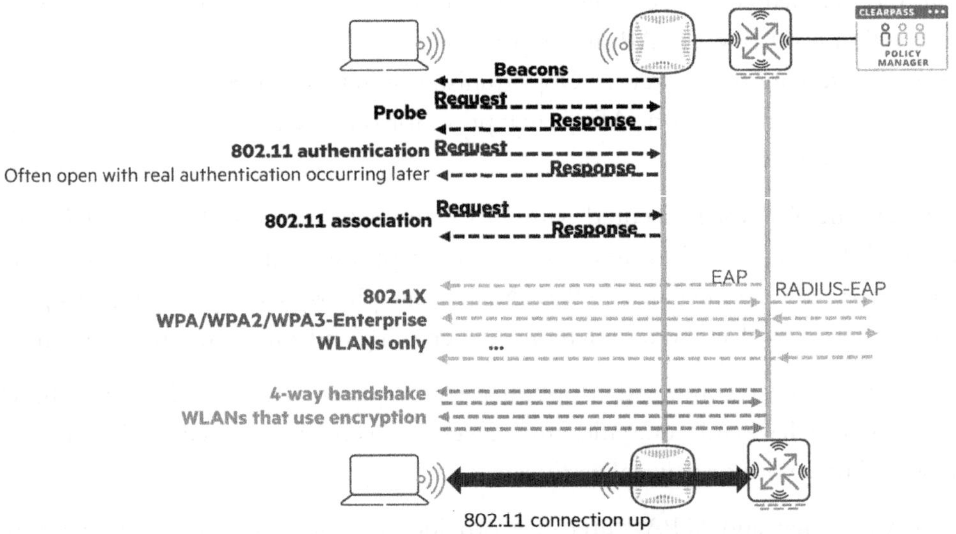

Figure 6-22: Overview of 802.11 Connection Steps

An AP periodically sends out beacons to advertise the WLANs that it supports. The beacons include information about the network name, or SSID, and the security options. A client typically also sends a probe request for the specific SSID to which it wants to connect. If the AP supports that SSID with compatible options, it will send a probe response.

The client then sends an 802.11 authentication request. 802.11 authentication most often operates in open mode, which means that the AP permits all clients to pass 802.11 authentication. The real authentication will occur later.

Next 802.11 association occurs. The client sends an association request, and the AP responds.

If the WLAN is using Enterprise mode, 802.1X authentication follows the 802.11 association. When a WLAN in a controller-based solution uses tunnel mode, the MC is the 802.1X authenticator. As you learned earlier, the 802.1X process can involve many packets, but in the end the client is authenticated.

The final step for all WLANs that use encryption is an exchange called the 4-way handshake. The client and AP/MC use this handshake to derive the keys used to secure communications; this handshake might also provide implicit authentication, as you will see.

After the handshake, the 802.11 connection is up, and the client can send and receive data on it.

WPA Key Hierarchy

To understand the 4-way handshake and data encryption/authentication keys used for all WLANs that use encryption, you need to understand the WPA key hierarchy. At the top of the hierarchy is the Pairwise Master Key (PMK), as shown in Figure 6-23. The PMK itself does not encrypt or decrypt data. Instead, wireless devices derive a Pairwise Temporal Key (PTK) from the PMK. In addition to the PMK, the PTK takes as inputs a random number used once (nonce) from the AP, a random nonce from the client, the AP's MAC, and the station's MAC. The ability to derive different PTKs from the PMK at different times keeps encryption keys from being reused too much.

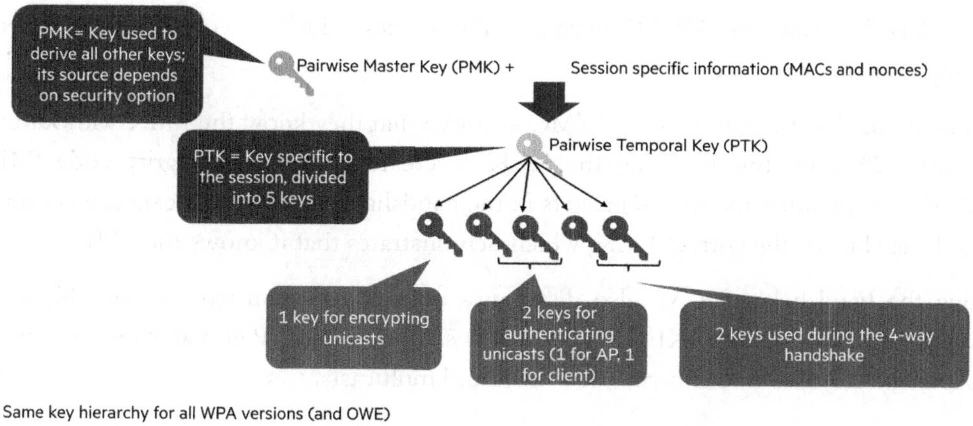

Figure 6-23: WPA Key Hierarchy

Wireless devices then derive five more keys from the PTK. You do not need to memorize the names of these keys, but you should understand their basic purpose:

- Temporal Key (TK), which the client and AP will use to encrypt unicasts in this session
- MIC transmit key, which the AP will use to provide integrity for data that it transmits in this session (client will use this key to check the integrity)
- MIC receive key, which the client will use to authenticate data that it transmits in this session (AP will use this key to check the integrity)
- Key Confirmation Key (KCK), used during the 4-way handshake
- Key Encryption Key (KEK), used during the 4-way handshake

WPA Four-Way Handshake

The four-way handshake, illustrated in Figure 6-24, provides several functions.

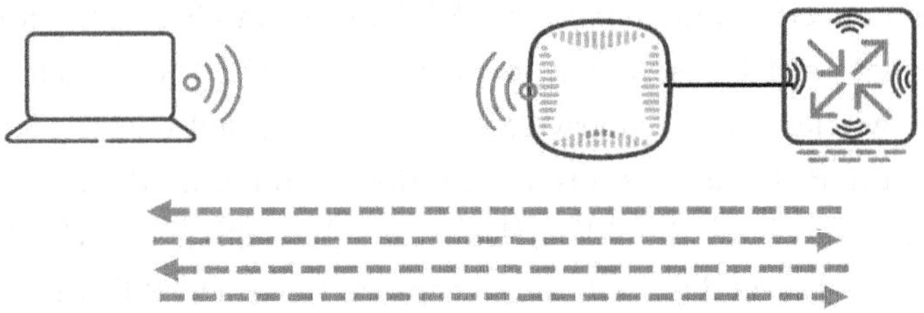

Figure 6-24: WPA Four-Way Handshake

First, it lets the client and AP/MC exchange the nonces, which, as you learned, are necessary for deriving the PTK from the PMK.

Second, it enables the station and AP/MC to prove that they know the PMK without transmitting the PMK. They do this by using the KCK to create a message integrity code (MIC), which authenticates the third and fourth packets in the handshake. The authentication will only check out if the device knows the correct KCK, which demonstrates that it knows the PMK.

During the handshake, the AP also distributes a groupwise transient key (GTK) to the client, encrypting this key with the KEK. The GTK is shared by the AP and all clients connected to the WLAN. The AP uses it to encrypt broadcasts and multicasts.

Finally, the handshake enables the AP and client to agree on the first initialization vector (IV) for their encrypted messages. Remember from Chapter 2 that encryption algorithms like AES use IVs to help prevent key reuse.

Krack: Attack on the Handshake

While this four-way handshake worked well for many years, it had one fatal flaw that was detected by Manthy Vanhoef in 2017: the Krack attack. The standard did not protect stations from replay attacks. Stations were supposed to accept any handshake messages from the AP that had a valid MIC even if the connection was ongoing. Let's look at the issue that this behavior can create.

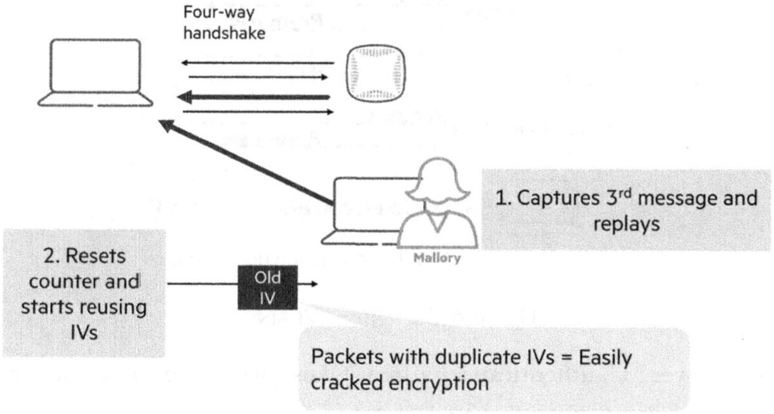

Figure 6-25: Krack: Attack on the Handshake

Mallory captures valid handshake packets between an AP and station, as shown in Figure 6-25. She replays the third packet in the handshake to the station.

This packet has a random counter that sets the initial value for the IV. When Mallory replays the packet, the station starts reusing IVs. As you learned in Chapter 2, reusing IVs with the same key opens major holes in encryption, so Mallory can start decrypting traffic. With some operating systems, the behavior was worse. Instead of installing the old IV, the reused counter triggered the OS to set the key to 0. Now the station is transmitting data in plaintext.

After the Krack attack was published, vendors created patches. Patched clients drop handshake messages that use a previously used counter, protecting them from the attack.

Open WLANs

With this brief background in wireless encryption, you are ready to look at the three categories of security in more detail. Begin with open WLANs.

An open WLAN simply allows clients to connect. As shown in Figure 6-26, the client passes through 802.11 open-system authentication, which permits all clients, and then 802.11 association. No further steps occur; the 802.11 connection is up, and the client can receive an IP address and start communicating—in plaintext.

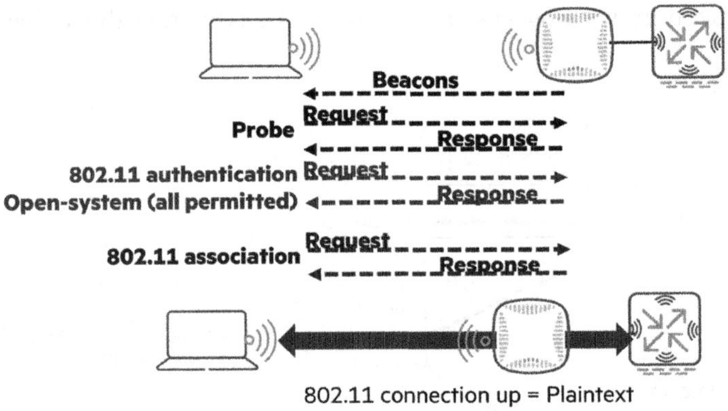

Figure 6-26: Open WLANs

After that, though, Layer 3 authentication often takes place. For example, companies often implement captive portal authentication on open networks.

Companies also sometimes combine open WLANs with MAC-Auth to provide minimal authentication for devices with no user interface or 802.1X support.

Opportunistic Wireless Encryption (OWE)

Now you will examine how Opportunistic Wireless Encryption (OWE), defined in RFC 8110, enhances open WLANs. OWE is not technically part of WPA, but instead rolled into the 802.11 standard. However, OWE uses WPA concepts like the key hierarchy and 4-way handshake.

An AP's beacons for an OWE WLAN indicate that it supports OWE.

When a station connects to the WLAN, it passes through open-system authentication as usual, as shown in Figure 6-27. But the 802.11 association frames include extra TLVs, which carry an unauthenticated Diffie-Hellman exchange to derive a PMK. The station and AP then conduct the four-way handshake to derive all the keys for encrypting and authenticating future data. From the user viewpoint, an OWE SSID works like earlier forms of open SSIDs. The user clicks Connect, and the connection happens without the user having to submit any credentials. However, the traffic is encrypted and authenticated.

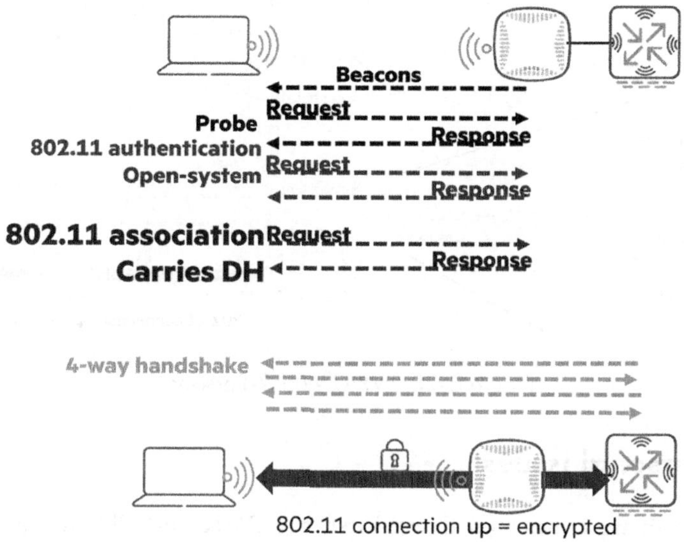

Figure 6-27: Opportunistic Wireless Encryption (OWE)

OWE provides best effort security—not true security. Remember that unauthenticated Diffie-Hellman exchanges are subject to MITM attacks, so a savvy hacker can still use a honeypot AP to lure clients to connect to it and generate a PMK with it. However, OWE does protect traffic against more casual snooping. Even though anyone can connect, the PMK for every session is unique and protected.

WPA/WPA2-Personal

You will next look at WPA/WPA2-Personal WLANs.

As shown in Figure 6-28, a client that wants to connect to a WPA/WPA2-Personal WLAN passes through open-system 802.11 authentication and then 802.11 association. It then conducts the 4-way handshake. Remember that part of the 4-way handshake is proving knowledge of the PMK. For WPA/WPA2-Personal, the PMK is expanded out in a predictable way from the passphrase configured for the WLAN. Thus the 4-way handshake proves that the client and AP are using the same passphrase, providing implicit authentication.

If the 4-way handshake is successful, the connection comes up.

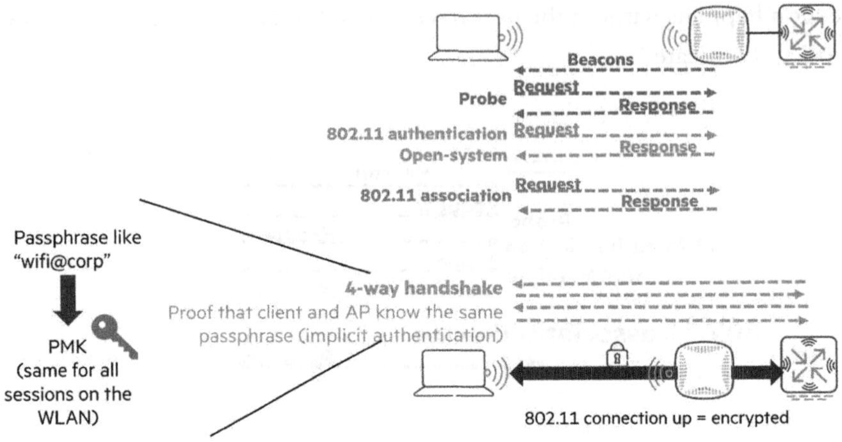

Figure 6-28: WPA/WPA2-Personal

WPA/WPA2-Personal Issues

WPA2-PSK has two main, connected flaws. First, the PMK and all other keys ultimately derive from a human-selected passphrase. Humans rarely select good passphrases because they cannot remember good passphrases. For this reason it is important that any key that derives from a passphrase cannot be subject to a brute force dictionary attack. However, WPA2-PSK is.

Mallory can capture a valid handshake between a station that knows the password and the AP. Two of the handshake messages includes a MIC, which is a hash of the plaintext message with the KCK (one of the keys derived from the PMK). Mallory takes this data offline and uses the massive parallel processing power now available with GPUs and hacking farms to simply try every possible passphrase until one produces keys that creates the right MIC.

Even if the passphrase is not a common word and includes special characters, if it is 8 characters or under, processing power is high enough now that brute force attacks can relatively quickly crack them. While companies could attempt to use very long and random strings, doing so is not feasible in practice.

Another issue with WPA2-Personal is that the password is often freely shared. While every client technically has its own PTK, in practice, anyone can easily derive another client's PTK. You just need the client's and AP's MAC addresses and the nonce values which are transmitted in plaintext during the handshake. This means that anyone who knows the passphrase can decrypt other devices' traffic. While you could hope that no one trusted with the passphrase would do this, that is often not the case.

WPA3-Personal

Instead of using a passphrase as the source of a PMK, WPA3-Personal uses Simultaneous Authentication of Equals (SAE) to generate a different PMK each time that a station connects.

The AP advertises SAE as the authentication method for a WPA3-Personal SSID in its beacon and probe response frames. This is a new 802.11 authentication method (value 3); WPA3-Personal is the only security option that uses an 802.11 authentication method other than open-system. (Shared-key is a deprecated option that is insecure and should never be used.)

The 802.11 SAE process involves four frames, as shown in Figure 6-29. You do not need to know the precise exchange but should understand the high-level concepts. SAE is based on the Dragonfly Key Exchange defined in IETF RFC 7664. This key exchange process uses some of the same cryptographic concepts as Diffie-Hellman and can use either Finite Field Cryptography (FFC) or Elliptic Curve Cryptography (ECC). HPE Aruba Networking mobility devices use group 19, a 256-bit Elliptic Curve group. The added twist, distinguishing SAE from Diffie-Hellman, is that the peers can only derive the same shared key—the PMK—if they also know the same passphrase.

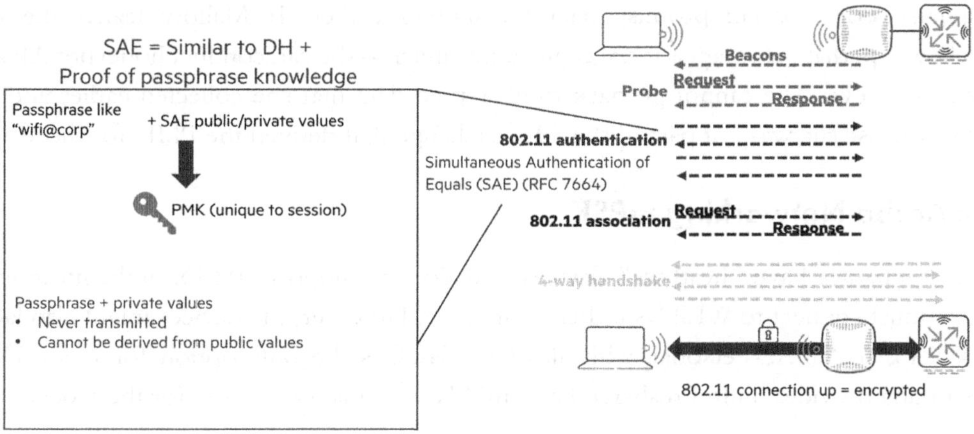

Figure 6-29: WPA3-Personal

After going through SAE, the client and AP have authenticated each other and derived a PMK that is unique to the session.

The client associates. The familiar 4-way handshake then occurs.

WPA3-Personal Benefits

WPA3-Personal offers many improvements over WPA/WPA2-Personal.

The SAE exchange is not subject to offline dictionary attacks because it is computationally infeasible to derive the passphrase from the values transmitted in the SAE exchange. While Mallory can still take the 4-way handshake offline, the PMK is no longer directly derived from the passphrase, so it is more resistant to a dictionary attack. And cracking the PMK will not be useful because it is different for each session. If Mallory wants to launch a dictionary attack, she must actually go through the SAE process and handshake for each guess of the passphrase, which makes the attack infeasible. Therefore, admins can choose weaker passphrases while still maintaining good security. In fact, SAE is designed to work with a large pool of passphrases known to hackers. (However, it is still best practice to choose good passphrases.)

WPA3-Personal also protects users' communications from eavesdropping by other authorized users. Every session has a unique PMK, known only to the client and AP/MC that established the session.

WPA3-Personal further provides forward secrecy, which means that a future compromise of the passphrase does *not* compromise data transmitted earlier. If Mallory learns the passphrase somehow—perhaps through a social engineering attack—she can conduct honeypot AP and MITM attacks. However, she cannot go back to encrypted data that she collected earlier and decrypt it. This is because she was not part of the SAE exchange that derived the PMK for those exchanges.

HPE Aruba Networking mPSK

Some companies have "headless" devices that do not support 802.1X authentication, so these devices must connect to WLANs in Personal mode. However, the devices might also lack support for WPA3, so WPA2-Personal, with all of its flaws, is the only option for them. HPE Aruba Networking provides multi Preshared keys (mPSK) to enhance security for these devices.

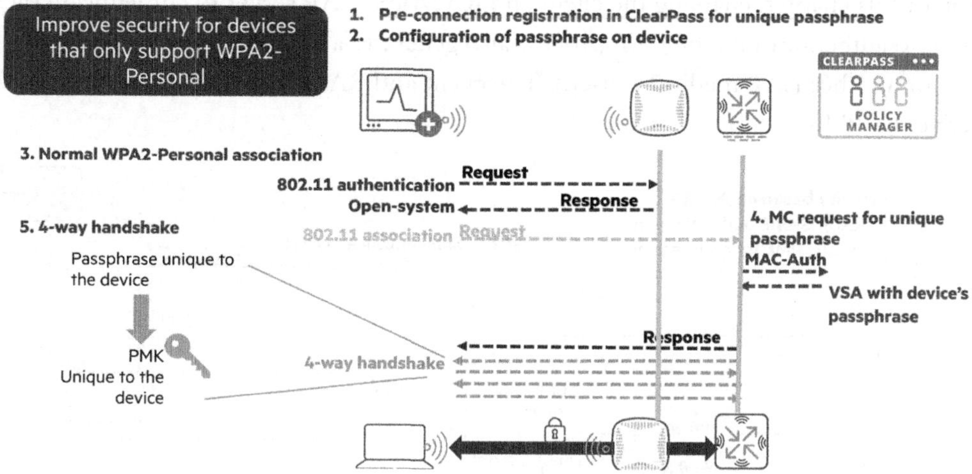

Figure 6-30: HPE Aruba Networking mPSK

A detailed explanation of mPSK configuration is beyond the scope of this study guide. At a high level, the solution typically works as follows (see also Figure 6-30):

1. Before attempting to connect the headless device, a user connects to the ClearPass Self Service Registration page on a different connection. In that page, the user authenticates and specifies the device's MAC address. ClearPass assigns the device a passphrase, or PSK, and displays that passphrase to the user.
2. The user then configures the client device to use the passphrase.
3. The client can now connect to the WPA2-Personal WLAN. From the client's point of view the connection proceeds like a typical WPA2-Personal connection.
4. However, at the point at which the client sends its 802.11 association request, the IAP or MC performs MAC-Auth to CPPM. CPPM responds with the passphrase assigned to the client in an Aruba-MPSK-Passphrase VSA.
5. The HPE Aruba Networking device then sends the 802.11 association response and performs the 4-way handshake using the PMK derived from the passphrase individual to the client device.

WPA/WPA2-Enterprise

You will now move on to WPA/WPA2-Enterprise.

A connection to a WLAN that enforces this form of security begins with open-system 802.11 authentication and an 802.11 association, as shown in Figure 6-31. 802.1X then shuts down the connection to all but EAP traffic. EAP authentication proceeds as you examined in the previous section of this chapter, enabling the client and RADIUS/EAP server to authenticate each other. In addition to authentication, the EAP process also generates a TLS master secret (whether EAP-TLS, PEAP, or another EAP method is used). The client and RADIUS/EAP server each use this secret to derive the PMK.

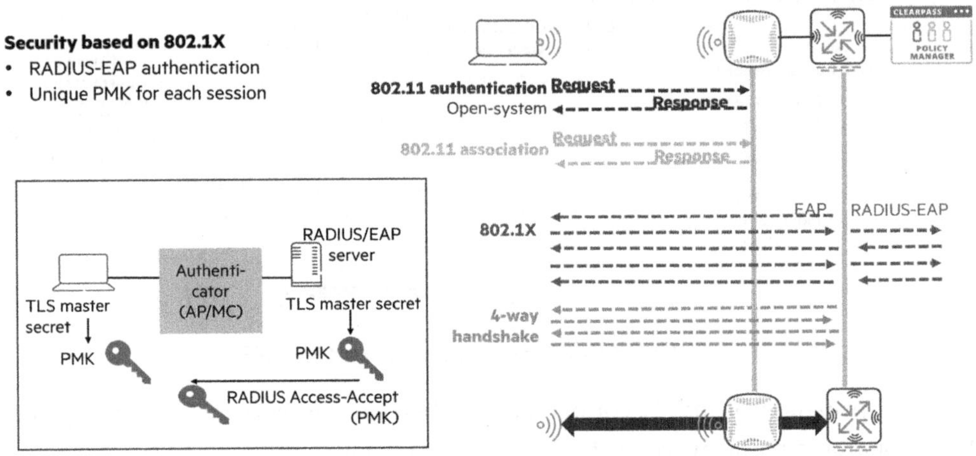

Figure 6-31: WPA/WPA2-Enterprise

Because it is the authenticator—the AP or MC—that needs the PMK for encrypting and decrypting wireless traffic, the RADIUS server sends the PMK to the authenticator in the RADIUS Access-Accept message. It encrypts the PMK with the RADIUS shared secret. Now both the client and the authenticator know the PMK for that session.

The client and AP then complete the 4-way handshake, and the connection is up.

WPA/WPA2-Enterprise Considerations

WPA/WPA2-Enterprise is not subject to dictionary attacks on the PMK because the PMK derives from the TLS master secret, which is long and more or less random. If a hacker takes the 4-way handshake offline, the hacker would need to launch a brute force attack on the 256-bit PMK, which is not computationally feasible. In addition, cracking the PMK gives the hacker less benefit because it would crack the PMK for one session only. By enforcing reauthentication periods, a solution can ensure that a client's PMK is renewed periodically.

Also keep in mind that simply by using 802.1X authentication, WPA/WPA2-Enterprise is generally more secure than *all* versions of WPA-Personal, including WPA3-Personal. 802.1X authenticates individual clients, so the solution can track and control clients individually based on the user and device identity. WPA3-Personal also remains susceptible to passphrase leaks, but with 802.1X, the most that individual users can do is expose their own credentials.

The main security issue in WPA/WPA2-Enterprise is that the security of the PMK rests on the security of the TLS exchange. This could be very secure. However, TLS 1.2, which is what EAP uses, does have some flaws. Highest on the list is that it supports a wide variety of key signing, encryption, and authentication algorithms from which servers and clients can choose; some of these algorithms are outdated and insecure. If the server and client agree on insecure algorithms that hackers can exploit, the TLS master secret is a risk and so is the PMK. While it is possible on some devices to lock down the set of supported algorithms to only secure ones, very few administrators struggle through the complexities of doing so.

In addition, WPA2 uses AES Counter CMAC Protocol (CCMP), which is no longer the fastest data encryption/authentication option. AES CCMP, properly implemented with securely derived keys and no repeat IVs, remains secure. However, the algorithm is aging. It is time for longer key lengths and a more modern, faster mode of operation.

WPA3-Enterprise

WPA3-Enterprise continues to work much like WPA2-Enterprise. Stations use 802.1X with EAP to authenticate, and the PMK is derived during the exchange. However, WPA3-Enterprise has added support for new data encryption and authentication algorithms. WPA2-Enterprise used AES CCMP with a 128-bit key for encryption and a 64-bit key for ensuring integrity. WPA3-Enterprise still supports this mode, but also supports AES-GCM-256, which simultaneously encrypts and ensures integrity for data with 256-bit keys.

The key new feature is Commercial National Security Algorithm (CNSA) mode. When an AP broadcasts this operation mode, it forces devices to commit to using the newer, more secure algorithms to secure data after they connect. In addition, the station must use secure TLS options during the EAP exchange too. CNSA locks down the TLS options to these:

- TLS_ECDHE_ECDSA_WITH_AES_256_GCM_SHA384 (mandatory)
- TLS_ECDHE_RSA_WITH_AES_256_GCM_SHA384 (optional)
- TLS_DHE_RSA_WITH_AES_256_GCM_SHA384 (optional)

The PMK generated during the key exchange must be at least 384-bits.

CNSA mode further specifies 256-bit Galois/Counter Mode Protocol (AES-GCMP-256) to secure unicasts and to secure group-addressed (broadcast/multicast) data sent between APs and stations.

APs secure group addressed management frames with 256-bit Broadcast/Multicast Integrity Protocol Galois Message Authentication Code (BIP-GMAC-256).

The AP that implements WPA3-Enterprise CNSA mode advertises these requirements in beacons, probe responses, and 802.11 association responses:

- AKM Suite Selector as 00-0F-AC:12
- Pairwise Cipher Suite Selector as 00-0F-AC:9
- Group data cipher suite selector as 00-0F-AC:9
- Group management cipher suite (MFP) selector as 00-0F-AC:12

A station can only complete authentication and connect if it supports all of these suites. Note that the AP is the other side for negotiating the suites. However, the authentication server is the other side for EAP-TLS. For this reason, HPE Aruba Networking mobility devices signal to CPPM that it must use Suite-B compatible EAP-TLS.

Review of WPA3 + OWE Enhancements

Use Figure 6-32 to review what you have learned throughout this section.

OWE, WPA3-Personal, and WPA3-Enterprise provide a similar user experience to an open network, WPA2-Personal, and WPA2-Enterprise. However, the emerging standards improve security.

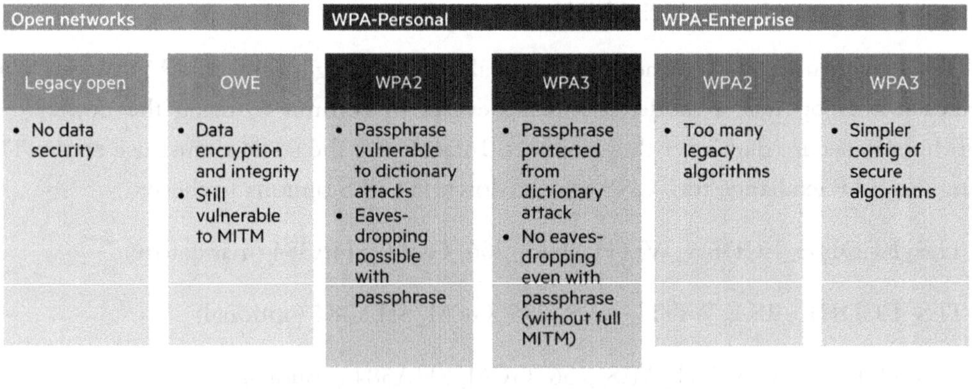

Figure 6-32: Review of WPA3 + OWE Enhancements

OWE provides best effort protection for communications. The station and AP use unauthenticated Diffie-Hellman to create a shared secret to encrypt and authenticate communications. However, because the Diffie-Hellman exchange features no data signing, a hacker could still use a rogue AP to launch an MITM attack.

WPA3-Personal protects passphrases from dictionary attack and creates a unique PMK for every client session. Even if a hacker learns a passphrase, that hacker cannot decrypt other client's communications without launching a full MITM attack.

WPA3-Enterprise adds support for new encryption modes and longer keys. When run in CNSA mode, it makes it simple for admins to ensure that devices use secure algorithms during the 802.1X EAP process and after the establishment of the 802.11 connection.

Additional Security with Management Frame Protection (MFP)

You will conclude this section by examining an additional layer of protection for 802.11 management frames. 802.11 uses management frames to regulate sessions and manage the cell. If hackers can forge these frames, they can cause trouble for clients. As one simple example, a hacker can send a disassociation frame to disconnect a client from the WLAN, causing a DoS attack on that client.

To protect clients from such attacks, WPA2 and WPA3 can enforce Management Frame Protection (MFP), also called Protected Management Frames (PMF), a feature based on 802.11w. This protocol protects the management frames listed in Figure 6-33 from being forged by unauthorized devices. Unicast management frames are further protected from eavesdropping. The management frames are protected by the same keys that protect data frames.

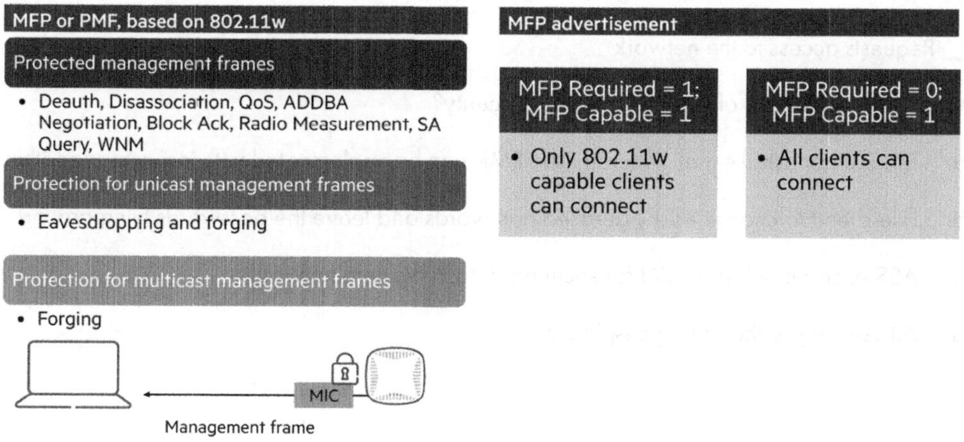

Figure 6-33: Additional Security with Management Frame Protection (MFP)

An AP that supports PMF advertises the capability in the RSN information element (IE) in beacons and probe responses. (In RSN IE, PMF is referred to as MPF.) The AP can indicate whether the

SSID is MFP capable and MFP required, in which case only 802.11w capable clients can connect, or MFP capable but not MFP required, in which case a mix of 802.11w and non-802.11w clients can connect.

Summary

You now understand the technologies that secure wired and wireless networks. Over the next chapters, you will learn how to put these technologies into action to protect your network.

Learning checks

1. Match each role to the actions associated with that role.

   Name of role

   a. Supplicant

   b. Authenticator

   c. Authentication server

   Actions associated with role

   __ Grants access to the network

   __ Determines whether or not to grant access

   __ Requests access to the network

2. What is the main flaw of WPA2-Enterprise security?

   a. The TLS handshake that generates the PMK can be implemented with insecure algorithms.

   b. Users tend to choose easily guessed passwords and leave the passwords lying around.

   c. AES encryption is not valid for securing data in modern networks.

   d. All users share the same passphrase.

3. Match the security enhancement to the security mode to which it applies.

   Security enhancement

   a. Unauthenticated Diffie-Hellman to negotiate a PMK

   b. Simultaneous authentication of Equals (SAE), which protects weak passwords from attack

   c. CNSA mode to ensure secure algorithms

   Security mode

   __ WPA3-Enterprise

   __ WPA3-Personal

   __ OWE

4. Consider these questions about implementing 802.1X with EAP-TLS.

   - Assume that decision makers at your company have asked you about certificate-based versus password-based authentication for the wireless and/or wired networks. They want to know if certificates are really more secure. What would you explain?
   - Assume that your company has decided to implement certificate-based authentication with EAP-TLS. What would you take into account as you make a plan for implementing the new security method?

Note

For answers to all learning checks, refer to Appendix B.

---

7 Edge Security with HPE Aruba Networking

EXAM OBJECTIVES

✓ Deploy WPA3-Enterprise to secure wireless networks

✓ Deploy 802.1X to secure the wired edge

Assumed Knowledge

- Basic IP addressing
- Basics of Ethernet switching, including VLAN operation
- Basics of 802.11 wireless networks

Overview

An HPE Aruba Networking network infrastructure provides the ideal solution for implementing the technologies that you examined in the previous chapter.

This chapter begins with a scenario introduction. You will then learn how to enforce WPA3 and 802.1X over a wireless network. Next, you will learn how to enforce 802.1X on a wired network. Finally, you will complete another lab applying 802.1X on your AOS-CX switches.

Introduction

This section introduces the scenario.

First Steps in Protecting the Edge

In this scenario, outlined in Figure 7-1, the company wants to begin enforcing 802.1X with EAP-TLS authentication for both wired and wireless access. The eventual goal is to customize access based on user identity and other factors. First, though, you will be focusing on getting authentication working. So, at this point all authenticated users will receive full access to the network. In addition, switches will forward authenticated user traffic without any tunneling to HPE Aruba Networking Mobility Controllers (MCs).

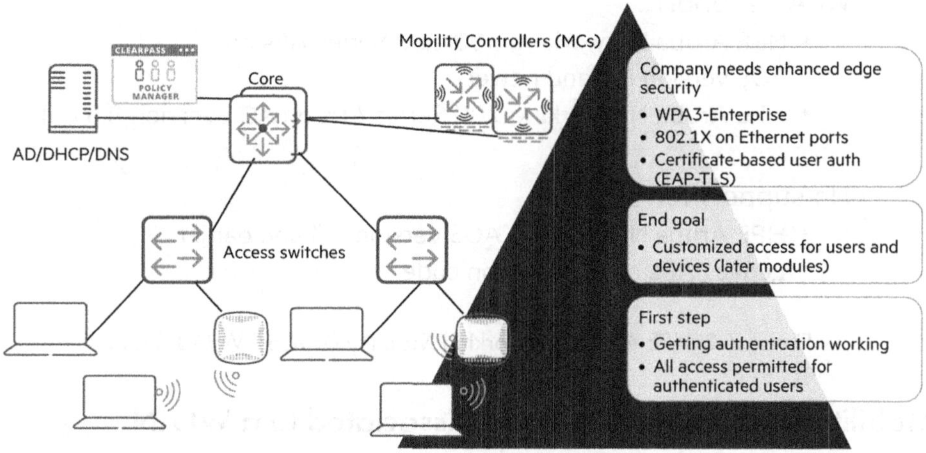

Figure 7-1: First Steps in Protecting the Edge

Throughout this chapter, you will learn how to apply this level of security with HPE Aruba Networking mobility devices and AOS-CX switches.

Enforce WPA3- Enterprise

You will now learn how to enforce WPA3-Enterprise.

HPE Aruba Networking Mobility Devices' WPA3 Support

Before you do so, you should understand how the deployment's software and hardware affects support for WPA3. In version 8.3 and earlier, HPE Aruba Networking Wireless Operating System

(AOS) did not support WPA3, which was not yet a standard when those versions were released. AOS 8.4 added WPA3 support. In this version and version 8.5, 3xx and 5xx Series APs can implement WPA3 operating modes, but 9x, 1xx, and 2xx Series APs cannot. If you set up a WLAN that uses WPA3 on an MC or IAP cluster with a mix of AP models, the 3xx and 5xx Series APs will implement WPA3 while the 2xx Series APs will implement the corresponding WPA2 modes. For example, instead of implementing WPA3-Personal with Simultaneous Authentication of Equals (SAE), the 2xx Series AP will implement WPA2-Personal.

AOS version 8.6 added support for WPA3 on 2xx Series APs. Legacy APs (prior to 2xx Series) continue to support WPA2 only. Again, if you deploy a WPA3 WLAN on APs with a mix of capabilities, the WPA2-only APs will implement the corresponding WPA2 modes.

These points are summarized in Figure 7-2.

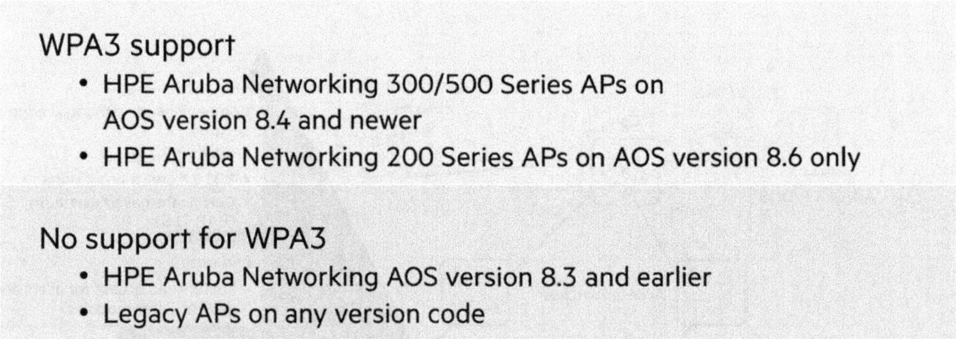

Figure 7-2: HPE Aruba Networking Mobility Devices' WPA3 Support

How Mobility Devices Handle Users Associated to a WLAN

To understand how to deploy a WLAN, you need to understand just a little about the concept of AOS roles because these roles affect how the AOS firewall treats every client that associates to a WLAN.

Whenever a client associates to a WLAN, the AOS firewall assigns the client a role and a VLAN.

The role is where the power of AOS really reveals itself. Based on the individual client's role, the AOS firewall will treat that client's traffic differently from other clients in the same WLAN or VLAN. The role determines which firewall access control policies apply to the user traffic. It can also impose bandwidth contracts to rate limit the traffic. And it can dictate the application of more advanced inspection and control policies. For example, deep packet inspection (DPI) enables the firewall to look beyond packet headers to make more informed choices about the traffic application while web content classification (WCC) enables the firewall to filter HTTP/HTTPS traffic based on the URL and sites' reputations.

You will learn more about how MCs assign client to roles a bit later in this course.

The VLAN determines how the HPE Aruba Networking mobility device that is responsible for forwarding the client traffic (an MC or AP) does so. The HPE Aruba Networking mobility device either bridges the traffic out this VLAN to a core device for routing (preferred), or the mobility device can be the default gateway for the VLAN.

You specify the VLAN as part of the WLAN settings, and it is perfectly valid to have all users connected to a WLAN be assigned to the same VLAN. Remember: it is the role that determines the user rights. If the company chooses, though, the RADIUS server that authenticates a client can override the VLAN assignment with an Aruba-User-VLAN VSA.

A role can also specify a VLAN that overrides other VLAN assignments. However, role-based VLANs are deprecated for wireless clients.

Simple Example of Role Assignment

Now look at a simple example of how a client can receive its role assignment when it connects to a WPA3-Enterprise WLAN. When a client first associates to the WLAN, the MC assigns the client to the "logon" role as a placeholder, as shown in Figure 7-3. It then immediately assigns the client to the initial role defined in the WLAN settings (more precisely, the WLAN's AAA profile, which you will examine in a moment). In this example, that initial role is set to "logon," so the client stays at that role. In a WPA-Enterprise WLAN (of any version), 802.1X is blocking the connection at this point, so the client's role does not really matter; however, the client still has it.

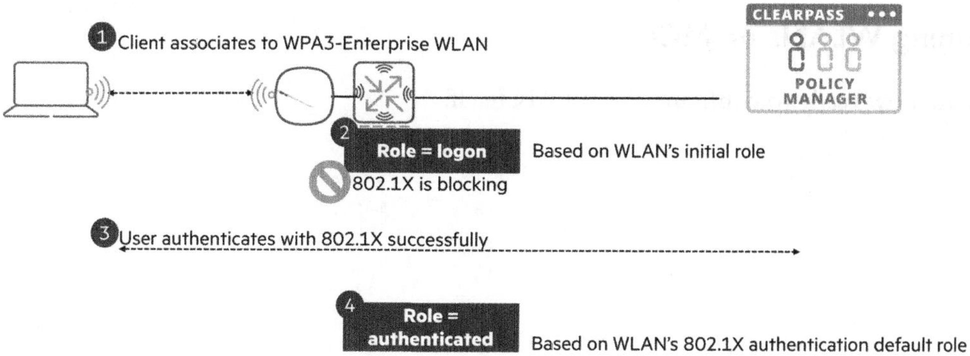

Figure 7-3: Simple Example of Role Assignment

The user then authenticates with 802.1X. If the authentication is successful, the MC updates the initial role with a role derived from the authentication. In this simple example, the MC has not received any information from the authentication server that enables it to assign a granular role. So the MC applies the 802.1X authentication default role defined in the WLAN (again, actually the WLAN's AAA policy). In the example shown in Figure 7-3, that role is "authenticated," so the user's role becomes "authenticated." This is a default firewall role that grants all access.

This simple setup fulfills the initial goal of getting users authenticated and granted full access to the network. In the next chapter you will learn how to implement more complex user role assignments to micro-segment clients.

You can check a wireless client's role in an MC's Web UI, shown in Figure 7-4. To do so, click the number next to the wireless icon under Clients in the top bar.

Figure 7-4: MC Web UI Top Bar

You will then see a list of wireless clients. As shown in Figure 7-5, the page lists clients by the name with which the user on that client authenticated. You can see the client's role, as well as other information such as IP address.

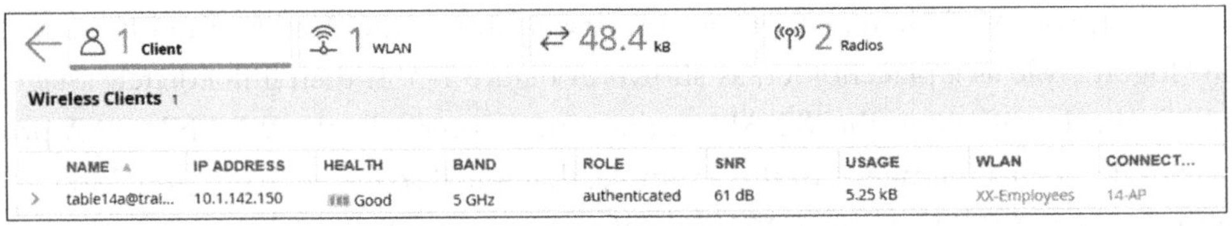

Figure 7-5: MC Clients List

Defining WLANs on MCs

You are now ready to configure the WLAN itself.

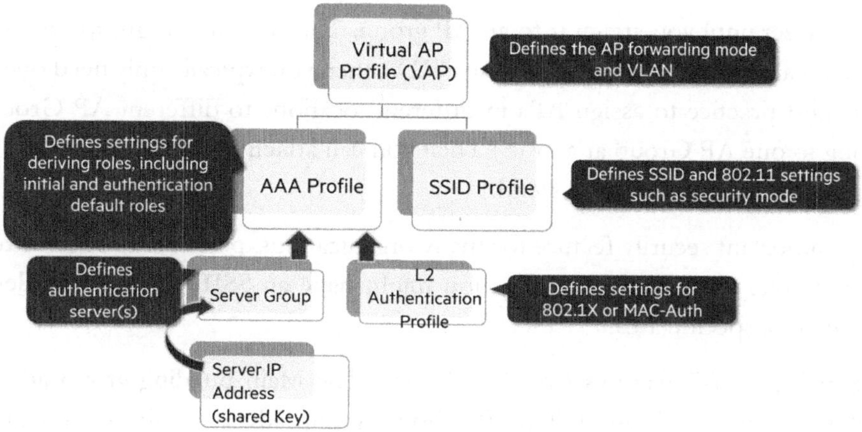

Figure 7-6: Defining WLANs on MCs

This chapter focuses on defining WLANs on MCs. A WLAN is defined in a virtual AP (VAP) profile, as shown in Figure 7-6. The VAP acts as the overarching container for the WLAN. It also defines the AP forwarding mode, which determines whether the AP decrypts traffic and whether the AP bridges the traffic locally or tunnels it to the MC. The VAP further defines the VLAN in which wireless client traffic is bridged by either the AP or the MC.

The VAP also references other profiles, including an AAA Profile and an SSID Profile.

The AAA Profile defines the settings that the MC uses to derive the client's role. These settings include the names of the initial role as well as the authentication default roles for any types of authentication implemented on the WLAN.

The AAA Profile also references a server group for each type of authentication. The server group holds one or more authentication server definitions. For example, a RADIUS server group could have two RADIUS servers, each with its IP address and shared key.

The AAA Profile also references L2 Authentication Profiles, which include 802.1X Authentication Profiles and MAC Authentication Profiles. These profiles define how the MC implements that form of authentication. For example, the 802.1X profile can set a maximum number of 802.1X failures or a reauthentication internal. Note that a WLAN can implement multiple forms of L2 authentication, but often implements just one form such as 802.1X for a WPA-Enterprise WLAN. The AAA profile for that WLAN will still have profiles for other L2 authentication methods; however, they will be default profiles and not actually put into use.

The SSID Profile for a VAP defines the WLAN's SSID and other 802.11 settings such as the security mode.

Attaching WLANs to APs

A VAP has no effect until you attach it to an AP group. The APs in that group then support that SSID. If you want all APs to advertise the same SSIDs, then you typically only need one AP Group. However, it is best practice to assign APs in different locations to different AP Groups. (An AP can only belong to one AP Group at a time.) Then you can attach the VAP only to the specific AP Groups where you want the SSID to extend.

This can be an important security feature for many organizations, particularly ones with specialized devices. For example, a healthcare organization might have an SSID for medical devices that it wants to confine to a specific area.

The example in Figure 7-7 features two AP Groups. The Main building group advertises three SSIDs: Employee, Voice, and Guest. The Warehouse group advertises the same employee SSID and a scanner SSID, which is unique to that location. The Warehouse group does *not* advertise the voice or the guest SSID, which are not required in that location. Preventing APs from advertising unnecessary SSIDs makes the deployment cleaner and simpler for users, who do not need to scan through as many SSIDs. It can also reduce the vulnerability surface. While you should be properly securing WLANs, every WLAN does represent an opening into your network, and there is no need to extend them unnecessarily.

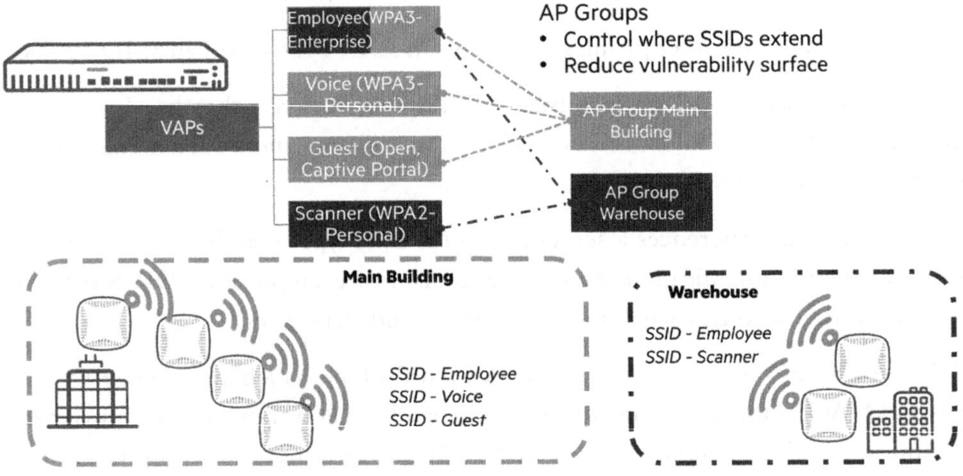

Figure 7-7: Attaching WLANs to APs

Using the WLAN Wizard to Define WLANs

The WLAN Wizard in the Web UI walks you through the WLAN configuration process.

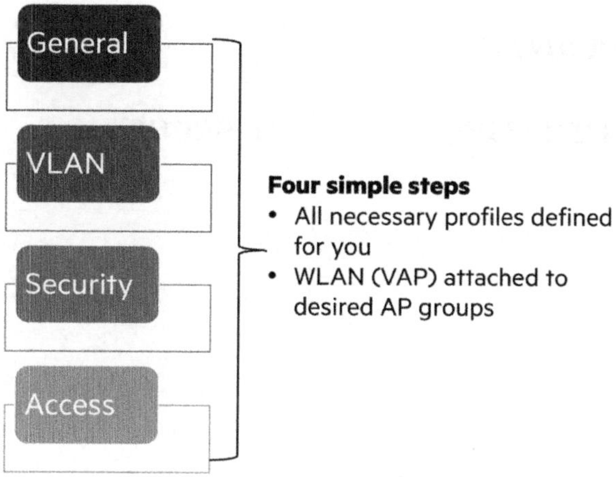

Figure 7-8: Using the WLAN Wizard to Define WLANs

The WLAN wizard is the easiest method for creating a WLAN. It has four pages, as shown in Figure 7-8. After you complete those pages, the wizard automatically creates all the pieces of the WLAN that you would otherwise have to create manually, including the VAP, the AAA profile, the SSID profile, and the L2 authentication profiles. The WLAN wizard also attaches the VAP to the desired AP groups and guides you through other configuration steps that require manual effort.

This section will guide you through the wizard, giving you insight into the meaning behind the settings and the reasons for selecting particular settings.

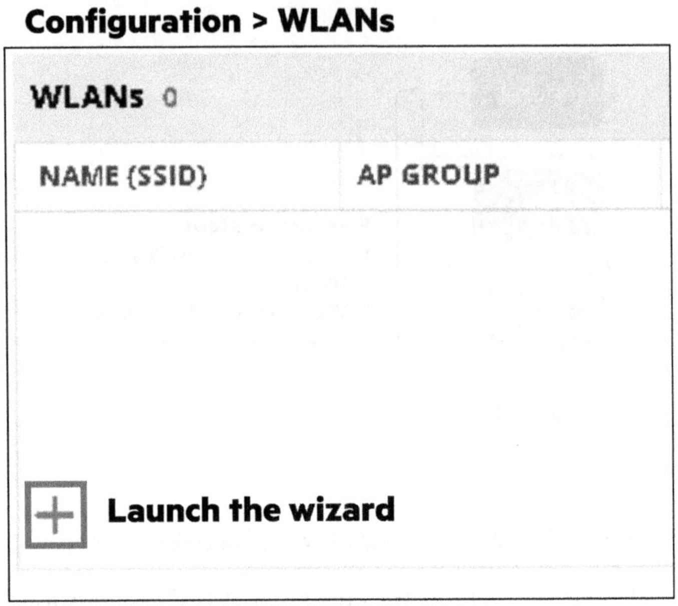

Figure 7-9: Using the WLAN Wizard to Define WLANs

To launch the wizard, navigate to **Configuration > WLANs**, and click the plus icon, as shown in Figure 7-9.

WLAN General Settings

Begin with the general settings, which are shown in Figure 7-10.

You must enter an SSID, which determines the SSID that APs broadcast for this WLAN. In addition to ASCII characters, AOS 8.4 and later supports Chinese characters with UTF8 encoding, which you can enter in the Web UI. The max SSID size is 32 bytes, which supports up to 10 Chinese characters. Only iOS, MacOS, Android, Windows 8 and Windows 10 clients support Chinese characters.

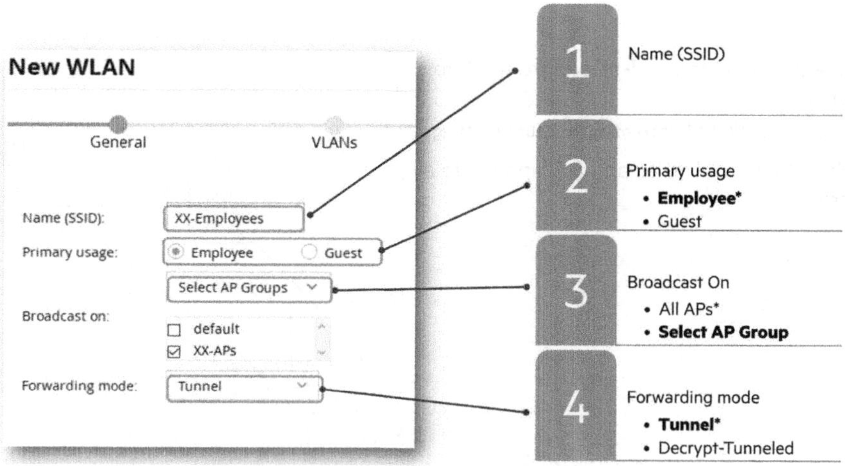

Figure 7-10: WLAN General Settings

You must set the primary usage for the WLAN to either **Employee** or **Guest**. The default setting is Employee, which is the correct setting for this scenario. Selecting this setting will prompt the wizard to give you WPA security options in the Security step while selecting **Guest** would prompt it to create an open network and give you captive portal options.

By default, the WLAN is set to broadcast on All APs. As you learned earlier, it is best practice to select the **Select AP Groups** setting and choose the specific groups that require the WLAN.

For the forwarding mode, choose **Tunnel**.

Security Benefits of Tunnel Mode for the WLAN

Consider the security benefits of keeping the WLAN's default forwarding mode, which is Tunnel. This mode and its benefits are illustrated in Figure 7-11.

Tunnel mode ensures that all traffic is encrypted with the form of encryption applied in the WLAN security settings, end-to-end between clients and MCs. In other words, campus APs (CAPs) forward traffic from clients over the Generic Routing Encapsulation (GRE) tunnel to their MC with encryption intact. In some competing vendors' solutions, APs decrypt and then re-encrypt traffic for transmission to their controller. But because HPE Aruba Networking APs never decrypt the traffic at all, they present less of a point of vulnerability to exploit, and fewer regulatory policies apply to them. The MC, typically deployed in a secure core location, is responsible for decrypting the traffic. Similarly, the MC encrypts traffic destined to a wireless client before tunneling to the client's AP.

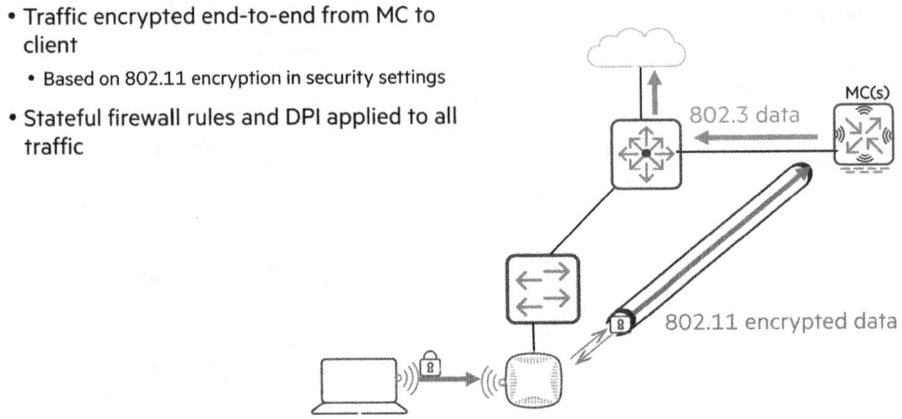

Figure 7-11: Security Benefits of Tunnel Mode for the WLAN

Tunnel mode also ensures that the powerful AOS firewall and deep packet inspection (DPI) capabilities apply to all wireless user traffic. While you are not using all of these features in this initial setup, the end goal is to be able to protect the internal network with them.

The alternative forwarding mode supported with WPA3-Enterprise is Decrypt-Tunneled, which has APs decrypt traffic before tunneling it to the MC. This can reduce the burden on MCs while still enabling the AOS firewall to control traffic. However, Decrypt-Tunnel is not typically necessary in a properly scoped deployment. Implementing this mode also means that traffic will pass through the network between APs and MCs in plaintext. While the network might be trusted enough for this to be acceptable for some customers, the emergence of the zero trust network model will be pushing some customers to want encryption as much as possible even within their campus LAN.

Other forwarding modes include Bridge and Split-Tunnel, which are not supported for WPA3-Enterprise. You can learn more about those modes in other training.

WLAN VLAN Settings—Recommendations

The next step in the wizard is defining the VLAN. Before you look at this step, stop to consider best practices.

The VLAN assigned to the WLAN determines the VLAN in which the wireless client traffic is bridged into the LAN. Because a VLAN is associated with a subnet, this in turn determines the IP address that the client receives with DHCP.

HPE Aruba Networking recommends dedicating a VLAN to the WLAN, as shown in Figure 7-12. You can then maintain a logical boundary between the wired and wireless network and between different wireless networks.

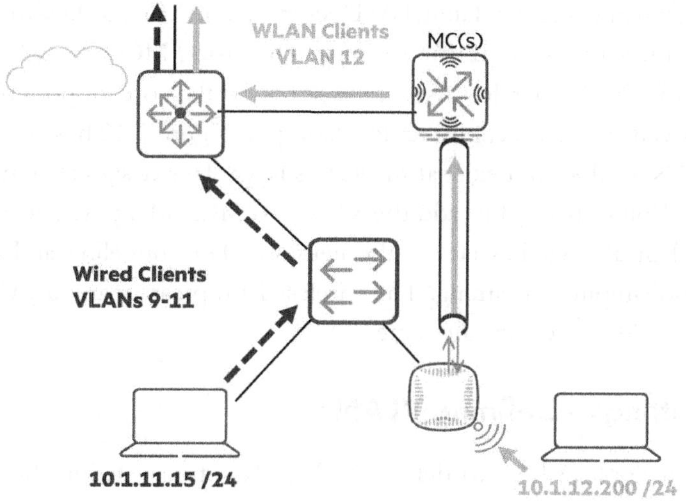

Figure 7-12: WLAN VLAN Settings—Recommendations

You should also remember that roles, not VLANs, will be your ultimate solution for segmenting traffic. In other words, you can have all the users in a WLAN assigned to the same VLAN and still create different firewall policies for those users, based on roles. Using fewer VLANs promotes simplicity. If you are familiar with traditional VLAN design for wireless networks, you might be concerned about assigning all wireless users to the same VLAN. Traditionally, this approach could cause problems due to the high number of broadcasts generated by the wireless users, which would then propagate across the WLAN and all wireless clients. However, MCs have mechanisms to suppress unnecessary broadcasts, so you can safely place many wireless clients in the same VLAN.

You do have the option, however, of placing clients connected to the same WLANs to different VLANs. HPE Aruba Networking ClearPass Policy Manager (CPPM) can assign clients to different VLANs with Aruba-User-VLAN VSAs, which override the WLAN's VLAN. The company might want to take this approach if the WLAN includes much less trusted clients, to which the company does not want to expose the corporate IP addressing.

However wireless users receive their VLAN assignment, HPE Aruba Networking recommends configuring the wireless user VLANs on MCs at Layer 2 only. In other words, do not configure IP settings on the VLAN on the MC. Instead make sure that the link between the MC and its switch supports the VLAN in question with the same tagging. The MC will then bridge the traffic out the link in the appropriate VLAN. The core routing switches or firewall will provide the default routing services for the VLAN. Similar recommendations hold true for IAP deployments, in which case the VLANs to which wireless users are assigned must be tagged on the IAPs' ports and the switch ports to which the IAPs connect.

Also remember that, when using the Tunnel or Decrypt-Tunnel forwarding modes in a controller-based solution, APs tunnel wireless traffic to MCs. Therefore, MCs, not APs, assign traffic to a VLAN. This allows VLANs for wireless users to exist only at the core, or possibly aggregation layer, where MCs are deployed. As you see, the wireless laptop in Figure 7-12 has an IP address in VLAN 12 even though VLAN 12 does not exist at the access layer. This design also makes it easier to add or change a VLAN. You only need to add the VLAN to MCs, MC ports, core switches, and core switch ports, instead of all switches across the network. The tunneling and centralized wireless WLAN approach also simplifies roaming. The wireless laptop can roam anywhere in the campus, and its VLAN and IP address remain the same.

WLAN VLAN Settings—Defining VLANs

You are now ready to look at how to define the VLAN settings within the wizard. (See Figure 7-13.)

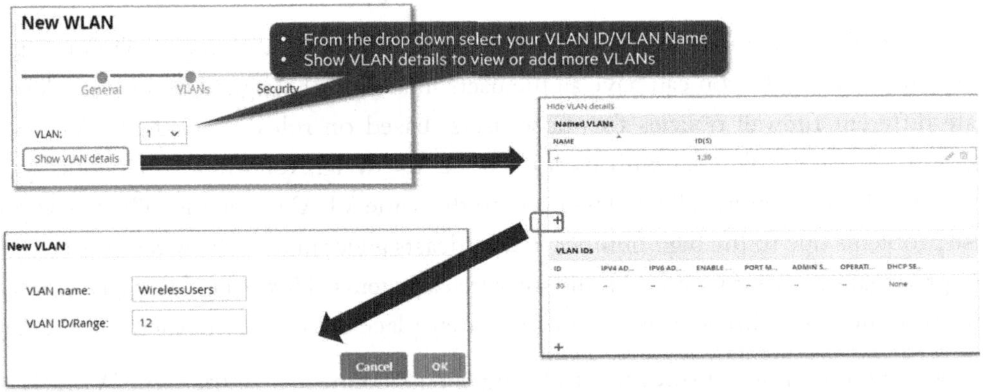

Figure 7-13: WLAN VLAN Settings—Defining VLANs

You select the VLAN from the drop-down menu by ID or by name. If the correct VLAN does not yet exist, you can click **Show VLAN details** to view the existing VLANs and add more VLANs. To add a new VLAN, select the + below the Named VLANs table, then enter a VLAN name and a VLAN ID or range of IDs for the new VLAN. To edit a named VLAN, select the VLAN from the table and select the edit (pencil) icon.

To create a range of multiple VLAN IDs, specify the beginning and ending VLAN IDs separated by a hyphen. For example, specify 55-58.

As mentioned earlier, you also need to verify that the proper MC port or ports carry the VLAN. You can select a port on an MC in the Web UI, as shown in Figure 7-14.

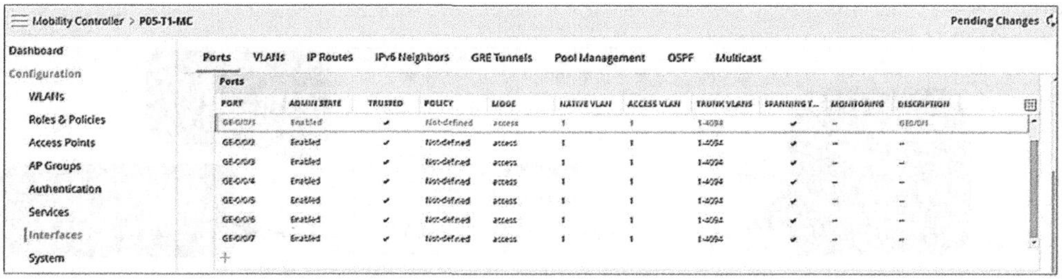

Figure 7-14: MC Ports List

You can then scroll down through the settings to view the VLAN configuration. In the example shown in Figure 7-15, the port's mode is Trunk. The trunk's native VLAN 1 is, and it allows all VLANs. This means that the port will carry all VLANs on the MC as tagged VLANs, except for VLAN 1 which it will carry untagged. In the real world, however, you should often select the exact VLANs that your company uses for the list of allowed VLANs.

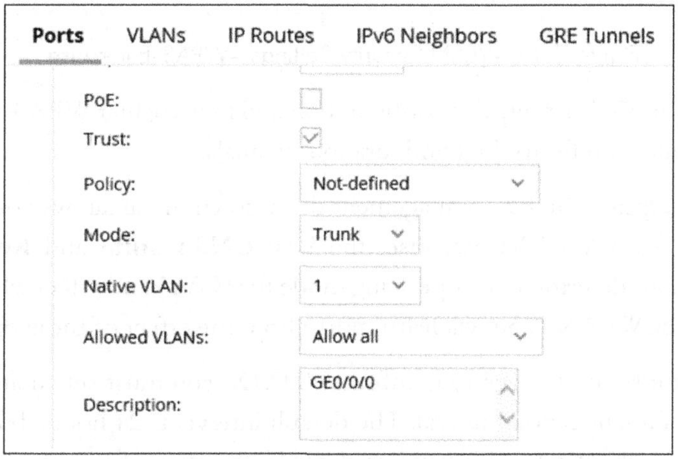

Figure 7-15: Defining VLANs on an MC Port

WLAN Security Settings—WPA3-Enterprise

The next step in the wizard is defining security settings.

When you have selected "Employee" for the primary usage in the General settings, the Security section includes the options shown in Figure 7-16. You can select Enterprise, Personal, or Open. As you learned in the previous chapter, Enterprise provides the highest level of security, including 802.1X and various forms of encryption based on the WPA version and mode. WPA/WPA2-

Personal, as you learned, uses pre-shared key (PSK) authentication while WPA3-Personal uses SAE. Open networks can be legacy open or Opportunistic Wireless Encryption (OWE) and are the least secure.

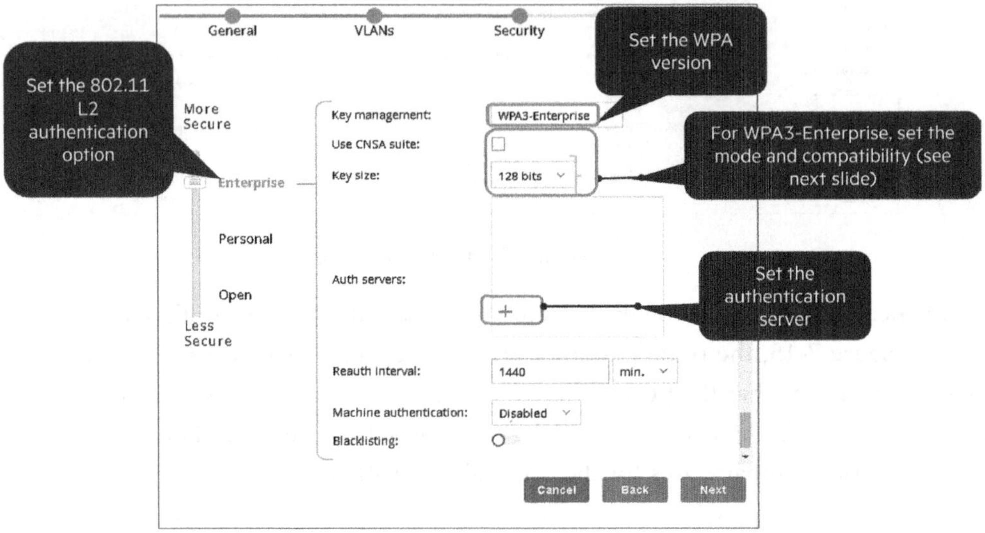

Figure 7-16: WLAN Security Settings—WPA3-Enterprise

This course focuses on WPA-Enterprise configuration and in particular WPA3-Enterprise. To learn more about other options, refer to the configuration manuals.

When you select Enterprise, the Key management drop-down menu allows you to select the WPA version. When you select WPA3-Enterprise, the **Use CNSA suite** and **Key size** options are displayed. These options determine the operating mode for WPA3 and affect whether WPA2 clients can also connect to the WLAN. You will learn more about the effect of these options in a moment.

Because WPA-Enterprise (of any version) enforces 802.1X, you must set an authentication server. You can also set a reauthentication interval. The default interval is 24 hours, but you can change it. If the user fails to re-authenticate with valid credentials, the state of the user is cleared. Requiring clients to periodically re-authenticate improves security because it prevents the status from becoming stale—proving that a legitimate user is indeed still using the client—and because it renews the Pairwise Master Key (PMK).

As you see, the wizard provides a couple more options. Machine authentication is a more advanced feature outside the scope of this training. It forces the device to authenticate before the user authenticates, allowing the MC to assign different roles to the client based on whether only machine authentication has succeeded, only user authentication has succeeded, or both types of authentication have succeeded. In this way, domain computers can gain limited access to the network before the user logs in, for example. The HPE Aruba Networking mobility devices can further grant different types of access to users on domain computers versus users on non-domain devices. However, CPPM can also implement policies that provide similar functionality, with even more options and flexibility—without requiring Machine authentication to be enabled on the HPE Aruba Networking mobility device. Therefore, when you are using CPPM as the authentication sever, you should typically leave this setting disabled and allow CPPM to make the decisions.

The feature at the shown bottom of Figure 7-16 has been renamed "denylisting." If it is enabled on a WLAN, a client can be denylisted in several ways. For example, if you enable the feature, and the client fails authentication a specified number of times, the MC prevents the client from associating to the WLAN at all.

WLAN Security Settings—WPA3-Enterprise Mode

When you choose WPA3-Enterprise for the WLAN's key management setting in AOS version 8.6, you have three choices, summarized in Figure 7-17.

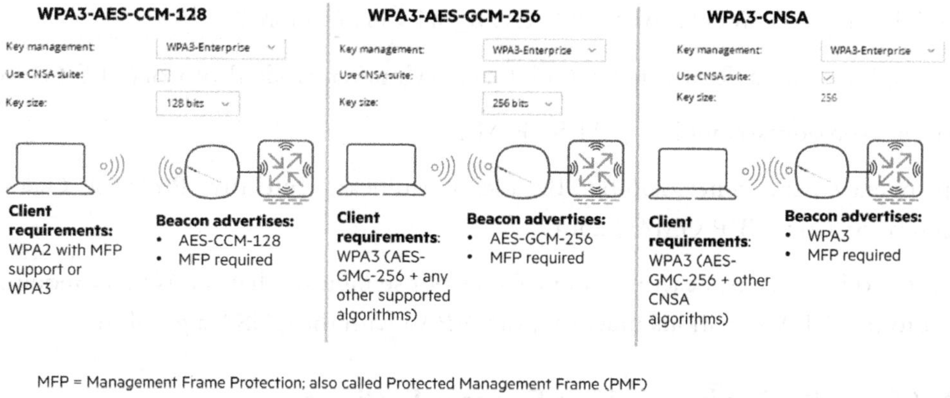

Figure 7-17: WLAN Security Settings—WPA3-Enterprise Mode

You can leave the **Use CNSA suite** box clear and select 128 bits for the key size. This sets the WLAN to use WPA3-AES-CCM-128 operating mode. In this mode, AP beacons and probe responses advertise WPA2 compatible algorithms for authentication and key management (AKM), pairwise encryption, and other algorithms defined within WPA. For example, the beacons advertise AES-CCM-128 for pairwise encryption. Because WPA2 clients support this form of encryption, they can connect to the WLAN—as long as they also support Management Frame Protection (MFP), which is required in WPA3. (Note that SSID Profiles include settings for setting the WLAN to "MFP capable" or "MFP required;" when you set the WLAN to WPA3-Enterprise, though, these settings are enabled automatically.) The WPA3-AES-CCM-128 option is the best choice if you need to provide support for both WPA2-capable and WPA3-capable clients.

The second option for the WPA3-Enterprise settings is leaving the **Use CNSA suite** box clear but setting the key size to 256 bits. This sets the WLAN's operating mode to WPA3-AES-GCM-256. In this mode, APs advertise AES-GCM-256 as the pairwise encryption standard for the WLAN, which means that only WPA3-capable clients will be able to connect. However, APs in this mode advertise support for a broader range of algorithms than they do in CNSA mode, potentially permitting more clients to connect to the WLAN, but also decreasing security as compared to WPA3-CNSA.

When you select the **Use CNSA suite** box in the WLAN settings, the WLAN operates in CNSA mode. In this case, in addition to indicating WPA3, beacons and probe responses mandate support for CNSA algorithms. To review, these algorithms are:

AKM: Suite B-compatible EAP-TLS, which generates at least 384-bit PMK

- TLS_ECDHE_ECDSA_WITH_AES_256_GCM_SHA384 (mandatory)
- TLS_ECDHE_RSA_WITH_AES_256_GCM_SHA384 (optional)
- TLS_DHE_RSA_WITH_AES_256_GCM_SHA384 (optional)

Pairwise cipher suite (unicast data): 256-bit Galois/Counter Mode Protocol (AES-GCM-256)

Group suite (broadcast/multicast): AES-GCM-256

Group management suite: 256-bit Broadcast/Multicast Integrity Protocol Galois Message Authentication Code (BIP-GMAC-256)

As with the other modes, beacons and probe response indicate that the WLAN requires MFP. To connect to the WLAN, stations must support WPA3 with the CNSA algorithms.

WLAN Security Settings—Auth Servers Settings

Now that you understand the WPA3 operating modes, move on to looking at the authentication server options. Click + to open the Add Existing Server window and select a preconfigured server from the list of servers, as shown in Figure 7-18. To define a new server, click + in the Add Existing Server window. The Web UI lets you define a RADIUS or LDAP server. RADIUS is the correct choice for this scenario.

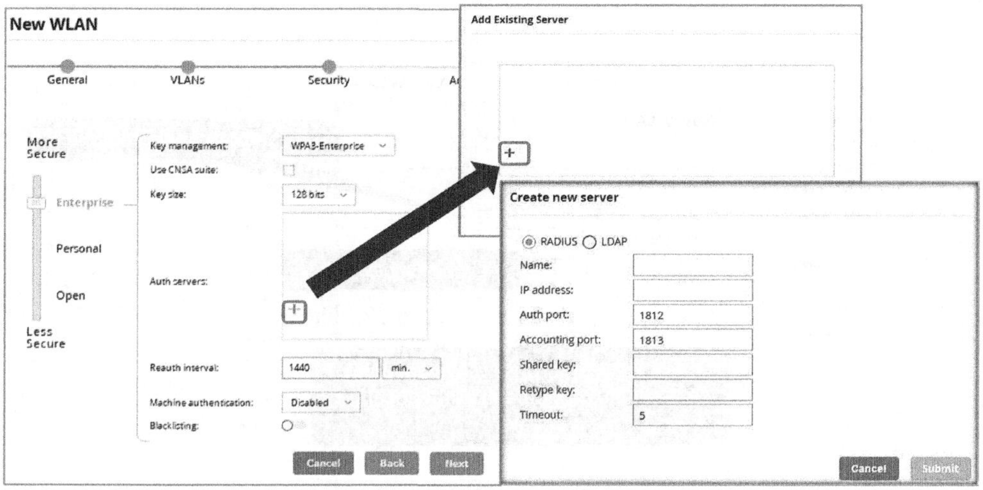

Figure 7-18: WLAN Security Settings—Auth Servers Settings

In version 8.6, WPA3-Enterprise does not support EAP termination, which means that the MC cannot use its internal RADIUS server for this WLAN; instead it must use an external authentication server like CPPM. For WPA3-CNSA mode, the MC communicates with this server, telling it which options it can use for TLS.

WLAN Access Settings

For the final step, you configure access settings, as shown in Figure 7-19. Here you can select the default role. For a WLAN that implements WPA2- or WPA3-Enterprise security, this menu sets the 802.1X authentication default role in the AAA profile created for the WLAN in the background. The MC assigns clients that pass 802.1X authentication to this role—unless another role overrides it in various ways about which you will learn in the next chapter. Because a client's role is associated with firewall policies and other settings, the role is very important for determining the client's access rights.

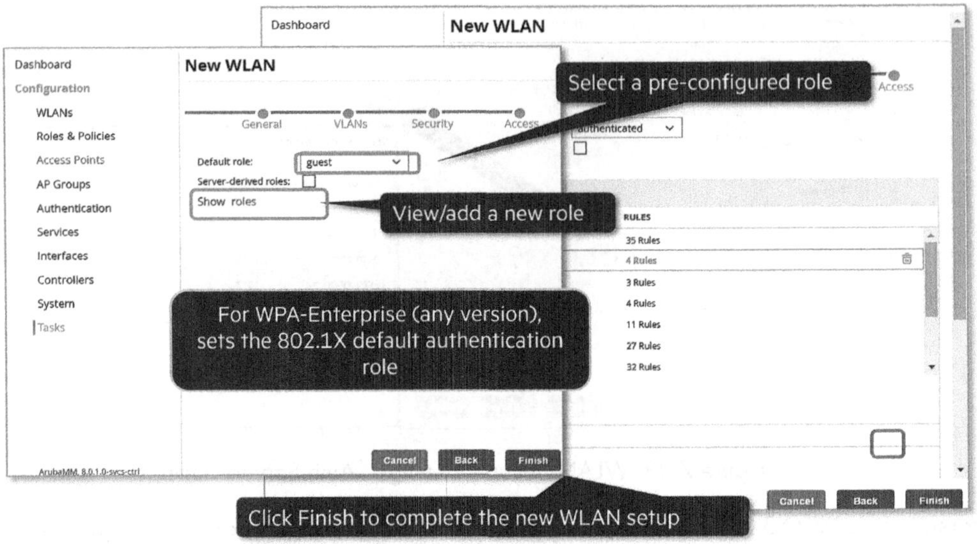

Figure 7-19: WLAN Access Settings

You can select an existing role from the menu. Or you can define a new role for the WLAN by selecting **Show Roles** and selecting the + below the Roles table.

The Server-derived roles check box enables you to create additional rules that an MC uses to assign different roles to different clients. However, learning how to set up these rules is beyond the scope of this course. Instead, you will focus (in the next chapter) on how the mobility device receives the role assignment from an authentication server like CPPM, which is what HPE Aruba Networking typically recommends.

Verifying the WLAN

When you click **Finish** in the Access section, the WLAN is complete. You can view the WLAN in the WLAN list. Remember to click **Pending Changes** and to deploy the changes to make them take effect, as shown in Figure 7-20. In an Mobility Conductor (MCR)-managed solution, deploying the changes pushes them through the group and subgroup hierarchy starting at the level in the hierarchy at which you configured the WLAN.

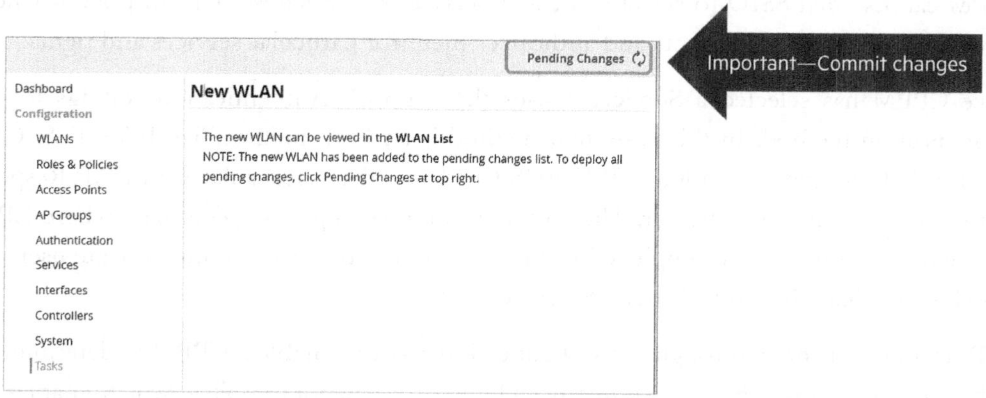

Figure 7-20: Verifying the WLAN

High-Level View of WPA-Enterprise Authentication to CPPM

In this course, you are using CPPM as your RADIUS authentication server for the WPA3-Enterprise WLAN. You should understand just a little about how CPPM is interacting with the solution.

As you recall, when a client associates to a WLAN that implements WPA-Enterprise (in any version), the authenticator transmits an EAP Request/Identity frame to initiate the 802.1X process. The authenticator then forwards the client's EAP Response/Identity frame in a RADIUS Request to CPPM. CPPM uses attribute value pairs (AVPs) in the RADIUS request to match the request to a Service, as shown in Figure 7-21.

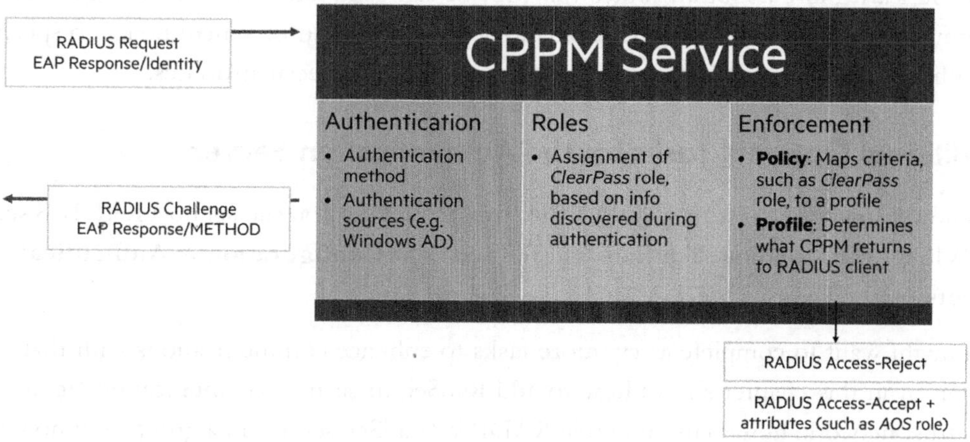

Figure 7-21: High-Level View of WPA-Enterprise Authentication to CPPM

For example, an AVP in the request might indicate that the client is accessing a specific SSID, and CPPM can use that SSID to select a specific Service. A Service is a CPPM policy, which defines how CPPM authenticates clients and authorizes them for particular services and rights.

Once CPPM has selected a Service, it uses the Service's Authentication settings to choose an authentication method. In this case, that method is an EAP method like EAP-TLS. CPPM then initiates that method by sending a RADIUS Challenge that encapsulates an EAP-Response using that method to the authenticator. The Authentication settings also determine where CPPM looks for the user account. For example, CPPM can use a local database or authenticate users against an LDAP server like Microsoft Active Directory (AD).

CPPM and the client go through the authentication process, enabling CPPM to determine the user's authentication status. The Service might include a role mapping policy, which enables CPPM to take what it has learned about the user and device during authentication and assign a ClearPass role based on that information. It is important that you understand that the ClearPass role is not the same as the AOS role—although CPPM can use this role to affect the role that the HPE Aruba Networking mobility device eventually assigns.

CPPM finally uses the Service's Enforcement Policy to decide what to do with the user's request. This policy includes rules that take inputs such as the ClearPass role, as well as other criteria, and outputs a profile. The profile tells CPPM whether to send an Access-Reject or Access-Accept. For an Access-Accept, the profile further defines attributes, which CPPM uses to communicate to the MC (or IAP cluster) more nuanced information about how to control the client's access. For example, the Aruba-User-Role vendor specific attribute (VSA) communicates the name of an AOS firewall role to the RADIUS client.

You will learn more about attributes that CPPM can send later. For this scenario, CPPM simply determines whether a user is authenticated. It uses a simple enforcement policy to specify that, for all authenticated users, it sends an Access-Accept with no special attributes.

Additional Optional Tasks for the Authentication Server

You now know how to create a WPA3-Enterprise WLAN. You can set up a RADIUS server during the WLAN configuration. You can find the server at **Configuration > Authentication > Auth Servers**, as shown in Figure 7-22.

You might want to complete a few more tasks to enhance communications with that server. This section of in this chapter shows how to add RadSec to secure communications in an insecure or zero trust network, how to use multiple RADIUS/RadSec servers in a group, and how to specify a dynamic authorization server.

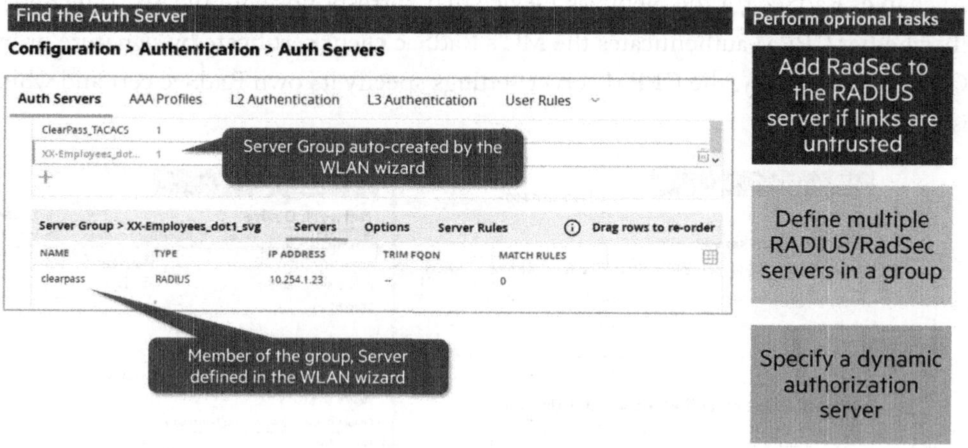

Figure 7-22: Additional Optional Tasks for the Authentication Server

Adding RadSec to Secure RADIUS on Untrusted Links

HPE Aruba Networking mobility devices such as MCs and IAPs support RadSec. To set up RadSec on MCs, create a RADIUS type authentication server in the Web UI or CLI and specify the server's IP address or hostname. You should also load the certificates that you plan to use for RadSec. These include:

- The trusted CA certificate for the certificate that CPPM (or another RADIUS server) uses for RadSec—Alternatively, if the RADIUS server uses a self-signed certificate, you can load that certificate as a public certificate.

- A server certificate for the MC that supports Client Authentication—To create the CSR for this certificate, you should communicate with the ClearPass administrators. CPPM can authorize a network device like an MC for RadSec in two ways. It might simply check the CA signature, in which case you simply need to ensure that you obtain a certificate for the MC signed by a CA that CPPM trusts. However, CPPM can also check both the certificate and the common name (CN) or subject alternative name (SAN) in the certificate. In the latter case, you must make sure to create a CSR in which the subject name or SAN matches what CPPM expects.

After you verified that the MC has the proper certificates, you can use the CLI to configure the device to use RadSec for the RADIUS authentication server. You specify the name of the trusted CA certificate or the server self-signed certificate. You also specify the name of the client certificate, which is the MC's own certificate. Finally you enable RadSec.

Shown in Figure 7-23, for your reference, are the corresponding settings on CPPM. As you see, CPPM enables RadSec for the Network Device that corresponds with the MC. The RadSec settings specify whether CPPM authenticates the MC's RadSec client certificate by signature or by signature and CN or SAN. Finally, the CPPM server settings specify its own RadSec cert and which root CA it trusts for RadSec.

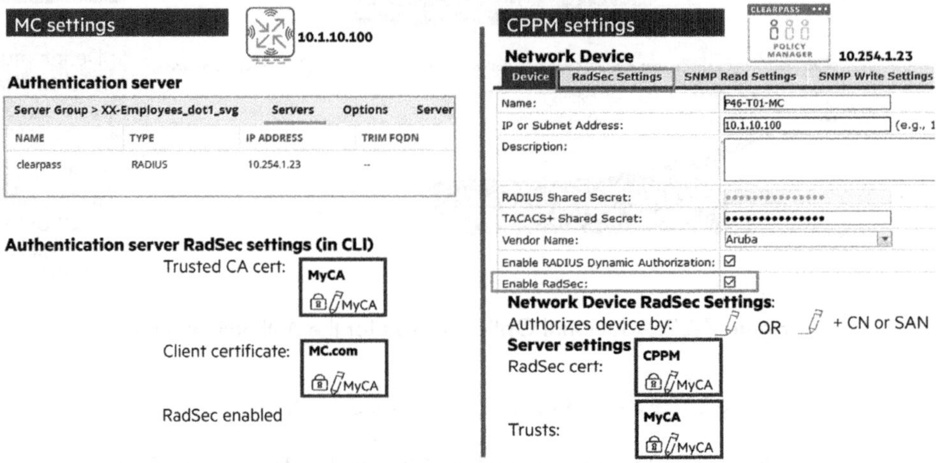

Figure 7-23: Adding RadSec to Secure RADIUS on Untrusted Links

Note that when you enable RadSec on CPPM and HPE Aruba Networking devices, the RADIUS shared secret (key) automatically becomes "radsec." The RADIUS standard requires a key, so the devices continue to use it in their RADIUS messages. However, it does not matter that the key is weak and publicly known because the key no longer fulfills the role of authenticating messages between the server and client. Instead the vastly more secure certificates fulfill this role.

CLI Commands: MC Configuration for RadSec

For MCs, you must configure RadSec for a RADIUS server from the CLI. The commands are provided in Figure 7-24 for your reference. These steps assume that you have already loaded the proper certificates on the MC (see instructions in a previous chapter).

In configure context

Step	AOS command
Access the RADIUS authentication server context.	aaa authentication-server radius <*name*>

In RADIUS authentication server context

Step	AOS command	Notes
Specify the name of the Trusted CA certificate, which is the root CA cert for the RadSec server's certificate.	radsec-trusted-cacert-name <*name*>	Install the Trusted CA certificate in advance
Specify the name of the MC's certificate, which it will use for RadSec.	radsec-client-cert <*name*>	Install the certificate in advance; it must have a Client Authentication EKU.
Enable RadSec.	enable-radsec	
Save the configuration	write memory	

Figure 7-24: CLI Commands: MC Configuration for RadSec

Note that instead of specifying the Trusted CA certificate, you can specify a trusted server certificate. You should do this when the RadSec server uses a self-signed certificate, which you should install as a Public Certificate in advance.

Defining Multiple RADIUS/RadSec Servers in a Group

You can define multiple redundant RADIUS servers within an authentication server group. With load-balancing disabled, the MC sends all requests to the first server in the group unless that server fails. Enabling load balancing distributes the load across all available servers, as shown in Figure 7-25.

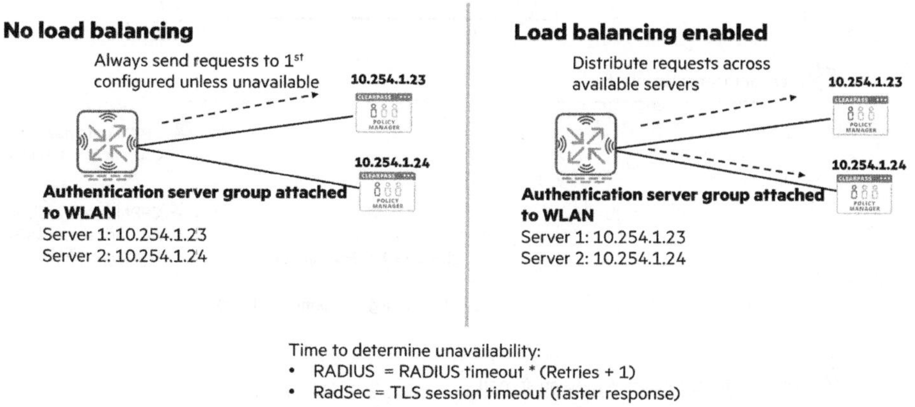

Figure 7-25: Defining Multiple RADIUS/RadSec Servers in a Group

Load balancing helps to promote good response times during times at which many users log in, such as at the beginning of the day.

You should also consider how the MC determines if an authentication server is available. RadSec is useful here, as an MC can more quickly detect that a RadSec session has failed. Standard RADIUS uses UDP, which has no concept of a session. The MC waits for the length of a RADIUS timeout to receive a response to its request. If the response does not come, the device sends another RADIUS request and waits for the timeout again. It continues this process for the number of retries defined. RadSec, on the other hand, uses TCP, and the device can determine that a server is unavailable as soon as the TCP session times out.

What is Dynamic Authorization?

A RADIUS server like CPPM is capable of making decisions about the correct authorization level for a user or device beyond the initial authentication. For example, ClearPass OnGuard can detect clients' security status and tell CPPM to disconnect a client that is infected with malware. Or CPPM might find out after the initial authentication that the client is an IoT device, which has particular security policies associated with it. Many other examples of such events exist.

So that CPPM can enforce the new policies, it must force the client to disconnect and re-authenticate. As shown in Figure 7-26, CPPM sends a Change of Authorization (CoA) message to the RADIUS client—here an MC. By default, it sends the message on standard port UDP 3799. If it is configured to use a different port, you must make sure to match that configuration on the HPE Aruba Networking mobility device acting as RADIUS client. But if CPPM and the RADIUS client are using RadSec, CPPM sends the CoA messages in the RadSec tunnel.

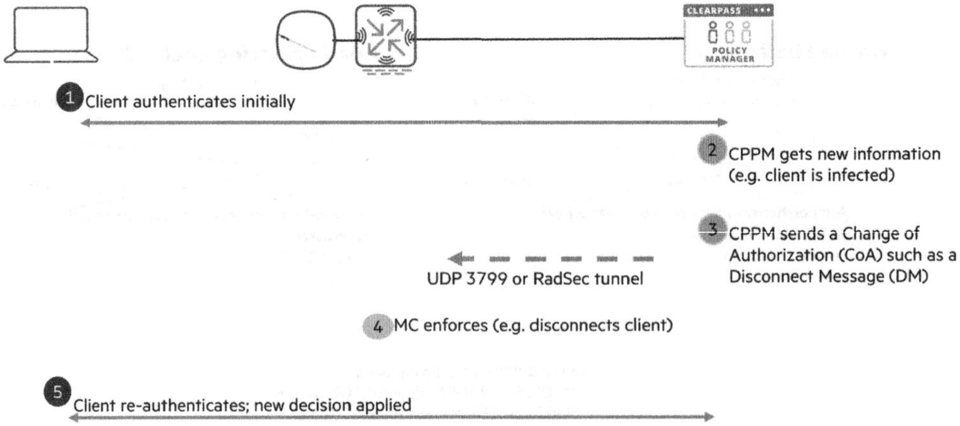

Figure 7-26: What is Dynamic Authorization?

When the HPE Aruba Networking mobility device receives a CoA from an authorized dynamic authorization server, it enforces the message and disconnects the client. The client will typically try to re-authenticate, and CPPM can then apply its policies to make new decisions about the client's access.

Specifying a Dynamic Authorization Server

If CPPM enables RADIUS Dynamic Authorization for the HPE Aruba Networking mobility device, you also need to set up that device to accept CoA messages.

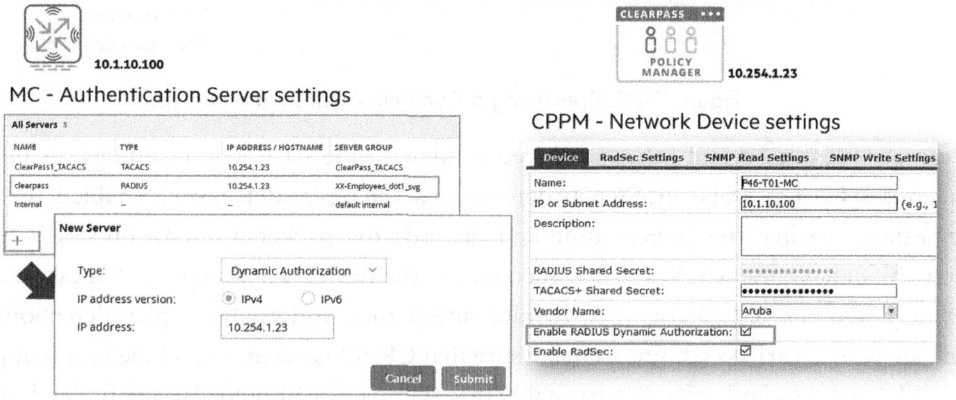

Figure 7-27: Specifying a Dynamic Authorization RFC Server

In Figure 7-27, you see how to specify a dynamic authorization server on an MC in the Web UI. You must add CPPM as both a normal RADIUS server and a dynamic authorization server. Note that you specify the dynamic authorization server by IP address, not by fully qualified domain name (FQDN).

You must then select the dynamic authorization server, also called an RFC 3675 server, and set the key, or shared secret, as shown in Figure 7-28. Make sure to match the shared secret specified for the corresponding network device on CPPM. If you are using RadSec, the shared secret is always "radsec," and you must also enable RadSec on the dynamic authorization server from the CLI.

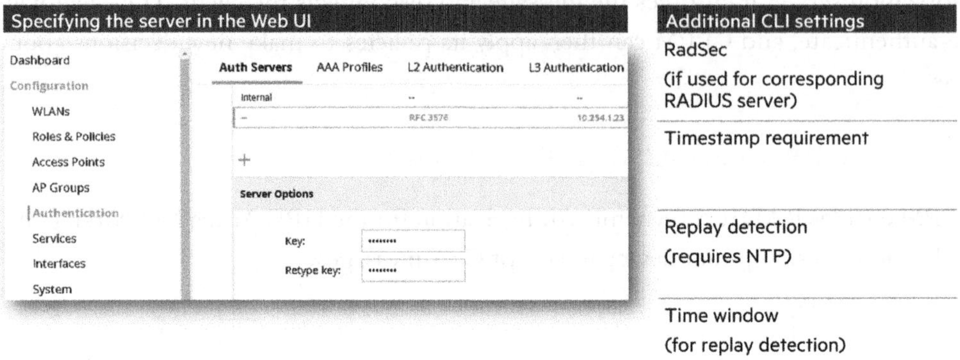

Figure 7-28: Specifying a Dynamic Authorization Server

You can also enable several other advanced settings in the CLI. These settings include requiring timestamps in CoA messages and replay detection. When replay detection is enabled, the MC checks the timestamp against the current time and discards the packet if it falls outside a configurable window. By default, the window is 300 seconds, so the device will accept CoA messages that are as much as 5 minutes old. Before you require timestamps and enable replay detection, you must coordinate with ClearPass admins to make sure that CPPM is set up to include timestamps in CoAs. If the CPPM admins authorize you to enable timestamps, you should also verify that both the HPE Aruba Networking mobility devices and ClearPass are taking their time from an authenticated NTP server to ensure that legitimate messages do not fall outside the window.

Activating the Dynamic Authorization Server on WLANs

To enable the MC to accept the CoA messages for wireless user sessions, you must also specify the RFC 3576 server on the WLAN to which the wireless users connect. You do so from the AAA profile associated with the WLAN. Go to the **Configuration > Authentication > AAA Profiles** window, expand **AAA** and the profile in question. Select **RFC 3576** server and click the + icon to add the server, as shown in Figure 7-29.

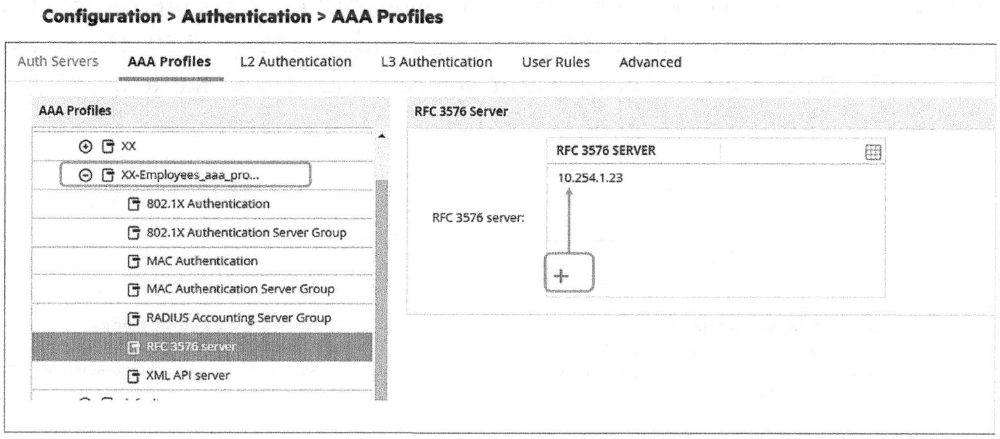

Figure 7-29: Activating the Dynamic Authorization Server on WLANs

Enforce Basic 802.1X on the Wired Network

In this section, you will learn how to enforce basic 802.1X on the wired network.

802.1X Configuration Process on AOS-CX Switches

Setting up basic 802.1X on AOS-CX switches is simple.

1. You must specify the RADIUS server.
2. If the links between the switches and the server are considered untrusted, you can add RadSec on AOS-CX switches.
3. You can also configure support for multiple servers for better availability.
4. You can configure the switches to accept dynamic authorization messages from the RADIUS server.
5. You will then be ready to enable 802.1X.

The next sections explain the process in more depth.

Specifying the RADIUS Server

Begin by configuring CPPM (or another RADIUS server) as a RADIUS host on the switch, specifying CPPM's IP address or fully qualified domain name (FQDN). Remember, if you use an FQDN you must configure the switch to communicate with a DNS server. You must also specify either an individual key or use a global key that applies to all RADIUS servers. If you define both a global key and an individual key for a server, the individual key takes precedence. The key must match the key configured on the RADIUS server, as shown in Figure 7-30. The exception is that, if you are using RadSec, covered in a moment, you do not have to set the key.

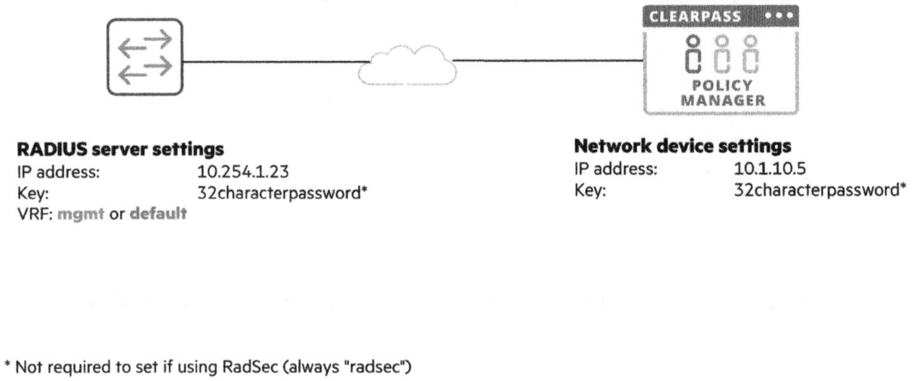

Figure 7-30: Specifying the RADIUS Server

You must also set the VRF. Use the mgmt VRF if the RADIUS server is deployed on the OOBM network. But often users need to access some ClearPass components, so CPPM is often deployed on the production network. In that case, the "default" VRF is usually correct.

Configuring RadSec on AOS-CX Switches: Certificate Common Name

If network connections in your environment are untrusted, it is important to secure RADIUS communications beyond the limited security offered by RADIUS. As of software version 10.05.0001, AOS-CX switches support RadSec. RadSec establishes a secure TLS tunnel in which to send all RADIUS messages (including authentication, accounting, CoA, and DM messages).

To set up RadSec on AOS-CX switches, you must verify that the switch has a trust anchor (TA) profile with the root CA certificate for CPPM's RadSec certificate. You also need to be careful in how you specify the RADIUS server on the switch so that this setting correlates with the subject name and SANs in CPPM's RadSec certificate.

You can specify the RADIUS server by IP address or FQDN. Make sure that you choose the option that matches the subject name in the CPPM RadSec certificate. For example, that certificate often uses the FQDN as the subject name. You must then specify the RADIUS server by FQDN on the switch. As shown in Figure 7-31, even though the FQDN resolves to an IP address, you cannot define the RADIUS server with that IP address because the certificate uses the FQDN.

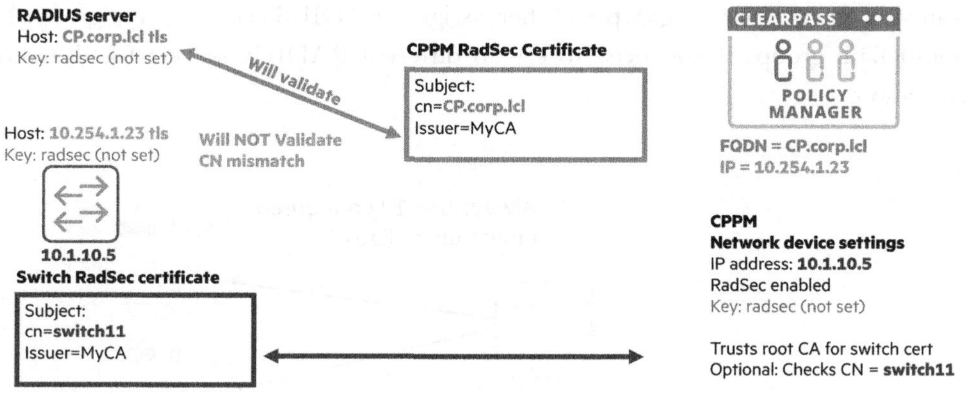

Figure 7-31: Configuring RadSec on AOS-CX Switches: Certificate Common Name

The switch must also have a local certificate set to the 'all' or 'radsec' usage that is signed by a CA trusted by CPPM and that has a Client Authentication EKU. You should coordinate with the CPPM admins as you create the CSR for this certificate online. Depending on the CPPM setup, CPPM might simply check that the switch has a certificate with a valid CA signature. Or it might check the CN, in which case you must make sure that you set the CN to a value that CPPM expects.

When you configure RadSec on an AOS-CX switch, you do not need to set the key. It is always "radsec."

Configuring RadSec on AOS-CX Switches: CLI Configuration Warning

When you specify a RadSec RADIUS server, you must specify the TLS option for every setting that you enter for the server. These commands correctly configure the server at FQDN cp.corp.lcl for RadSec and enable dynamic authorization on that server:

radius-server host cp.corp.lcl tls

radius-server host cp.corp.lcl tls dyn-authorization

The next example commands configure the RADIUS server incorrectly. The first command configures the server at FQDN cp.corp.lol for standard RADIUS with a PSK. The second command creates a second RADIUS server that uses RadSec and has dynamic authorization enabled on it. However, the switch will try to use the first, incorrectly configured server, first.

radius-server host cp.corp.lcl

radius-server host cp.corp.lcl tls dyn-authorization

Supporting Multiple RADIUS Servers

On AOS-CX switches, RADIUS servers are automatically added to the default group, named radius. You can also create a named group and then assign a RADIUS server to that group. You should use non-default groups if you need to set up different RADIUS servers for different types of authentication.

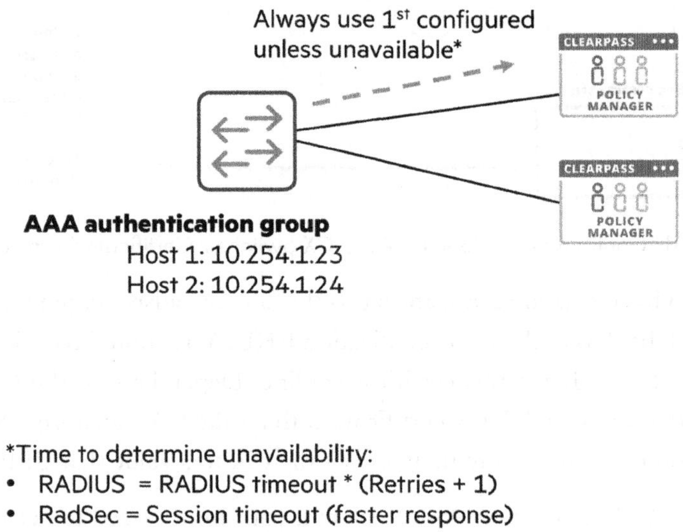

Figure 7-32: Supporting Multiple RADIUS Servers

You can define multiple servers in a group, but the switch sends all requests to the first server configured, as shown in Figure 7-32. It switches over to the other server only if the first becomes unavailable. As with HPE Aruba Networking mobility devices, how the switch determines whether a server is unavailable depends on whether the server is defined as a RADIUS or RadSec server. For RADIUS, timeout and retry settings determine the time.

Ensuring Switches Support Dynamic Authorization

AOS-CX switches support receiving CoA messages, but you must enable dynamic authorization for the RADIUS server.

As shown in Figure 7-33, if dyn-authorization is enabled, the switch accepts these messages on the default UDP port 3799, which is also the default port on CPPM. (You can change that setting if necessary, in your environment.)

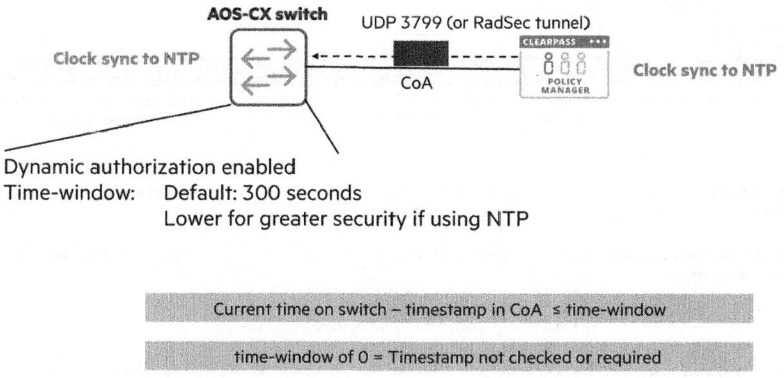

Figure 7-33: Ensuring Switches Support Dynamic Authorization

Even when an AOS-CX switch is configured to allow CoA messages, it checks these messages to determine if they are current. The switch only accepts the message if the time between the message's time stamp and the current time on the switch falls within its RADIUS time window. The default time window is 300 seconds, which means that the switch will not accept messages that are older than 5 minutes. This can help to protect against replay attacks. If you set the time-window to 0, the switch does not check the timestamp or care if the timestamp is not included at all. For better security, you should lower this window to a value such as 30 seconds. Make sure, however, that the switch can accept the CoA messages by ensuring that the switch has the correct time. It is best practice to configure NTP on the switch and the RADIUS server and have them take their time from the same server.

CLI Commands: AOS-CX Switch RADIUS Configuration

The table in Figure 7-34 includes the commands for configuring a RADIUS server on AOS-CX switches.

In configure context

<table>
<tr><th>Step</th><th>AOS-CX command</th><th>Notes</th></tr>
<tr><td>Specify the RADIUS server or servers.</td><td>radius-server host <<i>IP</i> | <i>hostname</i>> key <ciphertext | plaintext> <<i>secret</i>> [vrf <<i>vrf</i>>]</td><td>When you have redundant servers, remember to specify the primary one first.</td></tr>
<tr><td>Create a RADIUS server group and add the server(s) to it.</td><td>aaa group server radius <<i>group name</i>>
server <<i>IP</i> | <i>hostname</i>></td><td>Only required if you want to use the non-default group</td></tr>
<tr><td>Enable dynamic authorization and set the time-window.</td><td>radius dyn-authorization enable

radius dyn-authorization client <<i>IP</i> | <i>hostname</i>> secret-key <ciphertext | plaintext> <<i>secret</i>> [vrf <<i>vrf</i>>] [time-window <1-65535>]</td><td>The default time window is 300 seconds.</td></tr>
<tr><td>Enable RADIUS accounting to a central server.</td><td>aaa accounting <all-mgmt | port-access> <stop-only | start-stop> [interim [1-525600]] group <<i>name</i>></td><td>The switch can send accounting for management users (all-mgmt) and for network users (port-access).
The interim option only applies with start-stop. You can select a number of minutes for the interim updates or leave the default (60).</td></tr>
</table>

Figure 7-34: CLI Commands: AOS-CX Switch RADIUS Configuration

You can check the RADIUS server settings on an AOS-CX switch with this command:

```
AOS-CX# show radius-server

Unreachable servers are preceded by *

******* Global RADIUS Configuration *******

Shared-Secret: None

Timeout: 5

Auth-Type: pap

Retries: 1

Initial TLS Connection Timeout: 30

TLS Timeout: 5

Tracking Time Interval (seconds): 300

Tracking Retries: 1

Tracking User-name: radius-tracking-user

Tracking Password: None

Status-Server Time Interval (seconds): 300

Number of Servers: 1

AAA Server Status Trap: Disabled
```

```
---------------------------------------------------------------------
SERVER NAME                                     | TLS   | PORT | VRF
---------------------------------------------------------------------
clearpass.example.com                           |       | 2083 | default
---------------------------------------------------------------------
```

Below you see an example of checking the RADIUS server configuration on an AOS-CX that is using RadSec. Note that the TLS Connection Status indicates that the RadSec connection is up.

```
AOS-CX# show radius-server detail
******* Global RADIUS Configuration *******
<-output omitted->
****** RADIUS Server Information ******
<-output omitted->
Server-Name                : clearpass
Auth-Port                  : 2083
Accounting-Port            : 2083
VRF                        : default
TLS Enabled                : Yes
TLS Connection Status      : tls_connection_established
<-output omitted->
```

Enabling 802.1X on the AOS-CX Switch

There are several steps for enabling 802.1X on AOS-CX switches.

First, specify the RADIUS group for 802.1X on AOS-CX switches if you are using a non-default one. If you are using the default "radius" group, you can skip this step.

You must enable 802.1X on a per-interface basis.

And you must enable 802.1X globally. For 802.1X to operate on a port, it must be enabled both on the port and globally.

Additionally, best practice is to enable RADIUS accounting. You can select start-stop, in which case the switch sends a start accounting message to the RADIUS server when a session begins. Both start-stop and the stop-only options send a message when the session ends. Stop messages describe the user's identity, location, and bandwidth consumption, as well as other information. Finally, you can also configure switches to send periodic accounting messages.

AOS-CX Port Access Roles

When you enable any form of port-access control, including 802.1X, on an AOS-CX switch interface, the switch begins to implement roles on the port. It assigns each client that connects to the port to a role based on the user's authentication status, as shown in Figure 7-35.

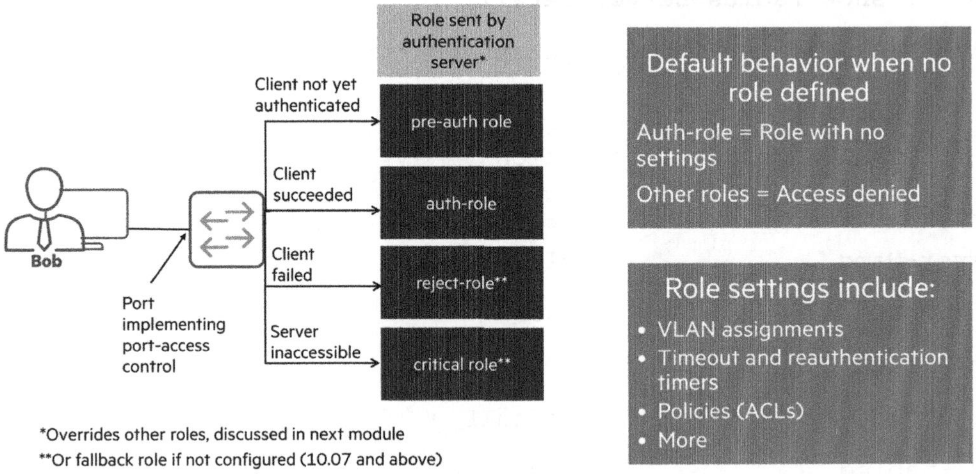

Figure 7-35: AOS-CX Port Access Roles

Before the client authenticates, it is assigned to the pre-auth role. If no pre-auth role is assigned, the switch denies access to the user. (Or, for switches running version 10.09 and above, after authentication times out, a fallback role can be applied.)

After the client authenticates successfully, the switch assigns the client to the auth-role. If no auth-role is defined on the port, the switch assigns a role with no settings; the settings on the port will take effect for the user. In addition, the authentication server can send a role for a user who passes authentication. This role overrides the auth-role, if it is a valid role that the switch can apply. (You will learn more about this option in the next chapter. For this chapter, your goal is to simply get authentication up and running with every user receiving full access.)

If the client fails authentication, the switch applies the reject-role. If the client is unable to complete the authentication process because the RADIUS server is inaccessible, though, the switch assigns the client to the critical role. If the port does not define either of these roles, a fallback role, if configured on the port is applied. (Note that the fallback role was added in AOS-CX version 10.07.) If none of these roles are configured, the client is denied access in these circumstances.

CLI Commands: AOS-CX Switch 802.1X Configuration

The table in Figure 7-36 shows the commands for configuring 802.1X on AOS-CX switches. For AOS-CX switches you need to either create port-access roles and apply them or have the RADIUS server send the role (covered in the next chapter).

In configure context

Step	AOS-CX command	Notes

Configure 802.1X to use RADIUS and set the RADIUS server group.	`aaa authentication port-access dot1x authenticator auth-method eap-radius`
	
	`aaa authentication port-access dot1x authenticator radius server-group <name>`
		EAP-RADIUS is the default method on AOS-CX.
		
		If you do not set a group, the default group ("radius") is used.

Enable 802.1X on ports.	`interface <ID>`
	
	`aaa authentication port-access dot1x authenticator enable`
	
	`exit`
		For AOS-CX, you must access the interface context to enable 802.1X on the interface.

Enable 802.1X globally.	`aaa authentication port-access dot1x authenticator enable`	
Create a port-access role and assign a VLAN to it.	`port-access role <name>` `vlan access <ID | name>`	
Assign a port-access role as the auth-role on a port.	`interface <ID>` `aaa authentication port-access auth-role <name>`	Use if the RADIUS server is not sending a role.

Figure 7-36: CLI Commands: AOS-CX Switch 802.1X Configuration

Summary

Congratulations! You can now configure a basic WLAN than enforces WPA3-Enterprise security. You can also configure AOS-CX switches to enforce 802.1X to protect the wired network.

Learning checks

1. True or false: Campus APs (CAPs) in Tunnel mode forward authenticated user traffic into the VLAN assigned to the WLAN.

   a. True

   b. False

2. You have campus APs and an MC. How can you control where an SSID is deployed?
   a. Assign location tags to APs. Specify the location tag within the New WLAN wizard.
   b. Assign APs in different locations to different groups. In the New WLAN wizard, choose the specific AP Groups for the WLAN.
   c. Within the SSID profile, choose the names of the APs that should broadcast the SSID.
   d. Create a location tag for areas of your network. Within the WLAN's AAA profile, select the location tag.
3. Which are steps for enabling 802.1X on an AOS-CX switch? (Select multiple options.)
   a. Creating a tunnel VLAN
   b. Setting a non-default admin password
   c. Enabling 802.1X on a per-interface basis
   d. Enabling 802.1X globally
4. What additional setup do you need to perform if you want to use RadSec between a network infrastructure device like an MC or switch and CPPM? (Take notes on this open-ended question.)

Note

For answers to all learning checks, refer to Appendix B.

---

8 Enforce Role-Based Authentication and Access Control

EXAM OBJECTIVES

✓ Define and deploy basic user roles for wireless access

✓ Define and deploy basic user roles for wired access

Assumed Knowledge

- Basics of IP addressing
- Basics of Ethernet switching, including VLANs
- Basics of 802.11 wireless networks
- Basic authentication concepts (from Chapter 6 and Chapter 7)

Overview

You will begin by examining role-based firewall policies. You will learn how HPE Aruba Networking ClearPass Policy Manager (CPPM) can assign users to a role, and how HPE Aruba Networking Mobility Controllers (MCs) and HPE Aruba Networking Instant Access Points (IAPs) learn the role. You will then learn about creating firewall policies for different roles. Finally, you will learn how AppRF 2.0 gives admins more visibility and control over the types of traffic that different users are allowed to send.

You will then examine how AOS-CX switches enforce role-based policies to dynamically adjust the switch port to the connected device. You will see how users authenticate and are assigned a role, and the types of settings that the role can apply. You will also learn how switches can download these roles directly from CPPM. You will also learn about Dynamic Segmentation, which allows an AOS-CX switch to use roles to determine that it should tunnel a user's traffic to the MC, which further applies role-based policies. And you will learn how to set up such tunneling on both the switch and the MC side.

Role-Based Firewall Policies

The first section of this chapter covers role-based firewall policies.

Scenario: Protecting the Edge with Role-Based Policies

The company in this scenario is now ready to unlock the power of the HPE Aruba Networking infrastructure. Building on the foundation of 802.1X with EAP-TLS across the complete wireless and wired edge, the company will now implement role-based access policies.

As indicated in Figure 8-1, CPPM will assign wireless users to different roles. The MC will then use that role to apply the correct firewall and other policies to the users' traffic. To expand the protections and ensure consistency of access, CPPM will also assign wired users to roles. The AOS-CX switches will use these roles to determine how to handle the wired clients' traffic. Sometimes the role will call for tunneling the traffic to the MC. The MC will then apply its firewall and other security policies to that traffic.

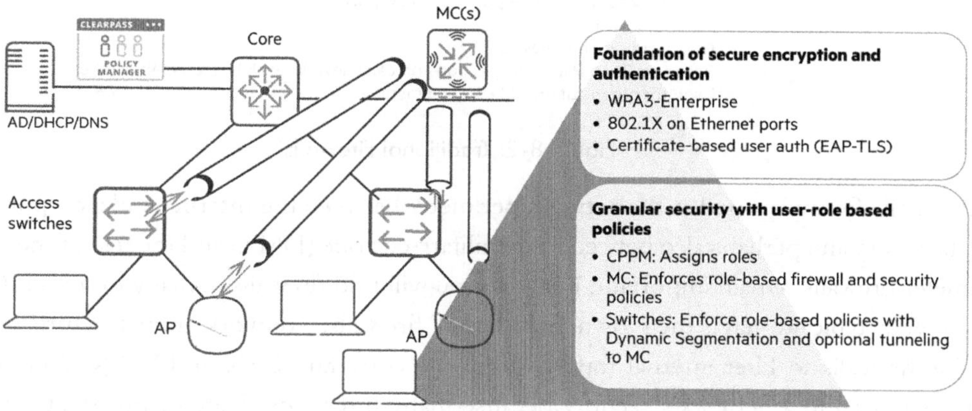

Figure 8-1: Scenario: Protecting the Edge with Role-Based Policies

Traditional Firewalls

To understand how the HPE Aruba Networking approach to security differs from the traditional approach, you need to look at how the AOS firewall differs from a traditional firewall.

A firewall is a device that stands between two network segments and filters traffic between them. As indicated in Figure 8-2, basic firewall functionality includes packet filtering based on rules, which are sometimes called access control lists (ACLs). For example, a rule might filter out traffic on all TCP ports except 80 and 443. In common terminology, you would say that the firewall has "closed" all TCP ports except that ports 80 and 443 are "open."

Modern firewalls often implement some basic protections against denial of service (DoS) attacks. For example, because older DoS attacks could involve sending packets with invalid flags, firewalls check TCP flags and ensure that they are valid and logical for the current point in the session.

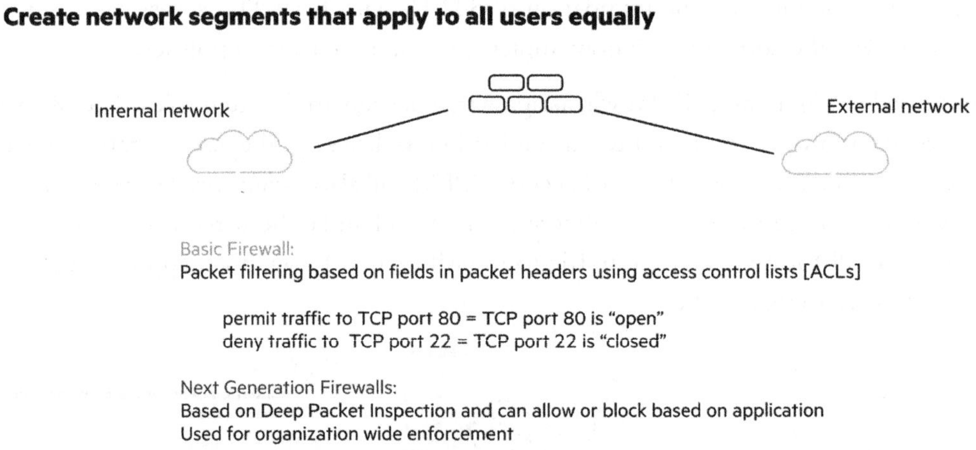

Figure 8-2: Traditional Firewalls

Traditionally firewalls stood at the network perimeter between the Internet and a corporate campus or data center, and perhaps also between a demilitarized zone (DMZ) and the rest of the data center. Perimeter firewalls remain important, but now you will find firewalls in many locations. Computers and servers typically have their own host-based firewalls. Network infrastructure devices also provide firewalls to filter internal traffic within the corporate LAN and WAN. Filtering internal traffic is crucial to a network's security because many threats originate or propagate from within. For example, if a company offers wireless guest access, some of those guests might try to abuse their access to the network. Or a user might innocently connect a device that malware has compromised, giving a hacker a door into the network.

AOS Firewall

The AOS firewall, provided by MCs and IAP clusters, gives organizations the firewall that they need for a zero-trust network. The AOS firewall is a fully functional Next generation firewall. Because of its unique location in the network it creates micro segments in the network based on client identity. This firewall filters all traffic to and from wireless users and, when you implement Dynamic Segmentation, even traffic to and from wired clients.

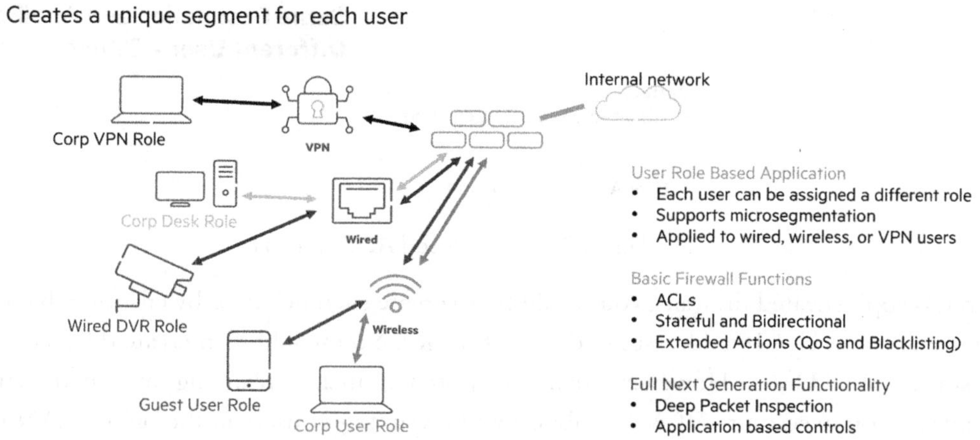

Figure 8-3: AOS Firewall

The AOS firewall is stateful and role-based. As shown in Figure 8-3, by treating each client differently based on its role, the AOS firewall can micro-segment traffic within the same VLAN.

Role-Based AOS Firewall

As shown in Figure 8-4, the role-based AOS firewall can enforce different rules based on user identity and assigned role, rather than simply on features in packet headers. For example, the firewall can restrict contractors to sending Web traffic to limited ranges of IP addresses while it allows managers' traffic to flow more or less unrestricted.

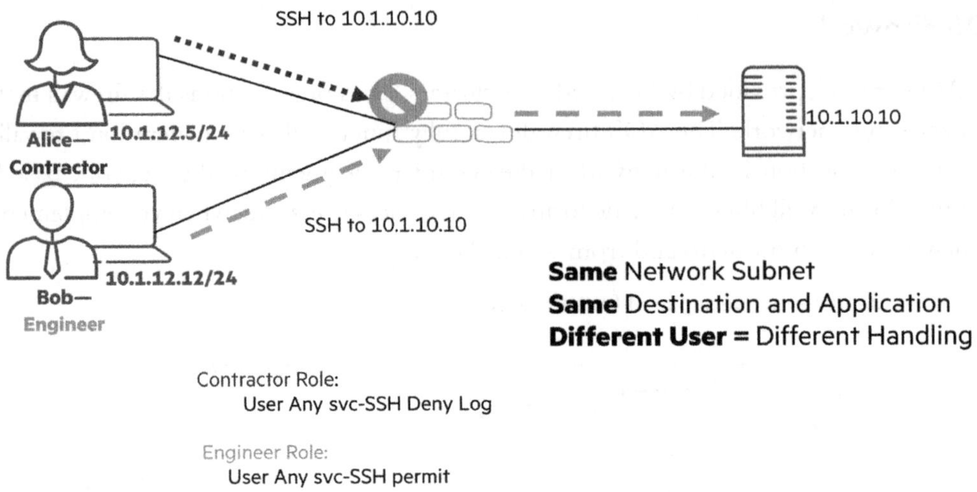

Figure 8-4: Role-Based AOS Firewall

With less sophisticated firewalls, you would try to create such behavior by ensuring that contractors and managers receive IP addresses in different ranges. You would then create different rules based on source IP address. However, such an approach makes planning user connections more complex—you need more VLANs/subnets and a way to get users in the right VLAN every time. With the AOS firewall, you can place users in the same subnets, but easily create different rules for different users.

The AOS firewall determines the user role associated with a device as the device authenticates and accesses the network. It might assign this role entirely on its own, or it might have help from CPPM.

By sending all internal user traffic through the role-based AOS firewall, you can change from a model in which you protect a single perimeter to a model in which you place every user within their own unique and properly customized firewall.

Review: What is a Role?

What is the role that controls a client's access? To review, the role bundles together multiple settings that enable the AOS firewall to control the client's traffic. (See Figure 8-5.)

Some of the most important settings for a role are access control rules, grouped in firewall policies. These rules filter traffic based on packet headers from Layers 2-4. They let you do things like control the destination IP addresses and TCP/UDP applications accessed by clients assigned to the role.

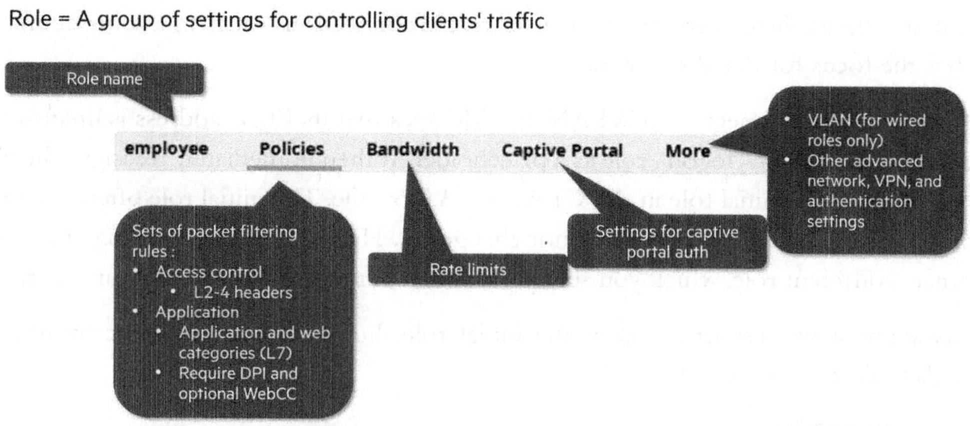

Figure 8-5: Review: What is a Role?

However, the AOS firewall has far more capabilities. A role's policies can also include application rules, which let you control clients' access to specific applications—such as social media or gaming—or specific web site categories, based on URL or reputation. These policies are enabled by deep packet inspection (DPI) and Web Content Classification (WebCC).

The policies are the focus of this training. However, you should also be aware that the role can implement bandwidth controls to rate limit clients' traffic.

And, if the client connection is enforcing captive portal authentication, the role can further specify how the client is directed to the captive portal.

Finally, roles define a variety of other settings, which are covered in more advanced training, such as virtual private network (VPN) settings and other forms of Layer 3 authentication. The **More** section setting that you will examine in this chapter is VLAN. A role-based VLAN overrides the VLAN assigned in another way such as through the WLAN settings. For wireless users it is best practice to assign the VLAN through the WLAN and leave the role-based VLAN non-specified. However, for tunneled wired users, a role-based VLAN assignment works best.

How HPE Aruba Networking Mobility Devices Assign Roles

Because the power of the AOS firewall partially derives from its role-based nature, it is important that you understand how the firewall assigns clients to roles. The device assigning the role could be an IAP in a cluster or an MC. The discussion below will refer only to MC for simplicity, but the process is similar.

You looked at a simplified version of this process for clients that connect to a WPA-Enterprise WLAN earlier. Now you will examine the process in more depth. (See Figure 8-6.)

The MC assigns a role to every client on an untrusted connection—such as a connection to a WLAN, which is the focus for this discussion.

When a client first connects to a WLAN, the MC sees that its MAC address is unknown. The MC assigns the client to the "logon" role as a placeholder. It then immediately reassigns the client to the role specified as the initial role in the WLAN's AAA profile. The initial role often is set to "logon," which means that the actual role does not change yet. However, you can change the initial role to reference a different role, which you set up with the desired policies to suit your security purposes.

The assignment process can stop at the initial role; however, the MC typically uses additional methods to determine the role.

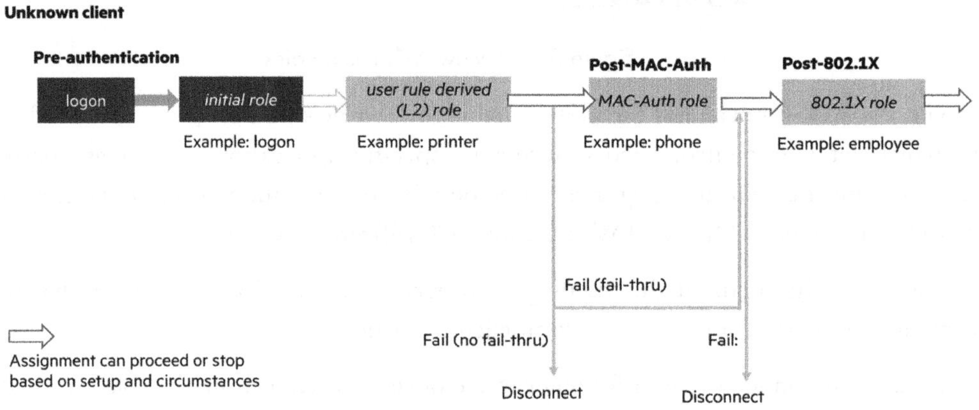

Figure 8-6: How HPE Aruba Networking Mobility Devices Assign Roles

First, the MC checks the client against the user rule set defined in the WLAN's AAA profile. User rule sets consist of one or more rules that map client attributes to roles. Rules with Layer 2 (L2) attributes, such as MAC address, apply before authentication. MCs might use these rules to assign clients that connect to open networks to different roles. For example, the MC might use MAC addresses to identify printers and assign them to a "printer" role.

If the AAA profile has no user rule set, or if no L2 rules apply to client, the client remains in the initial role at this point.

Often the WLAN will then implement a form of L2 authentication, MAC-Auth, 802.1X, or both. First consider a case in which the WLAN implements MAC-Auth. If the client succeeds MAC-Auth, it is assigned to a new role based on the authentication. The MC can derive this role in one of three ways, which you will examine in detail in a moment. If the client fails MAC-Auth, and L2 fail through is disabled, the client is disconnected. If it fails MAC-Auth, and L2 fail through is enabled, though, the client remains connected in the current role.

If the client is still connected, the MC next enforces 802.1X, if applied on the WLAN. Again, if the client succeeds 802.1X authentication, the MC assigns it to a new role based on the authentication. The role derived during 802.1X typically relates to the user identity, but the MC can derive it in many ways. The client is now connected.

If the client fails 802.1X, though, it is disconnected.

As you see, the last role to which the client is assigned is the client's role for the connection. In other words, the 802.1X role (if assigned) overrides the MAC-Auth role (if assigned), which overrides the L2 user rule derived role (if assigned), which overrides the initial role.

In the example in Figure 8-6, the client passes through both forms of authentication. However, if the WLAN does not enforce a particular method, the role derivation associated with that method simply does not apply to the client. For example, if a WLAN enforces 802.1X only, the MC goes straight from determining whether any L2 user rules apply to the client to enforcing 802.1X. Or if a WLAN only enforces a L3 authentication method, the client will stay in the initial or user rule derived role for the initial part of the connection until L3 authentication occurs (discussed in a moment).

As mentioned previously, the MC can derive the role for a client that has passed L2 authentication in one of three ways. Next, you will now look at those methods in more detail (see Figure 8-7).

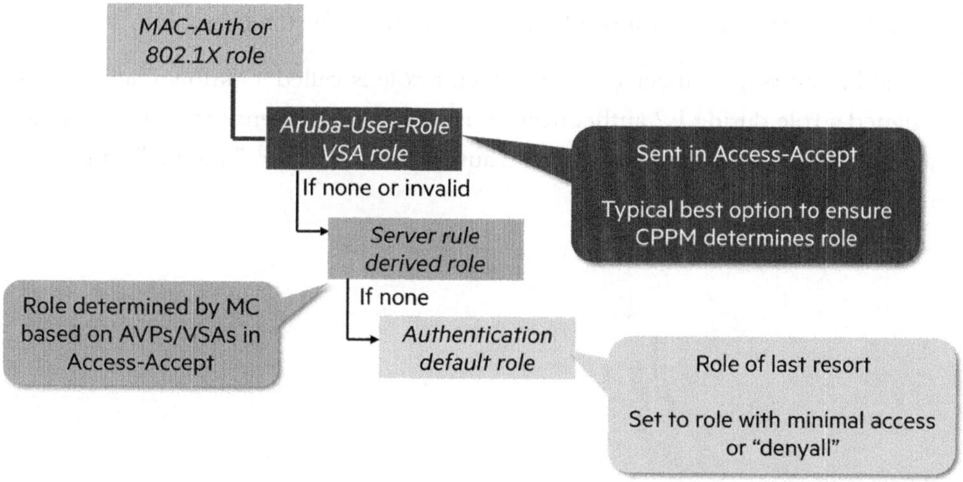

Figure 8-7: Assigning the MAC-Auth or 802.1X Role

First, the MC looks for an Aruba-User-Role vendor specific attribute (VSA) in the Access-Accept from the authentication server. If this VSA is present, and if the VSA specifies the valid role name—in other words, a role that exists on the MC—the MC assigns this role to the client. This is the recommended way to assign authenticated clients to roles, particularly when you use CPPM as the authentication server. This approach lets CPPM do what it does best—determine the right role for a device—while the MC focuses on what it does best—taking that role and enforcing settings based on it.

Only if there is no valid Aruba-User-Role VSA does the MC move on to the next type of role derivation, which are server rules. You can define these rules in the AAA Profile or in the Access section of the WLAN wizard. Server rules take attribute value pairs (AVPs) or VSAs sent by the authentication server in an Access-Accept, as well as client attributes such as MAC address, and map these criteria to a role. These rules allow the MC to make the ultimate decision about the authenticated client's role, but under advisement, in the form of AVPs/VSAs, from the authentication server. These rules are covered in more detail in advanced training.

Finally, if no valid Aruba-User-Role-VSA role or server rule applies to the client, the MC assigns the client to the default role for the authentication method that the client passed. This is sometimes called the role of last resort because you typically want to assign the client to a granular role based on information received from the authentication server. For security, choose a role with the lowest access level appropriate to authenticated users or even "denyall."

For example, you are expecting MCs to receive roles from Aruba-User-Role VSAs, so the WLAN has no server rules. However, CPPM sends a role name with a different spelling from the role name on the MC. The MC would then assign the client to the authentication default role. You want to ensure that this situation does not result in elevated access for the client.

A client that has already connected and received a role is called a "known" client. For example, a client assigned a role during L2 authentication is "known." A client that was assigned to the initial role in a WLAN that does not implement L2 authentication is also "known." In Figure 8-8, you see how an MC treats known clients.

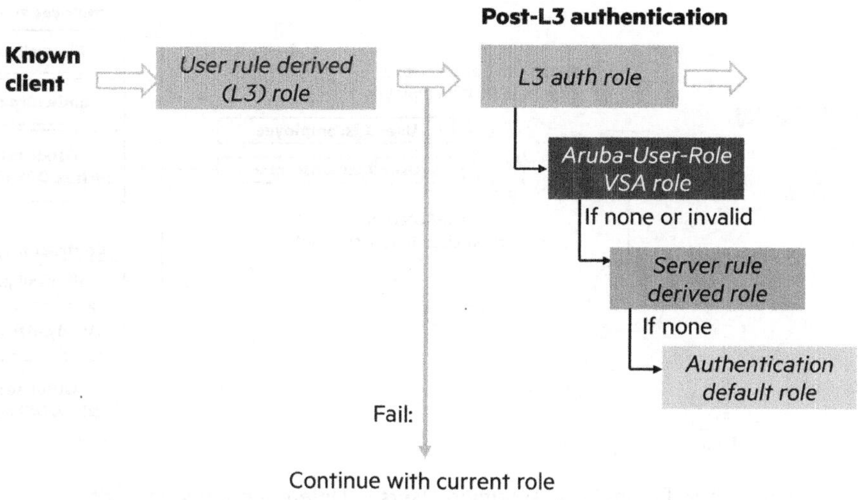

Figure 8-8: Assigning Post L3 Authentication Roles

The MC continues to check known clients against any L3 rules within the user role set in the clients' AAA profile. Rules for DHCP fingerprinting, about which you learn in the next chapter, are L3 rules. If the client matches any of those rules, the MC assigns the client to the role indicated in the rule. This role overrides the role that the client received earlier, including a role assigned during the L2 authentication process—even by an Aruba-User-Role VSA.

If the WLAN enforces L3 authentication such as captive portal, that process occurs after the initial connection as well. If the client succeeds authentication, the MC uses the same three methods to determine the client's role that it would for L2 authentication.

But with L3 authentication, a failure does not disconnect the client. Instead, the client remains connected with the existing role.

Almost always, when you use 802.1X, you do not implement L3 authentication, so a client that has passed 802.1X authentication simply retains its role assigned after passing 802.1X.

Example of Assigning Users to Different Roles with CPPM

While the role derivation can seem complex, it is often simpler in practice.

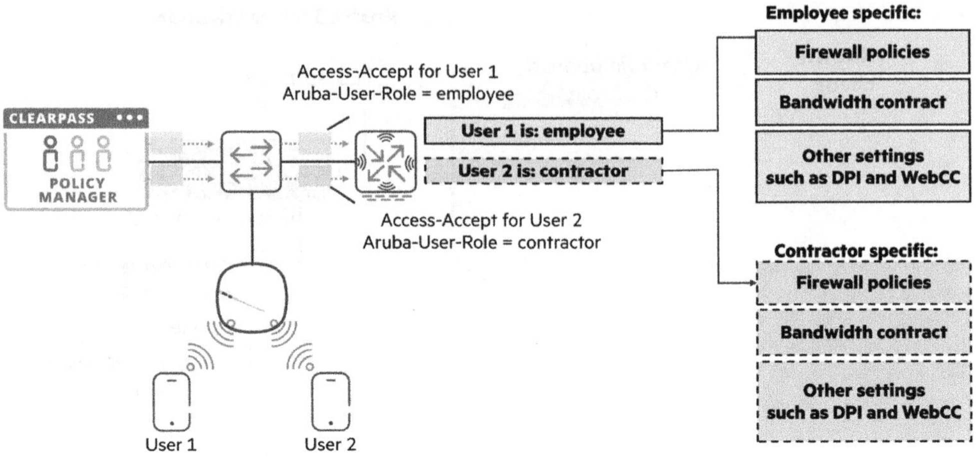

Figure 8-9: Example of Assigning Users to Different Roles with CPPM

In Figure 8-9, you see a simple example for our scenario. Both employees and contractors connect to a WLAN that enforces WPA3-Enterprise. The WLAN uses 802.1X authentication, but not MAC-Auth.

Therefore, clients connect to the WLAN and are briefly assigned to the initial role (but the connection is blocked by 802.1X). They then either fail authentication and are disconnected or pass authentication and receive a role based on their identity and perhaps other criteria uncovered during the 802.1X authentication process.

In this example, the company wants most employees to have the same rights, but to assign different rights to contractors. The MC receives the role name—"employee" or "contractor"—in the Aruba-User-Role VSA within CPPM's Access-Accept message at the user's successful authentication. In other words, you do not need to set up the WLAN with any user rules or server rules. CPPM simply decides the role to which each user belongs, and the MC enforces the decisions.

An Access Tracker record on CPPM is shown in Figure 8-10. As you see, CPPM has authenticated this user and assigned the user to the "contractor" role with an Aruba-User-Role VSA.

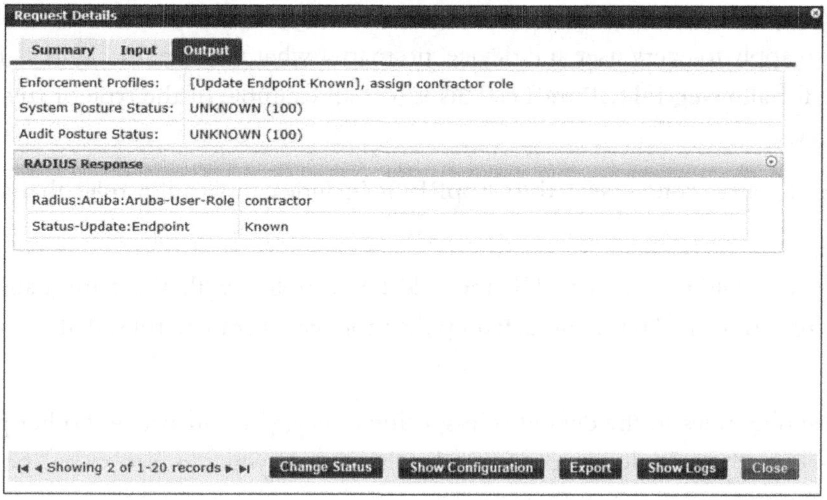

Figure 8-10: CPPM Access Tracker Record (Contractor Role)

Role-Based Firewall Policies

You are now ready to look at the role-based firewall policies in more detail, concentrating on MCs' firewalls.

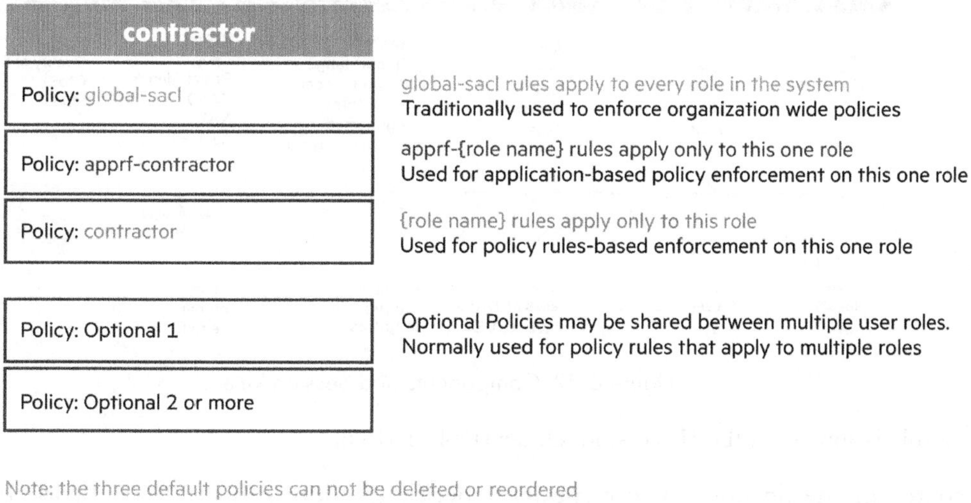

Figure 8-11: Role-Based Firewall Policies

An MC's AOS firewall implements a different set of policies for each role, as shown in Figure 8-11. Each policy contains a series of rules or access control entries (ACEs), which filter traffic for users assigned to that role.

At the top of every role's list of policies is the "global-sacl" policy, in which you can place any rules that you want to apply to every user and device, no matter what the role. In Figure 8-11, this policy includes a rule for allowing DHCP traffic. This is just an example of the type of rule that you can put in this policy.

The next policy in every role is related to AppRF. It includes application rules that are specific to the role.

When you create a role in the Web UI, the role has a policy with the same name as the role automatically created for it. This is the default policy for access control rules that are specific to the role.

In addition to adding rules to the default role-specific policy, you can also add other policies to the role. AOS comes with multiple pre-configured policies, and you can also create policies of your own.

Components of a Session Rule

Now examine the components of a rule in more detail. In Figure 8-12, you are looking at access control session rules.

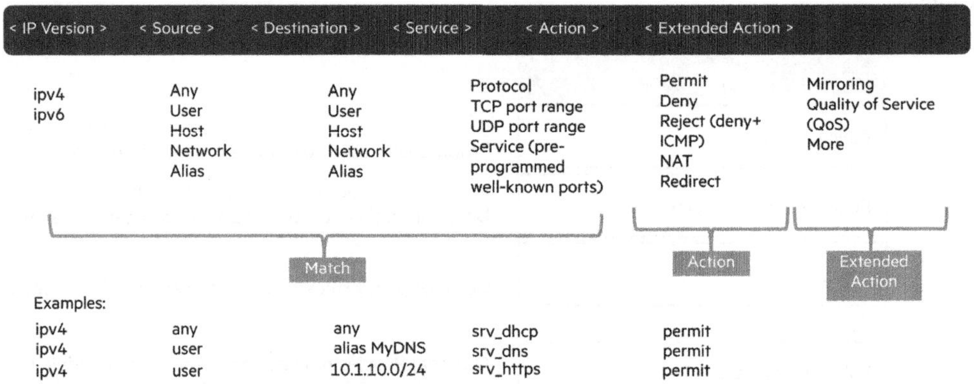

Figure 8-12: Components of a Session Rule

Such a rule begins with the IP version, either IPv4 or IPv6.

For source and destination, you can specify a specific IP address (host) or a network. Because the rules are policy-based, the easiest way to create most rules is usually to specify "user" for the source. "User" applies to any IP address assigned to a client in the client table that is assigned to this role. In some cases you will need to apply the rule to any IP address, rather than user. For example, a client needs to send DHCP traffic before receiving an IP address.

You can also specify an alias. An alias is a named object, which includes one or more host or network addresses. Aliases also make rules simpler to define and more flexible.

In addition to specifying the destination for an access control rule, you can also specify the service, which is defined by TCP or UDP destination port. You can define those manually or select a named service. You can also specify the service based on IP protocol.

Finally, you define the action. The basic actions are permit and deny. You can also specify reject, which denies the traffic and sends an Internet Control Message Protocol (ICMP) reject. More advanced actions include Network Address Translation (NAT), in which the MC forwards the traffic and translates its address, and redirection.

The MC can also add extended actions on top of the main action. For example, you can specify mirroring for traffic that matches the rule—perhaps mirroring the traffic to an intrusion detection system (IDS) for further analysis. Refer to HPE Aruba Networking user guides or more advanced training for more information on other extended actions.

As you can see, while these rules might look like traditional ACL rules at first glance, they can do much more than filter packets.

At the bottom of Figure 8-12, you can see a few simple examples of rules. The first permits any DHCP traffic. The second permits DNS traffic between clients assigned to the role and an alias named MyDNS, a DNS server. The third permits HTTPS traffic between clients assigned to the role and servers in 10.1.10.0/24.

Aliases

Aliases are very useful, so take a moment to examine them in more detail.

MCs support two types of aliases. Network aliases can be used for the source or destination of an access control rule; they represent one or more networks or host addresses. MCs come with several pre-defined aliases. For example, "controller" is an alias that includes any IP address of an MC controlled by this solution. "User" and "any" are very useful aliases, which you will examine in more detail in a moment. Admins can also define their own aliases such as "Domain_controller" or "Internal-network."

Service aliases represent a service or application based on IP protocol or TCP or UDP port list. MCs also come with many pre-defined service aliases for well-known applications. For example "svc_icmp" corresponds to the ICMP protocol, and "svc_https" corresponds to TCP port 443. If the company has custom applications that use defined ports, you can create your own aliases that similarly select traffic by IP protocol or TCP or UDP destination port or port range, as shown in Figure 8-13.

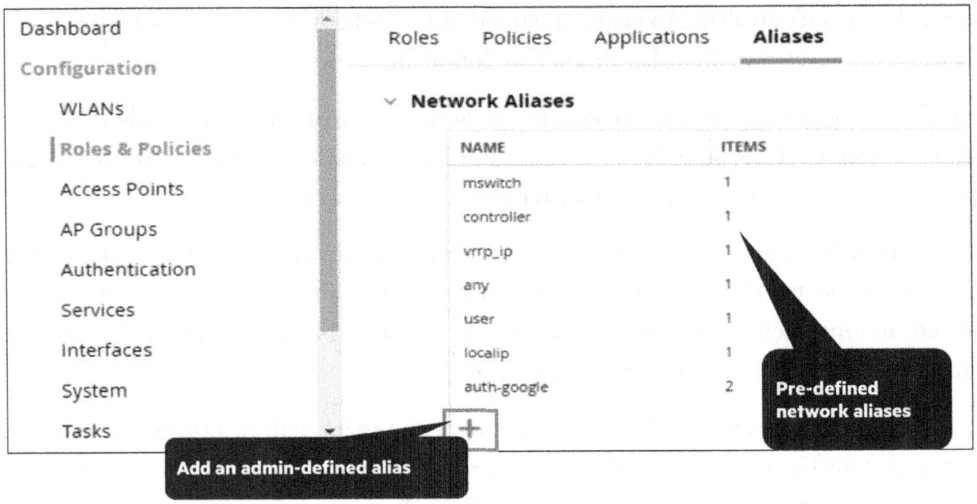

Figure 8-13: Aliases

A more complete list of pre-defined destination aliases is:

- Mswitch—the primary IP address of the Mobility Conductor (MCR)
- Controller—The IP address of the MC
- Vrrp_Ip—The VRRP address used by the MC
- Any—Any IP address
- User—IP address of a user assigned to this role
- localIP—the local IP address of a Remote AP
- Auth-google--An IP address used with Google authentication
- Auth-facebook—An IP address used with Facebook authentication
- Wificalling-block—A component of the UCC service for voice. This alias enables you to block voice calls.

Benefits of Aliases

Aliases have several benefits.

Aliases are self-documenting. For example, when you specify "Internal-network" rather than 172.16.0.0/16 in a rule, other admins can instantly understand the intent of the rule.

In addition, aliases make the solution more manageable and scalable. You can specify multiple subnets and ranges within an alias, so that you can create one rule rather than multiple rules.

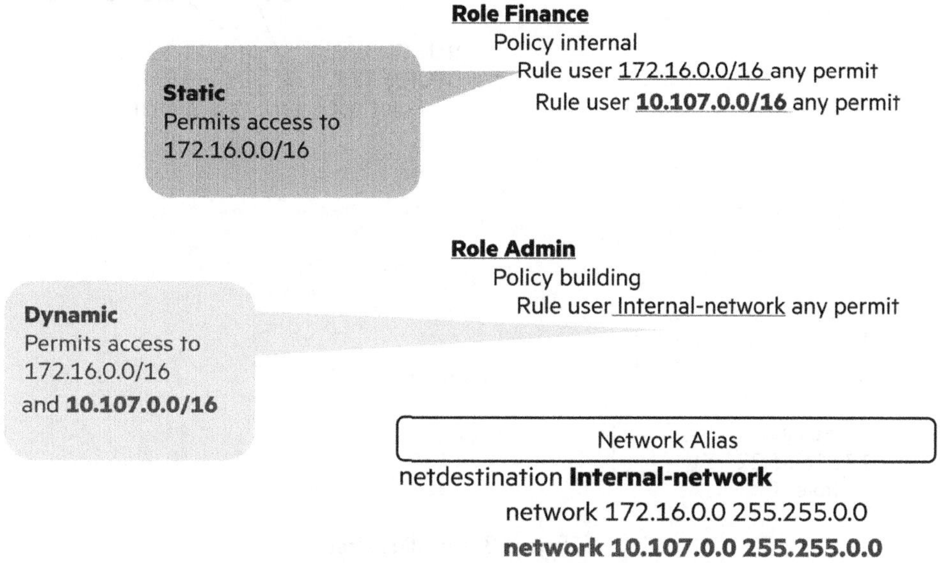

Figure 8-14: Benefits of Aliases

Aliases also update rules automatically. In the example shown in Figure 8-14, the Internal-network alias originally included 172.16.0.0 255.255.0.0. At that point, the policies in the Finance, Sales, and Admin roles all permitted users to access 172.16.0.0/16. However, the rule in the Finance role is static. It never updates. This makes the policy harder to manage when changes occur. For example, the network might expand into the 172.17.0.0/16 space. With aliases, you can easily add that network to the Internal-network alias. The rules that use this alias now automatically update, permitting sales and admin users to access 172.16.0.0/16 and 172.17.0.0/16.

Alias User

The pre-defined "user" alias is very useful and is best the source for many rules. This alias automatically includes the IP address of any client that is assigned to the role under which the rule is configured. The IAP or MC learns these IP addresses by snooping clients' traffic.

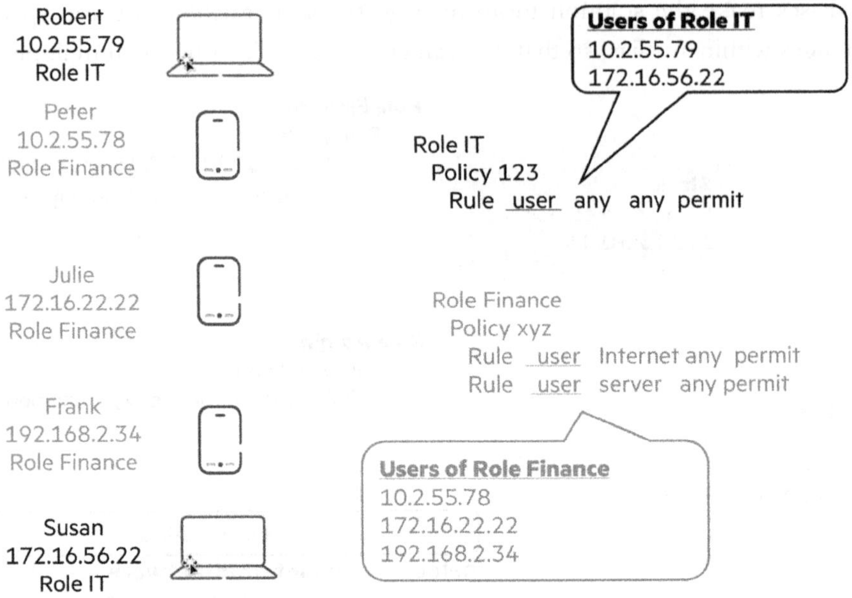

Figure 8-15: Alias User

The example in Figure 8-15 features two roles, each of which have rules with the "user" alias. In this example, Robert and Susan are at different sites. They have been assigned IP addresses in different subnets by DHCP. But both users are assigned to the IT role, and so their IP addresses are both automatically included in the "user" alias for that role.

Peter is in the same subnet as Robert, but he's assigned to the Finance role, so his client IP address is included in the "user" alias for Finance rules. Similarly, Julie's and Frank's IP addresses are included in the Finance rules' "user" alias, even though Julie and Frank are at different sites and in different subnets.

The AOS firewall always applies a role's rules only to traffic from or to clients assigned to that role. Therefore, the "user" alias can seem like the "any" alias in some ways. However, using the "user" alias in rules can help to prevent IP spoofing. For example, if Robert tries to spoof IP address 10.15.10.10, his traffic will not match the "user any any permit" rule, as "10.15.10.10" is not included with the IT users.

As you learn more about the bi-directional nature of the firewall, you will come to understand more differences between the "user" and "any" alias.

Session Rule Handling by Stateful AOS Firewall

You should also understand that the session access control rules are stateful and bi-directional.

As shown in Figure 8-16, a stateful firewall like the AOS firewall can track which packets belong to the same session and handle the traffic as a session, rather than only as individual packets. When a packet that is not part of an existing session arrives, the firewall checks that packet against its rules and permits or denies it. For a permitted packet, the firewall creates a session associated with the traffic. If more traffic in the same session arrives, including return traffic, the firewall permits that traffic. If you have set up a rule to permit clients to access a server on TCP port 443, as in this example, you do *not* have to create another rule to permit the server to access clients.

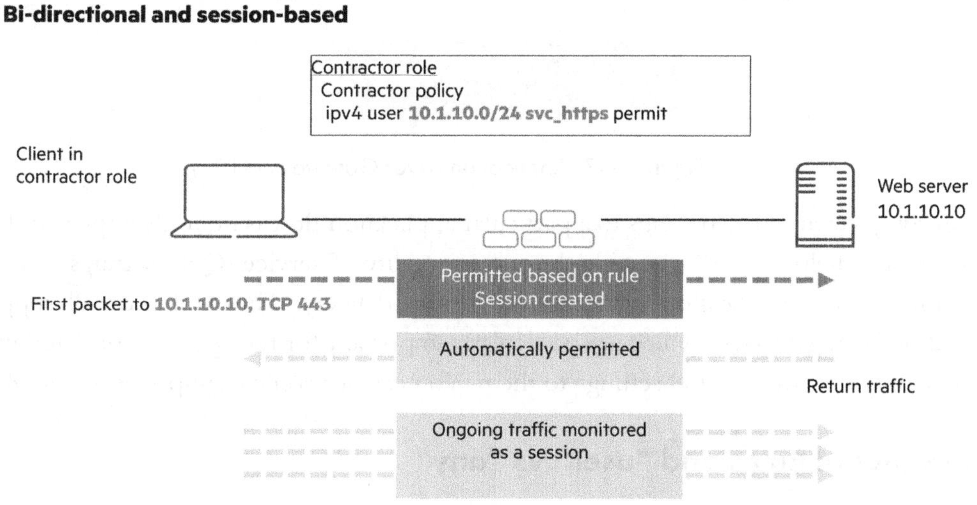

Figure 8-16: Session Rule Handling by Stateful AOS Firewall

The same holds true for a rule that permits UDP DNS traffic from the client to the DNS server. When the client sends a DNS request to the server defined in the MyDNS alias, the firewall permits the DNS response from that server without needing another rule for the return traffic.

Application Layer Gateways (ALGs)

As another key benefit, the stateful AOS firewall provides application layer gateways (ALGs), which help to secure and properly handle the many applications that use dynamic ports.

With these applications, a client first contacts the server on a well-known port. As shown in Figure 8-17, the client and server then negotiate a dynamic port for the session. If you were using simple rules, you would have to open up all the ports that might be used as dynamic ports, which could pose a security risk. An application layer gateway is specific to an application, and it gives the stateful firewall the insight that it needs into handling traffic for that application. For example, when a client first initiates a session for that application, the ALG notes the selected dynamic port and opens that port just for that session.

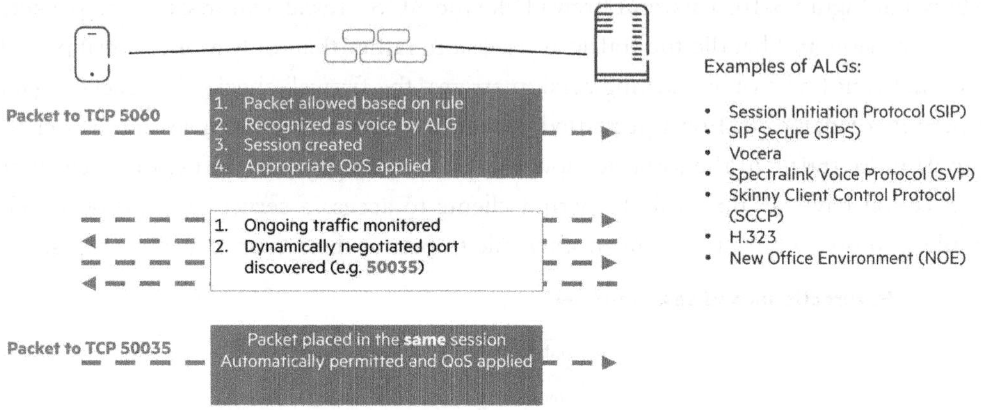

Figure 8-17: Application Layer Gateways (ALGs)

Recognizing which traffic belongs to a particular application flow is not only important for security, but it can also help the ALG provide the proper quality of service (QoS) settings to the traffic so that time-sensitive applications can receive priority delivery. The AOS firewall supports many application layer gateways , which are particularly important for recognizing voice and video traffic and assigning the correct QoS settings to them. You can see some examples in Figure 8-17.

Bi-directional Rules and "user" vs "any"

Now that you understand how the stateful firewall works, you can better understand the difference between the "any" and "user" aliases.

When you create a "permit" rule with the "user" source, you are not just preventing spoofing. You are also indicating that the controlled client must initiate the traffic flow associated with that rule. When a "permit" rule includes the "any" source, on the other hand, a device on the outside of the firewall can initiate the session.

In Figure 8-18, you can see the differences between the "any" and "user" aliases. The top rule has "any" for source and permits ICMP (pings). A client assigned to the role can ping the server. The reply is permitted by the bi-directional stateful firewall, but would be permitted in any case, because pings are permitted from any IP address. The server can also initiate the ping, and the rule permits this traffic because it permits ICMP from any IP address, within or outside of the firewall.

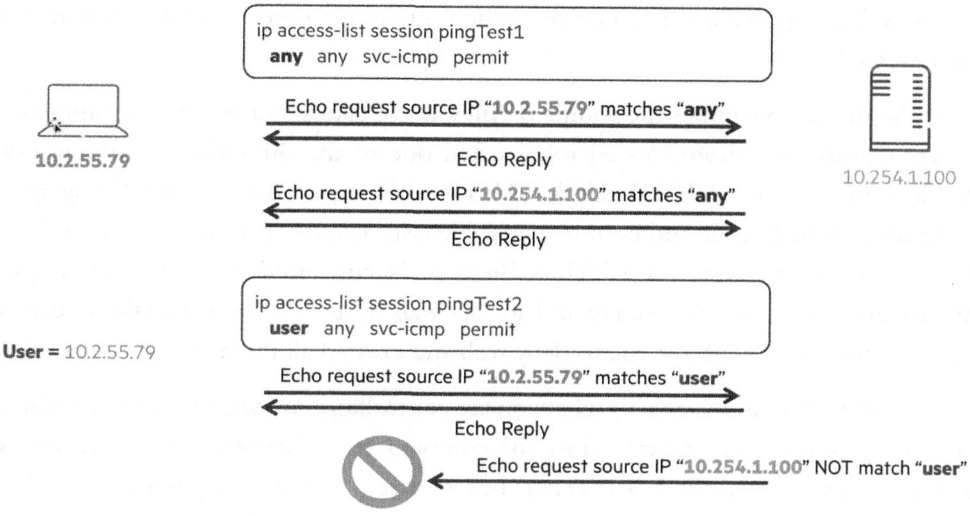

Figure 8-18: Bi-directional Rules and "user" vs "any"

The bottom rule, on the other hand, uses "user" for the source of the rule that permits ICMP. The client assigned to the role can ping the server and get a reply, because the bi-directional, stateful firewall permits return traffic. However, the server is not a user, so it is not allowed to initiate the ping.

Translating Security Needs into Firewall Policies

With this background in the AOS firewall, you can start to consider how to translate your security needs into firewall policies.

First you need to know how you will group users and devices into roles. As you work to define users on the MCs, you will need to coordinate with RADIUS server admins, as they are ultimately responsible for assigning users to roles. You need to match the names sent by the RADIUS server exactly.

Because a role corresponds with security policies, you should have a unique role for each set of users or devices with a different trust level. You should also group users and devices by how they use the network. The most secure approach to creating policies is granting each user only the level of access that they need to run their required applications. In other words, even if users belong to a trusted role, if the users do not need to use SSH to do their jobs, you should not allow the users to send and receive SSH traffic.

This approach to security is sometimes called allow listing. You deny access to all services by default and explicitly define which services to permit. This approach is ideal for less trusted users. It can also work well for specialized IoT devices, which often only need to send a specific type of traffic on documented ports.

On the other hand, sometimes the company will want to take a more permissive approach to setting up access controls for more trusted roles, often due to the difficulty in defining exactly which applications users in that role will require. This is sometimes called a block listing approach. You define various denied traffic and permit everything else. The AOS firewall has an implicit "deny all" at the end of all roles' policies, which is the most secure model; so, if a role requires a more permissive approach, you will need to add an "allowall" policy at the end. Make sure, if you do, that you have documented which traffic to deny well and created all the necessary deny rules.

You also need to define well exactly what you are controlling. Are you stopping users from reaching specific destinations? For example, does the company have different sets of servers dedicated to applications used by different departments? In that case, you can simply permit each department to access only its own set of servers. Are you defining which types of traffic users can send by TCP or UDP application such as HTTP, DHCP, and DNS? The most secure approach is to do both, defining precisely which services users can access on which devices.

As you are defining rules, you will probably run up against the challenge of defining permitted and denied applications based on UDP and TCP port, as many different applications use HTTP or HTTPS. You can look to AppRF 2.0 for help here, as you will learn a bit later.

Example AOS Role Configuration

In this section, you will follow an example of defining an AOS role with appropriate policies and rules.

Defining the Contractor Role

You begin by refining requirements so that the policy will permit only the traffic that a specific user role requires.

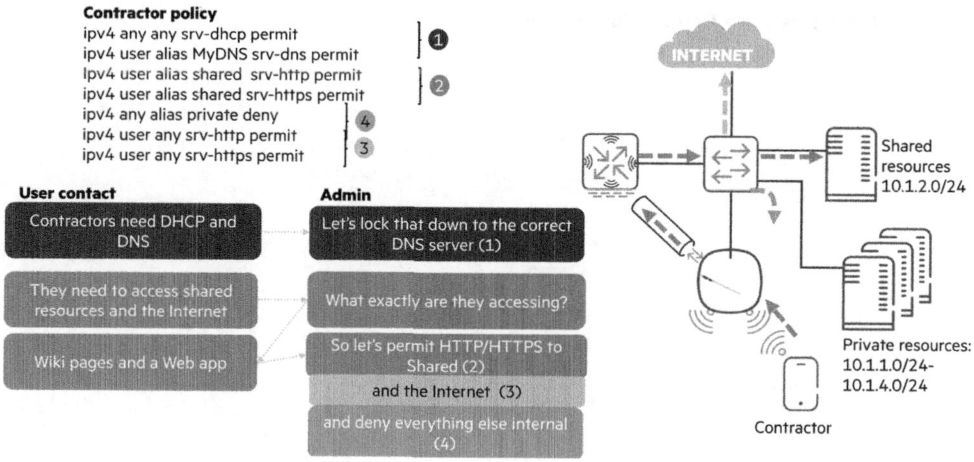

Figure 8-19: Example AOS Role Configuration

In the example shown in Figure 8-19, contractors need DHCP and DNS services, of course. But you can lock down the DNS services to the correct DNS server to prevent misuse of this protocol. Discussions with the department using the contractors reveal that the contractors need access to servers in the subnet labeled shared resources. Further discussion reveals that the contractors are only accessing Wiki pages and a single Web app. So you decide to permit access to the shared resources over HTTP/HTTPS only.

You then deny the contractors access to the rest of the private network, and you allow them HTTP/HTTPS access to the Internet.

Adding a User Role

You will now look at how you can create such a role in the MC (or MCR) Web UI.

As shown in Figure 8-20, you follow these steps:

1. Select **Configuration > Roles & Policies > Roles**. Click the plus icon.
2. Name the role and click **Submit**.

Configuration > Roles & Policies > Roles

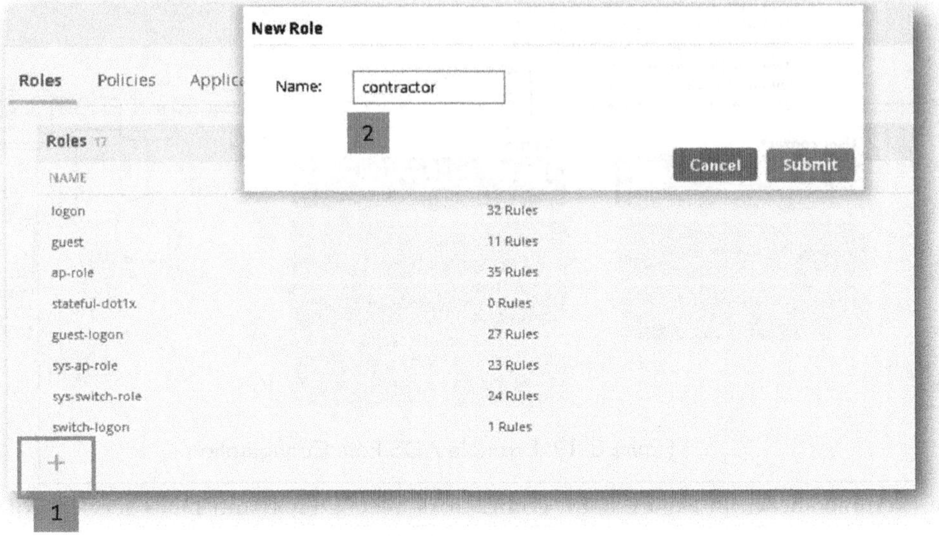

Figure 8-20: Adding a User Role

Viewing the Role

3. As shown in Figure 8-21, you can now select the role and see the rules that the firewall enforces for those users. Initially, no rules exist unless the global-sacl includes rules. No rules means no access. The policies end with an implicit deny.

Configuration > Roles & Policies > Roles

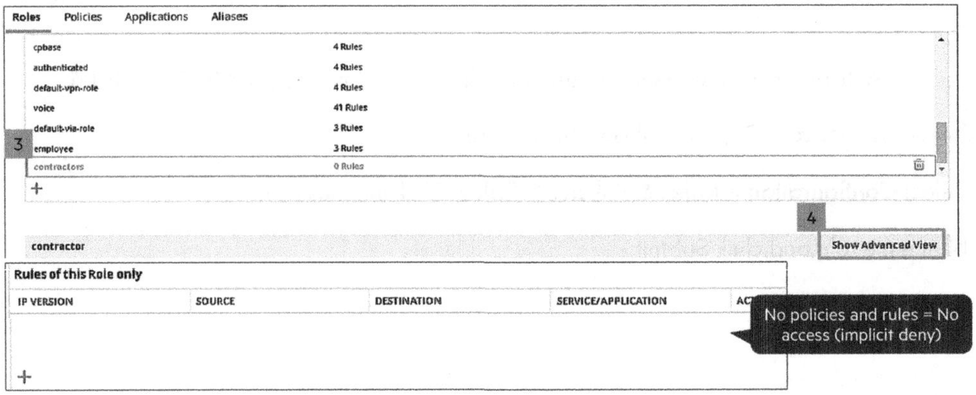

Figure 8-21: Viewing the Role

4. You can add rules by clicking the plus icon under **Rules of this Role** only. But in this example, you will click **Show Advanced View**.

Adding Access Control Rules to the Role Policy

In the Advanced view, you see the policies assigned to the role' as shown in Figure 8-22. At the top is a global policy, in which you can place any policies that you want to apply to every user and device, no matter what the role. The next policy is related to AppRF—you will look at that feature in a moment. Next is the policy automatically created for the role with the same name as the role. Any rules that you add in Basic view are added to this policy.

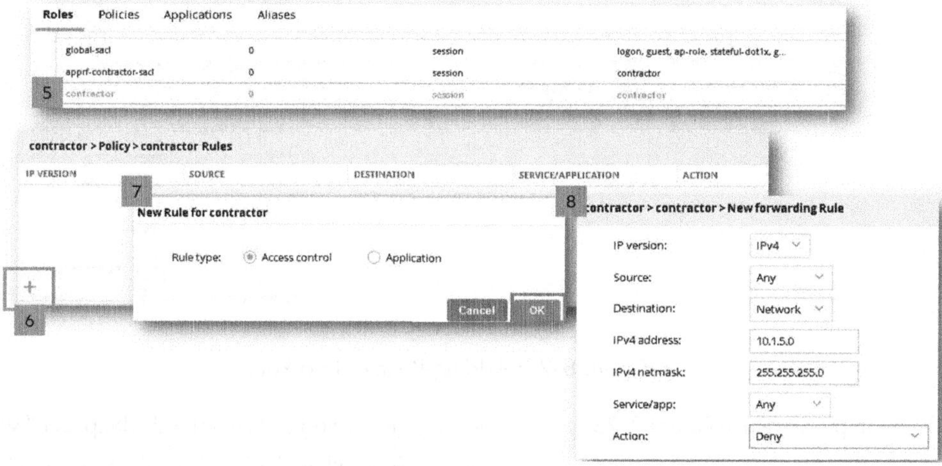

Figure 8-22: Adding Access Control Rules to the Role Policy

5. If you want to add a rule to the default policy, click it.

6. Then click the + icon.

7. Select **Access control** and click **OK**.

8. In Figure 8-22, you see how you can choose the IP version, source, destination, service, and action. For this example, you want to deny traffic to and from a specific subnet. You set the source to "any." This denies traffic originating from clients assigned to this role destined to this subnet, and it also denies traffic from this subnet destined to clients assigned to this role. For destination, you select Network and then fill in the network address and mask. You can alternatively create an alias for the network in advance. In this example, you want to deny all traffic, so you set Any for the Service/app. Then you select the Deny action. Refer to the appropriate user guide for more details if you are interested in learning about more advanced rules. When you finish configuring the rule, click **Submit**.

Adding Policies to a Role

9. You can also use the Advanced View to add policies to the role. Click the + icon in the policy list, as shown in Figure 8-23.

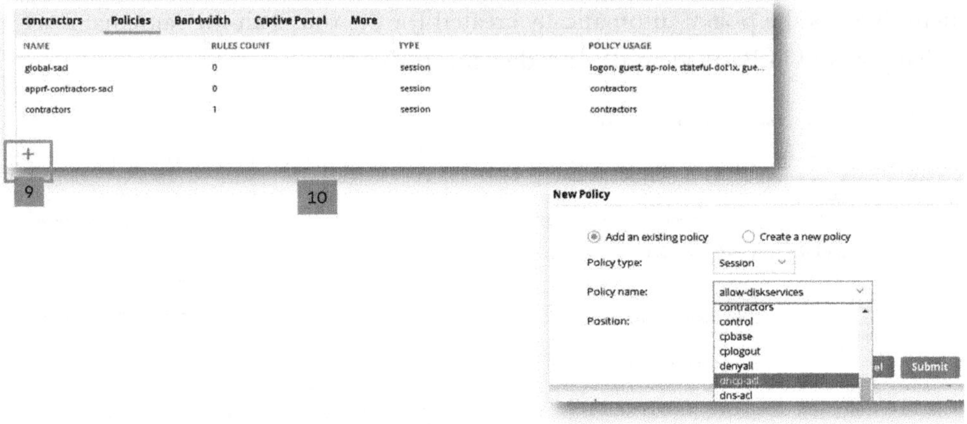

Figure 8-23: Adding Policies to a Role

10. In the example shown in Figure 8-23, you are adding an existing policy named "dhcp-acl," which permits DHCP traffic. Alternatively, you could create a new policy with multiple rules, which you could then apply in other roles too. When you add the policy, you can indicate the placement in the policy list for this policy. As you will see, order matters.

Viewing Rules and Examining Processing Order

11. When you are done adding policies and rules, you can return to the Basic View and see the complete list of rules that apply to the user. See Figure 8-24.

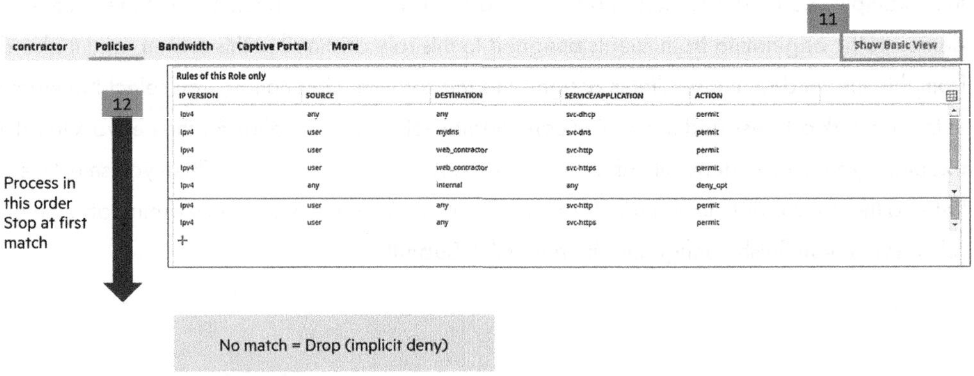

Figure 8-24: Viewing Rules and Examining Processing Order

12. It is important to pay attention to the order because the firewall processes the first rule first. If the traffic matches that rule, the firewall applies the action and does not match the traffic against any more rules. If there is no match, the firewall matches the packet to the next rule. It continues until it finds a match, and then it applies the action and stops.

In this example, a contractor might send HTTP traffic to 10.1.5.5, which is within the internal alias. This traffic hits the rule to deny all traffic to 10.1.5.0/24 before it hits the rule to permit HTTP traffic, so it is denied—the desired behavior.

The list ends with an implicit deny all. In other words, if the traffic does not match any of the rules, the firewall drops it. In this example, contractors are essentially locked down to Web traffic with a few required services, and contractors are locked out of some internal subnets entirely. Remember, though, that the firewall does automatically permit any return traffic for permitted traffic. In other words, the "ipv4 user web_contractor svc-https" rule, for example, permits contractors to *initiate* HTTPS sessions with servers included in the "web_contractor" alias. Devices within the "web_contractor" alias *cannot* initiate HTTPS sessions with contractor clients, but they can send return traffic within the sessions initiated by the contractors.

If you want to take a more permissive approach for a trusted role, you can create a few deny rules and then permit all other traffic; simply add the "allowall" policy to the end of the list.

AppRF 2.0

One of the challenges of controlling user access with a firewall is that so many different modern applications run over HTTP/HTTPS. For example, as shown at the top of Figure 8-25, a user could access social media, work in a shared Wiki, send chats over a company app, and even attend video conferences—all over Web. With traditional ACLs you must allow either none of the applications that use HTTP/HTTPS or *all* of these applications. But the HPE Aruba Networking AppRF 2.0 feature provides deep packet inspection (DPI), which gives visibility down to the application layer. This feature lets the firewall choose which applications to permit or deny more granularly, as shown at the bottom of Figure 8-25.

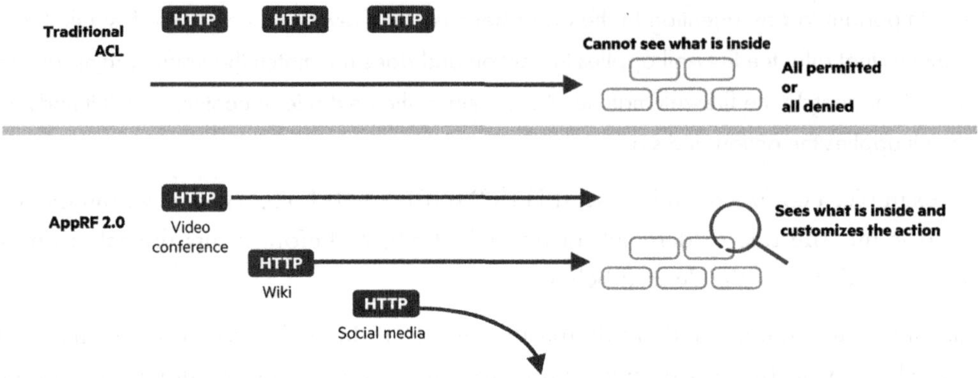

Figure 8-25: AppRF 2.0

To use AppRF, first enable DPI globally, which requires a reboot. You must also ensure that DPI is enabled on the role (it is by default). If you want to filter by Web category, as well as by application, you must also enable WebCC globally and on the role (it is enabled on the role by default).

Also make sure that you have the proper licenses for these features.

Example Creation of Application Rules

You can then create application rules within the role's AppRF policy. The policy is always second in the list of policies, directly after the global session ACL. It is called "apprf-*role-name*-sacl"—for example, "apprf-contractors-sacl."

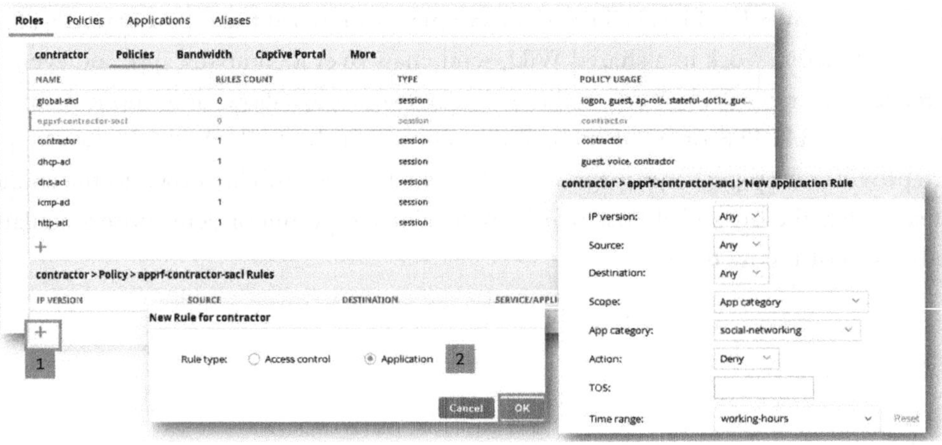

Figure 8-26: Example Creation of Application Rules

This policy only supports application type rules. As you see in Figure 8-26, an application rule specifies permitted applications, such as social networking, rather than permitted services.

Application Rule Components in More Detail

In Figure 8-27, you see the components of such an application rule in more detail.

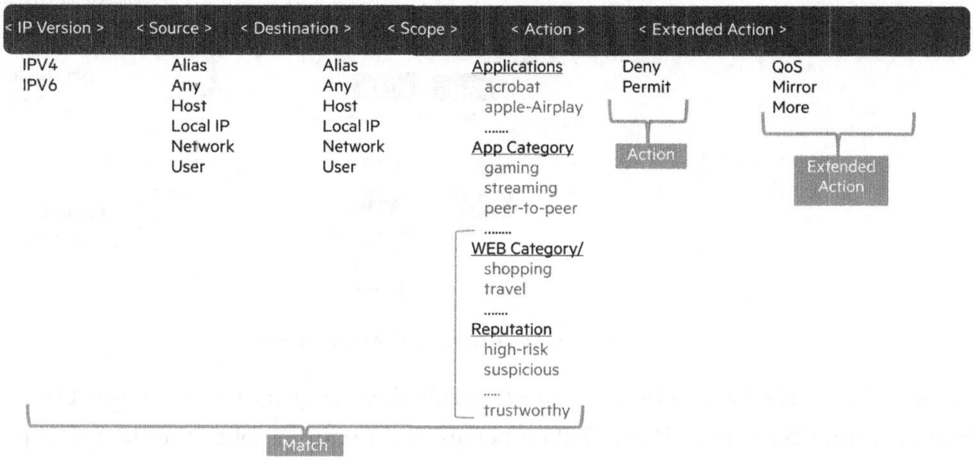

Figure 8-27: Application Rule Components in More Detail

The source and destination options are the same as those for access control session rules. However, instead of selecting services, you can select scopes.

The scopes fall into three categories. Application scopes deal with applications such as Acrobat, Apple Airplay, and many more. The Application category breaks down into categories such as gaming and streaming. When you select Web Category for the scope, you receive the Web Category and Reputation options. Categories include ones like shopping or travel. Reputation settings range from high risk up to trustworthy.

You can only configure two actions: deny and permit. You can then add extended actions if you want.

AppRF 2.0 enforcement

To enforce application rules, the firewall goes through two phases. In Figure 8-28, examine a situation in which a packet arrives for which no session yet exists. Phase 1, the application detection phase, begins. The firewall checks the packet's Layer 3 and 4 headers against all rules in the user's role, beginning at the top. Application rules should be at the top, so, if the packet uses the implicit ports associated with an application rule, it hits the application rule first. The firewall flags the packet as subject to that rule. (If the packet does not match an application rule, the firewall applies the first matching access control rule as normal.)

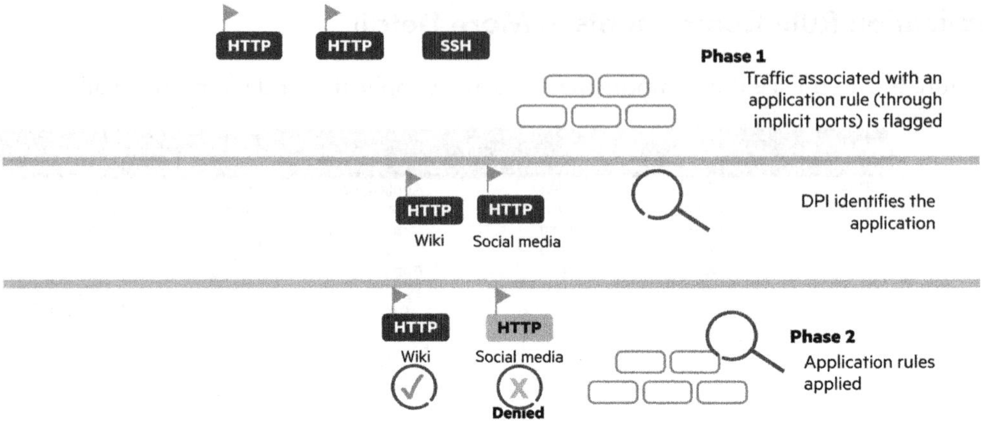

Figure 8-28: AppRF 2.0 enforcement

The firewall lets traffic flagged by an application rule flow temporarily, but it uses DPI to detect the application within the traffic. Phase 2 then begins; the firewall applies the flagged application rule or rules and permits or denies the traffic based on application category or Web category and reputation.

PEF and WebCC Licenses

Many of the features that you have examined in this section are enabled by the Policy Enforcement Firewall (PEF) license. The table in Figure 8-29 reviews features that this license adds to the AOS. To apply these AOS firewall features to wireless clients, you must have one PEF license per AP. As you will learn in more detail in a moment, MCs can also apply AOS firewall features to wired clients using a feature called "tunneled node." MCs also require one PEF license per switch implementing tunneled node.

License	Main features	License consumption

PEF	• Custom user roles
	• DPI
	• Global and role-based policies
	  • Access control
	  • Application rules (over 3,000 applications)
	• Bandwidth controls
	• ALGs
		1 per-AP
		1 per-tunneled node switch

WebCC	• Ability to filter traffic based on web category and reputation	1 per-AP; 1,3,5,7, or 10 year subscription

Figure 8-29: PEF and WebCC Licenses

If you want to use the WebCC options within application rules, you must obtain one WebCC license per-AP. WebCC licenses are subscriptions for varying numbers of years. WebCC licenses are not strictly required to implement WebCC for tunneled node clients; however, the MC must have at least as many WebCC licenses as AP licenses.

Colorless Ports and Dynamic Segmentation

In this section of the chapter, you will learn how you can enforce role-based access and micro-segmentation in the wired infrastructure.

Traditional Ports

Traditionally, admins have had to configure each switch port with the correct settings based on the user or device type that connects to the port. For example, admins might configure a port that connects to an employee as a member of the employee VLAN while they configure a voice over IP (VoIP) phone's port in the voice VLAN. They also give the VoIP phone's port a quality of service (QoS) priority of DSCP 46. What happens, though, if the employee accidentally swaps the jacks for the computer and VoIP phone? Now the devices are in the wrong VLANs and cannot function.

Even worse, errors can lead to security vulnerabilities. For example, a company might have a guest VLAN that provides minimal access as a courtesy for guests. Ports in meeting rooms might be assigned to this guest VLAN. However, if a guest connects to the wrong port, whether on purpose or accidentally, that guest could be placed in the employee VLAN and receive elevated access to the company's internal resources.

Traditional manual configuration of network ports also makes it difficult for companies to implement the role-based security that is so important for a zero-trust world. If the company wants to assign different departments to different VLANs, each of which has its traffic controlled in different ways, the administration becomes more complex. Every time that employees want to change seats, IT has to get involved. Particularly in modern offices that often use desk sharing, this is not feasible. Instead IT probably assigns all employees to the same VLAN and loses out on the benefits of segmenting users based on role. This ultimately makes the network less secure.

Finally, the traditional approach can cause a proliferation of unused ports. Each office might require multiple pre-configured phone ports, printer ports, and user ports, just in case they are needed. See Figure 8-30 for a summary of the disadvantages of traditional ports.

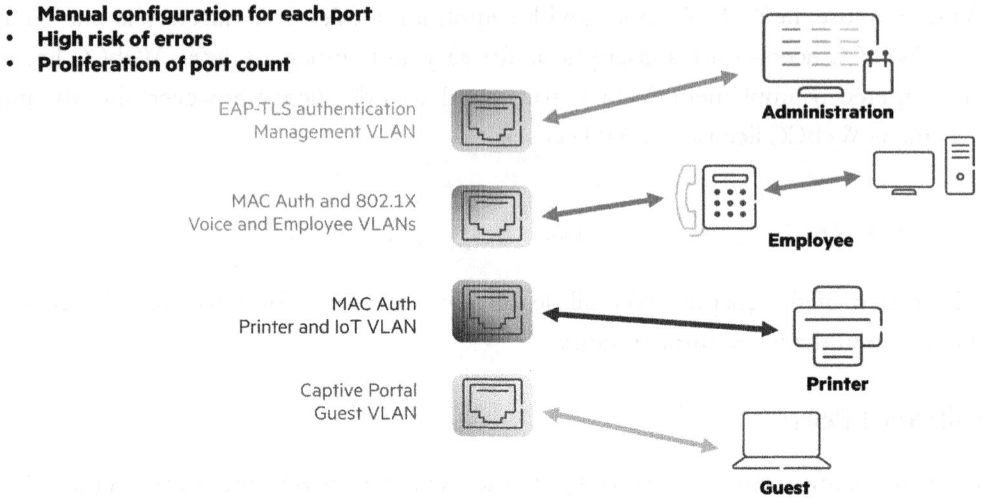

Figure 8-30: Traditional Ports

Colorless Ports

To deal with these issues, companies have long been searching for "colorless" ports. In other words, all ports have the same configuration, but they automatically adjust their settings based on which device connects. A VoIP phone connects to the port, and the port automatically takes on the correct VLAN and QoS. Next a user moves the VoIP phone to a new port and connects their laptop to the original port, and the port automatically changes to the correct VLAN. Similarly, the port adjusts itself with the correct settings for a guest user versus an employee. The connection settings follow the device automatically.

The benefits of colorless ports are summarized in Figure 8-31.

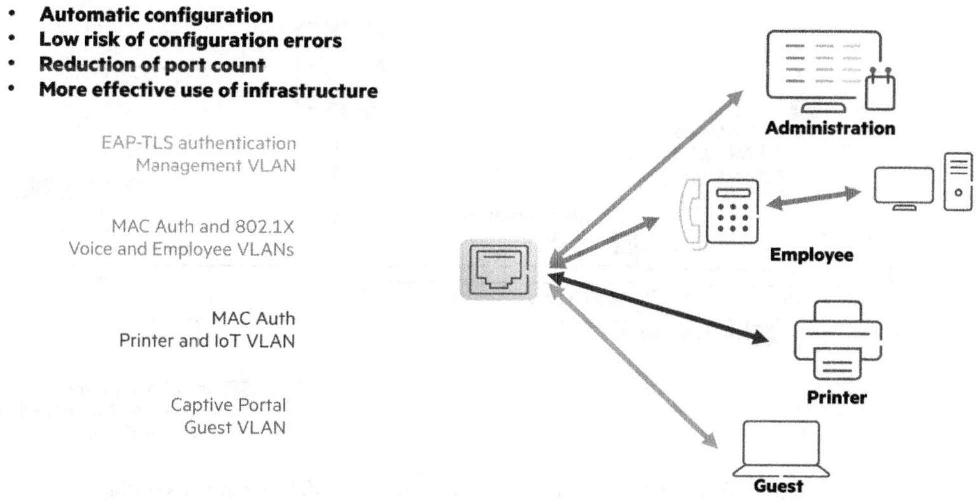

Figure 8-31: Colorless Ports

Colorless ports are easier to manage. They also support greater flexibility for users, who can now move their wired devices wherever required without IT involvement. Colorless ports also help to promote greater security because IT ca More feasibly create logical divisions between different types of users and devices and be confident that the assignment truly follows the client. And colorless ports help to conserve ports because one port can adapt to many uses depending on the device that currently connects to it.

AOS-CX Switch Colorless Ports With Local Forwarding

AOS-CX switches support colorless ports through role-based authorization. When a client authenticates to a port, the switch assigns the client to a role and then enforces particular settings based on that role. In this first scenario, the switch is applying a role that uses local forwarding.

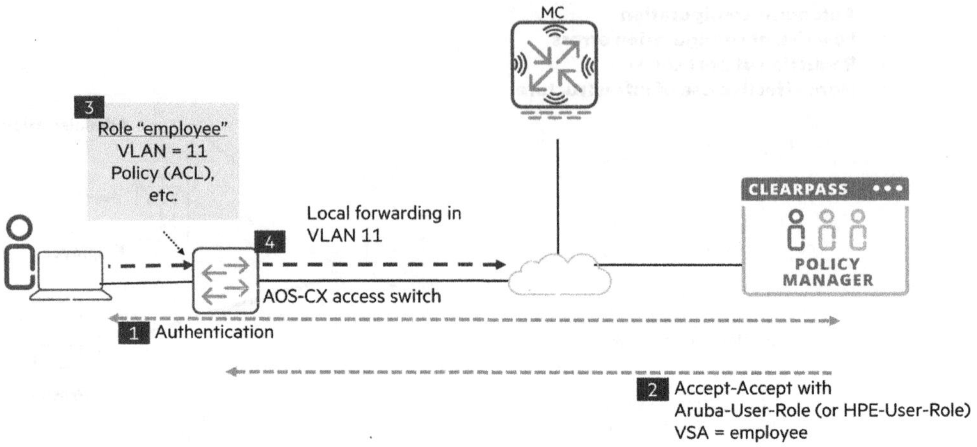

Figure 8-32: AOS-CX Switch Colorless Ports With Local Forwarding

As shown in Figure 8-32, the role-based authentication and authorization process, at a high level, is:

1. A client connects to an AOS-CX switch port, and the user or device authenticates.
2. By the end of the authentication process, the user is assigned to a role. In this example, CPPM sends an Aruba-User-Role VSA with the role name.
3. The switch uses the settings in that role to define settings for the client traffic. Role settings include VLAN settings, policies (ACLs), session settings such as reauthentication periods, and device settings such as power over Ethernet (PoE) settings for the port. A role can also enable tunneling of traffic to an MC, as you will see in the next section.
4. In the example shown in Figure 8-32, though, the role does not specify tunneling. Therefore, the switch forwards traffic as normal in the VLAN specified in the role.

Note that the VLAN assigned in the role must exist on the switch. However, software version 10.08, added a feature called role-based auto-VLAN (**port-access auto-vlan**), which automatically creates a VLAN assigned with a role if that VLAN does not already exist. Be aware, however, that the auto creation does not guarantee that the VLAN will be tagged on the appropriate uplinks. This feature would work best if the uplinks are trunk ports that permit all VLANs.

Dynamic Segmentation: Tunneling to an MC

Together with role-based authorization, AOS-CX switches support an innovative feature called Dynamic Segmentation. This study guide focuses on user-based tunneling (UBT) as the Dynamic Segmentation option. With UBT a switch "segments" clients' traffic by tunneling it to an MC, which then passes the traffic through the AOS firewall and applies further policies. The feature is "dynamic" because the switch dynamically chooses whether to tunnel a client's traffic based on the client's role. (With UBT switches can also tunnel traffic to AOS-10 gateways, but this study guide focuses on MCs.)

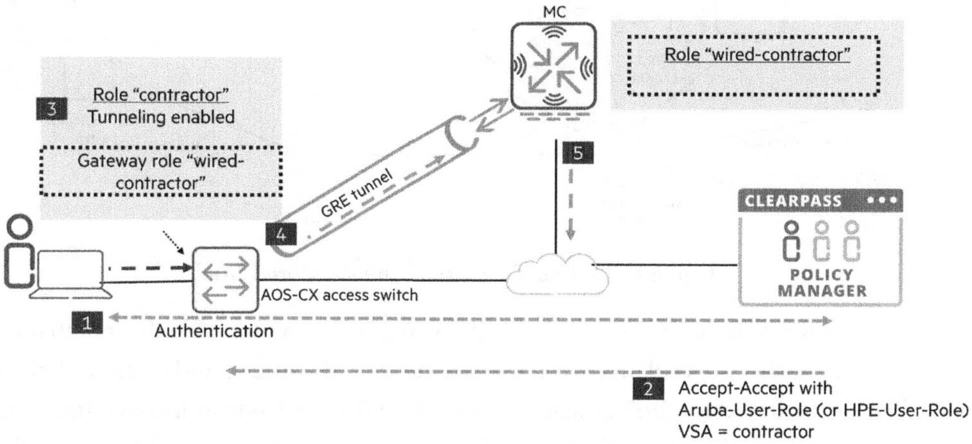

Figure 8-33: Dynamic Segmentation: Tunneling to an MC

Examine the process shown in Figure 8-33 in more detail:

1. The client connects and authenticates.
2. The client is assigned to a role. Typically, CPPM sends a VSA that specifies the role.
3. In this example, the role specifies that tunneling is enabled, as well as the name of the role to which the MC should assign the client. (On AOS-CX switches, this role is called the gateway-role.)
4. The switch sends the traffic to the MC in a Generic Routing Encapsulation (GRE) tunnel.
5. The MC applies the role communicated by the switch, which is "wired-contractor" in this example. This role can indicate firewall policies and other settings. The MC applies those settings and forwards the traffic.

An AOS-CX switch can simultaneously tunnel some devices' traffic and forward other devices' traffic locally, based on the settings in each device's role.

Use Cases for Tunneling Wired Traffic

You should tunnel wired traffic to MCs when you want to apply the same traffic control capabilities that these controllers traditionally provide for wireless traffic to wired traffic as well. These security features include a stateful, role-based firewall, DPI, and other capabilities.

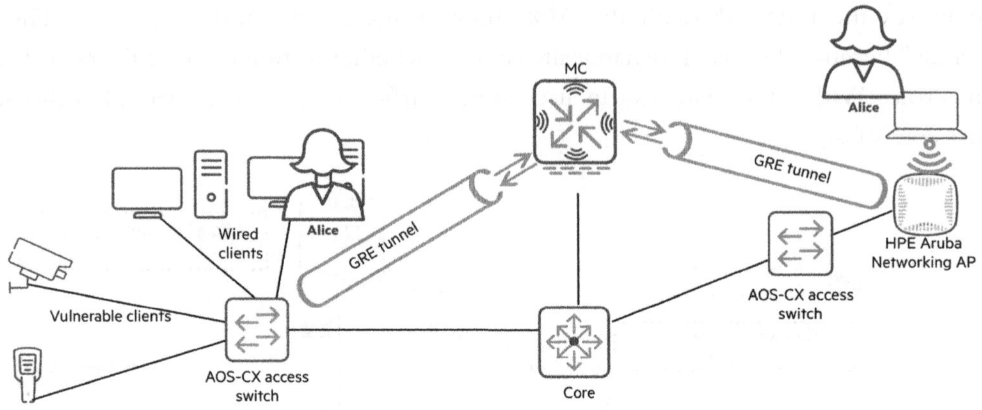

Figure 8-34: Use Cases for Tunneling Wired Traffic

For example, as shown in Figure 8-34, the company might have vulnerable devices that require extra filtering for their traffic. Surveillance cameras, payment card readers, and some IoT devices can be easily hacked or disrupted by DoS attacks because they do not have an internal firewall. By having the switch tunnel all the client's traffic to the MC, you can take advantage of the controller's security features to protect both the vulnerable device and the network.

Tunneled-node also provides tighter and simpler unification of wired and wireless access. Switches simply tunnel traffic to the MC, which handles the traffic much as it would handle traffic tunneled from an AP. An employee can connect through their desktop, log in, be assigned a role, and receive customized access. Later the employee can connect wirelessly using a laptop in a meeting room, log in, be assigned the same role, and receive the same access.

Do keep in mind that tunneling the wired devices' traffic to the MC, of course, increases the demands on that device. Architects will need to plan the HPE Aruba Networking solution's capacity and licensing accordingly.

Also note that in version 10.12 and above, some AOS-CX switch models can apply application-based policies on their own without tunneling traffic to an MC or gateway. In version 10.13 and above this support requires an Advanced Feature Pack. However, this study guide does not cover these capabilities.

AOS-CX Switch User Role Assignment (No Auth)

As you learned in the previous chapter, AOS-CX switches always apply role-based authorization on ports that implement a port access method such as 802.1X.

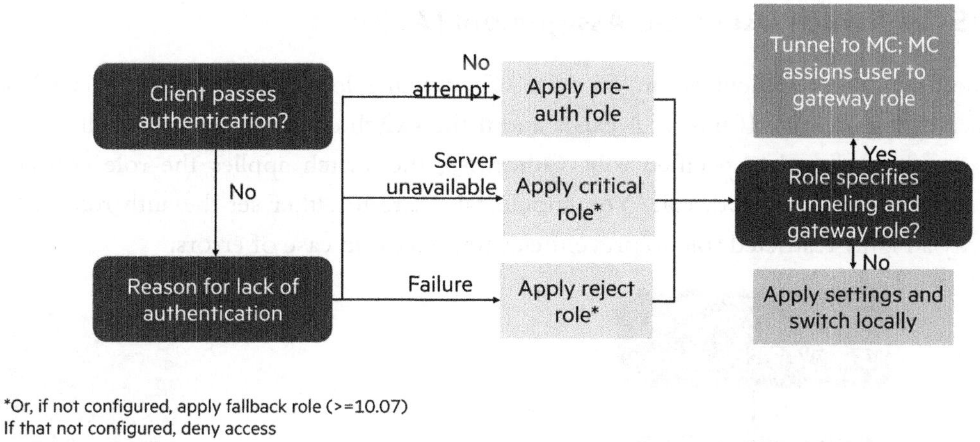

Figure 8-35: AOS-CX Switch User Role Assignment (No Auth)

As shown in Figure 8-35, these switches can assign unauthenticated clients to different roles based on the reason that the client is unauthenticated. If the client never attempts authentication, the switch assigns it to the pre-auth role. If the client attempts authentication, but the server is inaccessible, the switch assigns the user to the critical role. And if the user fails authentication, the reject role applies. These are all types of roles—not role names. If you wanted to use the role, you would create a role such as "limited_access" and apply it to that role type on a switch port. All role settings are per-port.

Note that AOS-CX version 10.07 and later add a fallback role. The client receives this role if it fails authentication, but no critical or reject roles is defined. Similarly, the client receives this role after authentication times out if the client never attempted authentication.

If an unauthenticated client cannot receive a role in one of these ways, the client is denied access entirely. For example, if the RADIUS server rejects the client, and no reject role nor fallback role is defined on the interface, the client is denied access. Similar considerations apply for when no critical, pre-auth, and fallback roles are defined.

Any of these roles can use local forwarding or specify tunneling to the MC and a gateway role, which the MC applies.

In the simplest and most secure set up, you leave all of the roles undefined, which would deny access to all unauthenticated users. However, if the company needs, for example, to support wired access in the case of loss of connectivity with the RADIUS server, you can use the other roles, such as the critical role, to meet the needs. Take care to configure settings for that role that are appropriate for unauthenticated users.

AOS-CX Switch User Role Assignment (Auth)

If the user passes authentication, an AOS-CX switches looks for the Aruba-User-Role VSA, as shown in Figure 8-36. If this VSA exists and if the switch can find the role locally or download it, the switch applies the specified role. Otherwise, the switch applies the role configured as the interface's port-access auth role. You should take care to either set the auth role to "denyall" or specify a highly restricted role to prevent elevated access in case of errors.

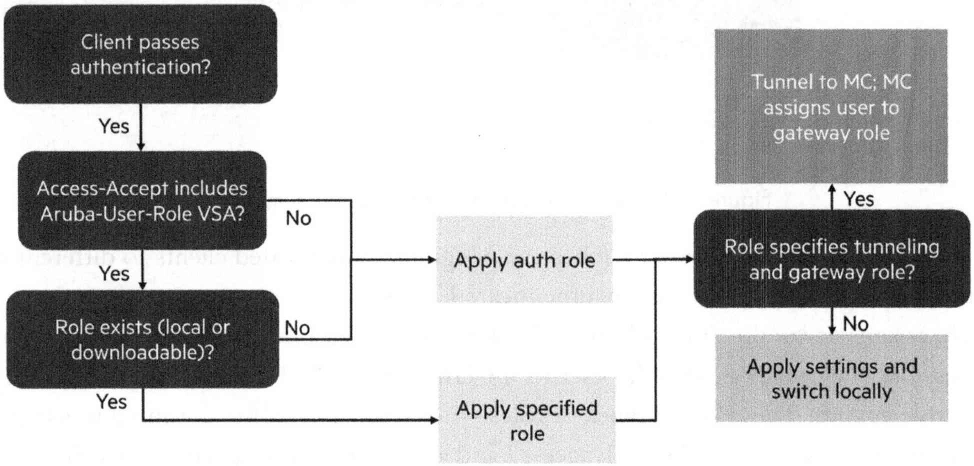

Figure 8-36: AOS-CX Switch User Role Assignment (Auth)

Again, any of these roles can use local forwarding or specify tunneling to the MC and a gateway role, which the MC applies.

Configuring Local Roles on AOS-CX Switches

For role-based authorization to work correctly, the switch must be able to obtain settings for the roles to which the RADIUS server assigns clients. You can configure local user roles on the AOS-CX switches manually, or the switches can download the roles from CPPM. First, you will example the local user role setup, as shown in Figure 8-37.

AOS-CX	Use RADIUS VSA - Aruba-User-Role = employee

```
port-access role employee
   vlan access 11
   reauth-period 7200
port-access role contractor
   gateway-zone zone MC gateway-role wired_contractor
```

Figure 8-37: Configuring Local Roles on AOS-CX Switches

When you are using local roles, you configure the roles in advance.

AOS-CX roles support settings such as untagged VLAN ID, untagged VLAN name, tagged VLAN IDs, tagged VLAN names, user policy, captive portal profile (used to redirect unauthenticated users to a captive portal for authentication), reauthentication period, cached reauthentication period, and a gateway zone (for UBT).

A role does not have to specify all of the settings. If the role does not specify a particular setting, then the switch keeps the default value that would apply to the user session based on other switch settings. For example, if the role has no untagged VLAN associated with it, the user's traffic is assigned to the untagged VLAN configured on the port.

As you see, if you want to apply tunneling to a client's traffic you specify that setting in the client's role. The secondary role or gateway-role setting indicates the role name to which the MC assigns traffic. If you do not specify it, the switch tells the MC to use the same role name as the primary role name.

After you have set up the roles, the switch is ready to enforce authentication. When a client authenticates, the authentication server, such as CPPM, sends the role name in an Aruba-User-Role VSA within the Access-Accept.

An example of CPPM sending the Aruba-User-Role VSA is shown in Figure 8-38.

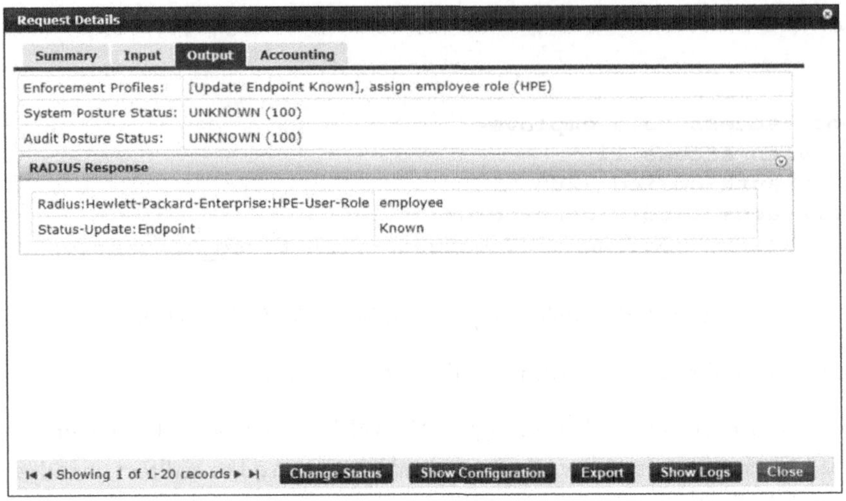

Figure 8-38: CPPM Access Tracker Record (Employee Role)

Using Downloadable User Roles

Alternatively, when you are using CPPM as the authentication server, you can have the switch automatically download the roles from CPPM. Downloadable User Roles (DURs) centralizes role administration and simplifies management. In addition to receiving the role *assignment* from CPPM, the switch also receives the role *configuration* from CPPM. This configuration can even include the user policies and classes associated with a policy specified in the role.

To apply a DUR, CPPM sends an Access-Accept with an Aruba-CPPM-Role VSA, as shown in Figure 8-39. The VSA specifies the role name. If the switch does not yet have this role, the switch contacts CPPM over HTTPS, logs in, and copies the role configuration. The switch then applies the role to the user's session. It also caches the role and can immediately apply it if CPPM sends an Aruba-CPPM-Role VSA with that role name again. If CPPM admins change the DUR settings, the number at the end of the role name changes. When a user is next assigned to the DUR, the switch sees that it does not have a role with that exact name, so it downloads the role again with the new settings.

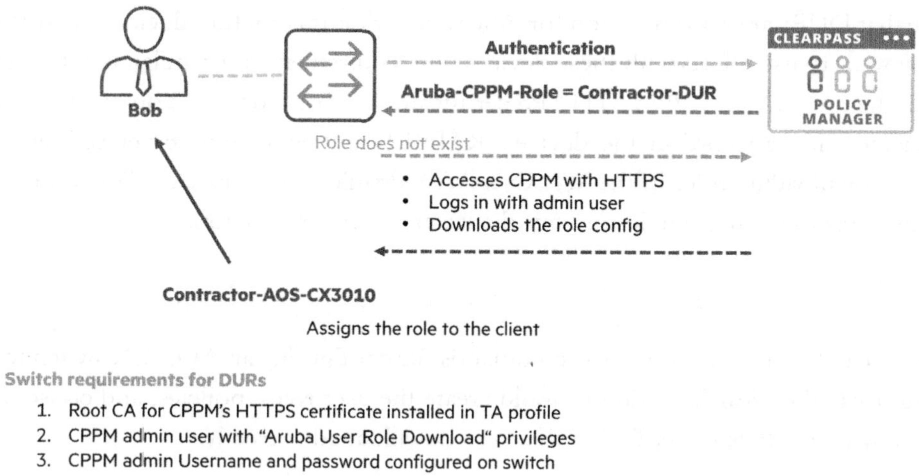

Figure 8-39: Using Downloadable User Roles

As shown in Figure 8-39, for DURs to work, the switch must trust the root CA for CPPM's HTTPS certificate. Make sure to load that certificate on the switch. The switch also needs the username and password for a CPPM account. That account must be in CPPM's Admin User repository. The typical privilege level is "Aruba User Role Download."

An example of what the record in CPPM's Access Tracker looks like when CPPM sends a DUR to an AOS-CX switch is shown in Figure 8-40.

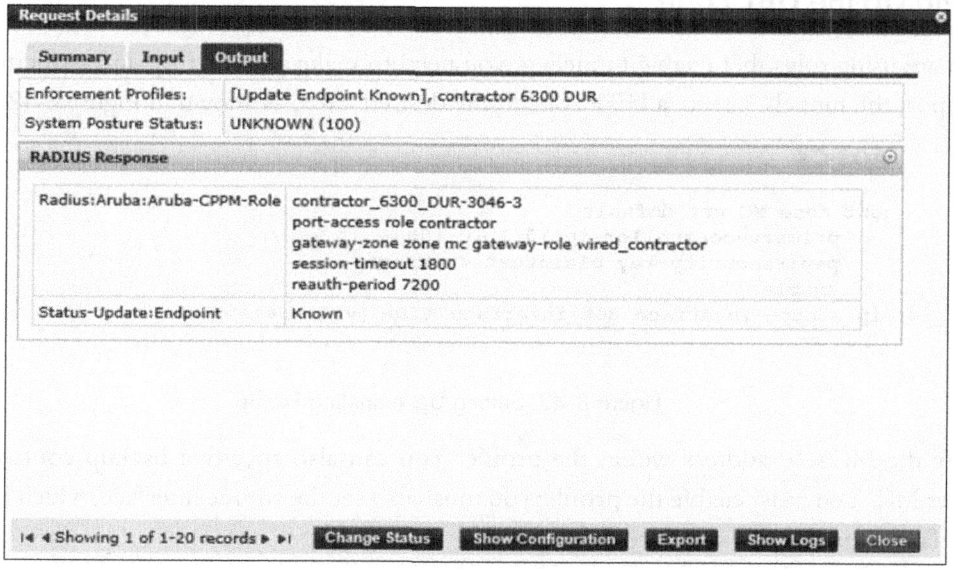

Figure 8-40: CPPM Access Tracker Record (DUR VSA)

Note that DURs are also an option for other network infrastructure devices, such as IAPs and MCs. If you want to use DURs with these devices, the requirements are similar as for switches. You must install the root CA for CPPM's HTTPS certificate as a Trusted Cert. You must also specify a CPPM username and password in the device's RADIUS authentication server settings. And you must enable downloading roles in the WLAN's AAA Profile or (for an IAP) WLAN settings. For more details, refer to the user guide for your device and software version.

CLI Commands: Enabling Downloading Roles on AOS-CX Switches

The table in Figure 8-41 shows the commands for configuring an AOS-CX switch to support DURs. In this case, the ClearPass admin should create the user roles, policies, and classes in advance. Use the show commands to verify that the roles downloaded correctly.

In configure context

<table>
<tr><th>Step</th><th>AOS-CX command</th><th>Notes</th></tr>
<tr><td>Set a ClearPass username and password</td><td>radius-server host <<i>FQDN</i> | <i>IP</i>> clearpass-username <<i>name</i>> clearpass-password <ciphertext | plaintext> <<i>password</i>></td><td>Use a ClearPass account that has rights for accessing the DURs.
Also make sure that the switch trusts CPPM's HTTPS certificate.</td></tr>
<tr><td>Verify the DURs.</td><td>show port-access role clearpass</td><td></td></tr>
</table>

Figure 8-41: Using Downloadable User Roles

Setting Up the UBT Zone

If you are using roles that enable tunneling, you need to make sure that the infrastructure is ready to support the tunnels. Create a UBT zone in the desired VRF, as shown in Figure 8-42.

AOS-CX

```
ubt zone MC vrf default
   primary-controller ip 10.1.10.100
   papi-security-key plaintext <string>
   enable
ip source-interface ubt interface <ID> [vrf <name>]
```

Figure 8-42: Setting Up Tunneling Profile

Specify the MC's IP address within the profile. You can also specify a backup controller IP for another MC. You must enable the profile; you must also set the source interface, which is the Layer 3 interface on which the switch reaches the MC.

With the profile configured and enabled, the switch and MC also establish a control channel that uses Provisioning API (PAPI), the same protocol that MCs use to send configs and other important messages to APs. You should set a PAPI password to provide MD5-based data integrity for messages sent across it. (While MD5 is a less preferred option, it is the only one available and better than no security.) The password can have a length from 10 to 64 characters; set the password as long as possible to prevent dictionary or brute force attacks.

The switch establishes a Generic Routing Encapsulation (GRE) tunnel to the MC to carry user traffic, using the same technology that encapsulates packets between HPE Aruba Networking APs and MCs. The switch sends all traffic from endpoints assigned to tunneled roles over this single tunnel. The MC separates the traffic out per-user and maintains one tunnel per user on its side.

Configuring MTU for Tunneling

In Figure 8-43, you see the commands for setting the MTU on Layer 2 and Layer 3 interfaces on AOS-CX switches. Be careful to always match or exceed the IP MTU when you set an L2 interface's MTU. When using UBT, set the MTU large enough to accommodate the extra headers.

In L2 interface context

Step	AOS-CX command	Notes
Set the MTU	mtu <46-9198>	**The L2 size must match or exceed the IP MTU**

In L3 interface context

Step	AOS-CX command
Set the IP MTU	ip mtu <68-9198>

Figure 8-43: Setting Up Tunneling Profile

UBT Client VLAN

AOS-CX switches support two UBT modes: local VLAN mode (also called reserved VLAN mode) or extend VLAN mode. (In version 10.04 and 10.05, local VLAN mode was the only option. In version 10.06 and later, you have a choice.) The mode determines whether the switch or the MC assigns clients to VLANs. This study guide focuses on local VLAN mode, which is the default mode. When you use this mode, the MC or gateway that receives the tunneled traffic assigns tunneled clients to VLANs. On the AOS-CX switch, you simply set up a UBT client VLAN, as shown in

Figure 8-44. You do not need to configure VLANs in switch roles that use tunneling, nor to coordinate those VLANs with the MC. Instead, the switch uses the UBT client VLAN for all tunneled traffic.

- Used to establish tunnel communication between Switch and Controller
- Must be unique and not used any other place in the network.

AOS-CX

```
ubt zone MC vrf default
    primary-controller ip 10.1.10.100
    papi-security-key plaintext <string>
    enable
ip source-interface ubt interface <ID> vrf <name>
vlan 2222
ubt-client-vlan 2222
```

Figure 8-44: UBT Client VLAN

You can choose any VLAN that is not used for another purpose. The switch communicates this VLAN with the MC for receiving the traffic at the other end of the tunnel. The VLAN basically acts as a placeholder in which the MC receives the user traffic. You do not need to add it to the MC or to the network. When the MC receives the client's tunneled traffic, it makes the decision on how to forward the traffic into the LAN.

Setting up the MC to Terminate Wired Tunnels

In addition to setting up the AOS-CX switches, you need to ensure that the MC is ready for terminating the wired user tunnels. First you should check the licensing. The MC requires similar licensing to support tunnels from AOS-CX switches as it does for controlling APs: one AP and one Policy Enforcement Firewall Next Generation (PEFNG) license per switch. And, if the MC has RF Protect (RFP) licenses for APs, the MC requires one RFP license per switch as well. A company might choose to deploy a dedicated MC for receiving traffic from the switches; in this case, the MC would not be using RFP and would only require an AP and PEF license per switch. Also remember that, if the MC is using WebCC, it needs as many WebCC licenses as AP licenses.

Remember to set the PAPI security key to match the key that you set on the switches. The key must use at least 10 characters. You can do this from the CLI with these commands:

(MC) [mm] (config) #**papi-security**

(MC) ^[mm] (PAPI Security Profile) #**enhanced-security**

(MC) ^[mm] (PAPI Security Profile) #**key ex@mple12!**

(MC) ^[mm] (PAPI Security Profile) # **write memory**

As you learned earlier, the switch communicates to the MC the role to which the MC should assign the client. This role is the "gateway-role" defined in the role setting that enables tunneling, or, if that is not set, it is the same role name to which the client is assigned on the switch. (In the example shown earlier, the "contractor" role, which enabled UBT, used "wired_contractor" as the gateway-role. You must create roles on the MC to match the gateway-roles that the switch sends. You set up these roles much as you do for wireless users and devices, assigning policies and rules based on how you want to control the traffic.

When you are using a UBT client VLAN, you must also configure the role on the MC to specify the VLAN for the wired users. It is best practice to use a different VLAN from the one used for wireless users. For example, if the MC uses VLAN 12 for wireless clients, you might have it use VLAN 13 for wired clients. You would then assign VLAN 13 to the roles created for wired clients, such as "wired_contractor." Much as for the VLAN associated with a WLAN, you should make sure that the MC's ports support the VLAN for wired users. Besides that, the VLAN only needs to exist within the network core (or at the aggregation layer, if that is where the MC connects).

Steps for completing this example setup are:

1. Go to the AOS device (MC or MCR) UI. Go to the **Configuration > Roles & Policies > Roles** page. Add the desired role.
2. Select the role new (**wired_contractor** in this example shown in Figure 8-45).
3. Select **Show Advanced View**. (As shown in Figure 8-45, the link then changes to **Show Basic View**.)

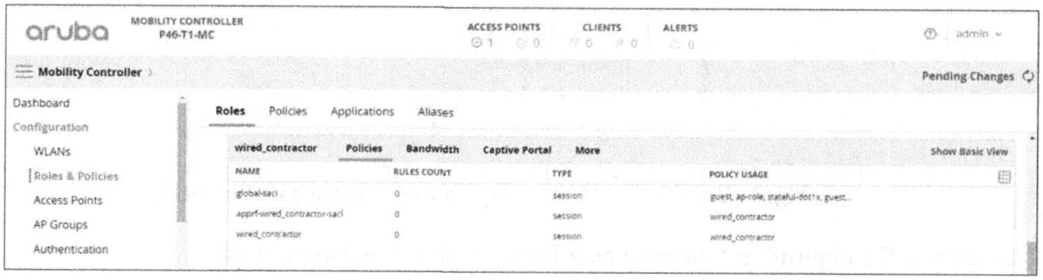

Figure 8-45: AOS UI (Configuration > Role & Policies > Roles)

4. Click **More**.
5. As shown in Figure 8-46, expand **Network.**
6. For **VLAN**, select the VLAN to which you want to assign wired users.

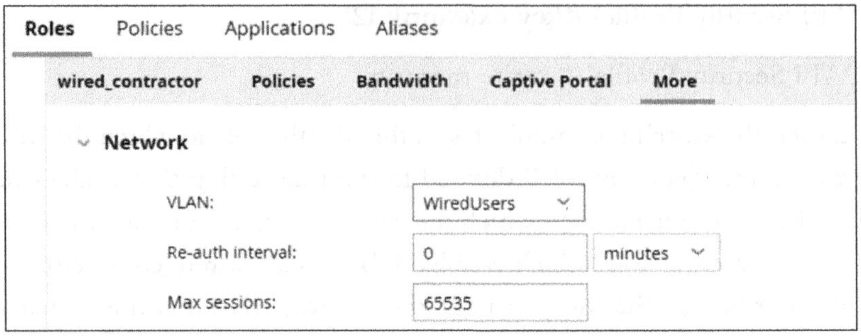

Figure 8-46: AOS UI (Configuration > Role & Policies > Roles > More > Network)

7. Click **Submit**.
8. Remember to deploy the pending changes.

It is also best practice to enable jumbo frames. Follow these steps in the AOS device's UI:

1. Navigate to **Configuration > Services > Firewall**.
2. Scroll down and select the **Jumbo frames processing** check box, as shown in Figure 8-47.
3. Click **Submit**.

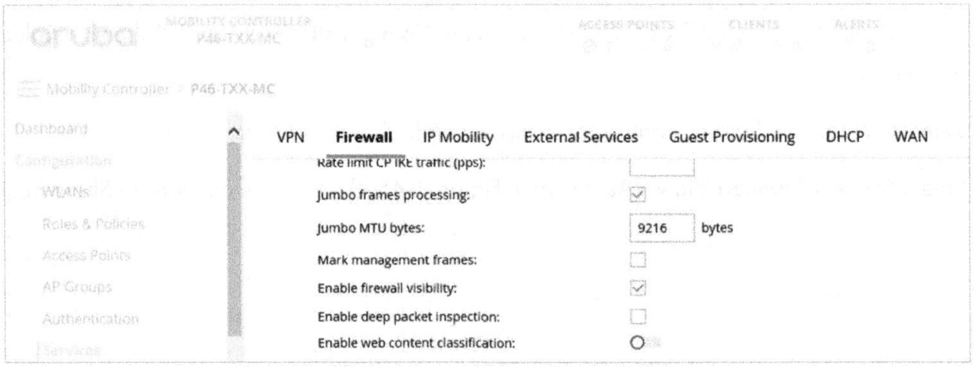

Figure 8-47: AOS UI (Configuration > Services > Firewall)

4. Navigate to **Configuration > Interfaces > Ports**, as shown in Figure 8-48.
5. Select the port on which the switches reach the MC or MCs.
6. Scroll down and select the **Jumbo MTU** check box.
7. Click **Submit**.

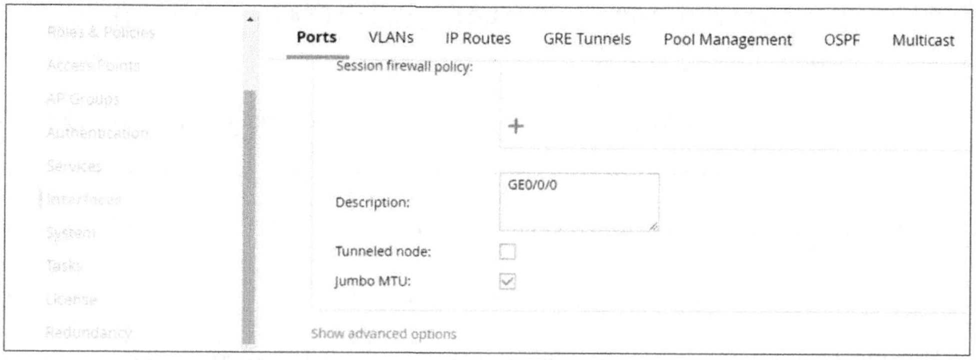

Figure 8-48: AOS UI (Configuration > Interfaces > Ports)

8. Remember to deploy the pending changes.

View Dynamic Segmentation Status

After you have set up dynamic segmentation on both the switch and MC sides, you can check the status.

Viewing UBT Status on an AOS-CX Switch

Below is the command viewing the tunnel status on an AOS-CX switch. The output shows what you should see if the switch has successfully established a tunnel with an MC, which has IP address 10.1.10.100 in this example. In this example, the switch has established a tunnel with a single MC; it could also establish tunnels with an MC cluster.

```
AOS-CX# show ubt state zone mc
======================================================================
Zone mc:
======================================================================
Local Conductor Server (LCS) State:
LCS Type       IP Address     State                    Role
----------------------------------------------------------------------
Primary      : 10.1.10.100 ready_for_bootstrap operational_primary
Switch Anchor Controller (SAC) State:
               IP Address      MAC Address          State
------------------------------------------------------------------
```

```
Active        : 10.1.10.100     00:0b:86:b4:ef:47     registered
User Anchor Controller(UAC): 10.1.10.100
User                 Port    State              Bucket ID  Gre Key  VLAN
----------------------------------------------------------------------
00:50:56:a5:da:f0    1/1/20 registered          143        14       2222
```

Viewing Tunneled Clients on AOS-CX Switches

When an AOS-CX switch has successfully established a tunnel to an MC, you are ready to let clients connect. The clients can connect to the switch, authenticate, and are assigned to roles that specify tunneling. The switch then tunnels those clients' traffic to the MC.

You can view the status for tunneled clients on an AOS-CX switch with this command:

```
AOS-CX# show ubt user all
======================================================================
Displaying All UBT Users for Zone: mc
======================================================================
Downloaded user roles are preceded by *
Port    Mac-Address     Tunnel Status     Gateway-Role             Failure
Reason
----------------------------------------------------------------------
1/1/20 00:50:56:a5:da:f0  activated     wired_contractor          ---/---
```

The output shows that a client on port 1/1/20 has an active tunnel in the MC UBT zone. This client's role on the MC is "wired_contractor."

Viewing Tunneled Clients on an MC

You can view wired clients for which the traffic is being tunneled to the MC in the Web UI. As shown in Figure 8-49, you can see the number of wired clients connected to the MC in the top bar of the UI. In this example, only one wired client is connected.

Figure 8-49: AOS UI Top Bar (Wired Client)

You can click the number to see more details about the clients. As shown in Figure 8-50, you can see information such as the client's IP address and role. Note that this is the role to which the MC assigns the client, taken from the secondary role or gateway role, communicated by the switch. You can also see the name of the switch to which the client is directly connected in the **Connected To** column.

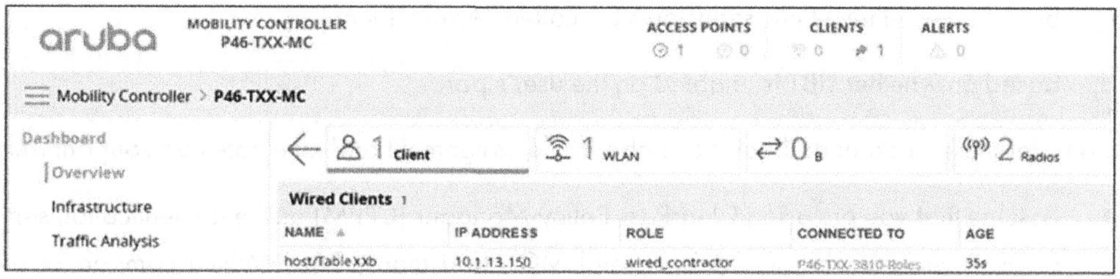

Figure 8-50: AOS UI Wired Clients List

Summary

Congratulations! You have learned how CPPM and HPE Aruba Networking infrastructure devices can work together to control individual users and devices based on their identity. You now understand how to integrate your HPE Aruba Networking infrastructure with CPPM to secure your network from the edge and enhance the user experience.

Learning checks

1. True or false: A role assigned with an Aruba-User-Role VSA takes precedence over a role derived from a server derivation role.

    a. True

    b. False

2. What is one feature of the AOS stateful firewall?

    a. It establishes sessions and can automatically permit return traffic in a session.

    b. It ensures that all users' traffic is controlled in the same way.

    c. It maintains good performance by offloading packet processing to a third-party firewall such as CPPM.

    d. It adds encryption to unencrypted traffic before forwarding it on the LAN.

3. An AOS-CX switch has a UBT zone profile configured on it. How does the switch decide when to tunnel a user's traffic?

   a. Based on whether the initial role is bound to tunneling globally

   b. Based on whether that zone is specified in the user role assigned to the user

   c. Based on whether CPPM sends an VSA called "Aruba-Tunnel"

   d. Based on whether UBT is enabled on the user's port

4. Consider these open-ended questions about role assignment and take notes on your answers:

   - Assume that you are using ClearPass Policy Manager (CPPM) as the authentication server, and you have a network with an MCR, several MCs, and many CAPs. Which component should be the decision maker for which roles users are assigned?

   - What approaches can companies take toward dividing users into different roles? When should users have the same role? When should they have different roles?

5. Consider these open-ended questions about defining policies on the AOS firewall. Take notes on your answers.

   - Describe allow list versus deny list approaches to defining firewall policies. Discuss advantages and disadvantages.

   - What strategies can you take toward defining the services that users need?

   - How can you distinguish between and control different types of applications that use HTTP/HTTPS?

6. What are the benefits of roles, policies, and aliases for the AOS firewall? Take notes on your answers.

7. You are defining a WLAN that uses WPA3-Enterprise security and authenticates users to CPPM. CPPM sends the roles for authenticated users in Aruba-User-Role VSAs. What should you set for the default role in the WLAN's Access settings?

8. This learning check refers to the policies listed in Table 8-1 and Table 8-2.

   a. Will contractors be able to send an HTTPS request to these IP addresses:

   - 10.254.1.15

   - 10.254.1.21

   - 203.0.13.3

b. Will this traffic will be permitted:

- HTTPS response from 10.254.1.15 after an HTTPS request from a contractor client
- HTTPS request from 10.254.1.15 (a server outside the firewall)
- DNS response from 10.254.1.21 after a DNS request from a contractor client

Table 8-1: Roles and Policies

Role	Rules
contractor	policy allowdhcpdns ipv4 any any srv-dhcp permit ipv4 user MyDNS srv-dns permit policy contractors ipv4 user Web_contractor srv-http permit ipv4 user Web_contractor srv-https permit ipv4 any Internal any deny policy webaccess ipv4 user any srv-http permit ipv4 user any srv-https permit

Table 8-2: Aliases

Alias	Items
MyDNS	Host 10.254.1.21
Web_contractor	Host 10.254.1.15
internal	Network 10.254.1.0 255.255.255.0 Network 10.1.10.0 255.255.254.0

Note

For answers to all learning checks, refer to Appendix B.

9 Identify and Classify Endpoints

EXAM OBJECTIVES

✓ Identify and evaluate discovered endpoints

✓ Compare passive and active classification methods

Assumed Knowledge

- Basics of IP networking
- Basics of Ethernet switching
- Basics of 802.11 wireless networks
- Basics of DHCP

Overview

In a zero-trust world, identifying users and controlling their access to resources based on their identity is crucial—but so is identifying devices. A laptop and a smartphone, even when used by the same employee, each represent different risks to the network, and the network should adapt rights to the device types. In addition, a modern network has many more devices than just desktops and laptops, and many of the devices are specialized, "useless" devices. Think of not just printers and security cameras, but all the "smart" devices introduced by the Internet of Things (IoT). The network needs to grant each of these devices the correct level of access.

This chapter introduces you to endpoint classification and explains passive and active classifying techniques. It also shows you the options that HPE Aruba Networking provides for endpoint classification.

In particular, you will explore the broad array of device profiling capabilities that are built into HPE Aruba Networking ClearPass Policy Manager (CPPM). You will focus on how you can configure HPE Aruba Networking infrastructure to take advantage of these capabilities.

You will then learn at a high level about other device profiling solutions offered by HPE Aruba Networking. about ClearPass Device Insight (CPDI), a separate solution that can integrate with CPPM and HPE Aruba Networking infrastructure, to provide the most advanced device profiling capabilities.

Endpoint Classification Introduction

This section introduces you to the benefits of endpoint classification.

Scenario Introduction

The company from previous scenarios now wants to build on the role-based policies with endpoint classification, as outlined in Figure 9-1. Working together with CPPM, the HPE Aruba Networking infrastructure will identify the type of device with which each user connects, including the device's broad category, such as PC or smartphone, and the device's OS, such as Windows Enterprise 10 or Android.

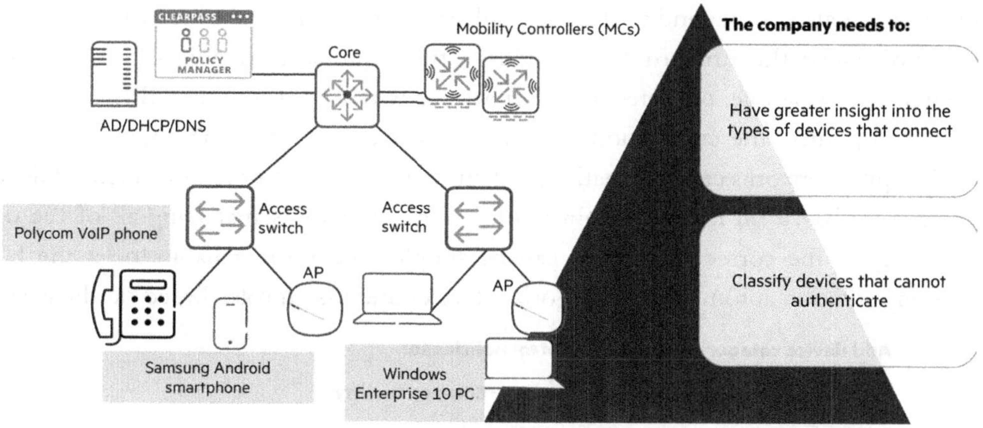

Figure 9-1: Scenario Introduction

This will give the company greater visibility into their devices. It will also help the infrastructure apply roles to devices that do not support 802.1X authentication.

Better Device Visibility = Better Security

Classifying devices in this way gives customers greater visibility into their network. The more information that a company can gather about types of devices users are actually using to connect to the network, at what times, and in what locations, the better equipped the company's security team is. Particular exploits often target specific operating systems. For example, Windows XP machines were susceptible to the WannaCry ransomware attack. (It might seem odd that XP was still broadly deployed in 2017, but organizations like hospitals often have very long refresh cycles for specialized devices.) While ideally companies have an inventory of the operating systems in their environment, these inventories can be out-of-date or unreliable. Endpoint classification gives companies the hard data that they need to assess how vulnerable they are to a particular attack.

Similarly, IT will be better able to plan upgrades or patches with better insight into how many devices run a particular OS and where those devices are used. When companies use CPPM to profile devices, CPPM can feed information into third party endpoint management systems.

Visibility into device types can also reveal misbehavior. For example, if an organization like a hospital has policies against smartphones, admins might want to know if, indeed, smartphones are connecting and who is connecting those phones. By carefully analyzing endpoint classification data, security teams can even detect more alarming misbehavior. Some networks use MAC authentication for devices like printers, and it is easy for a savvy hacker to spoof the printer's MAC address and gain unauthorized access. Through endpoint classification, you can see that a device changes from printer to laptop—a suspicious circumstance.

Companies can also go beyond visibility to enhancing control, as shown in Figure 9-2. Endpoint classification allows the infrastructure to adjust settings and security policies dynamically based on device type. For example, on detecting a wireless or wired VoIP phone, HPE Aruba Networking devices can optimize the connection with the correct quality of service (QoS) settings for voice traffic. Hospital networks can automatically adapt to different IoT devices or medical devices. While assigning roles based on identity might work if the company is managing all of the devices and placing them in the correct groups, it can be simpler and more reliable to let the HPE Aruba Networking solution automatically categorize devices and assign roles based on those categories.

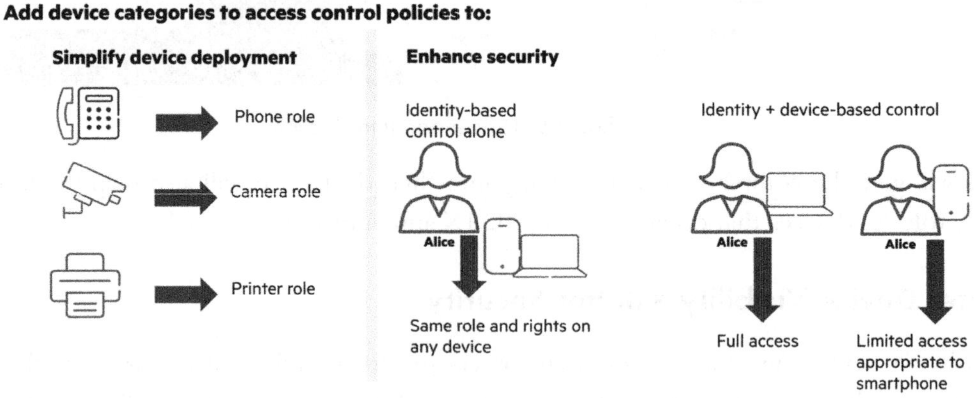

Figure 9-2: Better Device Visibility = Better Security

And sometimes identity-based roles do not give the company the granular control that it needs. Users often connect to the network on multiple devices. If you only use identity to assign roles, the same user will always receive the same rights no matter what the device. But smartphones tend to be less secure against some exploits and employees tend to use them for less demanding work applications—like browsing the Internet or checking email. That means that a more limited access profile could make sense. By combining identity-based and endpoint category-based criteria, the company can give an employee on a Windows PC unlimited access, but the same employee on a smartphone access to the Internet only.

Overview of Endpoint Classification Methods

As shown in Figure 9-3, endpoint classification methods fall into two broad categories: active and passive. With passive classification, the classifying device snoops traffic that endpoints already send in order to detect the device type. Common forms of passive classification include DHCP fingerprinting and HTTP User-based agent fingerprinting.

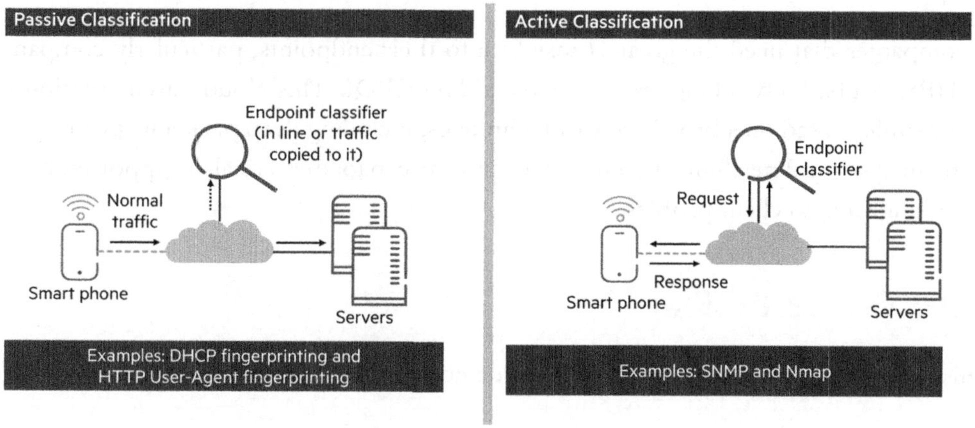

Figure 9-3: Overview of Endpoint Classification Methods

Active classification involves the classifying device sending out traffic to prompt responses. Examples of active classification include TCP/IP fingerprinting using Nmap and scanning with SNMP, SSH, and WMI.

HPE Aruba Networking Solutions for Endpoint Classification

HPE Aruba Networking offers several solutions for endpoint classification.

HPE Aruba Networking Mobility Controllers (MCs) and Instant AP (IAP) clusters support a degree of endpoint classification on their own. These devices support a technology called DHCP fingerprinting to identify a device and assign it to a role.

However, for greater capabilities, HPE Aruba Networking recommends that customers use CPPM, which can create a complete device profile for every endpoint and information about that device's hardware and software. Many vendors use only a MAC OUI to determine a device type. The MAC OUI, the first six hexadecimal characters in a MAC address, are specific to a NIC's vendor. However, they are easily spoofed and do not always give a reliable result. CPPM looks at the MAC OUI, but also uses DHCP fingerprinting for a more reliable result. And CPPM also supports other fingerprinting methods such as HTTP User-Agent strings, active scanning, and integration with other solutions. These capabilities allow CPPM to identify devices more reliably and also detect discrepancies—such as a device that has spoofed its MAC address to pose as a different type of device. CPPM consolidates the information for admins to view and search. Admins can also use whatever pieces of the profile that they want to create flexible policies to control user and device access.

For companies that need the greatest insight into their endpoints, particularly companies that use IoT, HPE Aruba Networking recommends adding CPDI. This cloud-based solution creates rich device profiles based on a broad array of techniques, including deep packet inspection (DPI), traffic pattern analysis, and machine learning. Many similar capabilities are also supported by HPE Aruba Networking Central client profiles.

CPPM Device Profiler

In this section, you will learn about CPPM's device profiling capabilities.

CPPM Device Profiler within Policy Manager

The CPPM Device Profiler delivers sophisticated endpoint classification. When this feature is enabled, CPPM assembles a device profile for every endpoint that it detects on the network, as shown in Figure 9-4. Detected endpoints include endpoints authenticated by CPPM, endpoints that attempted authentication but failed, and endpoints that CPPM detects through any of the passive or active methods that you will examine in this section.

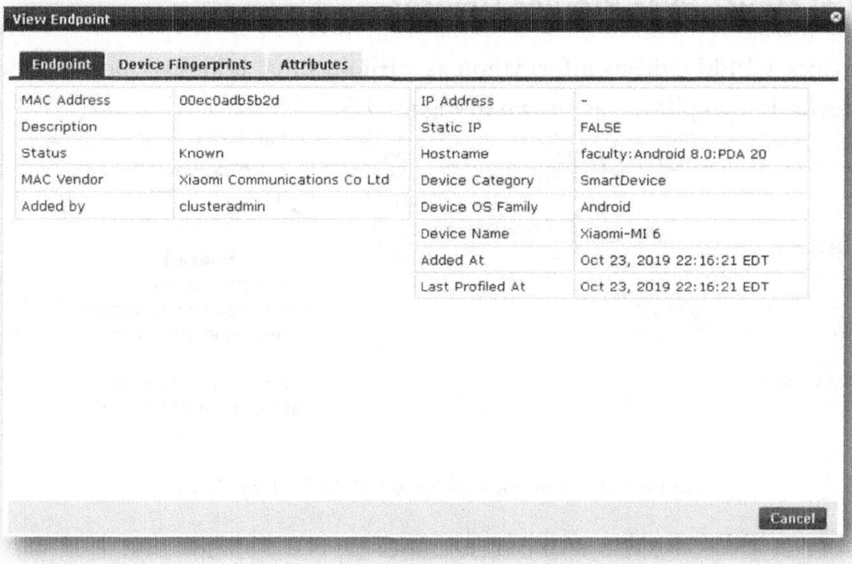

Figure 9-4: CPPM Device Profiler within Policy Manager

A device profile includes a hierarchy of information:

- Device Category, which identifies the device type such as Computer, Smart Device, or Printer
- Device Family, which further classifies the device by OS or vendor; for example, a Computer might be in the Windows, Linux, or Mac OS family
- Device Name, which further segments the family; for example, Windows computer might have Windows 10 for its Device Name

Other information in the profile includes:

- IP address
- Hostname
- MAC Vendor
- Timestamp when the device was first discovered
- Timestamp when the device was last seen

Overview of How CPPM Profiles Devices

To profile devices, CPPM collects information as attributes that it knows about the device. It then profiles the device in two phases, as shown in Figure 9-5.

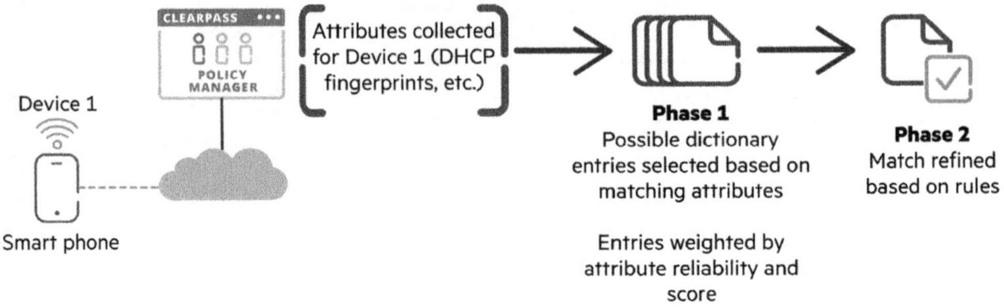

Figure 9-5: Overview of How CPPM Profiles Devices

In the first phase, CPPM matches attributes to entries in its fingerprinting dictionary. Sometimes attributes on the device might point in different directions. CPPM also assigns a reliability and score to each attribute type, which enables CPPM to choose the most likely entry.

CPPM refines the match in phase 2 using rules such as requiring the device to have a particular MAC OUI.

Ongoing Classification and Detection of Conflicts

Profile data does not expire in CPPM. After CPPM has profiled an endpoint, it knows the endpoint's profile when that endpoint connects to the network again later. However, CPPM also re-profiles endpoints whenever they connect or whenever CPPM runs an active scan. If new information causes the profile to change, CPPM changes the profile.

In this way, CPPM keeps the profiles up-to-date and accurate. A change in profile can also flag a security issue.

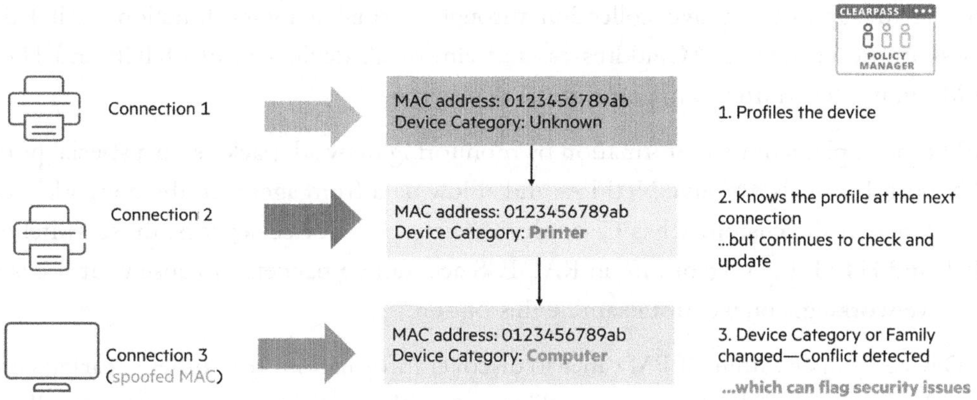

Figure 9-6: Ongoing Classification and Detection of Conflicts

In Figure 9-6, you see a device that was originally profiled as a printer has now become a computer. That indicates that a hacker might be spoofing the printer's MAC address to gain unauthorized access to the network.

CPPM detects that the endpoint's device category changed, so it indicates that it detected a conflict. CPPM admins can monitor conflicts and track down why they occurred. They can also create rules that tell the network infrastructure to block endpoints for which conflicts have occurred.

List of CPPM Collectors

Now that you understand the high-level process of device profiling, you can examine how CPPM collects device attributes in more detail. In Figure 9-7, you see a list of CPPM collectors as of version 6.8.

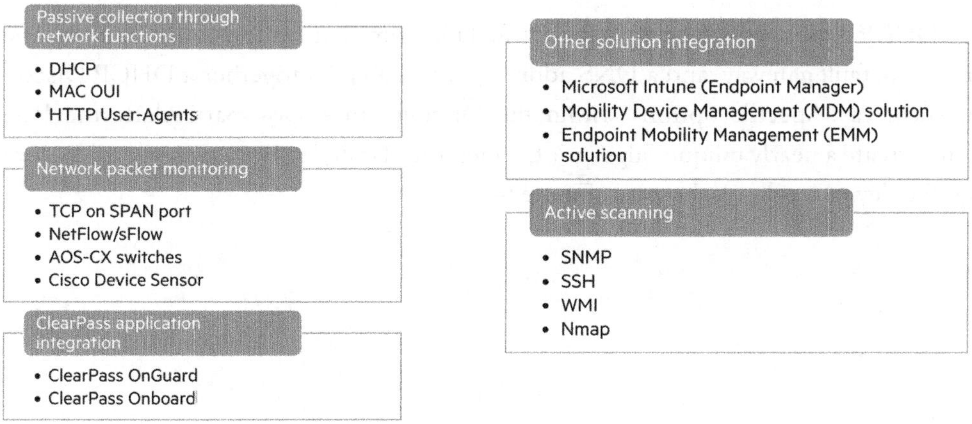

Figure 9-7: List of CPPM Collectors

First CPPM supports passive collection through normal network functions. All Ethernet and wireless traffic includes MAC addresses, and almost all devices send DHCP and HTTP traffic. CPPM can use this traffic to fingerprint the device type.

CPPM can supplement this information by monitoring network packets on a special port called the SPAN port. It can also receive NetFlow and sFlow data from agents in the network such as AOS-CX switches. If the customer has Cisco devices set up as device sensors, those switches can send DHCP and HTTP User-agent info in RADIUS accounting packets; because your focus is on HPE Aruba Networking, you will not examine this option.

The Device Profiler within CPPM can also discover information from other ClearPass applications. It can also integrate with third-party applications, such as Microsoft Intune (now called Microsoft Endpoint Manager). It can also integrate with other Mobility Device Management (MDM) and Endpoint Mobility Management (EMM) solutions.

Finally, CPPM can implement active scanning using SNMP, SSH, WMI, and Nmap.

Learning how to configure CPPM is beyond the scope of this study guide. Instead, you will focus on understanding the CPPM capabilities and how you can set up the HPE Aruba Networking infrastructure to support its collectors, as necessary.

DHCP Fingerprinting

You will first look at DHCP fingerprinting, one of the most useful forms of passive endpoint classification.

As you know, most endpoints must use DHCP to receive IP settings when they connect to a network.

While DHCP Discover messages typically include common options such as a request for an IP address, a default gateway, and a DNS address, every OS puts together a DHCP Discover message in its own way. Specific options within the Discover message—particularly the Parameter List option—create a nearly unique fingerprint, which the classifying device uses to discover the OS and assign the device a role, as shown in Figure 9-8.

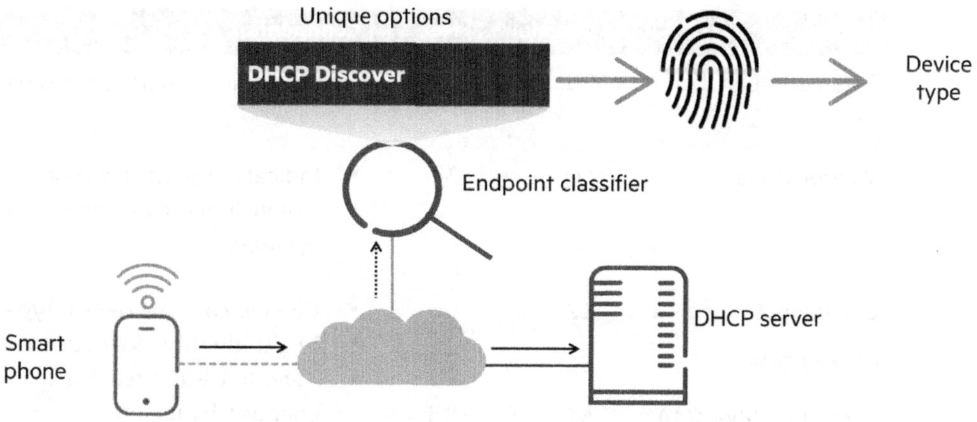

Figure 9-8: DHCP Fingerprinting

Common Options Used for DHCP Fingerprinting

One of the most important options for DHCP fingerprinting is DHCP Option 55—37 in hex—the Parameter List, which indicates the options that the client requests from the server. Different device operating systems will request different options, which already begins to distinguish device types. But the DHCP RFC also states that the options may be listed in the order of preference, which has led to every OS using a slightly different order. Together differences create a nearly unique fingerprint for the OS.

This option might be enough to create a DHCP fingerprint, and it is a great place to start defining a fingerprint.

A device profiler can also look for other options. DHCP Option 60, hex 3c, specifies the vendor class, which can point directly to the device type. However, not all operating systems include this option.

The other options listed in Figure 9-9 might help a device profiler like CPPM to identify individual devices. However, they can sometimes help to indicate device type. For example, some devices have a standard string as their initial hostname or the beginning of their hostname. Similarly the first three octets of a MAC address can point towards device types. But be aware that users can configure their machines' hostnames and easily spoof their MAC addresses, so these fields will not be as reliable as option 55.

<table>
<tr><th>Option</th><th>Decimal</th><th>Hex</th><th>Comments</th></tr>
<tr><td>Parameter List</td><td>55</td><td>37</td><td>Nearly unique ordering per device OS</td></tr>
<tr><td>Vendor Class</td><td>60</td><td>3c</td><td>• Indicates the device type
• Not included by all operating systems</td></tr>
<tr><td>Client Host Name</td><td>12</td><td>0c</td><td rowspan="3">• Can indicate the device type
• Might identify individual devices
• Tend to be less reliable (easily changed by users)</td></tr>
<tr><td>Client FQDN</td><td>81</td><td>51</td></tr>
<tr><td>Client Identifier (MAC address)</td><td>61</td><td>3d</td></tr>
</table>

Figure 9-9: Common Options Used for DHCP Fingerprinting

Examples of Parameter List Options

The packet capture in Figure 9-10 and Figure 9-11 show how a Windows Enterprise 10 PC and a Google Pixel Android device have different DHCP option 55 (hex 37). The devices both request different options and list options in different orders.

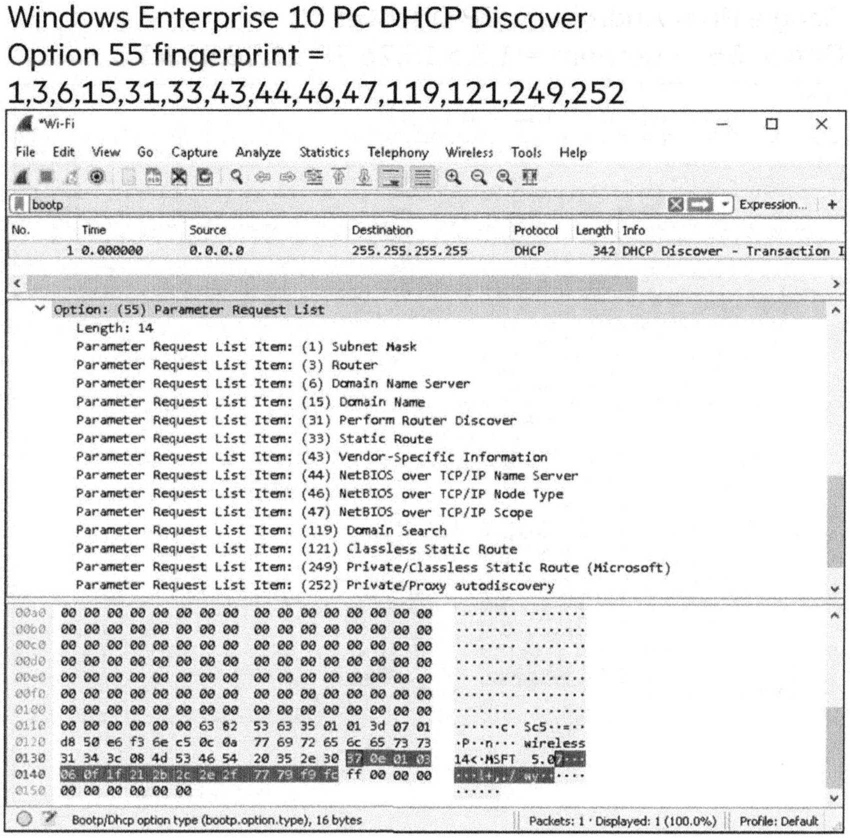

Figure 9-10: Windows Enterprise 10 DHCP Options

In the Android packet capture, you can also see how the Discover message includes Option 60 (hex 3C), the Vendor Class ID, set to "android-dhcp."

Google Pixel Android DHCP Discover
Option 55 fingerprint = 1,3,6,15,26,28,51,58,59,43

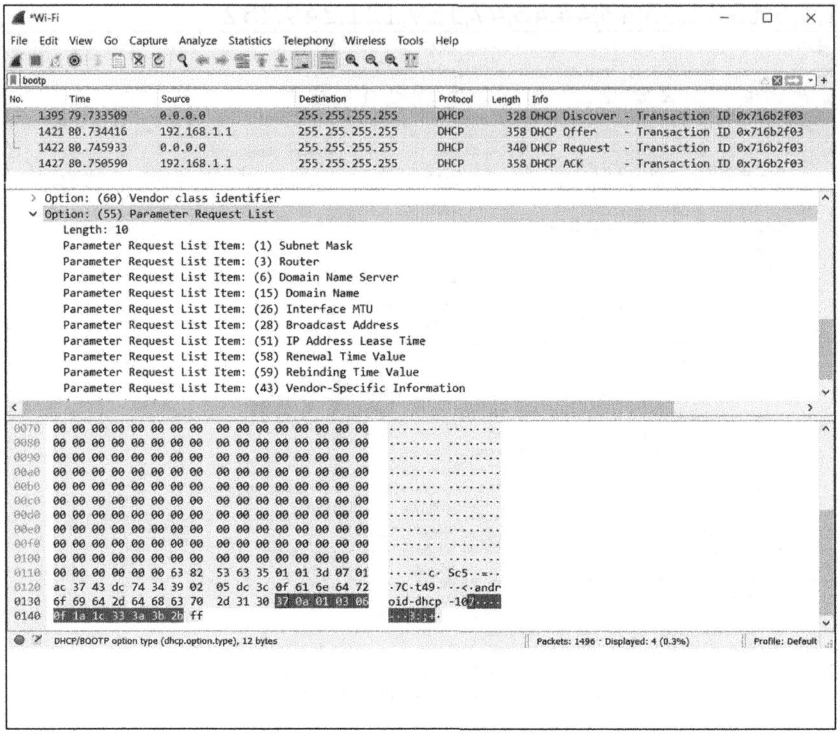

Figure 9-11: Android DHCP Options

Example Device Fingerprint Dictionary Entry

CPPM comes with an extensive dictionary that matches fingerprints to specific devices. In Figure 9-12, you see an example dictionary entry with a DHCP Option 55 attribute.

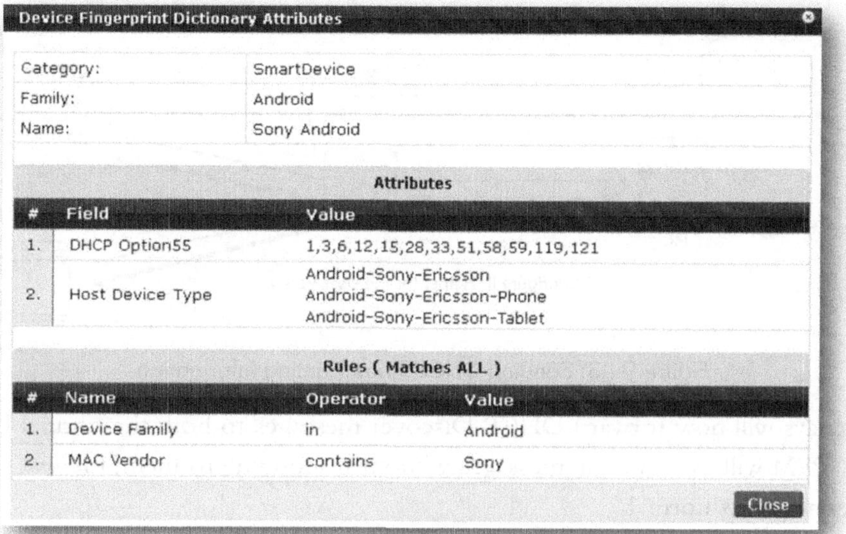

Figure 9-12: Example Device Fingerprint Dictionary Entry

As you see, the dictionary combines the DHCP fingerprint with other attributes that CPPM can collect. The dictionary also has rules for validating and refining the match. In this example, the entry for "Sony Android," the MAC OUI, which is the first three octets of the device MAC address, must indicate Sony as the vendor.

Sending DHCP Fingerprinting Information

For CPPM to use DHCP fingerprinting, it must receive DHCP traffic. Simply add CPPM as a DHCP helper address wherever the network currently defines helper addresses, as shown in Figure 9-13. This could be within MC VLAN settings. Or, if a core routing switch acts as the DHCP relay device, you could add the CPPM IP address to that switch's list of IP helper addresses on each endpoint VLAN.

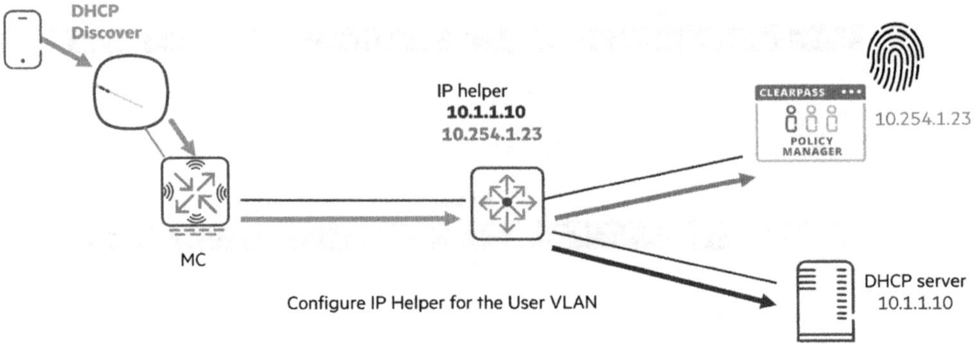

Figure 9-13: Sending DHCP Fingerprinting Information

The DHCP relays will now forward DHCP Discover messages to both the normal DHCP server and CPPM. CPPM will examine the messages without responding to them; the rest of the DHCP process will continue as normal.

HTTP User-Agent based Fingerprinting

Sometimes DHCP fingerprinting on its own does not provide enough information to fully classify an endpoint. As one example, the Apple OS produces DHCP messages with the same fingerprint on multiple types of devices. As shown in Figure 9-14, to distinguish an iPad from an iPhone, CPPM can look to another useful information source: HTTP User-Agent strings. As defined in RFC 1945, the HTTP User-Agent field identifies the agent that sends the HTTP request. While this field is not required, many applications send them so that they can receive tailored information. For example, if the HTTP server knows that it is sending information to a smartphone browser, it can adjust the page size.

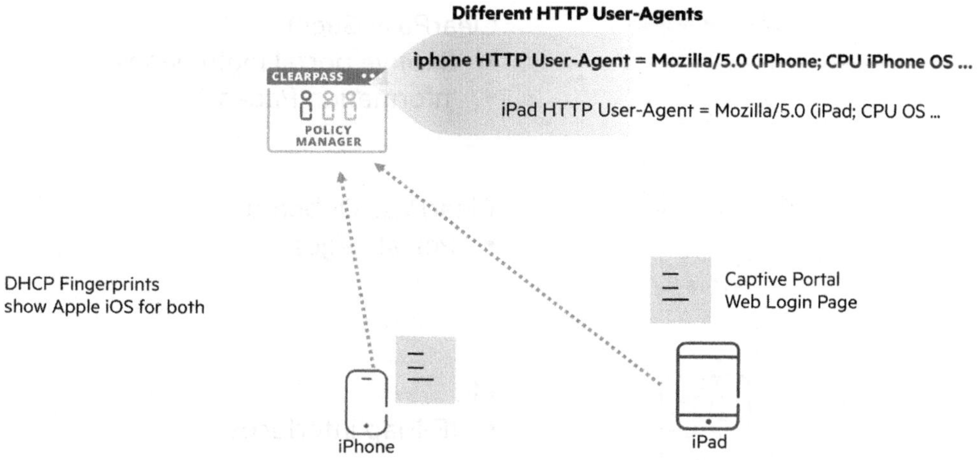

Figure 9-14: HTTP User-Agent based Fingerprinting

The application that generates the HTTP request—typically a browser—is responsible for generating the string. There is no set way to formulate strings. Depending on the application, it might include information such as the name of the application, the operating system, and the device—all of which can be very useful to a profiling device.

A repository of known User-Agent strings is available at http://www.user-agents.org/. CPPM has many User-Agent strings built into its default dictionary.

How CPPM Obtains HTTP User-Agent Strings

To examine HTTP User-Agent strings, CPPM needs to see devices' HTTP traffic. If the company is using ClearPass Guest, CPPM can obtain HTTP traffic from ClearPass Guest for devices that undergo captive portal authentication. Similarly, CPPM can obtain HTTP traffic from ClearPass Onboard for devices that are onboarded by this HPE Aruba Networking solution. To fill in the gaps that these sources might leave—and for companies that do not use Guest or Onboard—you can configure MCs to send HTTP traffic through an Interface for Metadata Access Point (IF-MAP) interface.

These options are summarized in Figure 9-15.

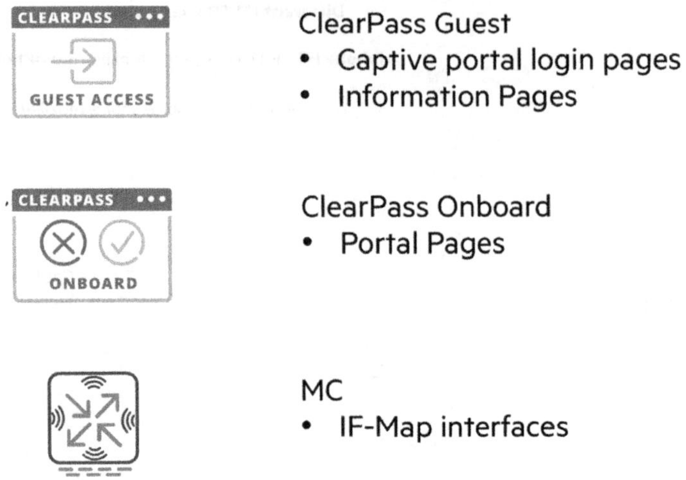

Figure 9-15: How CPPM Obtains HTTP User-Agent Strings

IF-MAP Profile Definition on AOS

IF-MAP enables devices to exchange any type of metadata. In this case, the metadata is information about wireless clients. An MC can snoop the http-user-agent strings in HTTP/HTTPS traffic that passes through it. It can then also snoop mDNS service announcements. The controller can then send this information to CPPM.

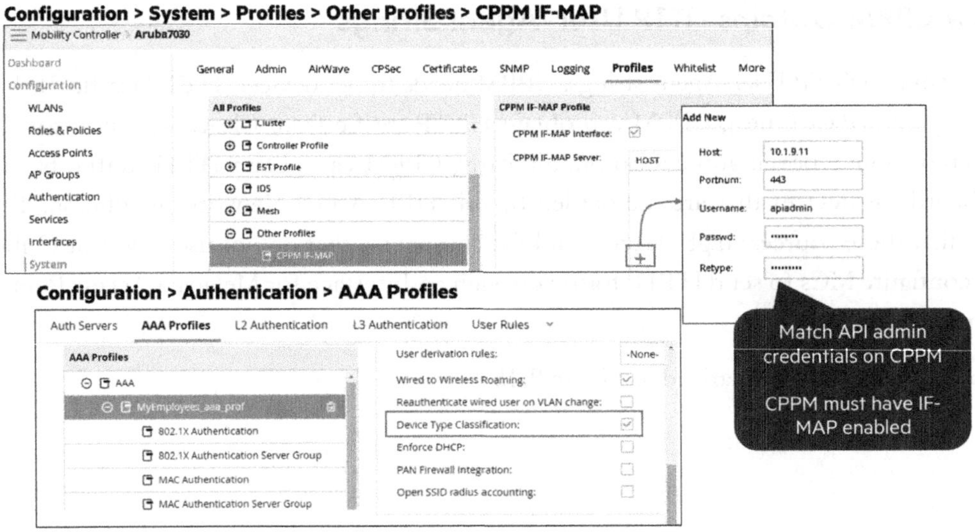

Figure 9-16: IF-MAP Profile Definition on AOS

As shown in Figure 9-16, to set up IF-MAP, go to the **Configuration > System > Profiles > All Profiles > Other Profiles** page in the AOS Web UI. Specify CPPM's IP address. You will also need to enter a username and password, which matches an apiadmin account on CPPM.

You also need to verify that Device Type Classification is enabled in the AAA profile that applies to the endpoints in question; it is by default.

This setup will require that you coordinate with ClearPass admins. They will need to give you the credentials for an appropriate account on CPPM. These admins will also need to enable CPPM to process IF-MAP data.

Sending Device Fingerprinting Information with AOS-CX Switches

As of software version 10.08, an AOS-CX switch can use device fingerprinting profiles to parse traffic from a client for information relevant to profiling, as shown in Figure 9-17.

This information includes the user-agent strings in the client's HTTP/HTTPS traffic. The profiles can also collect Link Layer Discovery Protocol (LLDP) Type Length Values (TLVs) and Cisco Discovery Protocol (CDP) TLVs. LLDP and CDP are industry standard protocols that devices use to advertise information about themselves, such as system name and port ID, to the other side of the link. A TLV is a specific attribute defined within the protocol, defined by both number and name. For example, LLDP TLV 6, system description, often includes useful information about the device type and vendor. Network infrastructure devices use LLDP and CDP, but some endpoints do as well, particularly VoIP phones.

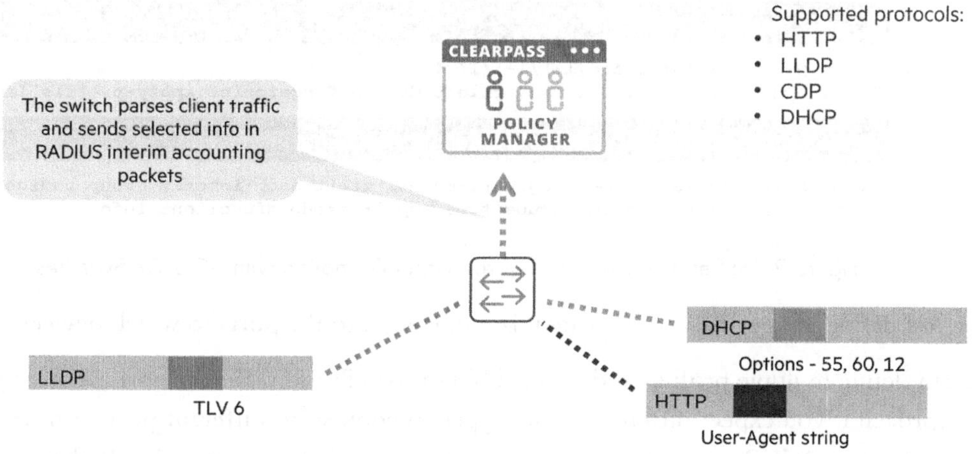

Figure 9-17: Sending Device Fingerprinting Information with AOS-CX Switches

A device fingerprinting profile can also collect DHCP options.

As of version 10.12, AOS-CX switches can send information gathered in device fingerprinting profiles to CPPM in RADIUS interim accounting packets.

Creating Device Fingerprint Profiles on AOS-CX Switches

To implement device fingerprinting on an AOS-CX switch, you must create a device fingerprinting profile, as shown in Figure 9-18.

The profile defines the protocols that the switch analyzes in user traffic, including DHCP, HTTP, LLDP, and CDP. For DHCP, you can define the specific option numbers that the switch parses and sends to CPDI. If you add DHCP to a profile without specifying an option, the profile uses the default options. These are 12, 55, and 60, and they are the options recommended for most use cases. If you do specify an option, the profile still parses the default options in addition to the one that you add. For LLDP and CDP, you can define the TLVs that the switch sends by name or number. If you add LLDP or CDP to a profile without specifying a TLV, the profile uses the default TLV, 6, which is typically recommended. If you do specify a TLV, the switch adds this one in addition to 6.

1. Create a device fingerprint profile.

```
Switch(config)# client device-fingerprint profile lab
Switch(lab)# dhcp
Switch(lab)# http user-agent
Switch(lab)# lldp
Switch(lab)# cdp
```

```
Switch# show client device-fingerprint profile lab
DHCP Attributes
Option Numbers : 12,50,55,60
Options List : Disable
HTTP Attributes
User-Agent : Enable
<- output omitted ->
```

2. Apply the profile to edge ports.

```
Switch(config)# interface 1/1/1-1/1/24
Switch(config-if-1/1/1-1/1/24)# client device-fingerprint apply-profile lab
```

3. Make sure accounting is enabled and send the profiling information

```
Switch(config)# aaa accounting port-access start-stop interim group radius
Switch(config)# vsa vendor aruba type avpair group dfp-client-info
```

Figure 9-18: Sending Device Fingerprinting Information with AOS-CX Switches

After you define the profile, you activate it by applying it to the ports to which devices connect.

You can define multiple profiles and apply different profiles to different ports. You might choose this approach if you expect different device types to connect to different ports. For example, you might configure DHCP and LLDP in the profile for ports that connect to VoIP phones and HTTP in the profile for ports that connect to computers. However, because you will often be using multi-use ports that apply the role dynamically as various types of devices connect, it can make more sense to create a single profile that includes all the protocols that devices might send and apply that to all edge ports.

To send the information to CPPM, you must ensure that interim accounting is enabled to that server. In the example shown in Figure 9-18, CPPM is defined as a RADIUS server in the default "radius" group. You must also enter this command to send the device fingerprint profiling information: **vsa vendor aruba type avpair group dfp-client-info**.

For more detailed command syntax for the device-fingerprint profiles refer to the table in Figure 9-19.

In configure context

Step	AOS-CX Command	Notes
Create a device fingerprint profile	client device-fingerprint profile <*name*>	

In device-fingerprint profile context

Step	AOS-CX Command	Notes
Collect DHCP options	dhcp [option-num <1-254>]	12, 55, and 60 are always included when you specify **dhcp**. You can add another option as well.
Collect HTTP User-Agent strings	http user-agent	
Collect LLDP TLVs	lldp [tlv-name <*name*> | tlv-num <1-8>]	6 is included by default. You can use the options to add another TLV.
Collect CDP TLVs	cdp [tlv-name <*name*> | tlv-num <1-100>]	6 is included by default. You can use the options to add another TLV.

In physical interface context

Step	AOS-CX Command
Apply a device fingerprint profile	client device-fingerprint apply-profile <*name*>

Figure 9-19: More Detailed Device-Fingerprint Profile Command Syntax

Additional CPPM Methods

In this section of the chapter, you will learn more about other CPPM profiling methods.

MAC OUI Fingerprinting

As shown in Figure 9-20, CPPM can use MAC OUI fingerprinting as well.

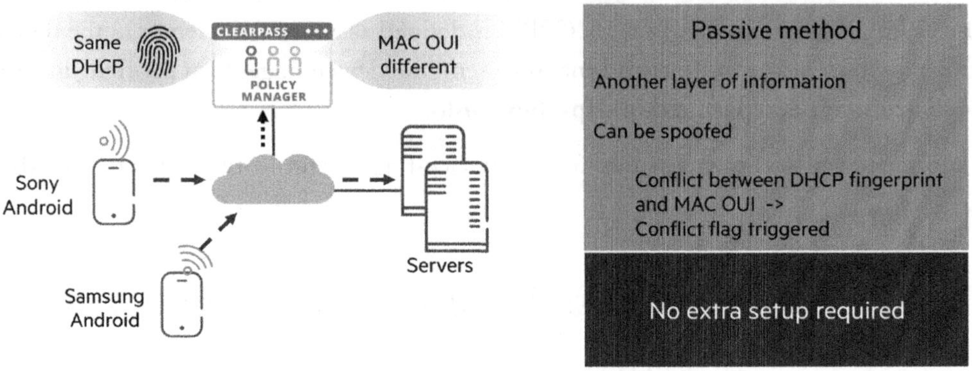

Figure 9-20: MAC OUI Fingerprinting

The first three octets of a MAC address are called the OUI. OUIs are assigned to hardware vendor, so the OUI can give valuable insight into the device type. For example, a DHCP fingerprint can identify an Android device while an OUI can detect the specific hardware for that device such as Samsung or Sony.

Keep in mind, though, that MAC addresses are easy to spoof, so MAC OUI fingerprinting does not detect when hackers are deliberately camouflaging their devices as different types. However, it can provide a valuable first layer of information to begin creating a device profile. If users and devices are behaving properly the MAC OUI fingerprint should be in line with the DHCP fingerprint. In this example, that is the case. Two smartphones were placed in the Android Device Family, and the MAC OUI gives the devices their types within this family.

CPPM can also detect conflicts when MAC OUIs place the device in one category or family while the DHCP fingerprint places it in another. It can then log the discrepancy as a conflict for admins to investigate or even change the device status to lock the device out of the network.

CPPM can learn devices' MAC addresses from a variety of sources, including during authentication. You do not need to configure anything extra in the network infrastructure to support this classification method.

TCP Fingerprinting

Using TCP fingerprinting, another passive classification method, CPPM analyzes the SYN and SYN-ACK messages in the TCP 3-way handshake that begins a session between a client and server. These messages include two important fields for identifying the client: the TTL and the time window. The TCP protocol does not mandate set values for these fields, and different operating systems have different values. Determining the values that the client uses can help CPPM pinpoint the client's OS.

To enable CPPM to receive the TCP traffic from wireless devices, you could set up mirror sessions on MCs' data path. However, often it is recommended that you mirror traffic at the core (or possibly aggregation layer), which catches traffic both from wireless and wired devices. Typically you would place the mirror source on a core link that connects to upstream resources such as a data center or the Internet, as shown in Figure 9-21. The mirror destination would then be the port that connects to the ClearPass server's span port.

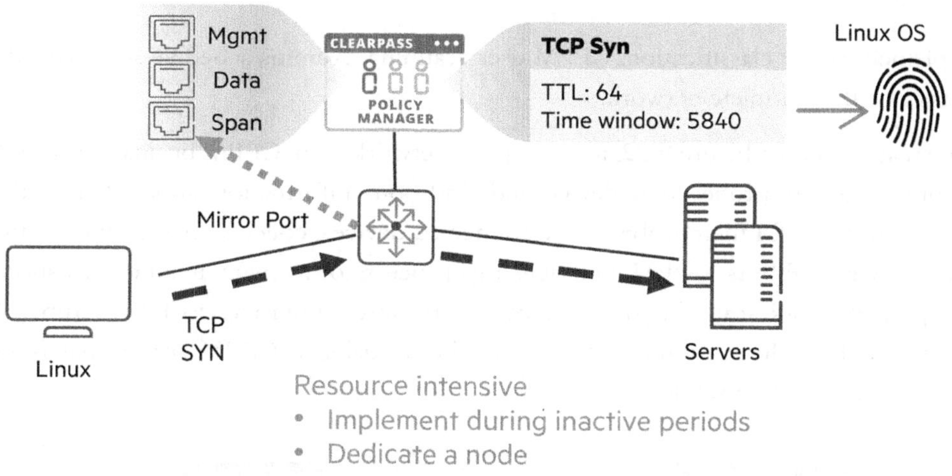

Figure 9-21: TCP Fingerprinting

Mirroring to the span port is a resource intensive function, so you should only implement it if CPPM needs to use TCP fingerprinting rather than other methods. A larger enterprise might dedicate a ClearPass node to the TCP fingerprinting functions.

ClearPass supports two configurations for a Span port. If ClearPass is using only its management port, it can use the data port as a span port. But if it is using both the management port and data port, a third port must act as the Span port. A hardware appliance has two extra ports, and the third will become the span port. However, for a virtual appliance, admins must use VMware to configure an additional promiscuous mode port to act as the Span port.

Integrations

You already learned how ClearPass can find the HTTP User-Agent strings in clients' traffic when the ClearPass solution includes ClearPass Guest or ClearPass Onboard.

ClearPass can also integrate with ClearPass OnGuard, which runs health checks on endpoints and can provide detailed information about the endpoint OS and status. ClearPass OnGuard already gathers detailed information about the device, so this solution can feed the device category, family, and name to CPPM, removing the need for ClearPass to perform further profiling.

When ClearPass integrates with an MDM/EMM, it polls the MDM/EMM server for information about endpoints. The MDM/EMM can tell CPPM about endpoints' device type, OS, security and health status, and encryption status. CPPM considers the information submitted by these solutions as definitive with no need to match against a dictionary.

Scanning

To implement active classification, CPPM uses scanning. Admins schedule scans of either specific subnets or of the complete network.

As shown at the left of Figure 9-22, in a complete network scan, CPPM begins at a seed device such as a core router. It probes that device and finds out information about it, including devices connected to it. CPPM then probes those connected devices, collects information about them, and discovers their neighbors. CPPM continues this process for as many iterations as specified in the scan depth. Network scans in particular can be resource intensive, so HPE Aruba Networking recommends that admins conduct them after hours using a CPPM server that is not actively providing authentication services.

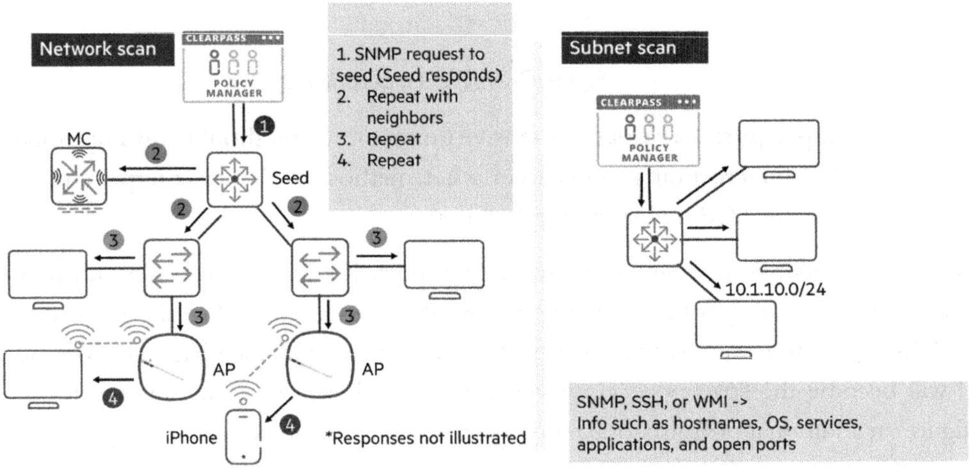

Figure 9-22: Scanning

In a subnet scan, show at the right on Figure 9-22, CPPM probes each device in the subnet.

During a scan, CPPM can perform three types of probes, based on whether admins have bound that type of probe to the device's IP address or subnet. The sections below describe these scans.

SNMP

CPPM sends SNMP Get requests to discovered devices and requests their SysDescr object. This object identifies the device type such as router. SNMP can also reveal the device's IP address and MAC address and device family. CPPM also requests three other objects, which helps it to find neighboring devices during network scans:

- cdpCacheTable, which lists neighbors that the device has discovered with Cisco Discover Protocol (CDP)
- lldpRemTable, which lists neighbors that the device has discovered with Link Layer Discover Protocol (LLDP)
- ARPtable, which lists all entries in the devices' ARP table

ClearPass admins can define multiple SNMP credentials for scans. These credentials must match read-write or read-only credentials on the scanned devices. If you have properly hardened your network infrastructure devices, they should be using SNMPv3, but many devices might be using SNMPv2 and the default "public" community.

SSH

CPPM can use SSH to log into a discovered device and discover its IP address, MAC address, device name, and hostname. This method can be useful for gathering information about Linux machines. If CPPM admins want to use this method, you and other admins should set up an SSH user account specific to this purpose on the devices that CPPM will probe.

WMI

Devices with Microsoft operating systems use Windows Management Instrumentation (WMI) to advertise information about themselves. CPPM can use WMI to discover device's MAC and IP addresses, OS information, hostnames, services, processes and applications, local users defined in the OS, and open ports. To conduct a WMI probe successfully, CPPM needs the correct domain name and credentials for a local administrator account on the machines.

Nmap Fingerprinting

Nmap is an open source network mapping tool. It sends out pings to identify the IP addresses for active devices. It then scans each device for open ports by sending TCP requests. While hackers can use Nmap for network reconnaissance, security teams can also use Nmap for benevolent purposes such as identifying vulnerabilities. CPPM also uses Nmap for legitimate purposes, taking advantage of its OS detection capabilities through TCP/IP fingerprinting, as shown in Figure 9-23.

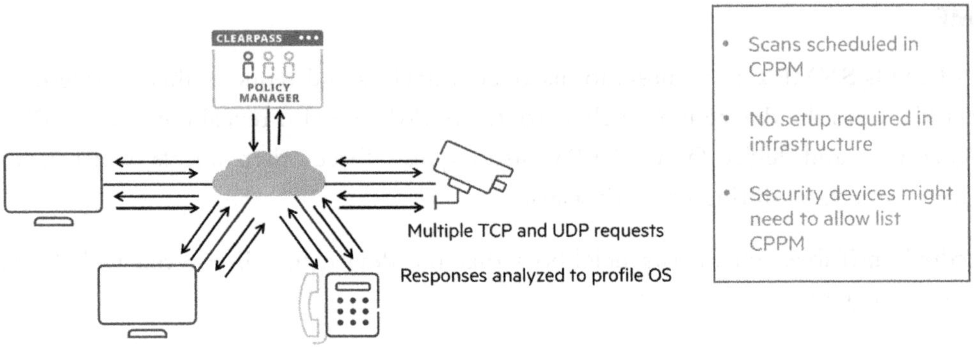

Figure 9-23: Nmap Fingerprinting

Nmap's active scanning differs from the passive TCP Syn observation because Nmap actively sends out TCP and UDP packets to determine how the device will respond. This lets Nmap identify the OS with greater certainty.

Admins can choose whether to use Nmap scans and when to schedule them. These scans can be resource intensive, so CPPM admins might choose not to use them if CPPM can gather information in enough other ways.

You do not need to set anything up in the infrastructure to support this fingerprinting type. You should be aware, though, that CPPM will be conducting scans. Some security devices, such as IDS/IPSes, will interpret CPPM's use of Nmap as a security threat. If ClearPass admins want to use this feature, they should work with security teams and make sure that they add CPPM to an allow list of devices authorized to run Nmap.

Viewing Device Profiles in CPPM

You now understand how CPPM can profile devices with the help of settings that you configure on HPE Aruba Networking devices. Admins can view device profiles in the CPPM Web UI and gain visibility into the devices on the network, as shown in Figure 9-24.

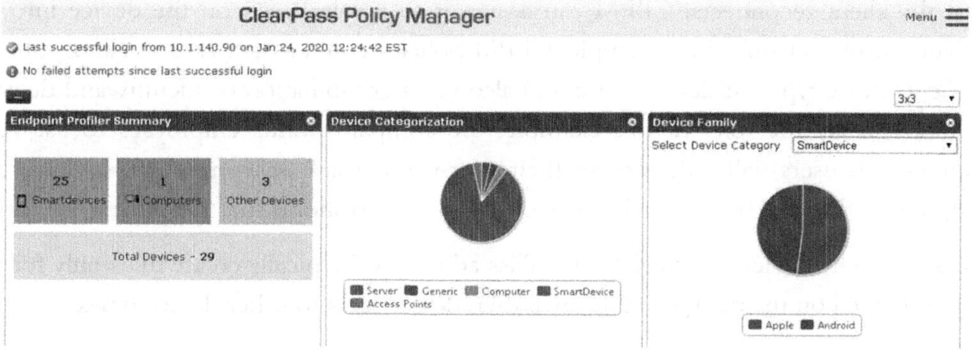

Figure 9-24: Viewing Device Profiles in CPPM

Using Device Information to Control Endpoint Access (Example)

Companies can also use the device profiles to adjust the network dynamically based on the device type.

As ClearPass admins create policies, they can specify a device category as one of the conditions for rules that assign authenticated clients to a role. The company can then more easily enforce policies about the types of devices on which users are accessing the network. The company can also improve security in a modern network with an array of IoT devices; CPPM profiles each device and assigns the device to a role based on device category.

As shown in Figure 9-25, the first time that a device connects, it will not yet have a profile. CPPM can detect that the device is not yet profiled and assign it to a "profiling role," which the network infrastructure associates with limited access. After a device is profiled or its profile changes, CPPM needs to reauthorize the device. It issues port bounce or disconnect messages to do so.

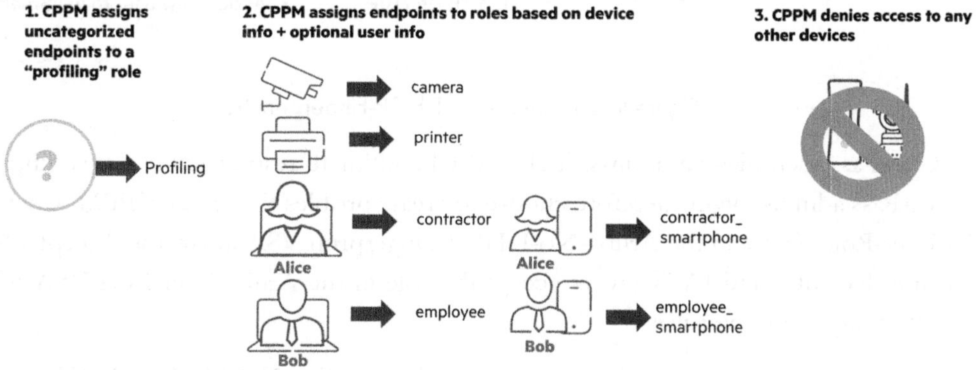

Figure 9-25: Using Device Information to Control Endpoint Access (Example)

When the client reconnects, CPPM can assign it to a role based on the device info, optionally combined with user info. For example, CPPM policies could assign cameras and printers to roles specific to those types of devices. It could also use a combination of identity and device category conditions to set the role. In this example, the company wants employees to use their proper computers, so users will only receive their user-based roles on computers. If users connect on smartphones, they will be assigned to a "Contractor_smartphone" or "Employee_smartphone" role.

Because so many device types exist, ClearPass admins will typically create rules only for the devices that are allowed on the network. They will then deny access to other device types.

Aruba-No-DHCP-Fingerprint VSA

Earlier you learned that MCs and IAPs can also use DHCP fingerprinting to classify devices and assign them to roles. They do so with user rules that use DHCP options. Such a rule is a Layer 3 rule, which takes effect after authentication. This means that the role overrides other roles, including one sent by CPPM with an Aruba-User-Role VSA.

However, as shown in Figure 9-26, an exception does apply. If the authentication server sends an Aruba-No-DHCP-Fingerprint VSA set to 1 in the Access-Accept, then the MC or IAP will no longer use its own DHCP fingerprinting rules to assign that client a role.

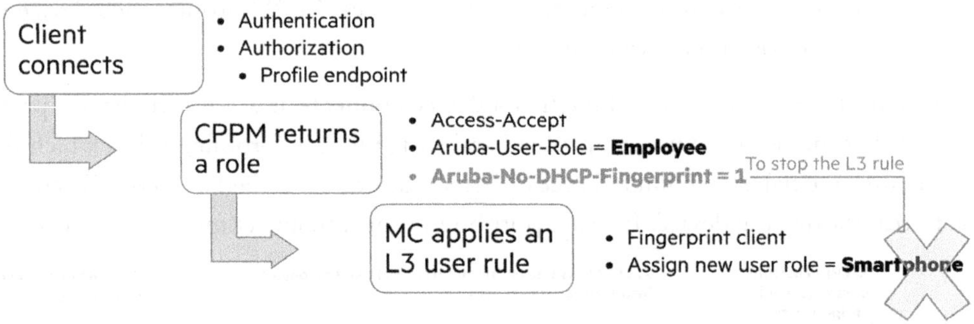

Figure 9-26: Aruba-No-DHCP-Fingerprint VSA

When ClearPass identifies endpoints, it should take ultimate responsibility for setting the device role. ClearPass admins might therefore choose to create profiles that cause CPPM to send both the Aruba-User-Role VSA and the Aruba-No-DHCP-Fingerprint VSA in Access-Accepts. In this way, you ensure that MCs and IAPs always accept the role in the Aruba-User-Role VSA despite their own DHCP fingerprinting rules.

However, in the long term, it is best to remove the conflict. If CPPM is classifying endpoints, remove any user rules on AOS devices that include DHCP options.

Wireless Profiling and Connection Process

You will now look at the process by which the HPE Aruba Networking solution classifies clients and assigns them roles.

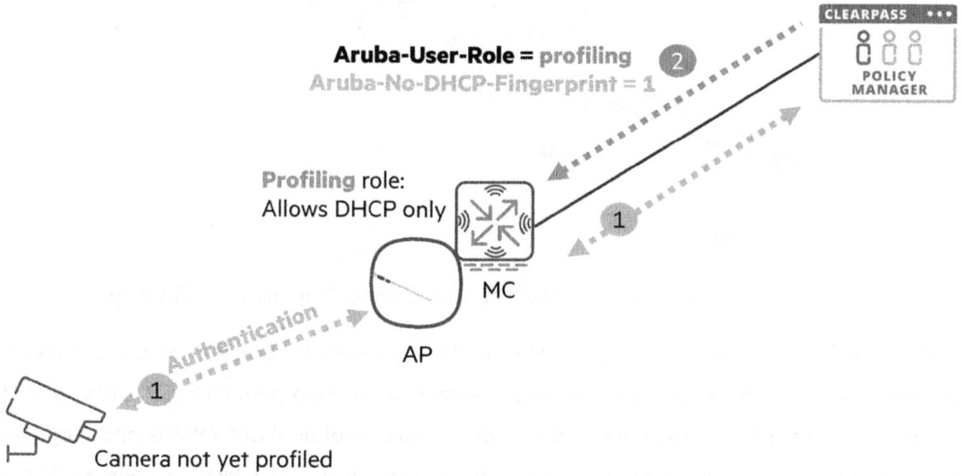

Figure 9-27: Wireless Profiling and Connection Process—Initial Connection

1. As shown in Figure 9-27, a wireless camera connects to the CompanyABC WLAN for the first time. It authenticates to CPPM.
2. CPPM detects that this device is not yet profiled, and it tells the MC to assign the device to the "profiling" role. On the MC, the "profiling" role has restrictive policies that permit DHCP traffic only.
3. As shown in Figure 9-28, the camera sends a DHCP Discover and receives an IP address. The DHCP relay also sends the DHCP Discover message to CPPM.

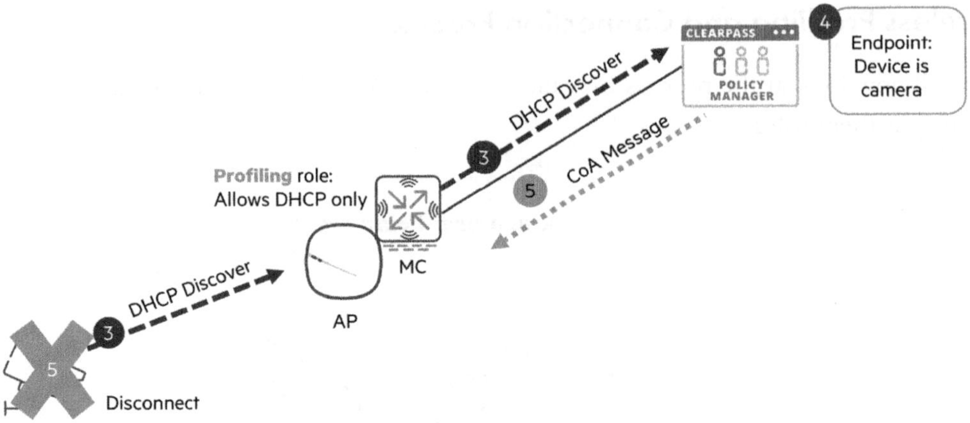

Figure 9-28: Wireless Profiling and Connection Process—Profiling

4. CPPM uses DHCP fingerprinting to identify the device category. Some companies prefer to set up the "profiling" role to implement captive portal. The role would then permit DHCP, DNS, and HTTP/HTTPS with a redirect to a ClearPass Guest page, which could explain that CPPM is profiling the device after which the user will receive network access. This approach allows CPPM to collect HTTP agent strings during the initial profiling. In either case, CPPM determines the category for the endpoint.

5. The CPPM service that originally authenticated the device has a rule that tells CPPM to trigger a CoA when a device's category is discovered or changes. CPPM sends a CoA message to the MC, telling the MC to terminate the session, which the MC does.

6. As shown in Figure 9-29, the camera now tries to reconnect and passes through 802.1X authentication again.

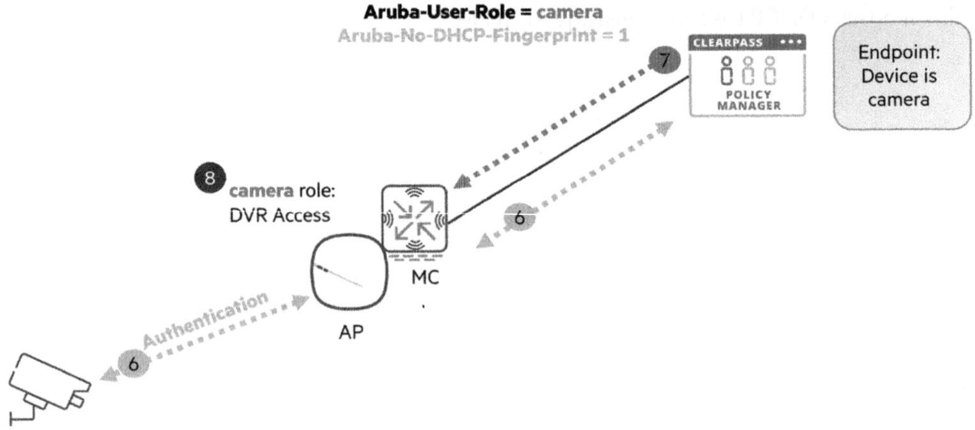

Figure 9-29: Wireless Profiling and Connection Process—Correct Role Assignment

7. This time, though, the policy enforcement rules match the rule that places cameras in the "cameras" role. CPPM sends this role to the MC in an Aruba-User-Role VSA.

8. The MC places the camera in the "camera" role. The camera will receive a DHCP address again. Now as the camera sends traffic, the MC will apply the correct firewall policies for the "camera" role and forward the traffic.

Wired Profiling and Connection Process

Now follow the process of endpoint categorization for a wired device.

1. A wired camera connects to the network for the first time. It authenticates to CPPM.

2. Because none of the rules in the profile apply to it—its category is unknown—the default profile applies. In this example, it is assigned to the "wired-profiling" role. In this example, the switch downloads this role from CPPM. The DUR has a port-access policy that permits DHCP traffic only. You could set up the role differently, though. For example, you could tunnel the traffic to the MC.

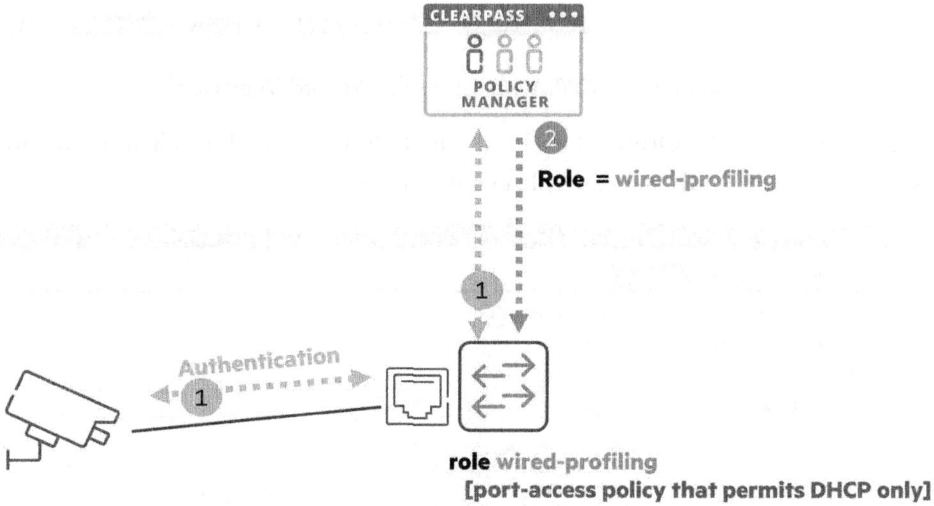

Figure 9-30: Wired Profiling and Connection Process—Initial Connection

3. The camera initiates DHCP. The switch broadcasts the DHCP Discover as usual. The wired infrastructure is configured to relay the traffic to CPPM. Alternatively, if the switch is tunneling traffic to the MC, the MC would apply the proper policies and forward the traffic in the correct VLAN. Again, CPPM must be an IP helper for that VLAN.

An example of what a CPPM Access Tracker record might look like at this point in the process is shown in Figure 9-31. The **Alerts** tab shows that CPPM does not yet know the device's profile. (While this record is for a wired client, you would see a similar record for wireless clients.)

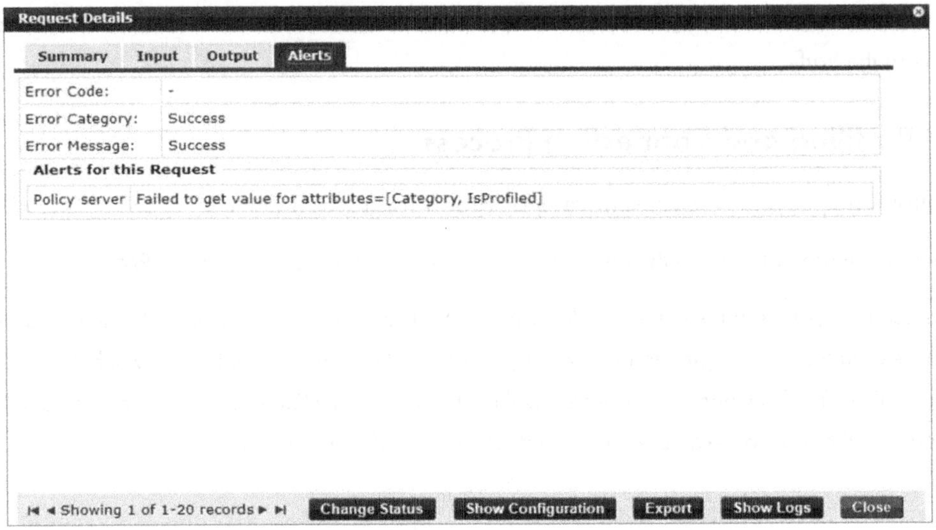

Figure 9-31: CPPM Access Tracker Record (Alerts tab)

As shown in Figure 9-32, the **Output** tab shows that CPPM assigned the client to the profiling role. In this example, this is a DUR, and the role enables tunneling.

Figure 9-32: CPPM Access Tracker Record (Output tab)

The process continues:

4. As shown in Figure 9-33, CPPM uses the DHCP fingerprint and other information that it is collecting to profile the camera.

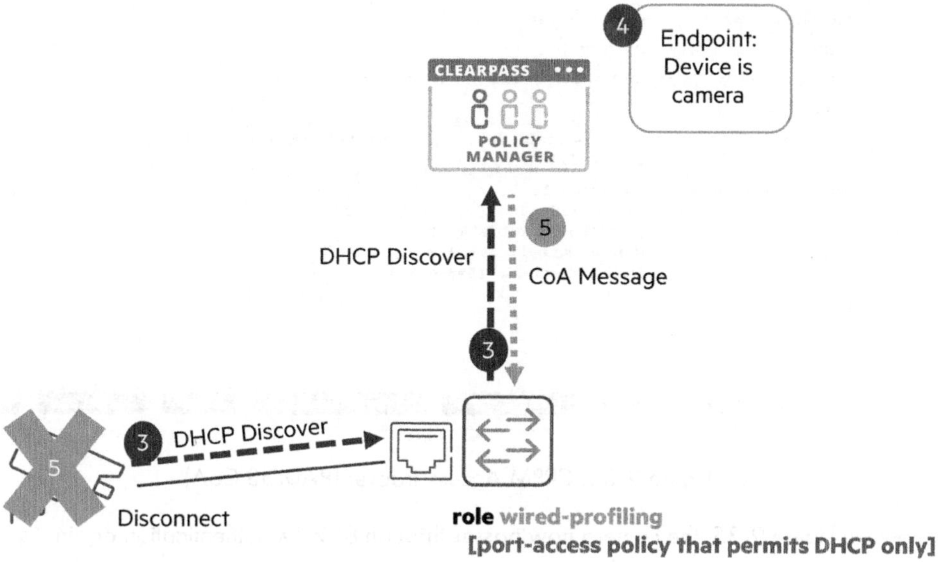

Figure 9-33: Wired Profiling and Connection Process—Profiling

5. CPPM sends a CoA to the switch. The CoA might tell the switch to terminate the session, or it might tell the switch to "bounce" the connection (terminate the session and keep it down for a few seconds).

When CPPM sends a CoA to change a client's status, it records that fact in the Access Tracker record created when the client first authenticated, as shown in Figure 9-34. As you see, in this example, the switch received the CoA, executed the termination action, and acknowledged the CoA.

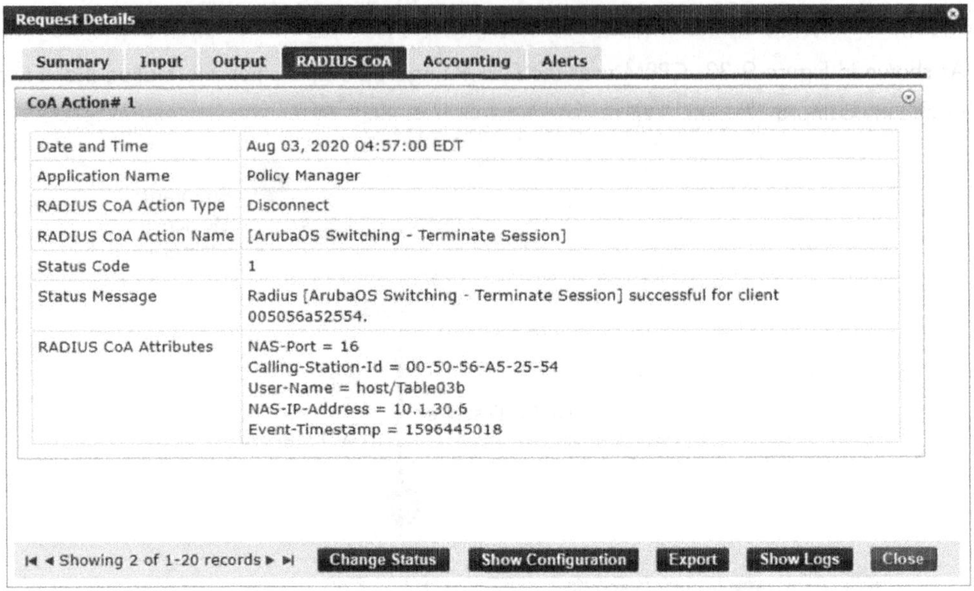

Figure 9-34: CPPM Access Tracker (RADIUS CoA)

6. As shown in Figure 9-35, the camera now passes through 802.1X authentication again.

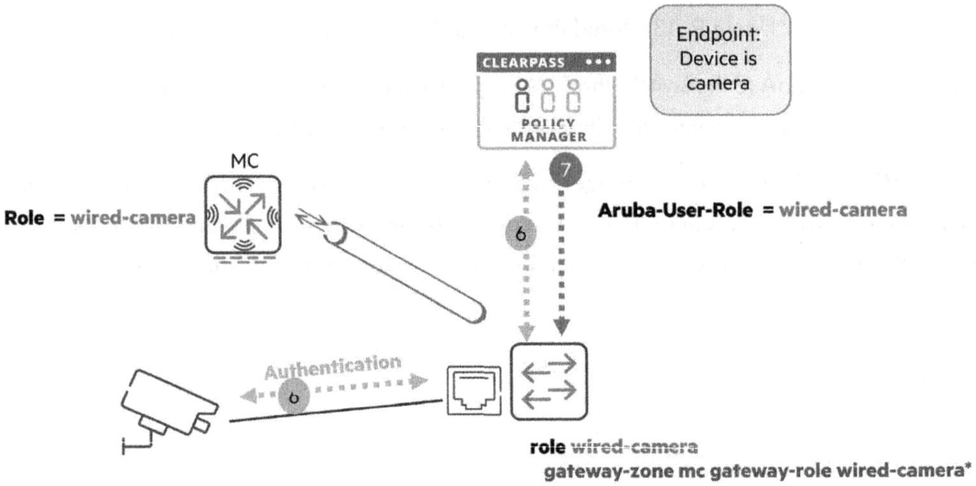

Figure 9-35: Wired Profiling and Connection Process

7. This time, though, CPPM matches the rule that places cameras in the "wired-cameras" role. The switch has a matching role, which, in this example, implements tunneling to the MC with "wired-camera" secondary role. The camera's traffic now passes to the MC, which applies "wired-camera" firewall policies. In this

example, the customer wants to unify traffic forwarding and security policies at the MC. But admins can choose local switching instead, if they want.

Summary of HPE Aruba Networking Infrastructure Setup

Conclude this section of the chapter by briefly reviewing what you have learned about setting up HPE Aruba Networking AOS network infrastructure devices to support the CPPM Device Profiler.

Configuring AOS Mobility Devices to Support CPPM Device Profiler

The configuration steps for AOS mobility devices are summarized in Figure 9-36.

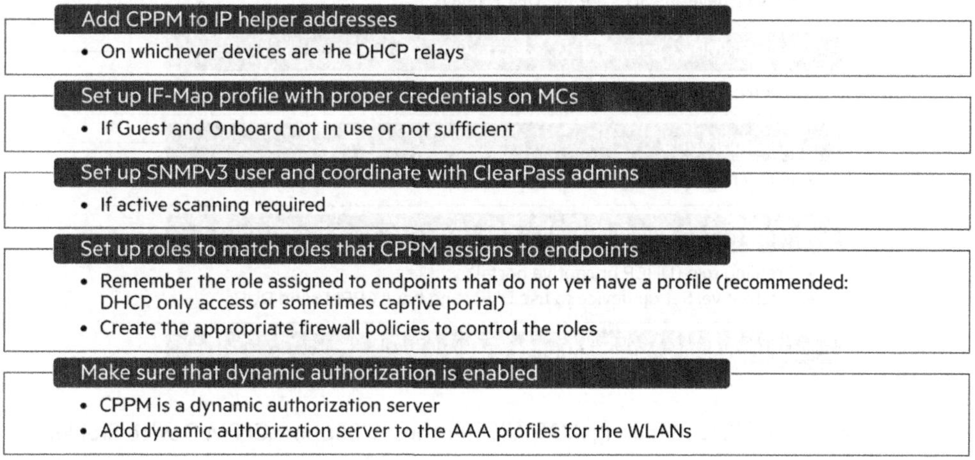

Figure 9-36: Configuring AOS Mobility Devices to Support CPPM Device Profiler

You must add CPPM's IP address to the IP helper list on whichever devices are currently acting as DHCP relays. These are typically routing switches, but sometimes MCs on some networks.

If the company is not using ClearPass Guest or Onboard, CPPM might need help learning the HTTP-Agent Strings that devices are using. You can set up an IP-Map profile on MCs to send this information.

If ClearPass admins plan to use active scanning, coordinate with them to set up the correct SNMP settings on MCs and APs. Recommend the use of SNMPv3 for security.

Remember one of the goals of endpoint profiling: CPPM assigning the proper roles to each user and device combination. Coordinate with ClearPass admins to determine which role names CPPM will send. Set up these roles on MCs or IAPs and configure the roles with the appropriate settings to implement the company's security policies. Remember to also set up the role assigned to devices that have not yet been profiled; typically, this role should only permit DHCP access.

Also make sure that the AOS device will accept CoAs from CPPM. Add CPPM as a dynamic authorization server on the device. If you are using RadSec for the RADIUS connection to CPPM, also enable RadSec on the dynamic authorization server. Also make sure to add the dynamic authorization server on the AAA profiles for your WLANs.

Configuring AOS-CX Switches to Support CPPM Device Profiler

You should also go through the checklist in Figure 9-37 to determine which steps to complete on switches.

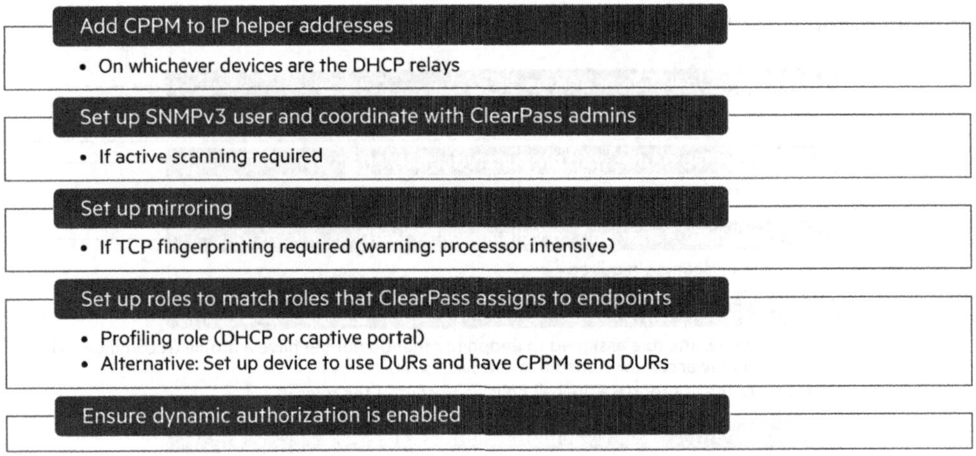

Figure 9-37: Configuring AOS-CX Switches to Support CPPM Device Profiler

You should add CPPM's IP address to the IP helper list on whichever devices are currently acting as DHCP relays. If ClearPass admins plan to use active scanning, coordinate with them to set up the correct SNMP settings on AOS-CX switches. SNMPv3 offers better security than SNMPv1/v2.

If the company has chosen to use TCP fingerprinting, you need to set up mirroring to the ClearPass server's span port on switches, based on the traffic that the ClearPass admins want CPPM to examine. Remember that this feature is less often used and very resource intensive.

Again, coordinate with ClearPass admins to determine which role names CPPM will send. Set up these roles on the switches and configure them with the appropriate settings to implement the company's security policies. Remember to also set up the role assigned to devices that have not yet been profiled with the appropriate rights.

As an alternative approach to configuring role settings on the network infrastructure devices, you can configure the devices to support downloadable user roles (DURs). Then ClearPass admins should set up DURs on CPPM, as well as create an admin account for the switches to use to download the DURs.

Finally, make sure to configure CPPM as a dynamic authorization server on the AOS-CX switches, following the process that you examined in a previous chapter.

In Figure 9-38, you see CLI commands that are relevant to setting up AOS-CX switches to support device profiling in conjunction with CPPM.

In configure context

Step	AOS-CX Command	Notes
Set up authentication to CPPM and user role authorization.	See Module 7. Make sure that the switch accepts CoA messages.	
Add CPPM as an IP helper.	`interface <L3_interface_ID>` `ip helper-address <IP>`	
Configure sFlow with CPPM as the collector.	`sflow` `sflow collector <IP> vrf <vrf>`	Optional

Mirror traffic (local).	`mirror session <ID>`
	`source interface <ID> <rx | tx | both>`
	`destination {tunnel <dst_IP> source <source_IP> | interface <ID>}`
	`enable`
		Optional.

Configure SNMPv3 user.		Optional. See Module 2.

Figure 9-38: CLI Commands: AOS-CX Setup to Support Device Profiler

As long as you have set up authentication and user role authorization on the switches, the switches can enforce CPPM's decisions. Make sure that you have enabled the switch to accept CoA messages so that the switch will bounce the port and re-authenticate the device when so ordered by CPPM.

If the switch acts as the DHCP relay, remember to add CPPM as an additional helper address to enable CPPM to implement DHCP fingerprinting.

The switch can also help direct other traffic to CPPM; commands are shown for your reference.

Other Device Profiling Solutions

Organizations have other ways to obtain advanced device profiling from HPE Aruba Networking. This section gives you a high-level look at this solutions. For more information, take the *Implementing HPE Aruba Networking Security* course or read the *HPE Aruba Networking Certified Network Security Professional Study Guide.*

Client Profiles built into HPE Aruba Networking Central

Client profiles are included within the HPE Aruba Networking Central Network Operations subscription. HPE Aruba Networking Central constructs these profiles using telemetry pulled from HPE Aruba Networking devices, such as APs, switches, and gateways. It uses machine learning (ML)-enhanced techniques to analyze that data and determine the client type, family, and category.

The client profile also includes rich context about the client such as the connections that it is establishing and the applications that it is using.

HPE Aruba Networking ClearPass Device Insight (CPDI)

Organizations can alternatively obtain CPDI as a separate subscription. CPDI uses Data Collectors, which run active and passive scanning methods to collect information about clients. CPDI can therefore profile clients in heterogenous networks that incorporate third-party network infrastructure while the client profiles in HPE Aruba Networking Central work only for clients connected to HPE Aruba Networking devices.

CPDI provides all the benefits of the client profiles in HPE Aruba Networking and more. CPDI supports security scanning, including posture and vulnerability assessments.

Refer to Figure 9-39 for a summary of differences between client profiles in HPE Aruba Networking Central and CPDI as a standalone application.

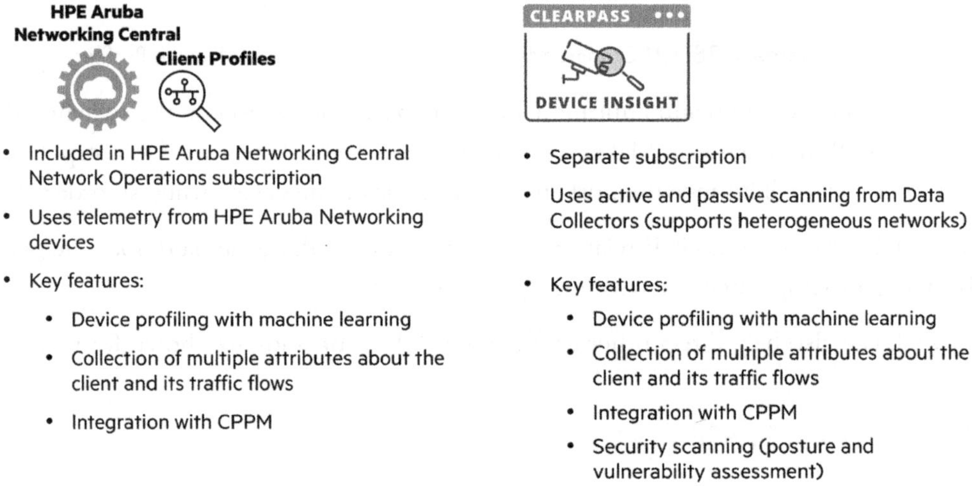

Figure 9-39: HPE Aruba Networking Central and CPDI

Note on CPPM Integration

If you integrate CPPM with Device Insight (either the client profiles on HPE Aruba Networking Central or CPDI), CPPM no longer supports its own profiling features. *Implementing HPE Aruba Networking Security* provides more information about how integration works and which features are disabled.

Summary

Congratulations! You have learned about how classifying devices based on their type can improve the security posture of a network. You learned about many passive and active classification methods supported by CPPM, and you learned how to configure HPE Aruba Networking infrastructure devices to feed information to CPPM. You also learned how CPPM can use device categories to determine role assignments, as well as how the HPE Aruba Networking infrastructure can enforce these roles.

Learning checks

1. What is a difference between passive and active endpoint classification?
    a. Passive classification collects data for admins to examine while active classification uses device types to assign roles.
    b. Passive classification uses traffic sent by endpoints on their own while active classification involves sending out probes or requests.
    c. Passive classification always occurs on-prem while active classification involves sending data to the cloud.
    d. Passive classification relies on agents installed on devices before they connect to the network while active classification occurs post authentication.
2. What should you configure in the network infrastructure to help CPPM perform DHCP fingerprinting?
    a. Configure CPPM as an additional IP helper address in the VLANs with endpoints.
    b. Configure MCs and switches to mirror all traffic to CPPM.
    c. Configure active scans with SNMP and DHCP on MCs.
    d. Configure user derivation rules with DHCP options on MCs.
3. Which purpose does an IF-MAP profile on an MC serve?
    a. It enables the MC to use Nmap to perform active classification on endpoints.
    b. It allows you to configure whether you want the MC to prioritize DHCP fingerprint roles from CPPM or the MC.
    c. It maps user roles sent by CPPM to user roles configured on the MC firewall.
    d. It sends HTTP User-Agent strings from the MC to CPPM.

4. Assign numbers to each criterion to indicate its precedence for assigning roles on MCs. The criterion with highest precedence should be 1.

   __ User Derivation Rule with L2 criteria

   __ Aruba-User-Role VSA only

   __ User Derivation Rule with L3 (DHCP option) criteria

   __ Aruba-User-Role VSA + Aruba-No-DHCP-Fingerprint VSA

5. What is one difference between profiling capabilities in CPDI and CPPM?

   a. CPDI supports active classification methods, and CPPM only supports passive classification.

   b. CPDI can use packet inspection, analysis of traffic patterns, and machine learning to identify IoT devices more precisely than CPPM.

   c. CPDI can use device categories as conditions for assigning roles while CPPM on its own cannot.

   d. CPDI runs as a module on MCs while CPPM is a standalone solution.

6. Consider these questions and take notes on your thoughts:

   - How can the profiling capabilities about which you learned in this chapter help companies improve visibility into their network?
   - How can the profiling capabilities about which you learned in this chapter help companies improve security?
   - What are two ways that you could use these capabilities in your or your customers' networks?
   - Assume that you are working with a company that already uses CPPM to authenticate users, but does not use device profiling. Now ClearPass admins want to add profiling. What would you discuss with the admins to help the implementation go smoothly?

Note

For answers to all learning checks, refer to Appendix B.

---

10 HPE Aruba Networking Unified SASE

EXAM OBJECTIVES

✓ Define Secure Access Service Edge (SASE)

✓ Describe HPE Aruba Networking unified SASE solutions

Assumed Knowledge

- Basics of IP networking
- Basics of Ethernet switching
- Basics of 802.11 wireless networks
- Basic definition of a wide area network (WAN)

Overview

The chapter begins by introducing you to trends and challenges that are driving companies to seek a new approach to securing branch and remote access. You will learn what SASE is and why it presents the best solution to these challenges. You will then explore at a high-level HPE Aruba Networking unified SASE solutions, including HPE Aruba Networking EdgeConnect SD-WAN, HPE Aruba Networking EdgeConnect SD-Branch, HPE Aruba Networking EdgeConnect Microbranch, and HPE Aruba Networking SSE.

Introduction to SASE

In this section you will learn about the challenges that customers face with traditional approaches to branch and remote connectivity. You will then learn how SASE helps customers overcome those challenges.

Key Trends in Edge Security

Three key trends are driving organizations to implement SASE.

Remote and hybrid workforce

The COVID-19 pandemic accelerated a trend that already begun pre-2020: the shift in the workforce away from the office and toward home offices. A Gallup Indicator poll on hybrid work, accessed July 17, 2024, surveyed US employees in "remote-capable" jobs (such as office jobs that can be performed offsite). As shown in Figure 10-1, the survey indicated that only 21% of these employees work in exclusively on-site. Over a quarter of (27%) work fully remotely. And most employees (53%) take a hybrid approach, working sometimes from home and other times in an office. While this poll surveyed US employees, the trends are similar worldwide.

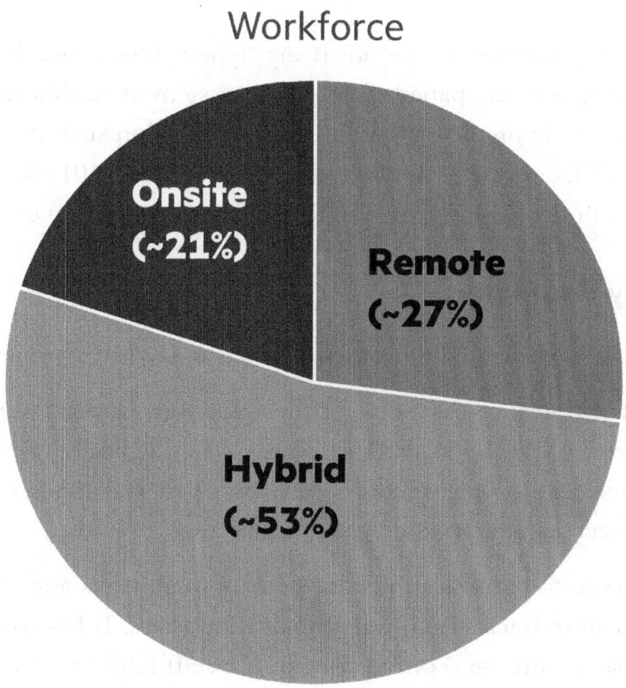

Figure 10-1: Hybrid Workforce

Technologies, such as virtual private networks (VPN), have made this shift possible. But as the balance tips towards a hybrid and remote workforce, those technologies are being strained to their limits.

Increased use of cloud

Much like users' locations have been shifting the location of the resources that they access has also changed. In addition to private apps running in companies' data centers, remote and branch users also need to access cloud apps. A DemandSage report reveals that by the end of 2023 95% of organizations used at least one Software as a Service (SaaS) application running in the cloud. Use of infrastructure as a service (IaaS) is also expanding ("17+ SaaS Statistics For 2024 (Trends, Growth & More)," DemandSage, Dec. 30, 2023). Gartner forecasted a 26.6% growth in spending in this area in 2024 ("Gartner Forecasts Worldwide Public Cloud End-User Spending," Gartner Press Release, Nov. 13, 2023).

Because remote and branch users need to access these new types of applications, their traffic patterns are shifting, and solutions designed for legacy patterns (in which most traffic flowed to a data center) no longer deliver the required performance.

Proliferation of IoT

Internet of Things (IoT) devices come in many types: smart building equipment, smart manufacturing equipment, sensors, patient monitoring system at healthcare facilities, and many more. A 101 Blockchain article predicted that more than 30 billion such devices will be connected to networks by 2025 ("IoT Connectivity Industry Forecast by 2030," 101 Blockchains, Mar 8, 2023). With so many new types of devices popping up, IT staff struggle to keep up.

Dissolving Security Perimeter

Together these trends have accelerated the dissolution of the security perimeter.

When most users work remotely at least some of the time, organizations can no longer maintain a strict delineation between internal and external access. They typically use some form of virtual private network (VPN) to regulate the remote access. But even with this security in place, opening the network to remote access increases the attack surface.

Organizations also struggle to enforce consistent policies to all users and devices in all locations. They might find it difficult to track who is connected from where. If Bob connects from site A on Monday, home on Tuesday, and site B on Wednesday, IT staff might find it difficult to ensure that Bob has the same rights each time he connects. The result can be frustration for Bob—if he does not receive the access he needs to do his job—as well as the unintended opening of security gaps.

- Increased attack surface
- Difficulty enforcing consistent policies to all users and devices in all locations
- Gaps in security

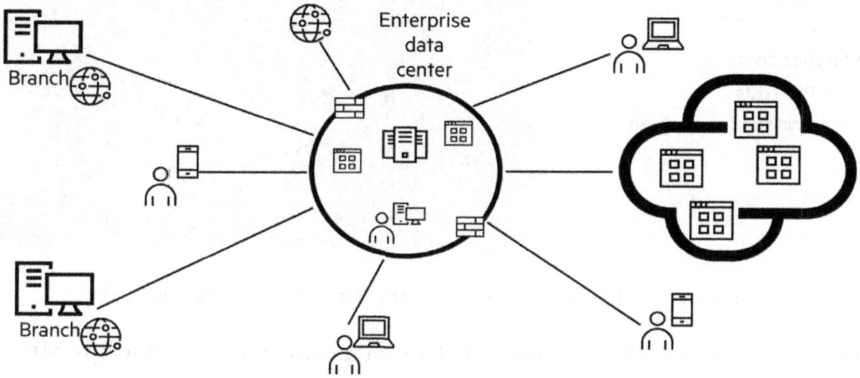

Figure 10-2: Dissolving Security Perimeter

As another example, companies often struggle to monitor IoT devices and apply appropriate access controls to them, particularly when those devices are deployed across many branch sites. But without proper oversight, the devices can become targets for hackers. After they have compromised an IoT device, hackers can use the device's connection to the internal network to spread laterally and attack other resources.

These challenges are summarized in Figure 10-2.

Challenges of Legacy Branch Networks (MPLS)

Legacy branch networks often use private Multi-Protocol Label Switching (MPLS) circuits to connect the branch office to the enterprise data center. As shown in Figure 10-3, often companies backhaul *all* branch traffic to that data center, pushing the traffic through firewalls before allowing it to flow to its ultimate destination. This traffic flow made sense when most branch traffic was destined to the data center itself. However, as you learned, much of branch traffic is now destined to the cloud. Backhauling that traffic from the branch to the data center to the cloud increases latency.

- Slow, unpredictable performance
  - Increased latency
  - Inability to consistently meet service level agreements (SLAs)
- Higher costs
  - MPLS costs
  - Administration overhead

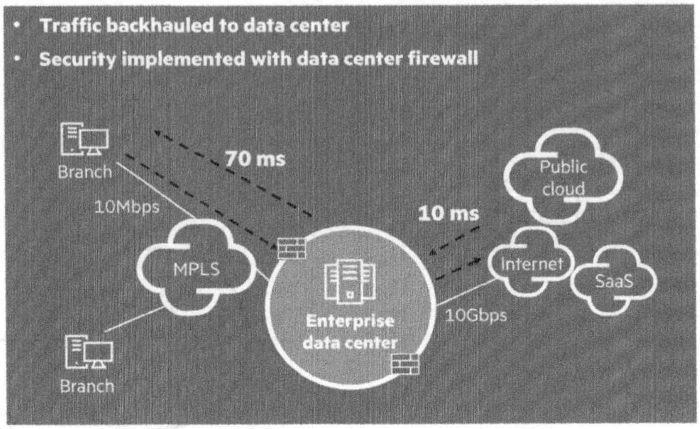

Figure 10-3: Challenges of Legacy Branch Networks (MPLS)

Consider, for example, a branch that connects to a data center over a 10 Mbps MPLS circuit. The data center has a 10 Gbps Internet connection. When a branch user accesses a SaaS application, the MPLS connection might impose 70 ms of latency while the connection from the data center to the SaaS application adds another 10 ms of latency. The total round-trip latency is 160 ms, significantly higher that it would be if the branch user could connect directly to the SaaS application from a branch Internet connection. This increased latency can make it more difficult for organizations to consistently meet service level agreements (SLAs) for performance. For example, Microsoft recommends that clients connect to Microsoft 365 without any backhauling.

Legacy branch networks can also be costly. As private lines, MPLS circuits are more expensive than simple Internet connections. Organizations must also devote resources to managing the solution.

Challenges of Legacy Branch Networks (VPN over Internet)

To reduce costs, some companies choose to connect their branches to the enterprise data center through the Internet, as shown in Figure 10-4. Because the Internet is an insecure, shared network, these companies must establish a VPN to protect the site-to-site traffic. Typically, they use IP Security (IPsec) tunnels. An IPsec tunnel encrypts traffic and ensures its data integrity.

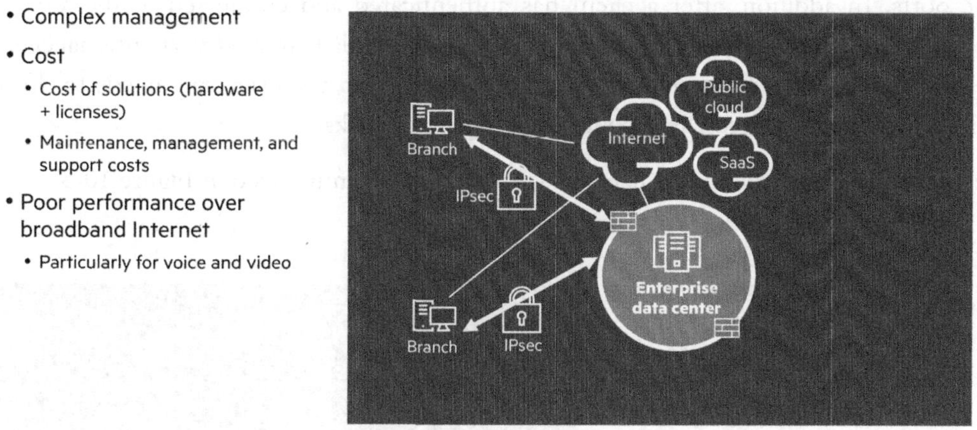

Figure 10-4: Challenges of Legacy Branch Networks (VPN over Internet)

While the Internet connections themselves are lower cost, the company must spend more time and effort deploying, managing, maintaining, and troubleshooting the VPN. Companies must pay for hardware and licenses for the VPN solutions. In addition to ongoing costs from their solutions' support contracts, companies must pay to maintain and manage the complex solutions.

In addition, broadband Internet connections often fail to provide adequate performance. Packet loss and jitter can reduce the quality of applications to an unacceptable degree, particularly for voice and video applications.

Challenges of Legacy Client-to-Site VPNs

To grant remote users access to the corporate main office or data center, companies have traditionally deployed client-to-site VPNs. These VPNs might use IPsec like site-to-site VPNs, or they sometimes use SSL.

As the size of the remote and hybrid workforce has increased, companies struggle more and more with maintaining these VPNs. Users' devices must have VPN clients installed and configured. Admins must either perform the setup for users or guide users through the setup. Both approaches add to admins' workload and can cause user frustration. Problems often occur, which take time to troubleshoot. As the demands on the VPN grow, the company might also need to deploy more gateways, incurring more costs in purchasing, deploying, and managing those gateways.

VPNs do impose security. They require users to authenticate to receive access. They encrypt traffic. However, opening up the private network to remote access—despite these security measures—tends to increase the attack surface. To allow clients to connect to the VPN gateway, the data center or main office must expose a port on that gateway to the Internet. Hackers often attempt to attack these ports. In addition, after a client has authenticated and connected to the VPN, the client receives an IP address on the internal network and is essentially treated as an internal client. Without careful controls, the client could receive more access than the user truly needs to do their job—access that a compromised client could use to launch attacks.

The disadvantages of traditional client-to-site VPNs are summarized in Figure 10-5.

- Complex management and user frustration
- Same costs as site-to-site VPNs
- Trouble scaling
- Split-tunnel VPN
  - Might address performance issues
  - But introduces security challenges
- Increased attack surface
  - Ports exposed
  - Entire private network exposed to remote users

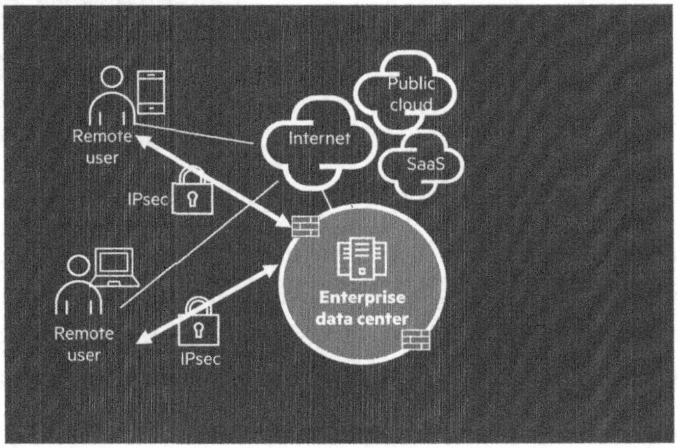

Figure 10-5: Challenges of Legacy Client-to-Site VPNs

Secure Access Service Edge (SASE) to Address the Challenges

To surmount these challenges, customers need a new framework for connecting branch and remote users to applications. Secure Access Service Edge (SASE) provides this framework.

Combining networking and security functions, SASE securely connects users or devices anywhere—branch offices, home offices, or others—to applications that run anywhere—from the private data center to the public cloud. SASE can also impose security controls to help companies protect sensitive data stored in the cloud.

As shown in Figure 10-6, SASE consists of two complementary components: a WAN edge, delivered by secure software-defined WAN (SD-WAN), and a Security Service Edge (SSE).

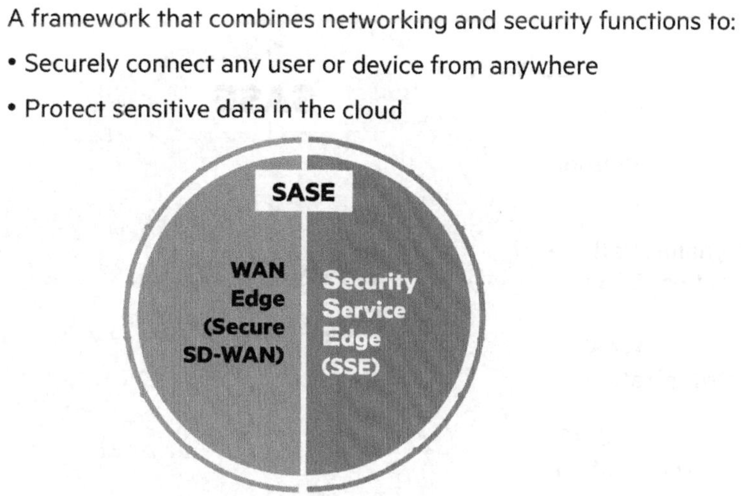

Figure 10-6: Secure Access Service Edge (SASE) to Address the Challenges

Secure Software-Defined-WAN (SD-WAN)

As an example of software-defined networking, SD-WAN is a centrally orchestrated solution that abstracts the underlying WAN connectivity technologies, such as MPLS and broadband. Rather than rely on traditional routing technologies, SD-WAN intelligently steers traffic across the correct path at any given moment based on business policies and current circumstances; this intelligent steering is called Dynamic Path Selection (DPS). Through automation and AI, SD-WAN removes the complexity of establishing secure connections from the branch to the data center and cloud.

Ideally SD-WAN should overcome performance shortcomings in the underlaying physical links. For example, the solution might repair packet loss and jitter. It might provide WAN optimization, reducing latency by compressing data and "deduplicating" transmissions. (Deduplication eliminates redundant data while still maintaining the integrity of transmissions.) It might also accelerate TCP applications through techniques such as variable window sizing and selective acknowledgements.

SaaS acceleration technologies improve performance for SaaS applications by—for example—intelligently steering traffic toward the correct point of access for each application.

Some customers choose to obtain SD-WAN hardware and software as a service. In other words, they pay for the solution on a periodic basis rather than upfront. They might also pay providers to manage the services for them.

See Figure 10-7 for a summary of key SD-WAN features.

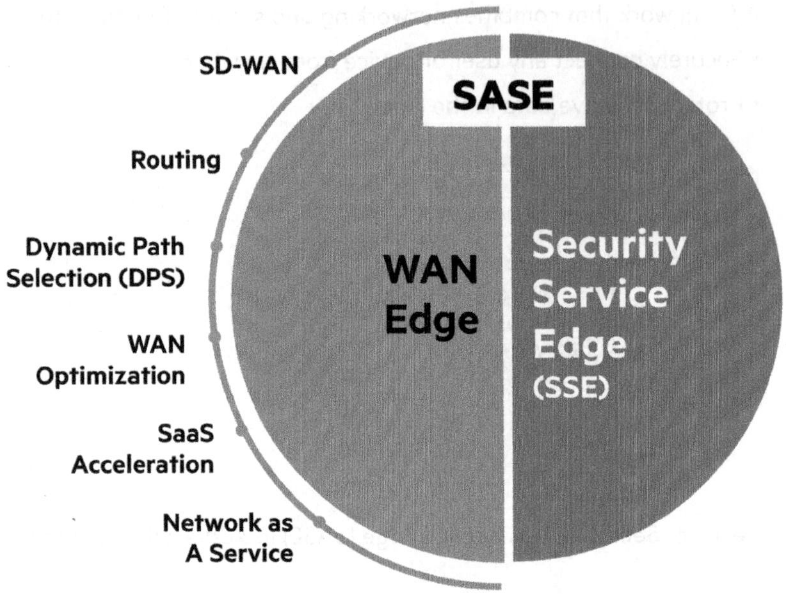

Figure 10-7: Secure Software-Defined-WAN (SD-WAN)

SSE

SSE consists of cloud-based services for securing user access to data and applications (see Figure 10-8). The sections below briefly describe services that industry analyst Gartner defines as part of SSE.

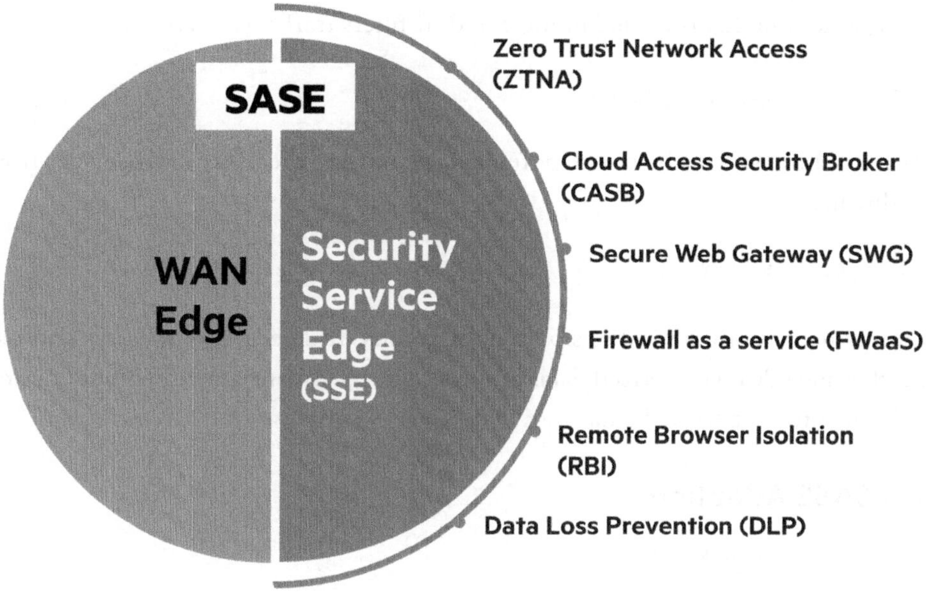

Figure 10-8: SSE

Zero Trust Network Access (ZTNA)

ZTNA replaces traditional client-to-site VPNs. Rather than tunnel a remote user into an internal network, ZTNA brokers access for each authenticated user to the precise private applications to which that user is authorized.

Cloud Access Security Broker (CASB)

CASB stands as a "broker" between users and SaaS applications. It can then enforce security controls, such as inspecting traffic and restricting certain actions (such as file transfers)—all based on the organization's security policies.

Secure Web Gateway

SWG provides security for users' Web browsing. It can perform actions such as blocking users from accessing untrustworthy sites or sites prohibited by the organizations' policies. It can also inspect traffic for malware or other threats.

Firewall as a Service (FWaaS)

An FWaaS is a firewall that is hosted in the cloud. It filters traffic as a service.

Remote Browser Isolation (RBI)

RBI hosts users' browsers in the cloud, potentially protecting the users' own devices from malware and other threats.

Data Loss Prevention (DLP)

DLP monitors users' communications with applications. It detects file transfers and blocks those transfers if they include unauthorized data. For example, the organization might not permit transfers that include customers' personal data.

Trends in SASE Adoption

Companies are embracing SASE.

In a recent analysis, Gartner forecasts that the SASE market will grow at a CAGR of 29% over a five-year period, reaching US $25 billion by 2027)"Forecast Analysis: Secure Access Service Edge, Worldwide," Gartner, Oct 10, 2023). In a Ponemon Institute study, 44% of respondents indicated that they either have deployed SD-WAN and a SASE security architecture or were planning to do so within the next year ("The 2023 Global Study on Closing the IT Security Gap: Addressing Cybersecurity Gaps from Edge to Cloud," Ponemon Institute, Mar 2023).

And more and more of these organizations favor a single-vendor approach to SASE. In the September 2023 Magic Quadrant for SD-WAN, Gartner forecasts that by 2026, 60% of SD-WAN purchases will be part of a single-vendor SASE offering, up from 15% in 2023 ("Magic Quadrant for SD-WAN," Gartner, Sept 27, 2023).

HPE Aruba Networking Unified SASE Solutions

You have learned what SASE means and why so many companies are choosing to implement SASE. This section provides you an overview of HPE Aruba Networking unified SASE solutions.

Overview of the HPE Aruba Networking Unified SASE Portfolio

HPE Aruba Networking provides a comprehensive, unified SASE portfolio for securing remote and branch users' access to the data center, public cloud, and SaaS applications. (See Figure 10-9.)

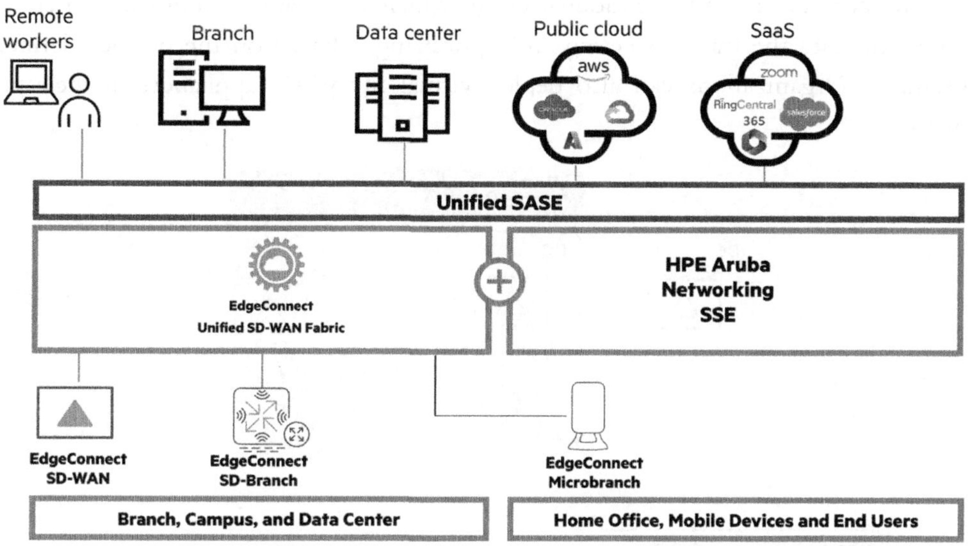

Figure 10-9: Overview of the HPE Aruba Networking Unified SASE Portfolio

For SD-WAN, connecting branches to the campus and data center, customers can choose from two options: HPE Aruba Networking EdgeConnect SD-WAN or HPE Aruba Networking EdgeConnect SD-Branch.

HPE Aruba Networking SSE fills out the unified SASE portfolio. These cloud services help users in home offices and other locations to securely connect to data center, public cloud, and SaaS applications. You might also offer customers HPE Aruba Networking EdgeConnect Microbranch, which provides users in home offices or very small, "pop-up" sites both wireless connectivity and a secure connection to a campus or data center.

HPE Aruba Networking EdgeConnect SD-WAN Architecture

You will now explore each solution in the portfolio, beginning with HPE Aruba Networking EdgeConnect SD-WAN. The architecture for this solution is shown in Figure 10-10.

Intended for branches of any size, this solution includes the HPE Aruba Networking WAN Orchestrator, which customers can deploy as a virtual machine (VM) in a data center or public cloud (AWS, GCP, or Azure). Or, rather than deploying the VM themselves, customers can purchase a subscription to the WAN Orchestrator service, which runs in the cloud. WAN Orchestrator provides the control plane for the solution. From its interface, admins orchestrate the solution centrally.

Each branch location requires an HPE Aruba Networking EdgeConnect SD-WAN appliance, as do the data centers or campus headquarters to which you want the branches to connect. These appliances handles the traffic, steering and optimizing it based on the policies defined in WAN Orchestrator. Organizations can also deploy virtual SD-WAN appliances in their public cloud instances.

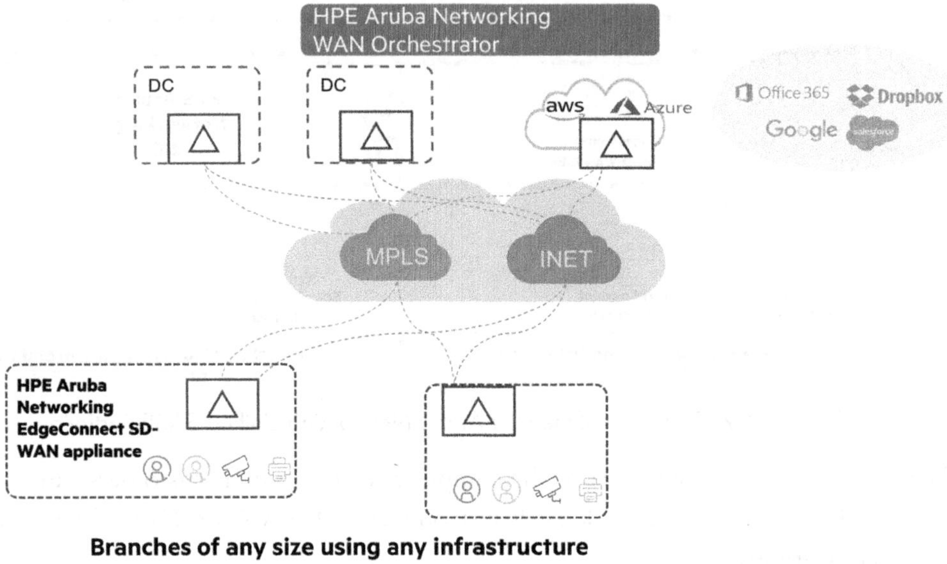

Figure 10-10: HPE Aruba Networking EdgeConnect SD-WAN Architecture

EdgeConnect SD-WAN is vendor agnostic. In other words, the company can use any vendor for the branch network infrastructure. It can overlay multiple connectivity options, such as MPLS and broadband Internet.

Key Features of EdgeConnect SD-WAN

You can deploy EdgeConnect SD-WAN appliances as physical or virtual appliances. These appliances seamlessly establish the SD-WAN, overlaying a variety of underlying media, to connect sites together. While they provide advanced routing, they go far beyond the capabilities of traditional WAN routers to deliver the highest performance and quality of experience for branch users, even over broadband Internet. With minimal admin effort, they optimize performance for applications of many types, including latency-sensitive, real-time voice and video. Their SaaS Express feature also accelerates performance for SaaS applications by automatically steering traffic to the best current IP address for accessing that application.

Cybersecurity features built into the EdgeConnect SD-WAN appliances help customers maintain the highest security across all branches. Appliances auto-establish secure IPsec connections. They also include a next-generation firewall, which, in keeping with other HPE Aruba Networking solutions, provides end-to-end, role-based segmentation. For example, appliances can segment IoT traffic from mission-critical traffic.

To simply deployment, appliances support zero-trust provisioning, automatically finding WAN orchestrator and receiving all instructions from it.

WAN Orchestrator provides single screen administration and orchestration of the complete SD-WAN fabric. It is designed for ease of use, allowing admins to configure desired functionality intuitively. For example, admins define Business Intent Overlays (BIOs) to control how the solution handles various types of traffic. Each BIO includes rules for selecting traffic, a desired experience profile, and various other choices for handling traffic. For example, a BIO might select voice traffic, ask for a high quality experience, and opt to use both Internet and MPLS links. The EdgeConnect SD-WAN solution then handles the details on its own, bonding links, establishing routing, and shaping traffic as required to meet the BIO's specifications. First-packet IQ app visibility ensures that the solution can immediately detect the correct BIO for each piece of traffic.

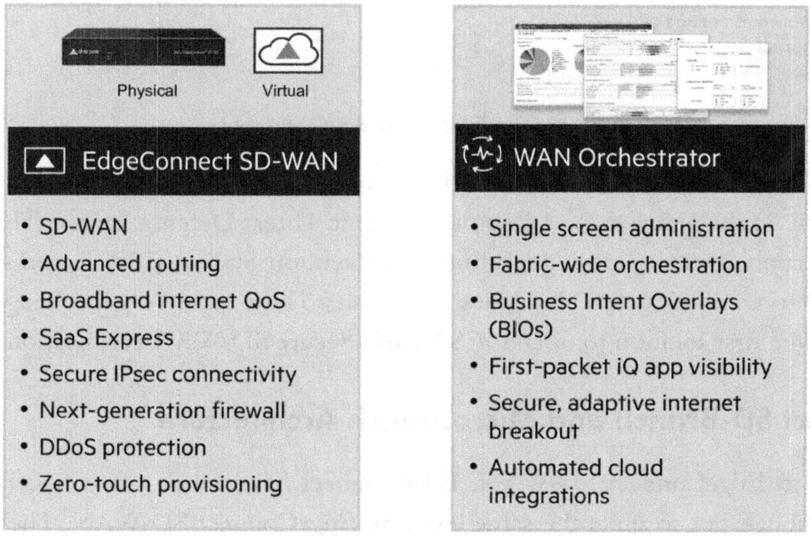

Figure 10-11: Key Features of EdgeConnect SD-WAN

As part of its flexibility, EdgeConnect SD-WAN allows customers to choose for each BIO whether they want to backhaul Internet traffic to the data center or break it out to a local Internet connection. Even when breaking traffic out locally, the solution maintains security with features such as

WAN Orchestrator also makes it easy to integrate with cloud services through automated cloud integrations.

Key EdgeConnect SD-WAN and WAN Orchestrator features are summarized in Figure 10-11.

Key Features of the Optional Licenses

Customers can add optional HPE Aruba Networking WAN Optimization licenses, which, as shown in Figure 10-12, add latency mitigation, data deduplication, and data decompression. The on-demand nature of this license makes it unique. Not every application benefits from WAN optimizations. With HPE Aruba Networking EdgeConnect, customers can choose to apply—and pay for—this feature only one the desired applications.

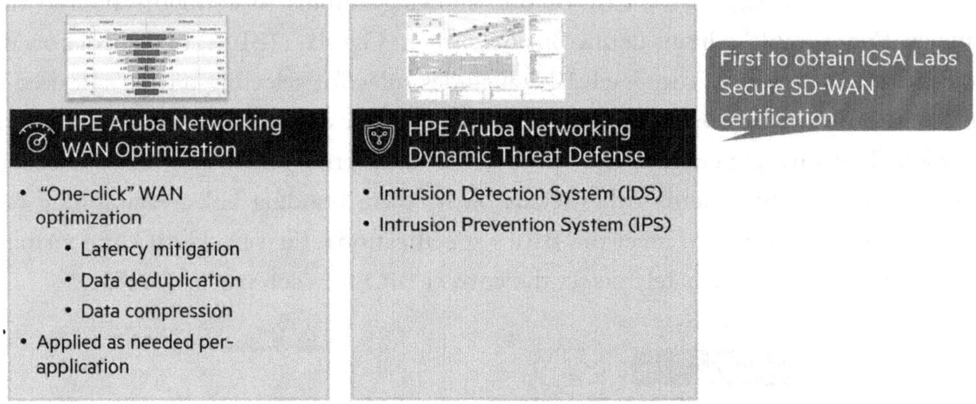

Figure 10-12: Key Features of the Optional Licenses

For customers who need even greater security than that built into all EdgeConnect SD-WAN appliances, HPE Aruba Networking offers the Dynamic Threat Defense license. This license adds Intrusion Detection System (IDS) and Intrusion Prevention System (IPS) features, enabling the appliances to detect and mitigate a broad range of threats. These features helped EdgeConnect SD-WAN become the first solution to obtain ICSA Lab's Secure SD-WAN certification.

EdgeConnect SD-Branch and Microbranch Architecture

An alternative to EdgeConnect SD-WAN, EdgeConnect SD-Branch offers excellent quality of experience, although not to quite the same levels as EdgeConnect SD-WAN. The EdgeConnect SD-Branch architecture is illustrated in Figure 10-13.

EdgeConnect SD-Branch is designed for maximum integration with HPE Aruba Networking infrastructure. It features HPE Aruba Networking gateways at large or small branches. These gateways connect to VPN concentrators (VPN-Cs), also called headend gateways, at headquarters or in the data center.

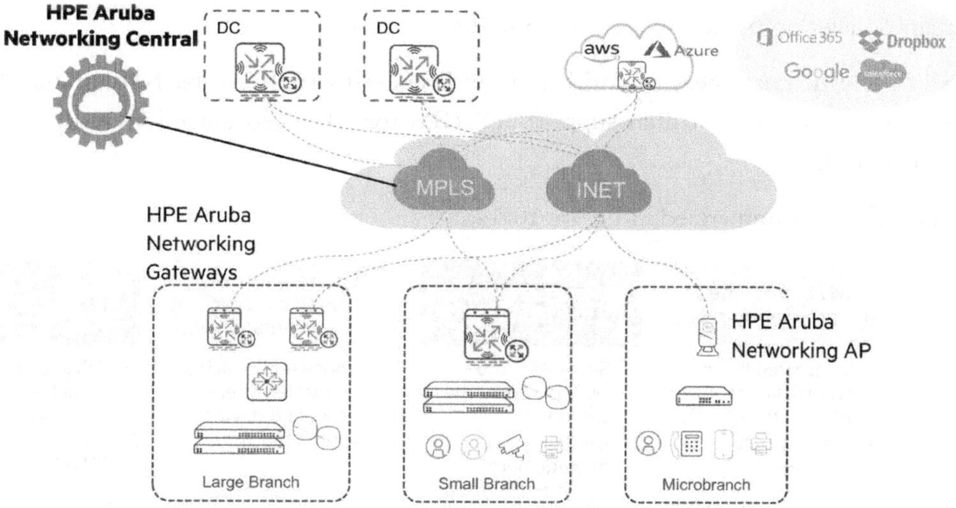

Figure 10-13: EdgeConnect SD-Branch and Microbranch Architecture

Organizations can deploy any HPE Aruba Networking AP at microbranches or home offices. Those APs too can automatically establish a secure connection to VPN-Cs at headquarters or the data center.

HPE Aruba Networking Central orchestrates the complete EdgeConnect SD-Branch and EdgeConnect Microbranch solutions.

Key Benefits of HPE Aruba Networking EdgeConnect SD-Branch

HPE Aruba Networking Central automates establishing secure IPsec tunnels between SB-Branch gateways at various sites. It also orchestrates routing and dynamic path steering across these tunnels and the underlying connections.

Customers can deploy the same types of APs and switches at the branches as at the campus, making the solution easy to manage. HPE Aruba Networking Central unifies management for both the SD-Branch solution and the HPE Aruba Networking switches, APs, and other network infrastructure deployed across the multi-site environment.

In addition, the gateways that establish EdgeConnect SD-WAN are designed to work with HPE Aruba Networking APs and switches deployed at the branch. For example, they can terminate tunnels from these devices as part of implementing end-to-end, role-based segmentation. In this study guide, you have learned how HPE Aruba Networking infrastructure can enforce role-based access controls based on roles sent by HPE Aruba Networking ClearPass Policy Manager (CPPM).

With EdgeConnect SD-Branch, the HPE Aruba Networking infrastructure can continue to enforce those same controls based on the same roles sent by CPPM. In short, it becomes easy to implement consistent policies no matter where users and devices connect.

For extra security, customers can add Security licenses to HPE Aruba Networking 9000 Series gateways. The gateways can then support IDS/IPS for advanced detection and prevention of a broad range of threats.

These benefits are summarized in Figure 10-14.

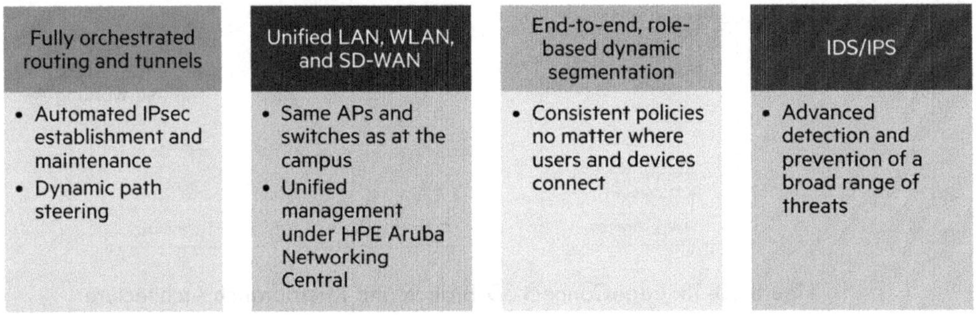

Figure 10-14: Key Benefits of HPE Aruba Networking EdgeConnect SD-Branch

Key Benefits of EdgeConnect Microbranch

When customers need a simple, all-in-one secure connectivity solution for home offices and tiny locations, they should consider EdgeConnect Microbranch.

An HPE Aruba Networking AP provides all the hardware required for the solution. When connected at the microbranch, that AP automatically discovers HPE Aruba Networking Central to receive its configuration. It can then automatically establish a VPN connection to a VPN-C at the main office, delivering the on-campus experience without requiring a branch appliance.

That on-campus experience includes consistent security controls. The AP, deployed in a home office or other microbranch, enforces the same end-to-end, role-based segmentation enforced at the campus. If the company has set up integration with SSE cloud services within HPE Aruba Networking Central, that security also applies to the microbranch.

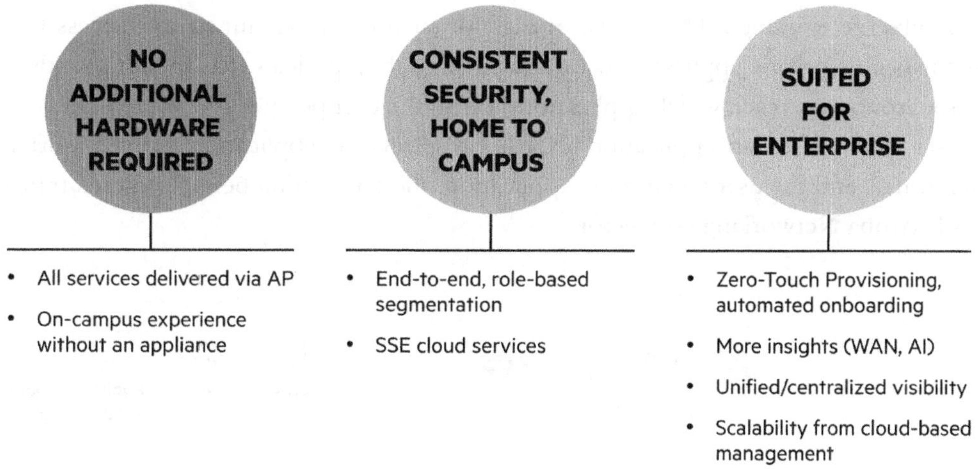

Figure 10-15: Key Benefits of EdgeConnect Microbranch

The simplicity and scalability of the solution makes it well-suited for enterprises. The company does not need to send IT staff to set up each home office or microbranch location. Instead, non-experts can easily connect the AP, which uses Zero-Touch Provisioning to become automatically onboarded by HPE Aruba Networking Central. From that solution's interface, admins can find AI-powered insights into performance and possible issues. With unified visibility into all sites, no matter how small, from a centralized location, admins can more easily manage and secure the network. Because HPE Aruba Networking Central is a cloud-based management solution, it scales well for enterprises with many sites.

The benefits of EdgeConnect Microbranch are summarized in Figure 10-15.

HPE Aruba Networking SSE Components

You will now learn about HPE Aruba Networking SSE, which combines with SD-WAN for a full SASE solution. This solution provides secure access for users anywhere to the Internet, SaaS applications, public cloud applications, and private applications running in the data center. Some customers might prefer it to EdgeConnect Microbranch for home office users because HPE Aruba Networking SSE, as a cloud-based solution, can work with home Internet setup users already have. This solution also offers many benefits as compared to a VPN-based solution.

As shown in Figure 10-16, HPE Aruba Networking SSE includes the key components of SSE about which you learned earlier.

ZTNA

HPE Aruba Networking ZTNA makes it easy for customers to grant remote access for authorized users to specific private applications. Admins easily set up policies that specify exactly which users and user groups can reach which applications. From the user perspective, the experience is seamless. The user browses to an application's URL or selects an application from a portal. The user authenticates, and the user reaches the application, the connecting being brokered transparently by an HPE Aruba Networking connector.

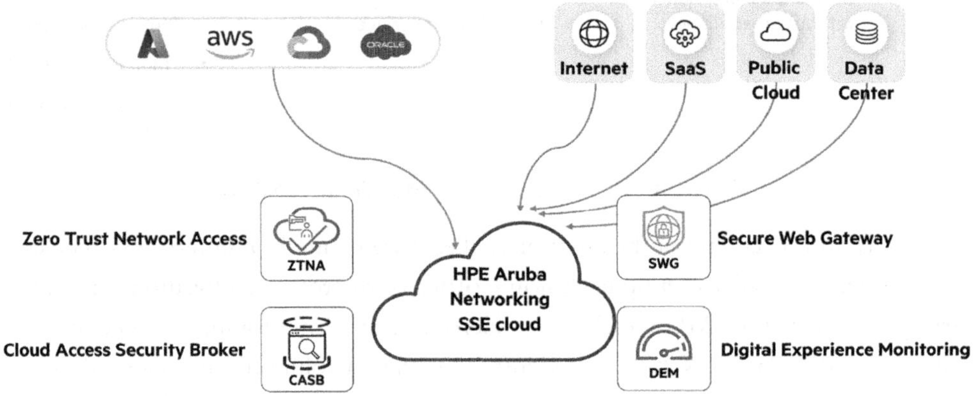

Figure 10-16: HPE Aruba Networking SSE Components

CASB

Admins can just as easily set up policies to regulate users' access to SaaS applications. They can choose precisely which actions specific users or user groups can perform with that application. When users access the SaaS application, a connector within the HPE Aruba Networking brokers the connection, inspects traffic, and enforces rules.

HPE Aruba Networking SSE also supports DLP together with its CASB and ZTNA capabilities; that is, it can inspect file transfers and block unauthorized ones. In this way, it can, for example, prevent the exfiltration of credit card numbers or personal information by malicious users or compromised devices.

SWG

HPE Aruba Networking SWG monitors users' browsing on the Internet. It can perform URL filtering, blocking traffic to various categories of suspicious or questionable sites, such as ones known for phishing. It can also inspect traffic for malware, including traffic encrypted by HTTPS.

Digital Experience Monitoring (DEM)

HPE Aruba Networking DEM continuously monitors sessions, granting admins insights. It helps admins detect performance issues and resolve problems more quickly.

HPE Aruba Networking SSE Benefits

HPE Aruba Networking SSE *replaces* client-to-site VPNs, offering several advantages over the traditional approach.

It shrinks the attack surface. Unlike a traditional VPN, HPE Aruba Networking ZTNA, with its connection brokering model, does not expose any ports. This solution also restricts users' access more tightly than many traditional client-to-site VPNs. HPE Aruba Networking ZTNA provides access to specific *resources* rather than to an entire network. Each time a user attempts to access a resource, that user must be authenticated, and HPE Aruba Networking ZTNA continuously monitors users' access.

With traditional client-to-site VPNs, companies face a choice. They either send only private traffic through the VPN tunnel, allowing users connect to the Internet freely, or they backhaul all traffic to the data center. The first approach can introduce vulnerabilities. Users might use cloud apps insecurely or accidentally compromise their devices by accessing untrustworthy sites. The second approach introduces all the inefficiencies of backhauling. With HPE Aruba Networking CASB, companies can easily control which actions users take with SaaS applications. And HPE Aruba Networking SWG applies URL filtering and other security measures to protect users on the Internet.

With traditional client-to-site VPNs, each new remote user adds more burden to the gateways. Eventually performance suffers, or companies must invest in more hardware to meet the rising demands. On the other hand, as a cloud-based solution, HPE Aruba Networking SSE can apply its security features, including traffic inspection, at scale.

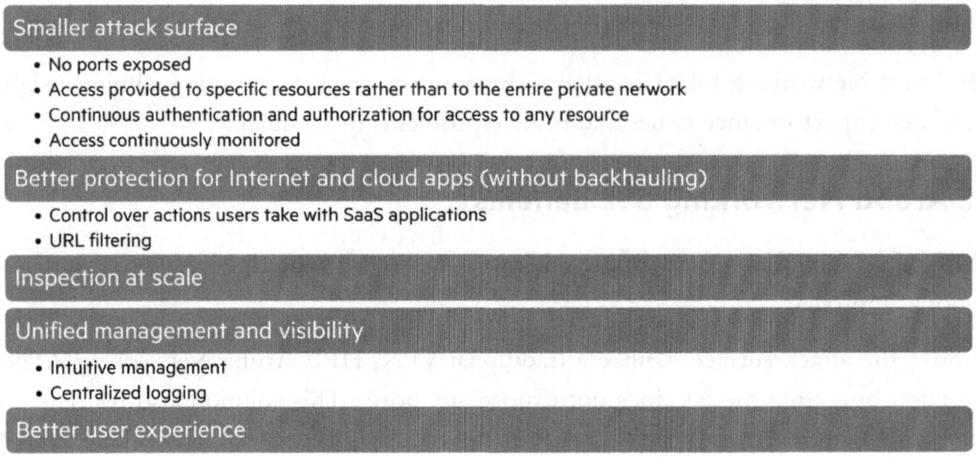

Figure 10-17: HPE Aruba Networking SSE Benefits

Admins benefit from unified management and visibility for the solution using an intuitive cloud console. The solution also offers centralized logging for better auditing.

Finally, users also benefit from HPE Aruba Networking SSE. Rather than struggle with a VPN client, users can easily access the applications they need from an intuitive user portal. (Some HPE Aruba Networking SSE features do require an agent, which is easily installed and managed.)

These benefits are summarized in Figure 10-17.

HPE Aruba Networking SSE Licensing

Companies can choose between various HPE Aruba Networking licenses based on the cloud services that they desire, as shown in Figure 10-18.

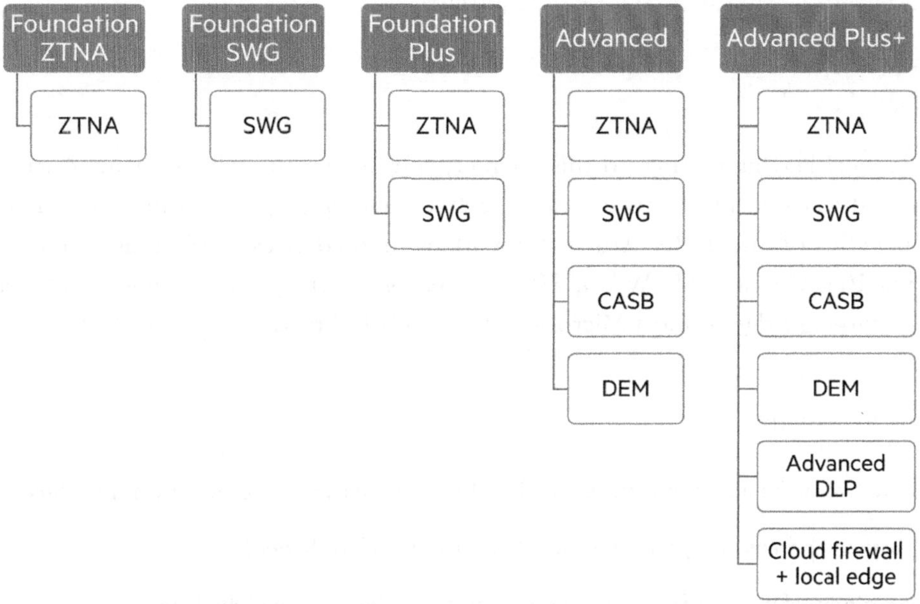

Figure 10-18: HPE Aruba Networking SSE Licensing

If they only want HPE Aruba Networking ZTNA or HPE Aruba Networking SWG, they select the Foundation ZTNA or Foundation ZWG license. The Foundation Plus license includes both of those services.

The Advanced license includes the four major services:

- ZTNA
- SWG
- CASB
- DEM

The Advanced Plus+ license also entitles companies to those four services. It also provides advanced DLP, which adds Optical Character Recognition (OCR) to DLP capabilities supported with ZTNA, SWG, and CASB. In addition, the Advanced Plus includes a cloud firewall. Its local edge feature brings the cloud services to a point closer to the company's users and devices.

Local edge is also available with other licenses as an add-on purchase. HPE Aruba Networking also offers several other add-ons for various features. For more information about those add-ons and the detailed features included with each licenses refer to this document: https://www.hpe.com/psnow/doc/a00141561enw.

Summary

Congratulations! Having completed this chapter, you understand why so many organizations are adopting SASE: a cloud-first, zero-trust approach to securing edge access anywhere. You also know the key benefits of the HPE Aruba Networking unified SASE solutions from HPE Aruba Networking EdgeConnect SD-WAN, HPE Aruba Networking EdgeConnect SD-Branch, HPE Aruba Networking EdgeConnect Microbranch, and HPE Aruba Networking SSE.

Learning checks

1. What are two of the trends that make it difficult for companies to close security gaps? (Select two.)
    a. Most companies lack perimeter security solutions such as firewalls.
    b. A majority of the workforce now works remotely at least some of the time.
    c. Most user clients lack support for advanced technologies such as 802.1X.
    d. Users are moving away from mobile connections toward Ethernet connections.
    e. The number of IoT devices is rapidly growing.
2. Which HPE Aruba Networking solution provides cloud access security broker (CABS) functions?
    a. HPE Aruba Networking EdgeConnect SD-WAN
    b. HPE Aruba Networking EdgeConnect SD-Branch
    c. HPE Aruba Networking EdgeConnect Microbranch
    d. HPE Aruba Networking SSE

Note

For answers to all learning checks, refer to Appendix B.

---

11 Troubleshooting and Monitoring

EXAM OBJECTIVES

- ✓ Collect logs on HPE Aruba Networking devices
- ✓ Collect and monitor data about historical network patterns
- ✓ Troubleshoot with Access Tracker and Event Viewer in HPE Aruba Networking ClearPass Policy Manager (CPPM)
- ✓ Describe the purpose and methods of packet captures

Assumed Knowledge

- Basics of IP networking
- Basics of Ethernet switching
- Basics of 802.11 wireless networks

Overview

Only by monitoring routinely when the company does not yet know a security incident has occurred, can a security team rapidly discover when an incident does occur. In this chapter, you will begin by examining several tools for monitoring the network on an ongoing basis, including logging tools, HPE Aruba Networking mobility dashboards, and ClearPass Insight.

You will next look at troubleshooting processes with a focus on troubleshooting authentication issues. You will learn about valuable tools for collecting information, including CPPM Access Tracker and Event Viewer. You will then take what you have learned and apply it to a few example scenarios.

You will then learn about how you can capture packets and provide those to a security or troubleshooting team who needs to analyze packet flows in more depth.

Logging and Monitoring

In this section, you will learn how to configure HPE Aruba Networking devices to log events to centralized solutions, helping organizations to better monitor their networks. You will also learn about the monitoring capabilities of HPE Aruba Networking ClearPass and other HPE Aruba Networking solutions.

Logging

Logging means to record an event, which can be viewed locally on the device that created the log or forwarded to a central solution. Setting up logging is critical for the purposes of auditing, ongoing security monitoring, and troubleshooting.

HPE Aruba Networking network infrastructure devices are capable of creating logs for a broad array of events that occur on the device. Just a few examples include a client failing authentication or an interface going down.

Logging Severity Levels

The system also assigns each logged event a severity level based on what the event is reporting. On HPE Aruba Networking Mobility Controllers (MCs) and HPE Aruba Networking Instant APs (IAPs), as well as on AOS-CX switches, the severity levels range from debugging at the low end to emergency at the high end, as shown in Figure 11-1.

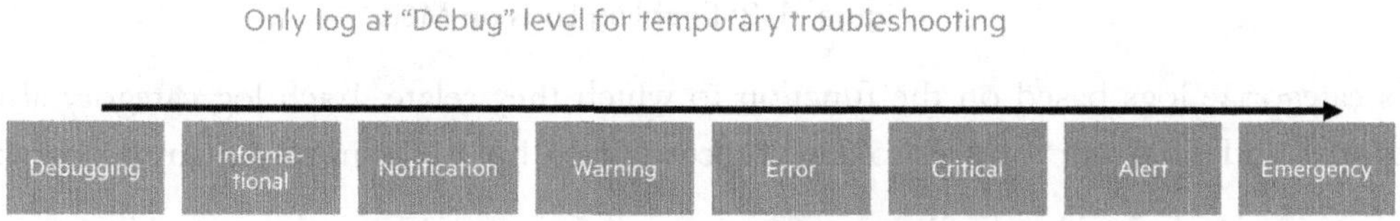

Figure 11-1: Logging Severity Levels

Debugging events give very low-level information about the processes, and you should only enable logging at this level for temporary troubleshooting purposes.

Informational and notification level events are routine. They provide information about what is happening on the system, but do not indicate events of particular concern.

Warning events and above indicate issues of varying levels of concern. Warning typically means that the event might cause an issue. Error means that the event is not normal and could be of concern, but probably is not causing widespread issues yet. Critical, alert, and emergency events indicate events that will cause immediate issues and should be addressed immediately; emergency events have the widest impact on the network.

You will need to work with your company's monitoring team to determine which logging level you should enable for each category or subcategory of events. Often warning events and above are the most important to collect, but your company might have policies about logging lower-level events for auditing or forensic purposes.

Enabling Logs on MCs

In Figure 11-2, you see how to set up logging on MCs (or MCRs [MCRs]) in their Web UI.

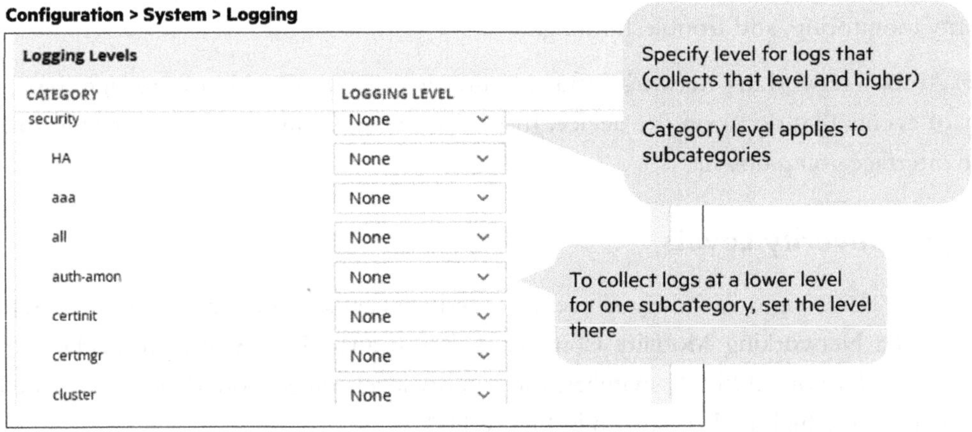

Figure 11-2: Enabling Logs on MCs

MCs categorize logs based on the function to which they relate. Each log category also has subcategories under it. Here you are looking at security logs, but MCs support many more types of logs.

You specify the desired level of logging for each category or subcategory to enable the device to log events locally. The MCs logs events at the desired severity level and above. The level that you apply to a category extends to all its subcategories. If you want to log events at a lower level on just one subcategory, you can specify a lower level on that subcategory.

Click **Submit** after you have made the changes.

Enabling User-Debug and AP-Debug Logging

Sometimes you might need to log or debug events on a particular user or AP. You can set up this targeted logging in the **Diagnostics > Logs > Log Settings** page (see Figure 11-3). Click the + icon.

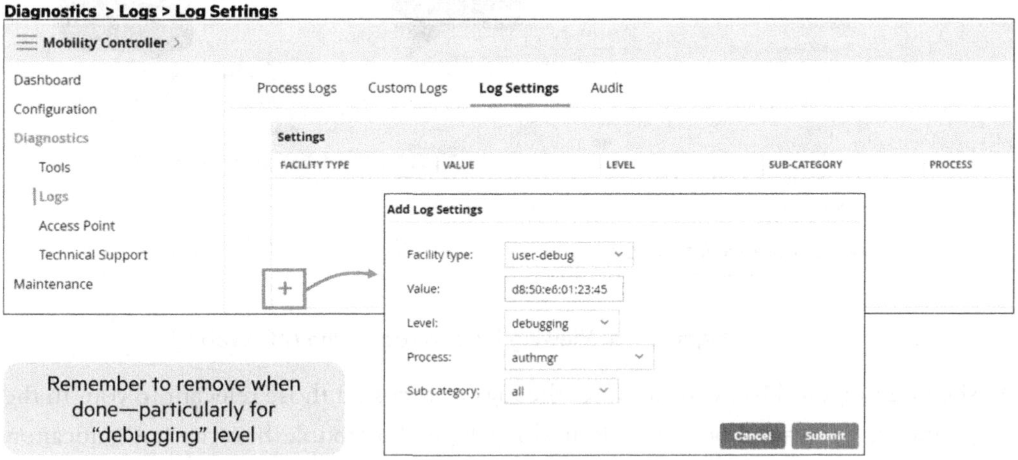

Figure 11-3: Enabling User-Debug and AP-Debug Logging

In this example, you want to log events for a particular client. Set the Facility Type to user-debug. You then set the value to the client's MAC address. Set the logging level. Remember that debugging is for temporary troubleshooting. You can narrow down the logging to a particular process. This example shows "authmgr," which is useful for troubleshooting issues in which the client cannot authenticate or is assigned to the wrong role. If you do not specify a process, events from all processes are logged. You can also specify, "all" or another subcategory.

If you want to collect logs from a particular AP, select ap-debug for the facility type and an AP name, IP address, or BSSID for the value. You can then set the desired level, and optional process and subcategory.

You can also log arm-debug events, but this training does not cover ARM.

Viewing Process Logs in the MC Web UI

You can use the MC (or MCR) Web UI to view the logs for events that have occurred. Go to the **Diagnostics > Logs > Process Logs** window, as shown in Figure 11-4.

If you are in the UI for a MCR, the page shows logs for all MCs, all MCs in a group, or an individual MC depending on where you are in the hierarchy. The process logs will include logged events based on the levels that you established in the **Configuration > System > Logging** window, as well as any user-debug, ap-debug, or arm-debug logs for which you have enabled collection. You can also see the ap-debug logs in **Diagnostics > Access Point**.

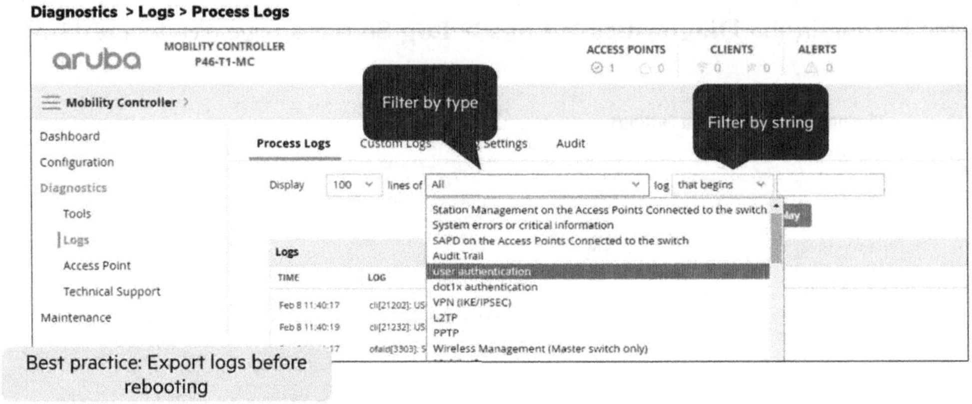

Figure 11-4: Viewing Process Logs in the MC Web UI

As shown in Figure 11-4, you can filter the logs to focus on those relevant to you. In the first drop-down menu, you can choose the desired log type. To troubleshoot an authentication issue, you might choose "user authentication" or "dot1x authentication." If you still have many logs to pore through, you can further filter by string. You can specify that the string begins the log, ends the log, or is included in the log. For example, you might want to find logs that include the MAC address of a client for which you are investigating an issue.

You can also export logs from the Process Logs list. On devices of all types, it is best practice to export logs before rebooting.

Viewing Logs on AOS-CX Switches

You can use the **show logging** command on AOS-CX switches to search those switches' logs for valuable information.

See most recent events first

Filter based on severity (shows that level and above)

Filter based on the category

```
AOS-CX# show logging -r -s info -c port_access
2020-05-21T11:50:04.802720-04:00 AOS-CX port-accessd[3040]:
Event|10503|LOG_INFO|MSTR|1|Port 1/1/16 is unblocked by port-access
2020-05-21T11:47:25.347940-04:00 AOS-CX port-accessd[3040]:
Event|10502|LOG_INFO|MSTR|1|Port 1/1/16 is blocked by port-access
```

Info and above logged by default

Figure 11-5: Viewing Logs on AOS-CX Switches

As shown in Figure 11-5, you can use -r if you want to see the logs with the most recent events first. Use the -s option if you want to filter by severity level. You can add the -c option to filter the logs by category. Some key categories for settings that you have examined in this study guide include "port_access," "aaa," and "role." You can use **?** to get a list of all categories.

You can also use **?** to see more options. For example, by default, the logging command shows events that occurred since the last time that the switch booted, but you can add -a to see events logged in previous boots as well. Or you can use the -n option to set the number of logs that you want to view.

Viewing Debug Information on AOS-CX Switches

Sometimes your managers will ask you to collect debug information so that they can troubleshoot a particular issue. Refer to Figure 11-6 for example commands for collecting this information on AOS-CX switches.

```
AOS-CX# debug destination <console | buffer | syslog | file> [severity
<level>]
AOS-CX# debug <category> [subcat] [severity <level>]
AOS-CX# show debug buffer
AOS-CX# no debug all
```

Figure 11-6: Viewing Debug Information on AOS-CX Switches

First, set the debug destination. You can send the debugs to the session or console, which will make new debug messages pop up in the terminal window of a console session as they occur. You can send the debugs to a buffer, which you can then view with the **show debug buffer** command. You can also send the debugs to a remote Syslog server. You will learn how to set up a Syslog server in a moment. You can also set the destination to **file**, which stores the debugs in a file on the switch. You can enter the **debug destination** command multiple times to specify multiple destinations.

AOS-CX switches defines severity level for debug events, and you can optionally filter the debug events that you send to a particular destination based on severity. However, often you are debugging because you want to see all of the events. Omitting the severity option leaves the default severity of "debugging" in place.

Next you need to enable debugging on the category and possibly subcategory of interest. For example, you could specify **debug portaccess dot1x all** to debug 802.1X functions and events. If you are logging to the console and you are accessing the command line interface (CLI) through the console port, you should see the debugs start to pop up. If you are logging to the buffer, you can see the debugs by entering **show debug buffer**. Or admins can view the debugs on the remote Syslog server if that is the destination.

Debugging can be processor intense, so always remember to turn it off when you have finished troubleshooting the issue.

Defining Remote Logging Requirements

Logging events locally gives you a limited view that can be useful for troubleshooting. But for ongoing monitoring, you need to forward logs to a central solution, as shown in Figure 11-7.

Because you cannot always anticipate what information you will need until after you need it, many companies have policies that require logs to be forwarded as they occur and archived on a central Security Information and Event Management (SIEM) server. The security team might treat these logs like the footage of a security camera. Much of the information might be irrelevant, but some could contribute to an investigation into an incident discovered later. Some companies might also copy the data so that they can have analytics-based security solutions look for patterns and suspicious anomalies. Your company might also have policies related to creating an audit trail by collecting and archiving logs.

Forwarding logs to a central server can also help to make the solution more resilient against malicious users. If hackers compromise a network device, they often delete the log history to conceal their activity. The hackers might also turn off remote logging. But if the network device has already been sending logs to a central solution, many of the logs will be preserved there, giving admins a better record and ability to detect the intrusion. Implementing remote logging can also help you to ensure that logs are preserved even if an admin reboots a device.

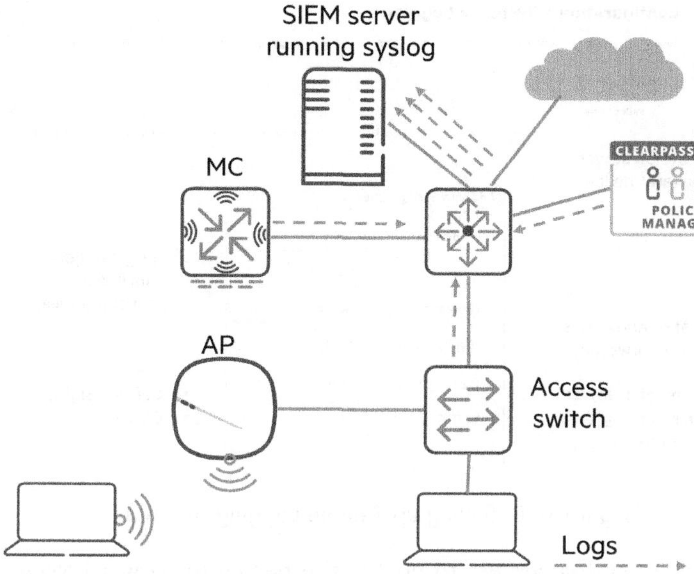

Figure 11-7: Defining Remote Logging Requirements

You should work with your company's IT monitoring managers to determine which logs they want to receive from the network infrastructure devices under your control. You need to discuss two factors. First, discuss which types of logs the network infrastructure devices should send. For example, HPE Aruba Networking mobility devices have security logs and wireless logs, among many other types. You can also select which logs to forward based on severity.

Syslog servers use the logging facility to identify the type of entity that sends the log. The server might be set up to handle logs from different facilities in different ways. You can configure the facility that HPE Aruba Networking devices use for their logs; ask your organization's monitoring admins which facility they prefer network infrastructure devices use.

With the types of logs that you want to collect defined, you are ready to set up the network infrastructure to send those logs securely.

Setting up Remote Logging on MCs

You configure the Syslog servers to which MCs forward logs in the **Configuration > System > Logging** page in the Web UI, as shown in Figure 11-8.

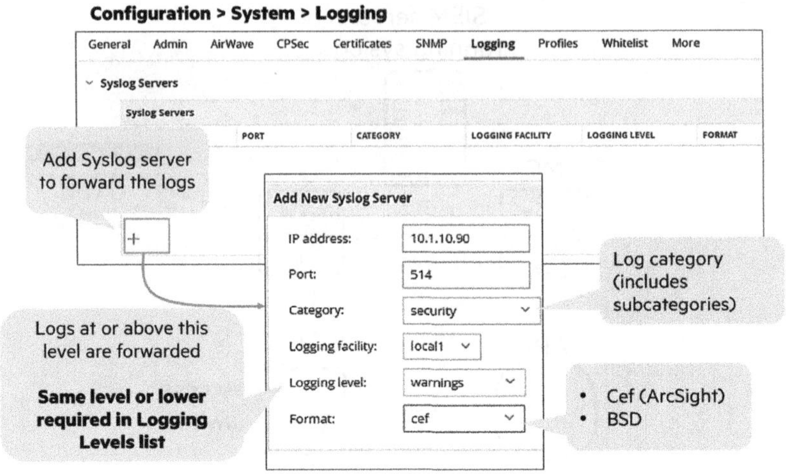

Figure 11-8: Setting up Remote Logging on MCs

You can add one or more syslog servers to enable the device to forward logged events to those servers. For each server you can define the IP address and port, as well as the category of logs to send. You also specify the logging level. The device forwards logs for events in the selected category that are at or above this level. However, to forward the logs, the MC also needs to be logging the events locally. You learned how to set that up earlier; make sure that the device is logging events within the selected category at the level that you want to forward or lower.

When you create the Syslog server, you can also set the format. MCs support two formats for logs: cef, which ArcSight uses to standardize logs, and BSD.

Securing the MC Logs

MCs devices send the logs with UDP, so by default the logs are not protected in transit, as shown on the left of Figure 11-9. In addition, the MC does not know whether the Syslog server is up and receiving events.

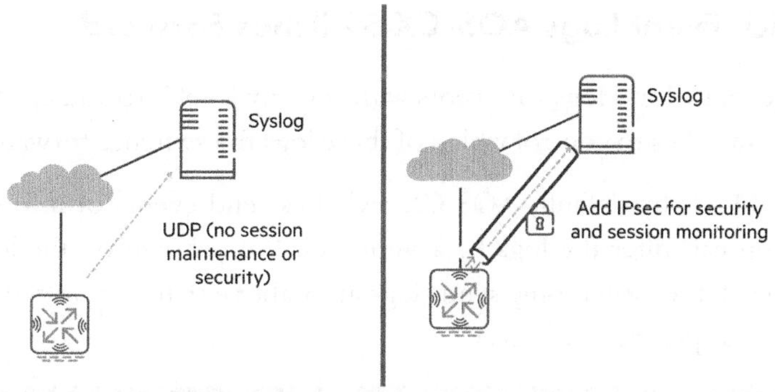

Figure 11-9: Securing the MC Logs

As shown on the right of Figure 11-9, you can establish an IPsec tunnel between the MC and the Syslog server, as long as the server is IPsec-capable as well. For more information on setting up IPsec connections, attend higher level training.

Setting up Secure Logging to a Syslog server on AOS-CX Switches

Now you will learn about setting up remote logging on AOS-CX switches (see Figure 11-10).

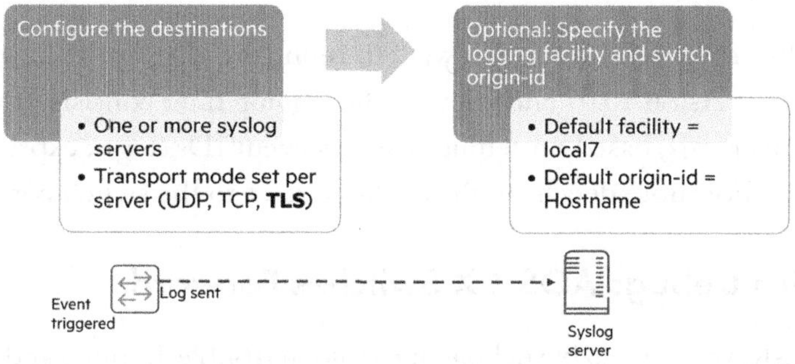

Figure 11-10: Setting up Secure Logging to a Syslog server on AOS-CX Switches

You can send an AOS-CX switch's log to a Syslog server by adding the Syslog server IP address as a logging destination. When you specify the server IP, you choose UDP, TCP, or TLS as the transport mechanism. If the company requires security, use TLS. Make sure that the Syslog server also supports TLS and has a certificate signed by a Certification Authority (CA) that the switch trusts. The switch will similarly require a certificate that is signed by a CA that the server trusts.

You can change the logging facility to match the type recommended by network monitoring admins. The default facility for AOS-CX switches is "local7." AOS-CX switches send their hostname for the origin ID.

Controlling Which Event Logs AOS-CX Switches Forward

AOS-CX switches generate local logs for events with severity level info and up. You use the logging settings on AOS-CX switches to control which of these logs the switches forward to a Syslog server.

As shown in Figure 11-11, by default, AOS-CX switches send events of info level and up to the Syslog server. But you can filter the logs by severity level. As shown on the left of Figure 11-11, when you specify a level, the switch only sends logs at or above that level to the server. You can set a logging severity that is specific to a server.

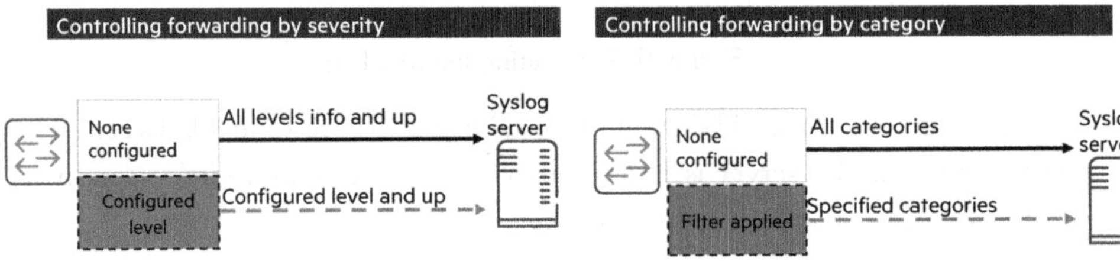

Figure 11-11: Controlling Which Event Logs AOS-CX Switches Forward

The default logging settings send all categories of logs with the configured severity levels and above to defined Syslog servers.

As shown on the right of Figure 11-11, if you want to refine the logs, create logging filters (available in AOS-CX in software version 10.07 and above). The logging filter consists of a set of rules, which permit or deny particular logs, based on settings such as event IDs, regex expressions for matching log strings, or severity. For more details on these filters, refer to the switch configuration manuals.

Controlling Which Debugs AOS-CX Switches Forward

As discussed previously, you can enable debugging to aid in troubleshooting and forward the debugs to a Syslog server. Controlling the remote logging of debug events is a bit different from controlling the remote logging of other events.

AOS-CX switches define multiple categories for debug events. As shown in Figure 11-12, when no categories are defined for debugging, the switch does not generate or forward any debug messages. You must enable debugging on one or more categories to see those debugs. (AOS-CX lets you filter the debug events by severity, but the default severity is "debugging.")

In addition, if you want to forward the debugs, you must set the debug destination to "syslog."

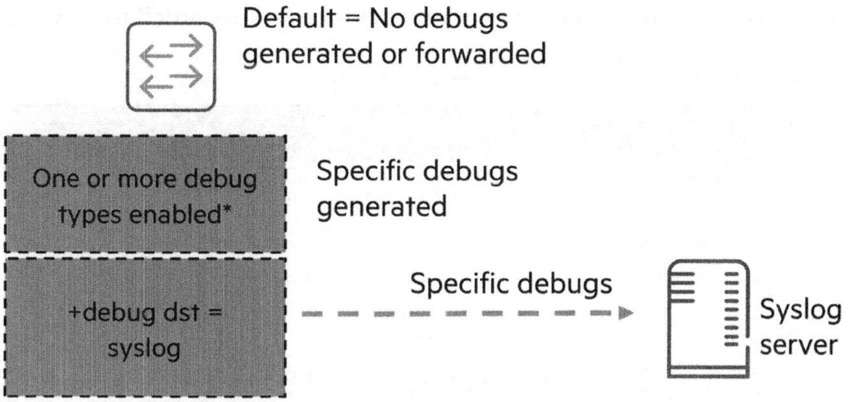

*Optional filtering of debug events by severity levels

Figure 11-12: Controlling Which Debugs AOS-CX Switches Forward

CLI Commands: Configuring Logging on AOS-CX Switches

The table in Figure 11-13 provides a reference for configuring event logging to a Syslog server on AOS-CX switches.

In configure context

Step	AOS-CX command	Notes
Specify the Syslog server and the transport protocol used to reach it.	`logging <IP> <udp [port] | tcp [port] | tls [port]> [vrf <VRF>]`	If you do not specify a port, the switch uses the default one: UDP = 514, TCP = 1470, TLS = 6514 If you do not specify a VRF on AOS-CX, "default" is used.
Set the logging facility.	`logging facility <facility>`	Use ? to see a list of valid facilities. Default: local7

Figure 11-13: CLI Commands: Configuring Logging on AOS-CX Switches

In Figure 11-14, you see how to set the severity level for logs forwarded to a Syslog server.

In configure context

<table>
<tr><th>Step</th><th>AOS-CX command</th></tr>
<tr><td>Set the severity for forwarded logs.</td><td>logging <<i>IP</i>> [include-auditable-events] severity <debug | info | notice | warning | err | crit | alert | emerg></td></tr>
<tr><td>Configure a logging filter.</td><td>logging filter <name></td></tr>
<tr><td>Apply a logging filter to a syslog server.</td><td>logging <<i>IP</i>> filter <<i>name</i>></td></tr>
</table>

In filter context

<table>
<tr><th>Step</th><th>AOS-CX command</th></tr>
<tr><td>Create a rule that permits or denies logs based on event ID.</td><td><<i>rule ID</i>> <permit | deny> event-id <<i>ID list or range</i>></td></tr>
<tr><td>Create a rule that permits or denies logs based on severity.</td><td><<i>rule ID</i>> <permit | deny> severity <eq | ge | gt | le | lt> <<i>severity</i>></td></tr>
<tr><td>Create a rule that permits or denies logs based on regex</td><td><<i>rule ID</i>> <permit | deny> includes <<i>regex string</i>></td></tr>
</table>

Figure 11-14: CLI Commands: Configuring Logging on AOS-CX Switches

You can alternatively configure logging filters to control which logs are forwarded in a very granular manner. You can define rules that permit or deny logs based on event ID, severity, or regular expression. You can also combine criteria. For example, you could permit logs with a certain event ID and severity. For more details refer to the switch configuration manuals.

Traffic Monitoring in HPE Aruba Networking Mobility Dashboards

Logs provide one important aspect of ongoing monitoring, but HPE Aruba Networking offers other tools as well.

You can use the MC (or MCR) dashboard to monitor the network at the current moment in time (see Figure 11-15). The dashboard gives you a great deal of information that you can use for a variety of purposes, including checking on the quality of wireless signals and how busy the network is. For security monitoring, you can check out the usage consistently at various points in the day to get an idea of your network's normal. You can then look out for usage spikes. (HPE Aruba Networking also has other tools that can help you look at patterns on an ongoing basis.)

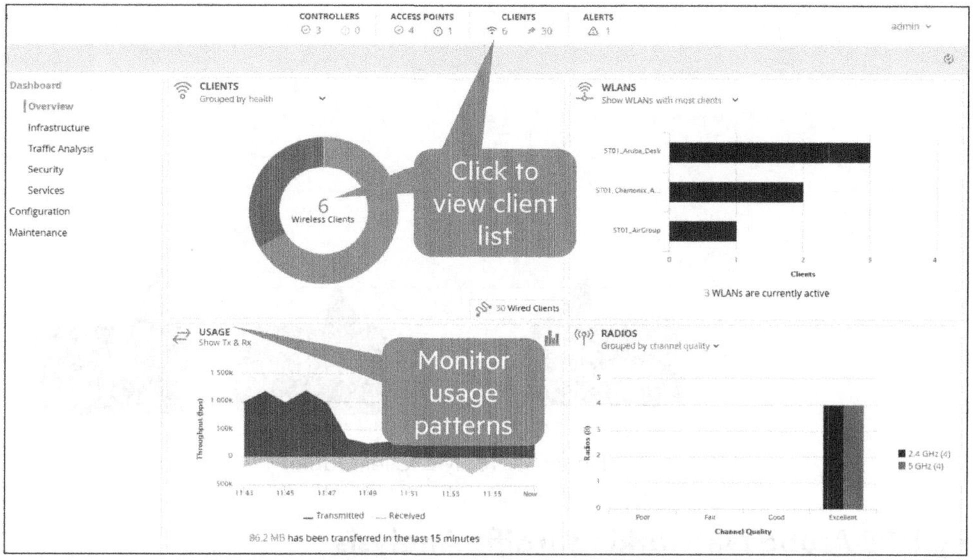

Figure 11-15: Traffic Monitoring in HPE Aruba Networking Mobility Dashboards

For security monitoring, the clients list can be particularly helpful. Click the clients link to open that list.

MC Client Table

You can search the MC client table, shown in Figure 11-16, to troubleshoot or monitor individual client's connections. It can be particularly useful to see which role has been assigned to the client so that you can understand which policies the firewall will apply. You can also check users' network usage. You can look for users with particularly high usage and begin to investigate whether that usage is legitimate.

Figure 11-16: MC Client Table

Using HPE Aruba Networking Traffic Analysis

You can use the Traffic Analysis section of the Dashboard to gain a deeper insight into trends that you see in the overview. For example, if you see usage spikes, you might want to investigate what applications are in use. The Traffic Analysis component of the MC (or MCR) dashboard can help you answer these questions.

For this dashboard to show any information, you must enable firewall visibility. As shown in Figure 11-17, go to the **Services > Firewall** window and find this setting in the global settings). To have the Traffic Analysis dashboard present information down to the application level, you must further enable deep packet inspection (DPI). If DPI is not already enabled on the device, you must reboot it.

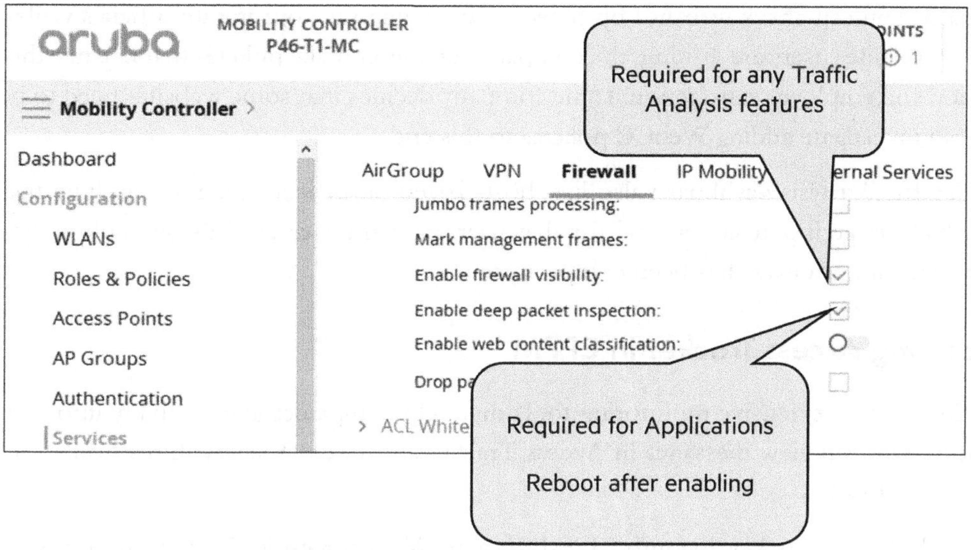

Figure 11-17: Using HPE Aruba Networking Traffic Analysis

Traffic Analysis Dashboard

View Figure 11-18 for an example of the information displayed in the Traffic Analysis dashboard. Often when you formulate firewall policies, you start with a more permissive approach. As you continue to monitor on an ongoing basis, you can see which applications users are actually running. You can work with your company to define policies for those applications. Should users be running them? If not, you can add them as denied applications to the appropriate roles' application access control lists (ACLs). If so, you can leave them permitted.

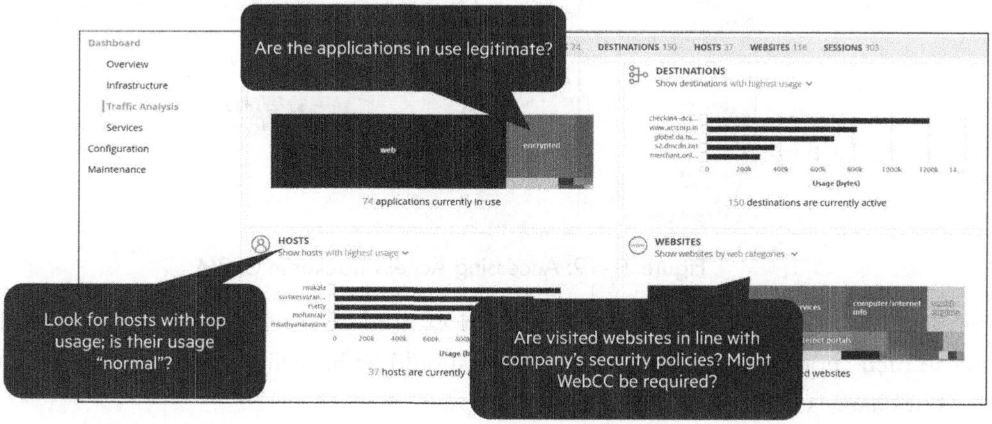

Figure 11-18: Traffic Analysis Dashboard

Similarly, you can check websites by category. By giving the appropriate officials visibility into the type of websites users are visiting, the company can better make policies that balance the company's mission and employee satisfaction. If the company decides that some websites need to be restricted, you can investigate adding WebCC policies to this end.

The Traffic Analysis dashboard also lists hosts based on their usage. You can investigate whether such high usage is actually normal for that user, or if the increased usage could indicate that the user's account or device has been compromised.

Accessing Access Tracker in CPPM

CPPM provides extensive monitoring tools that help companies gain visibility into their users and devices. You can view messages in Access Tracker and Event View with read-only (or read-write) access to CPPM.

To reach Access Tracker, log into CPPM. Expand **Monitoring** in the navigation tree, as shown in Figure 11-19. Then expand **Live Monitoring** and select **Access Tracker**. The Access Tracker displays logged access requests with information about the requests such as the ClearPass server that received the request, the source of the request, and the user associated with the request. It further shows to which service CPPM matched the incoming request and the action request status, which is ACCEPT, REJECT, or TIMEOUT.

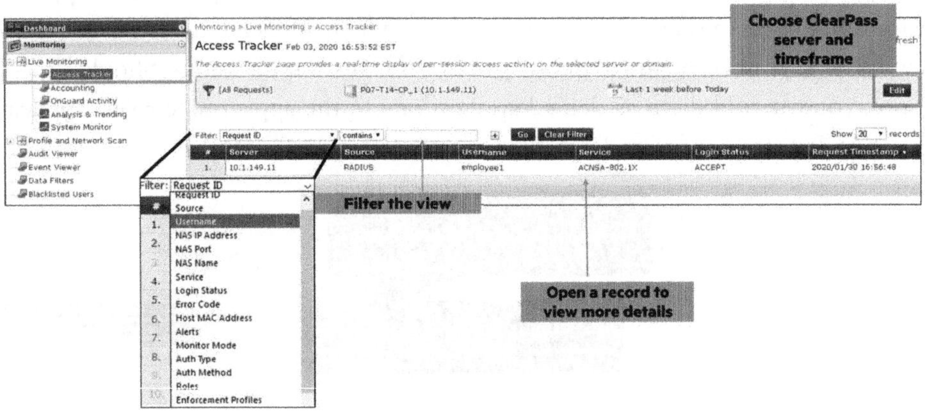

Figure 11-19: Accessing Access Tracker in CPPM

To ensure that you are seeing the records that you want to see, click **Edit** near the top of the window. You can then choose the ClearPass server that should be handling the request and the timeframe. Save your changes and return to the main window.

There you can use the Filter fields to narrow in on the records in which you are interested. For example, if you are troubleshooting an issue with authentication, you can filter for Username to find the record specific to your authentication attempt.

You can open a record to view more details.

Viewing Access Tracker Records

When you open the record, a pop-up window provides extensive details about the request and how CPPM processed it, as shown in Figure 11-20. The **Input** tab provides information about the RADIUS request (TACACS+ requests have a similar Request tab). This information can include the Attribute Value Pairs (AVPs) in the Access-Request. It might also include enforcement attributes that CPPM has calculated for the request. This information can help ClearPass admins understand why CPPM is matching a request to a particular service, ClearPass role, and enforcement policy.

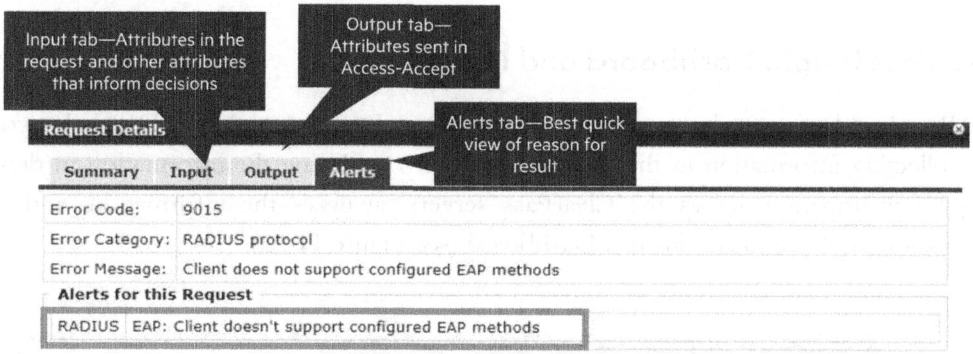

Figure 11-20: Viewing Access Tracker Records

If the user passes authentication, the **Output** tab (called **Policy** tab for TACACS+) shows the profile that CPPM applies to the user. If the user is not receiving the correct role, you might check this tab.

The **Alerts** tab often holds the most useful for troubleshooting purposes. For a rejected request, the alert gives information about why CPPM rejected it, and it can point you toward the issue. In this example, you can see that the client does not support the correct Extensible Authentication Protocol (EAP) method.

Accessing Event Viewer in CPPM

To access the Event Viewer, expand **Monitoring** and select **Event Viewer**, as shown in Figure 11-21. Again, you can filter the view. Events with source RADIUS or TACACS+ will probably be most useful for your purposes. You can open an event to view more details.

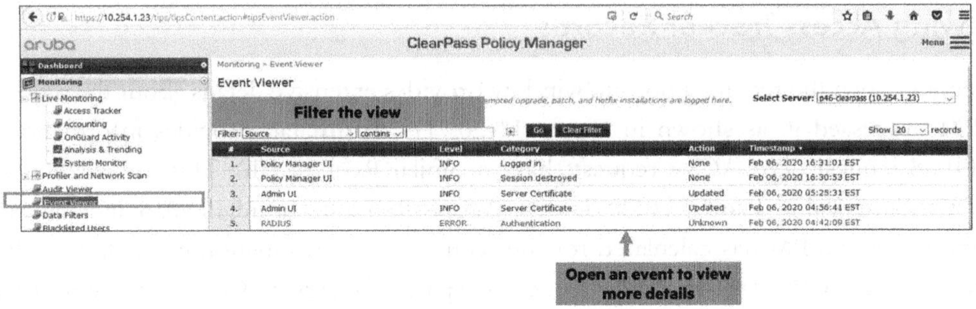

Figure 11-21: Accessing Event Viewer in CPPM

ClearPass Insight Dashboard and Inventory

ClearPass Insight (not to be confused with ClearPass Device Insight [CPDI]) helps you to move past collecting information in the moment and onto analyzing the information in depth. Insight compiles information across the ClearPass servers, analyzes the information, and makes that information available to admins in a Dashboard (see Figure 11-22).

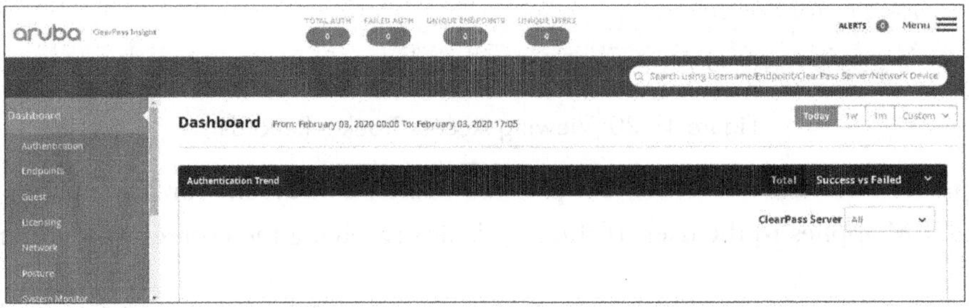

Figure 11-22: ClearPass Insight Dashboard and Inventory

ClearPass Insight Reports

As another key feature, ClearPass Insight makes the information that it collects and analyzes available in reports. Insight supports multiple report types, which companies can customize according to their needs. Admins can configure reports to run on schedule, and Insight can automatically send the reports to admins through email or text.

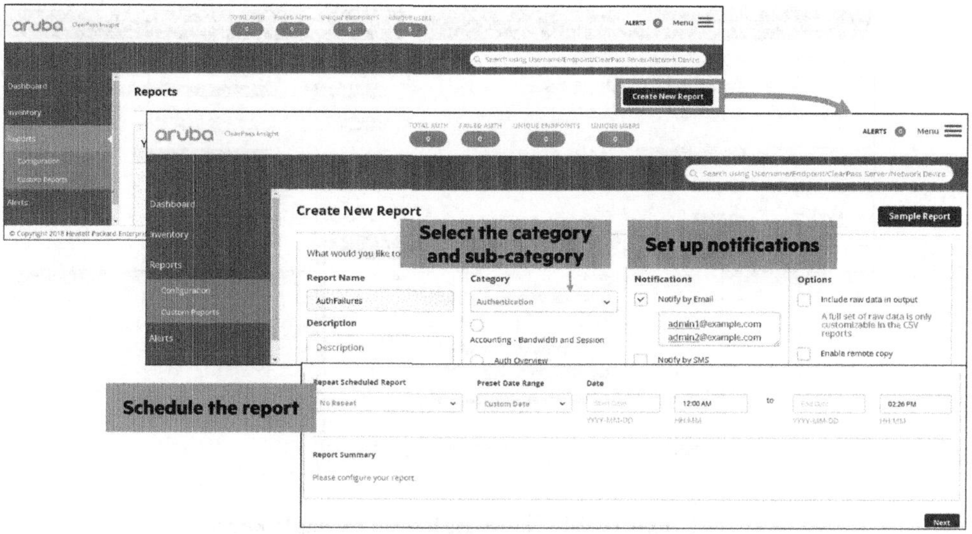

Figure 11-23: ClearPass Insight Reports

In Figure 11-23, you see the main settings that you would configure in a report. On the next page of the report configuration (not shown in Figure 11-23), you can also filter the fields that show in the report and customize the report with a logo.

ClearPass Insight Example Report

Example information provided by an Insight Report is shown in Figure 11-24. As you see, Insight analyzes the data to show relevant trends such as the top 10 Network Access Devices (NADs) with the most failed authentications or the top errors that are causing authentication failures. These reports can help admins see a bigger picture so that they can better investigate and troubleshoot.

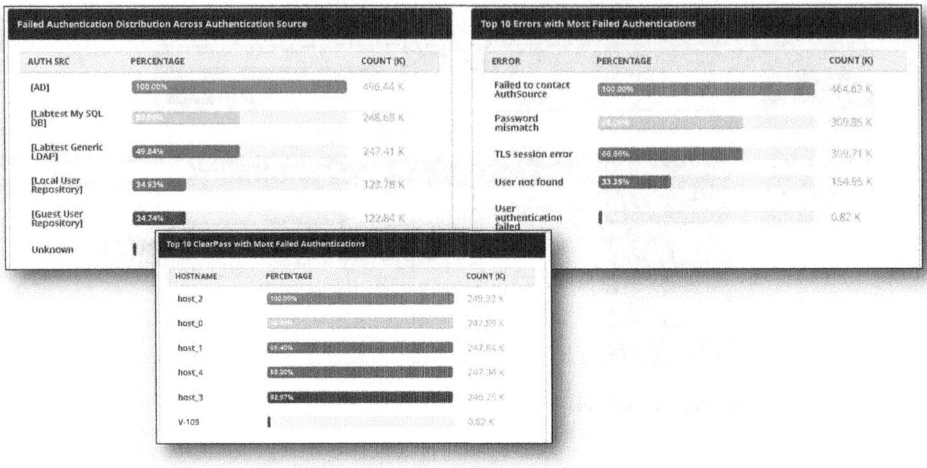

Figure 11-24: ClearPass Insight Example Report

Additional HPE Aruba Networking Monitoring Tools

HPE Aruba Networking offers three additional solutions in which companies might be interested to improve monitoring of their network (see Figure 11-25).

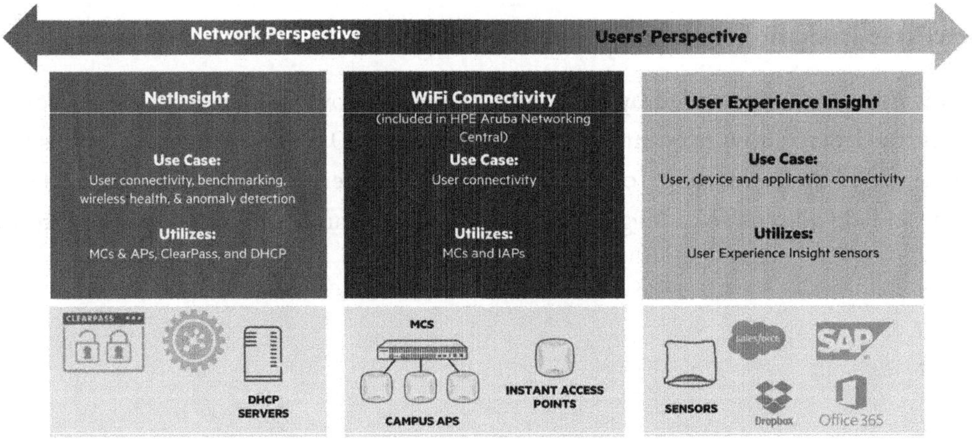

Figure 11-25: Additional HPE Aruba Networking Monitoring Tools

For more in-depth network analytics, HPE Aruba Networking offers NetInsight. The solution uses a data collector to gather info on network infrastructure, the DHCP server, and any other HPE Aruba Networking software, such as ClearPass and AirWave. (Note that Airwave is a legacy HPE Aruba Networking solution, which is currently end-of sale.) Customers then get information and analytics on user connectivity, benchmarking, wireless health, anomaly detection, and more.

WiFi Connectivity, included in HPE Aruba Networking Central and legacy Airwave, provides in-depth, phase-based connectivity awareness to help customers discover if a connectivity problem is due to an association, authentication, a problem accessing the DNS server, or more. It gathers data from MCs and instant access points.

HPE Aruba Networking also offers User Experience Insight. This solution helps companies to assess how good of an experience their network is providing for users, devices, and applications. Using User Experience Insight sensors, the solution provides user, device, and application connectivity analytics as well as information on wired and wireless performance.

Introduction to Troubleshooting Authentication Issues

This section provides a process for troubleshooting common authentication issues.

Defining the Problem

Many authentication issues begin with similar complaints. As shown in Figure 11-26, users might come to you saying things like "My smartphone won't connect to the network." Or "I can't get my log in to work." Because users often do not understand the root cause of the issue, many of their problems can sound the same. You will often find that you must separate RF signal issues from authentication issues.

If possible, you should have the user reproduce the issue for you. On a Windows machine, you can see a message indicating that authentication failed, tipping you off to an authentication, rather than a connection issue.

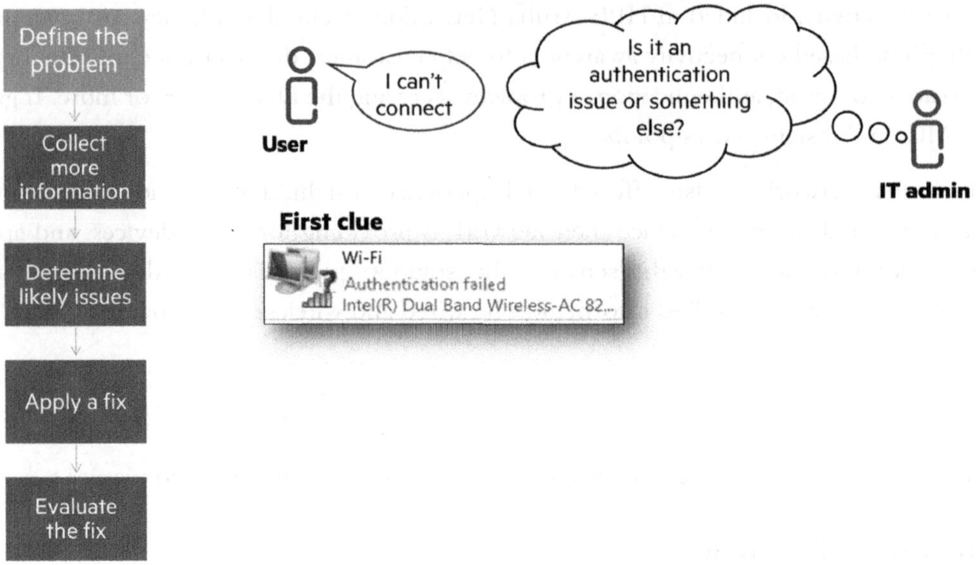

Figure 11-26: Reproducing the Issue and Checking the Client

Some devices do not show rich error messages, in which case it can be more difficult to determine whether a device is not connecting because it cannot get a good signal or because it failed authentication.

As shown in Figure 11-27, begin to collect as much information from users as you can so that you can tease out commonalities between the issues. For example, is it just one user or device that is having issues, or are the issues widespread across many users? If no one can connect to WLANs that implement authentication, the problem is more likely an authentication issue than a signal issue. For widespread issues, do the issues seem to be location-based? Location-based issues are often wireless signal based—in other words, a signal deadspot due to interference or inadequate AP planning. However, even location-based issues can relate to authentication. For example, if a switch that enforces 802.1X has the wrong RADIUS shared secret, no wired user in the area served by that switch will be able to connect.

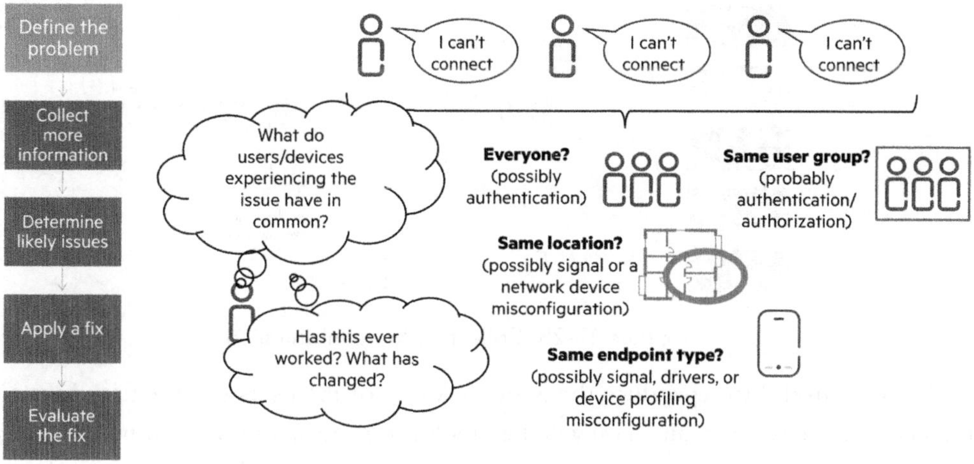

Figure 11-27: Finding Commonalities

Are all the users who are experiencing problems in the same user group? That could be an issue with authorization settings such as an error in the role to which the users are assigned. Do the users experiencing the problem use the same types of endpoint? That could be an issue with the signal or drivers, or it could possibly be an issue with device profiling/endpoint identification and the roles to which this feature assigns the devices.

Make a list of everything that you learn and begin to connect the dots.

It can also be very useful for you to find out whether this is a new problem and, if so, what has changed. For example, if users have been authenticated to a WPA3-Enterprise WLAN for months and suddenly none of them can connect, the issue probably is not that the MC is misconfigured—although it is always possible that an admin has accidently changed the configuration. More likely something else has changed. For example, maybe the ClearPass admins forgot to update the ClearPass RADIUS/EAP certificate before it expired.

Collecting More Information

Once you have defined the problem, you need to collect more information.

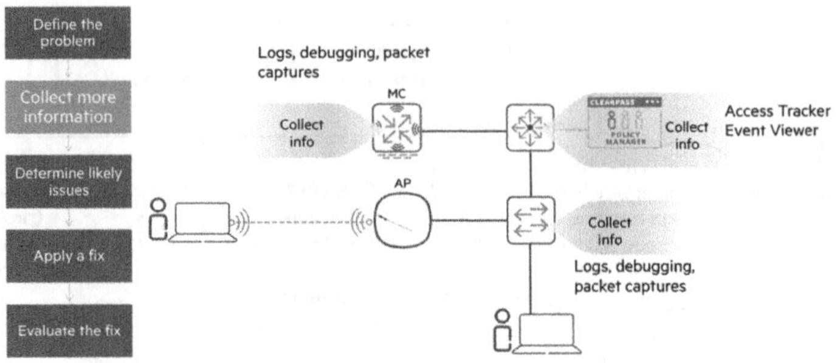

Figure 11-28: Collecting More Information

Again, it can be useful to reproduce the issue. Ask one of the users experiencing issues to attempt to connect to the network again. You will then look for information on your network infrastructure devices and on ClearPass, using the tools that you examined in the previous section (see Figure 11-28 for suggestions). Based on what you find, you will make a hypothesis about the error.

Determining Likely Issues

After collecting information, you should have a list of one or more potential causes of the issue.

Some common authentication issues include client configuration issues. For example, clients might be configured to use the wrong EAP method, or they might not have the proper certificates installed on them for EAP. The clients might also lack the root CA certificate for CPPM's RADIUS/EAP certificate, or there might be an issue with that certificate. The clients might also be configured with a RADIUS server name that does not match the subject name in CPPM's RADIUS/EAP certificate.

Another common issue is that one or more network devices have the wrong RADIUS shared secret, so CPPM drops their Access-Requests. If the network devices are using RadSec, there could be an issue with the certificates on either the network device or CPPM.

Sometimes you will encounter issues with the network infrastructure devices not applying the role that CPPM assigns. For example, the value that CPPM sets in the Aruba-User-Role VSA might not exactly match a role name configured on the device. Or, if CPPM is sending downloadable user roles (DURs), the network device might not be able to use HTTPS to log into CPPM and obtain the role. It might not have the proper root CA certificate installed, or it might have the wrong credentials.

This list, summarized in Figure 11-29, is non-comprehensive, but it gives you a starting place. The CPPM Access Tracker and network device logs that you collected will give you a better idea of which issue is at play in your network. In a moment, you will explore some example scenarios so that you can learn to correlate various symptoms with various issues.

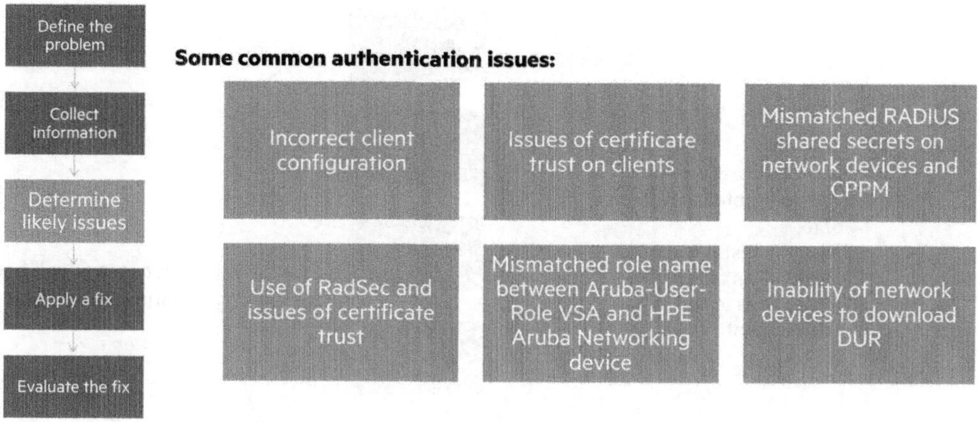

Figure 11-29: Determining Likely Issues

Applying and Evaluating Fixes

When you have determined the likely source of the issue, begin by fixing the most probable cause. For some simple fixes you might be able to make the change in a live network. For example, you might have clear indications of a mismatched shared secret between one network device and CPPM. No one is able to connect through this device in any case. You can fix the secret on that device. However, often you will be troubleshooting an issue that affects a segment of users in a solution that is functioning in many other ways. You do not want to make a change to this solution that could potentially affect other functions and make the situation worse. Instead set up a simple test environment that uses the same configuration as the actual solution but on a smaller scale. Apply the fix in that environment and evaluate the results.

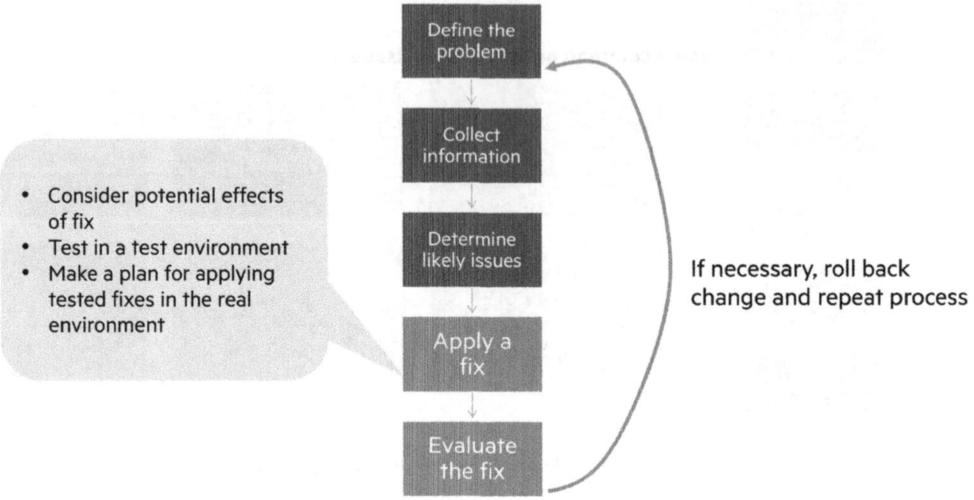

Figure 11-30: Applying and Evaluating Fixes

Keep in mind that you should make only one change at a time so that, when you evaluate the results, you know where the change in behavior comes from. As shown in Figure 11-30, if the issues persist, you can roll-back that change and make another change, or—if you suspect interacting factors, you can add a change.

If you find that the fix does work, you can make a plan for applying it in the actual environment. This plan should include consideration of the potential effects on other components of the solution, as well as a plan for when to apply the change and how to evaluate it. For some pressing issues, you might need to apply the change as quickly as possible. In other cases, the issue is less pressing, and you might need to apply the change during a scheduled network outage outside of business hours.

Authentication Troubleshooting Scenarios

You will now see the tools and processes about which you have learned put into action in several scenarios.

Scenario #1

In this first scenario, the network already implements 802.1X on wireless and wired connections. You have added a new switch and configured it to implement 802.1X as well. When you tested a connection, your authentication attempt failed. You have tried several client devices and user accounts and none of them are working.

This scenario could also apply to deploying a new IAP cluster.

Collecting More Information in Access Tracker

To collect more information, you try authenticating again. Then you go to CPPM's Access Tracker and look for a corresponding record by filtering for your username. In the example shown in Figure 11-31, Access Tracker indicates that no results match. You double-check the time period and selected server. You look without a filter too and cannot find a record that corresponds with your authentication attempt.

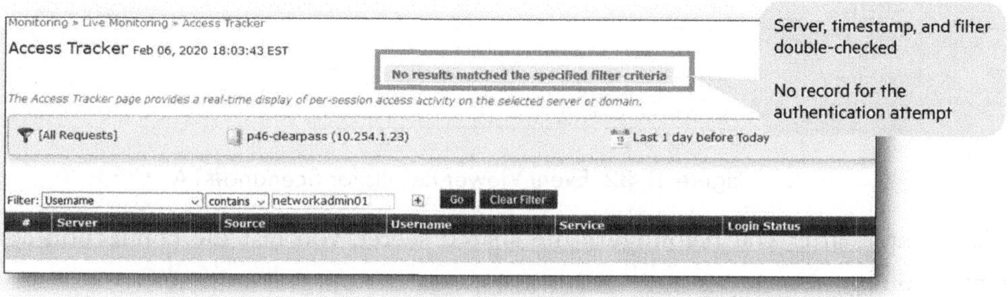

Figure 11-31: Access Tracker Results for Scenario #1

The most obvious issue that would cause this result is that the new switch's authentication requests simply are not reaching CPPM. This could occur if the switch has a routing or connectivity problem. It could also occur if the switch was configured with the wrong RADIUS IP address or if a misconfiguration causes the switch not to use RADIUS for the authentication attempt.

CPPM could also be receiving the requests but not logging them in the Access Tracker. This occurs when the shared secret on the Network Access Server (NAS) (here, the switch) does not match the shared secret on the RADIUS server. It also occurs if the NAS IP address does not correspond with the IP address of a network device configured on CPPM.

For your next step, you should check CPPM's Event Viewer, which can show you whether the requests are actually arriving on the CPPM server and, if so, why CPPM is rejecting them.

Collecting More Information in Event Viewer (A)

You go to the Event Viewer, and there you do get a result for RADIUS events, as shown in Figure 11-32. You open up the error message, and the message gives you a great clue. The shared secret is incorrect. Because, for security, you should not be able to view the actual shared secret on either CPPM or the NAS (AOS-CX switch), you cannot double-check it. Simply double-check with the ClearPass admins that you have the correct secret and reconfigure it. You might also want the ClearPass admin to reconfigure the secret in case the error was on that side.

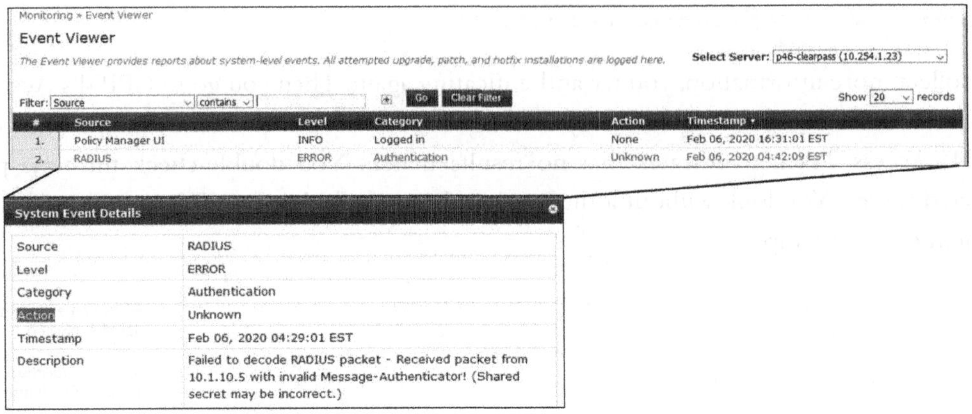

Figure 11-32: Event Viewer Results for Scenario#1A

In this example, after you reconfigure the shared secret, you test the connection again, and the authentication is a success.

Collecting More Information in Event Viewer (B)

In this alternate world, the event details were different, as shown in Figure 11-33. This time the event details indicate that the authentication attempt came from an unknown NAD IP address. It is possible that ClearPass admins accidently entered the incorrect IP address when configuring the new network device on CPPM. When a device has more than one IP address, unexpected issues can occur. For example, maybe you expected that the switch would use 10.1.20.8 on VLAN 20 as its NAS IP address, but the switch actually routes traffic toward CPPM out VLAN 10 with IP address 10.1.10.8. You can set the RADIUS source IP address manually on AOS-CX switches to prevent issues such as this.

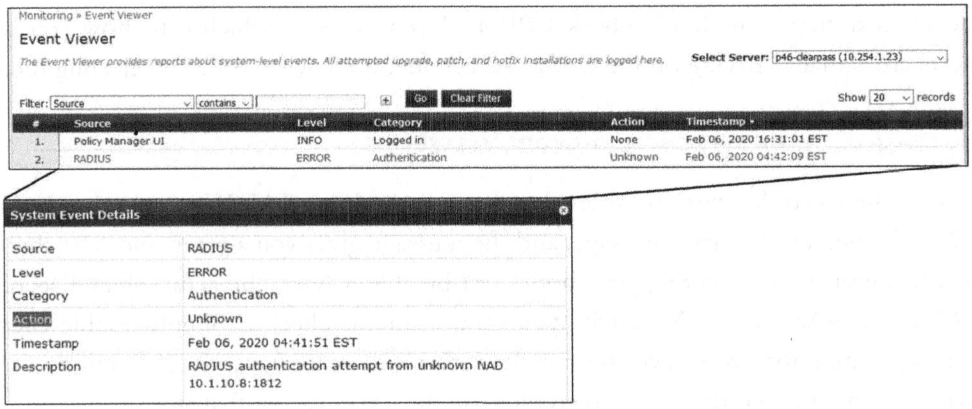

Figure 11-33: Event Viewer Results for Scenario #1B

Collecting More Information (C)

In the final alternative world, you cannot find any relevant RADIUS events in CPPM's Event Viewer, as shown in Figure 11-34. Now it begins to seem likely that the NAS's RADIUS messages are never reaching the CPPM server. You should return to the NAS in question, whether an AOS-CX switch, IAP, or MC.

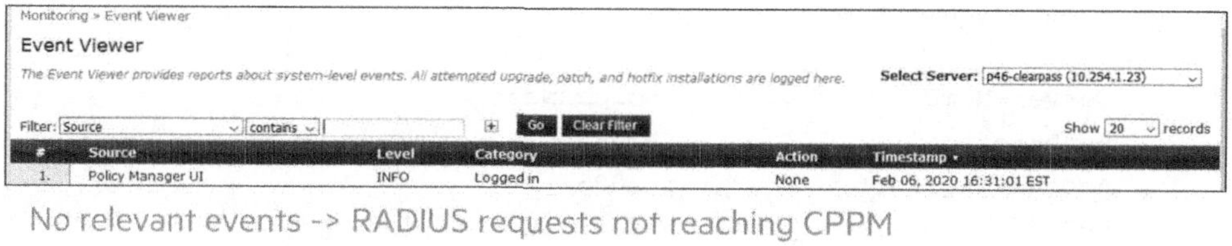

Figure 11-34: Event Viewer Results for Scenario #1C

Collecting More Information on NAS (C)

Start with a ping just to make sure that the device can reach CPPM. If not, you will need to examine the device ports, VLANs, and IP routes for issues.

If the switch can reach CPPM, you should look at the RADIUS server details, as shown in Figure 11-35. In this example, you can see an issue. The switch's TLS connection with CPPM has failed. You know that the switch can send pings to CPPM, so you should look for other issues. Perhaps the CPPM's certificate was updated and this switch no longer trusts the subject name in the certificate. Perhaps a control plane ACL or another security device between the switch and CPPM is allowing the pings but prohibiting the TCP 2083 traffic.

```
Switch# ping clearpass
<-output omitted->
5 packets transmitted, 5 received, 0% packet loss, time 4004ms
Switch(config)# show radius-server detail
******* Global RADIUS Configuration *******
<-output omitted->
****** RADIUS Server Information ******
<-output omitted->

Server-Name                  : clearpass
Auth-Port                    : 2083
Accounting-Port              : 2083
VRF                          : default
TLS Enabled                  : Yes
TLS Connection Status        : tls connection failed
<-output omitted->
```

Figure 11-35: RADIUS Server Details on AOS-CX Switch in Scenario #1C

If the switch is sending the requests, but CPPM is not getting them, you can use packet captures to find where the request is being dropped. You will look at packet captures in the next section.

You can take a similar approach for collecting information on HPE Aruba Networking mobility devices. These devices also have a useful feature for testing access requests. This feature can help you determine whether the issue is with the client making the request or with the request itself.

Scenario #2

In this next scenario, the company has a WLAN that enforces 802.1X with EAP-TLS. The network has been active for some time, and most users are able to connect without issues. However, one user is having issues connecting on a new device, and you have been assigned to help that user out.

Collecting More Information with Access Tracker (A)

You have the user attempt to authenticate with the same credentials and on the same device experiencing the issues. This time your search for an Access Tracker record associated with the attempt yields a result. The login status is reject, as shown in Figure 11-36. A reject often means that the client failed authentication. For example, the client might not have a valid certificate. However, when you open this record and check the Alerts tab, you see a different issue. The "EAP-TLS ??? Alert by client" indicates that the client closed the session because the server failed authentication. Remember EAP-TLS and PEAP implement mutual authentication, and the server failing authentication also shows up as a Reject. This issue occurs when the client does not trust

CPPM's RADIUS/EAP certificate. The client might not have the correct root CA certificate installed on it to validate CPPM's certificate. Or it might be configured to look for a particular subject name, and that name is misconfigured.

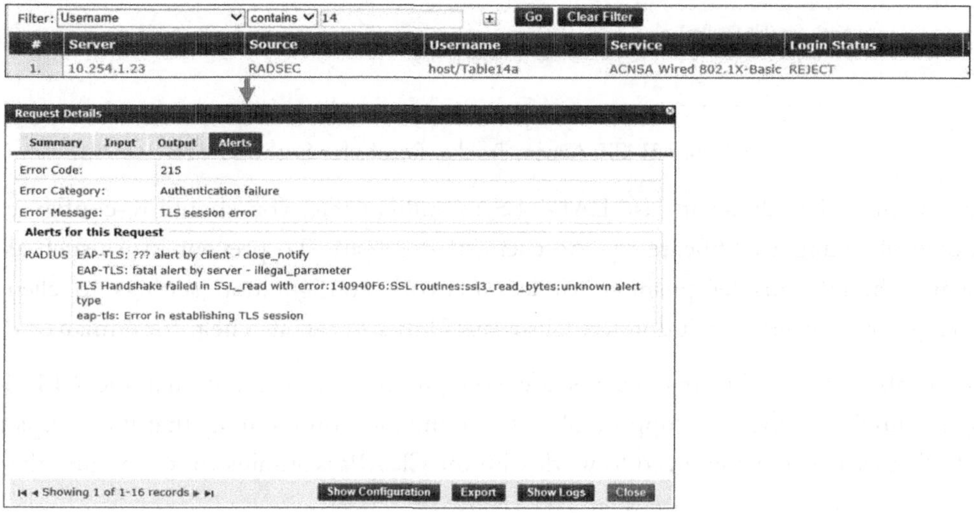

Figure 11-36: Access Tracker Results for Scenario #2A

To resolve this issue, you should check the 802.1X configuration settings on the client. (You learned how to configure those settings on a Windows client's Ethernet NIC in Chapter 6.)

If many clients experience the same issue, the ClearPass admins should check CPPM's RADIUS/EAP certificate. If it was recently renewed, perhaps an error occurred such as giving the certificate an incorrect subject name.

Collecting More Information with Access Tracker (B)

Next look at an alternative issue that you might find when one or a few individuals have trouble authenticating. The Access Tracker shows a reject record for the user experiencing the issue. As shown in Figure 11-37, when you open the record and click the **Alerts** tab, you see that the client does not support any of the authentication methods configured in the service. In this example, the service requires EAP-TLS, but the client is configured to use PEAP. This can happen if a client does not have an appropriate certificate, forcing it to a different option such as PEAP.

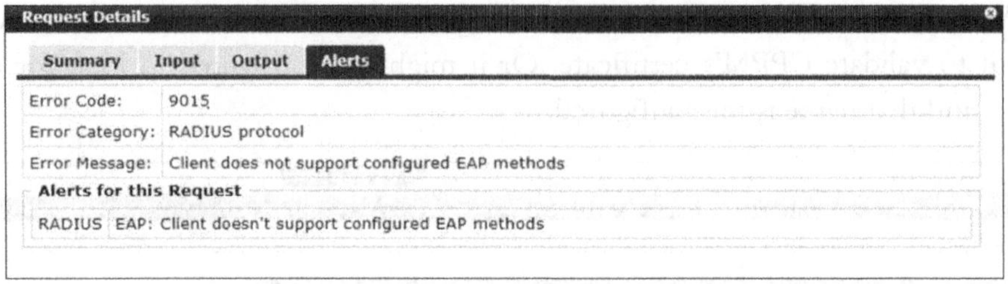

Figure 11-37: Access Tracker Results for Scenario #2B

If the intent is that all clients use EAP-TLS to authenticate, you should re-configure the client, including obtaining a certificate for the client, if necessary. To prevent issues such as these, the company should consider pushing wireless network configuration settings to clients through domain group policies. A solution like ClearPass Onboard can also help for unmanaged devices.

It could also be possible that clients are configured correctly, but that the CPPM service is misconfigured and does not support all of the authentication methods that the company wants to use. In that case, you would need to work with the ClearPass admins to reconfigure the service.

Scenario #3

In this scenario, the company has added some new users: contractors. The company wants contractors to connect to the WPA3-Enterprise WLAN but to receive different access rights from employees.

Domain admins have created accounts for contractors, and you have deployed user certificates to their devices. ClearPass admins have adjusted policies so that CPPM assigns contractors to the "contractor" role. You created the role on your MC and set up security policies for the role.

However, when you try to connect a Windows client to the WLAN using a test account in the contractor group, Windows reports "No Internet." The wireless NIC has not received an IP address.

Collecting More Information in Access Tracker

You go to the Access Tracker and see an Access-Accept for the user. When you look at more details for this record, you see that CPPM has sent the correct vendor specific attribute (VSA) for the "contractor" role, as shown in Figure 11-38.

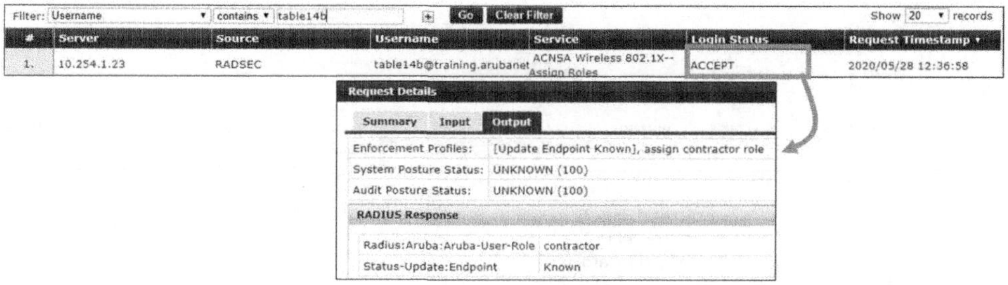

Figure 11-38: Access Tracker Results for Scenario #3

Something is occurring on the authenticator—a MC in this example—to prevent it from accepting the role. The most likely issue is that CPPM is sending a role name that does not match a role on the MC. If the MC *does* have a role that matches the one in this VSA, it will always assign the client to that role. The only exception is relatively rare. When the AAA profile has a user ruleset with Layer 3 (DHCP option) rules, those rules can reassign clients to new roles *after* authentication.

Another issue could be that the VSA is incorrect for the device type. Remember that HPE Aruba Networking mobility infrastructure, as well as AOS-CX switches, use the Aruba-User-Role VSA. (The HPE-User-Role is used for legacy switches that run AOS-S software.) Your best next step is to go to the HPE Aruba Networking device that is acting as authenticator and check the role that it is assigning to the client.

Collecting More Information on a MC

You go to the MC (or MCR) Web UI and view clients. As shown in Figure 11-39, you see that the MC has assigned the client in question to the "denyall" role, which is the 802.1X default role for the client's WLAN. This role explains why the client is not receiving an IP address. Assigning the client to the default role indicates that the MC did not accept the role in the Aruba-User-Role VSA as valid.

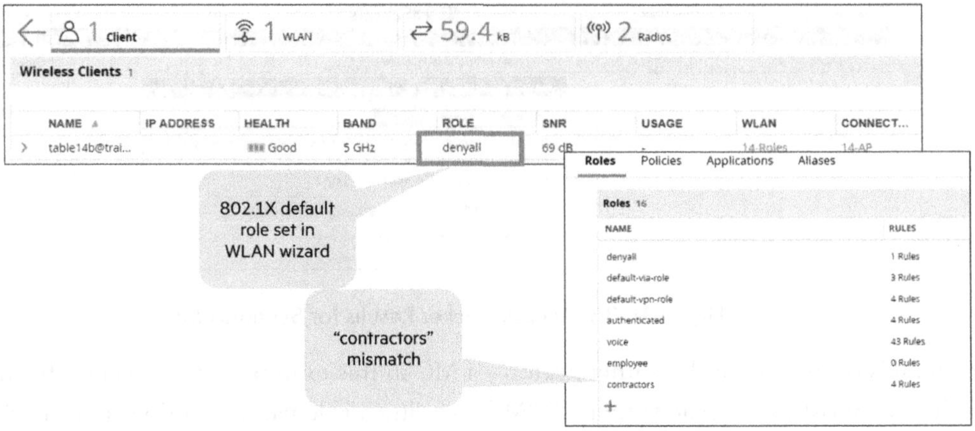

Figure 11-39: Collecting More Information on a MC

When you check the roles on the MC, you see that the role name is slightly different: "contractors" rather than "contractor." You either need to recreate the role with the name that CPPM is sending or have ClearPass admins change the name in their policies. The problem will then be fixed.

If the role seems to have the correct name to you, you should check that the role is applied to the correct hierarchy level so that the MC to which the client connects actually has the role.

Packet Captures

In this section, you will learn how to use packet captures to collect in-depth information about traffic flows.

Packet Capture and Traffic Mirroring

Packet captures can be useful tools for troubleshooting and investigating security issues. A packet capture intercepts and records packets in real-time as they flow through a network point. For example, a network infrastructure device can capture packets that flow through a particular port. In the example shown in Figure 11-40, a switch captures packets, but HPE Aruba Networking mobility devices can capture packets too. A client or server that runs a packet capturing application like Wireshark can also capture packets sent and received on its NIC.

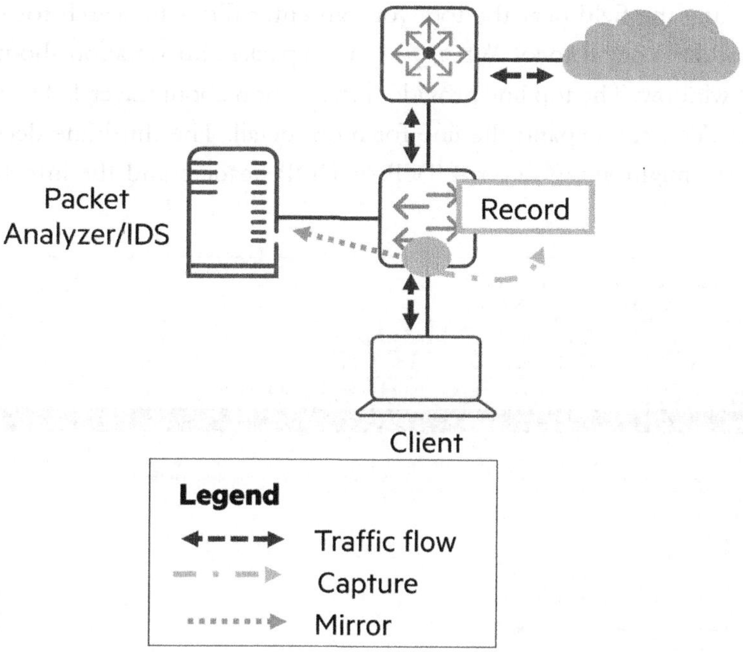

Figure 11-40: Packet Capture and Traffic Mirroring

Note that "interception" means "copying." The captured packets also continue to flow to their destination, as shown in Figure 11-40.

The device conducting the packet capture typically performs some basic packet decoding, converting the raw packet data into a more human readable form. It then typically stores the packets in a file for later analysis.

Somewhat related to the concept of packet capturing is traffic mirroring. Again, the network infrastructure device copies packets in real-time, but instead of recording them locally, the device sends the packets to another destination such as an Intrusion Detection System (IDS). The destination device can then conduct further analysis of the packets.

Example Packet Capture/Analysis Application: Wireshark

The idea of a packet capture might become a little more concrete for you with an example.

Wireshark is an application that can run on servers and clients. Wireshark captures packets on the device's NIC and puts the packets through a simple analyzer, presenting the results in an easy-to-use interface.

An example Wireshark packet capture is shown in Figure 11-41. The top section shows a chronological list of packets inbound and outbound on a device interface, enabling you to follow particular flows. Using the field near the top, you can enter filters to search for packets based on criteria such as IP address or protocol. When you click a packet, information about it appears in the bottom half of the window. The top line provides information about Layer 1. The next line decodes the Layer 2 header. You can expand the line for more detail. The third line decodes the Layer 3 header. Further lines might show Layer 4 TCP or UDP headers and the information contained within.

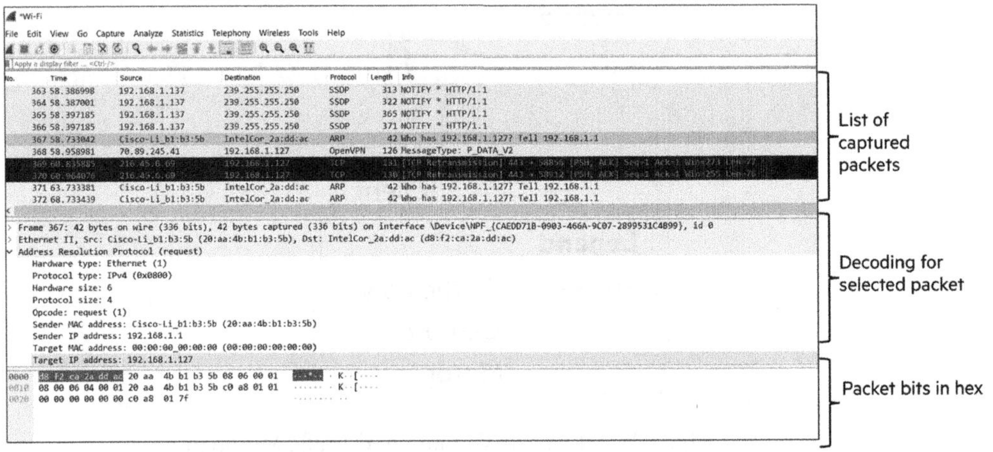

Figure 11-41: Example Packet Capture/Analysis Application: Wireshark

The middle section presents the data in a friendly format while the bottom section shows the actual hex characters in the packet.

This example gives you a basic idea of what a packet capture is and how useful it can be. You could spend a great deal of time learning how to use a program like Wireshark and how to analyze packets and packet headers. The focus for this study guide, though, is not on analysis. Instead you are focusing on how to obtain captured packets, which you can turn over to the company's security team for analysis.

Reasons to Capture Packets

Packet captures can help your company to investigate security issues, analyze traffic patterns, and troubleshoot.

Senior IT members at your company or support personnel might request that you perform a packet capture to provide more intensive analysis. Using CPPM and HPE Aruba Networking WIP, you might have pinpointed suspicious activity by a user or device that merits further investigation. You can capture packets to examine the suspicious traffic in more detail. You can submit the traffic to an IDS for further analysis against threat signatures.

Packet captures can also help admins to analyze network traffic in more depth and to see precisely how the traffic flows across the network. For example, if you capture traffic at two points in the network, you might see RADIUS packets from CPPM to a NAS leaving CPPM and then disappearing somewhere along the path. You can then pinpoint where the packets are lost and what is happening. Perhaps an ACL is misconfigured and dropping the packets.

In this and other ways, packet captures can help you and your organization to troubleshoot issues that resist resolution through other methods.

As you capture traffic, make are that you understand and follow your company policies. First, only capture traffic when necessary, as dictated by the policies. Second, follow policies that define what types of data that you can examine. For example, you might be allowed to analyze the types of traffic a user is sending but not actually look inside the application layer. Or you might have to run traffic through an intrusion prevention system (IPS) and find a credible threat before a human also examines the packets.

Where to Implement the Packet Capture

After you have decided to implement a packet capture, you need to think about the best place to capture the traffic, as indicated in Figure 11-42. For example, if you are trying to find more information about authentication communications between a wireless client and CPPM, you need to consider whether to capture packets at the client, the AP, the MC, or even on the link to CPPM. Similarly, for wired clients you have the choice of capturing traffic at multiple points in its journey.

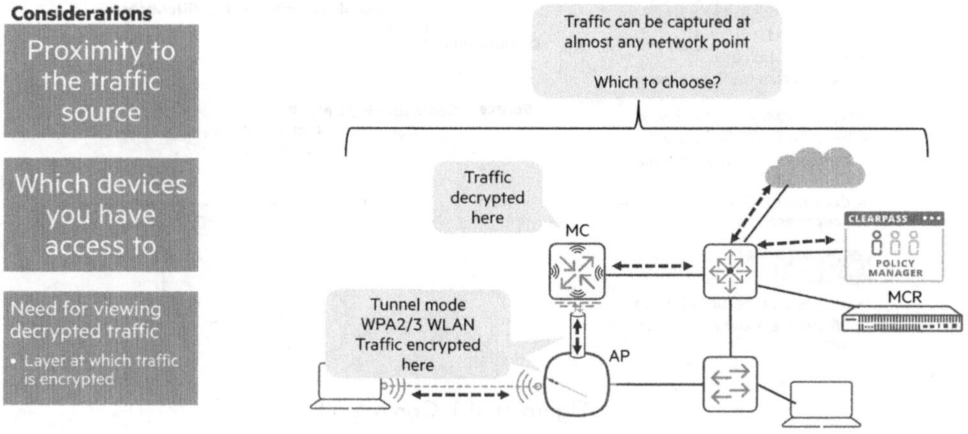

Figure 11-42: Where to Implement the Packet Capture

You might begin by capturing the packets on the client that you want to examine, as the most straightforward way to obtain all of the client's traffic and just the client's traffic. You should keep in mind whether traffic is encrypted by the application as well. Wireshark can output TLS keys, but it needs to be running on a device that has those keys—in other words the client or server in the session.

However, sometimes the client is not available to you. Then it can be useful to capture the packets as they flow through the MC, filtering for just the client in question. You could also try to capture packets on the APs, but if traffic is encrypted by a protocol like WPA3 and the WLAN operates in tunnel mode, the traffic will not be decrypted until it reaches the MC. For this reason, it is often a good idea to capture packets on the MC. Even if some traffic is encrypted at a higher level, such as by TLS, seeing the types of traffic can be useful in some situations.

You might even want to capture traffic at multiple points so that you can compare the results and look for lost packets.

Implementing a Packet Capture on a MC

You have several options as you capture packets on a MC. You can choose the controlpath or datapath as the source for the capture.

Controlpath

As you learned in Chapter 4, the controlpath includes traffic sent to and from the MC itself while the datapath includes traffic that flows through the MC to and from clients connected to the MC's APs.

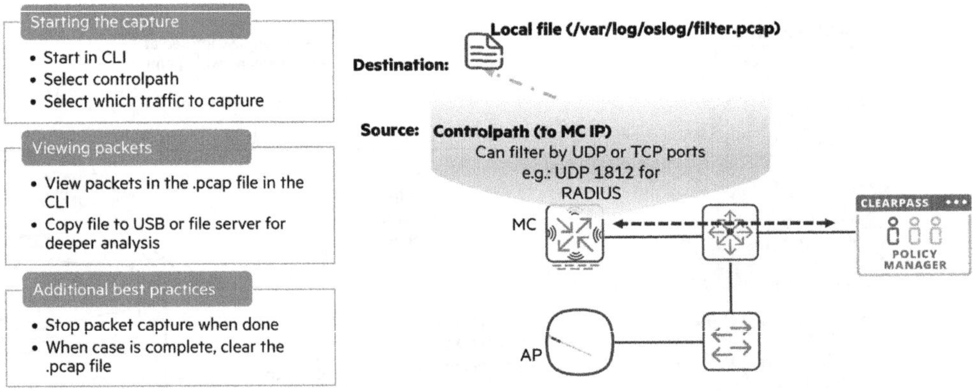

Figure 11-43: Controlpath

In Figure 11-43, you see guidelines for controlpath captures, which capture inter-process messages, as well as TCP, UDP, and other traffic that the MC exchanges with other devices.

When you begin the packet capture on the controlpath, you select the type of traffic to capture. For example, if your company is troubleshooting an authentication issue between the MC and its RADIUS server, your managers might ask you to capture UDP 1812 traffic.

The MC saves the captured packets to a local file on the flash at /var/log/oslog/filter.pcap. You can view the packets in that file in the CLI. However, for deeper analysis, you should copy the file off of the MC so that you or the security team can open it in a packet analyzer like Wireshark and subject the packets to deeper analysis. You can copy the file to a USB or an SCP, FTP, or TFTP file server. The security team can then take that file and open it in a tool like Wireshark.

When you are done capturing packets, you should stop the packet capture, which stops the MC from collecting more packets. The previously captured packets remain in the file.

When you start a new packet capture on the controlpath (or the datapath), the MC appends the new captured packets to the end of any packets already in the corresponding .pcap file. So that you can start new packet captures with a clean slate, it can be best practice to clear that file after you have copied out the file or closed the case.

Datapath

Sometimes your managers might ask you to capture traffic from a particular wireless client. In this case, you would want to conduct a datapath capture on the MC. As discussed earlier, if the wireless client is encrypting its traffic, the MC decrypts the traffic. A datapath capture lets the security team examine the unencrypted packets.

As shown in Figure 11-44, when you enable a packet capture on the datapath, you can select an IPsec tunnel or the MAC address of a user device within the MC's user table. The MC then selects all traffic to and from that device and copies it to the configured packet capture destination.

You can configure this destination as a file, which you would then typically turn over to the appropriate team for further analysis. (The local file for the datapath is /var/log/oslog/datapath.pcap.) You can also mirror the captured traffic out a particular interface. You would typically connect that interface to a packet analyzer or IDS. If the packet analyzer or IDS is on a non-directly connected device, you can direct the packet capture to an IP address, rather than an interface.

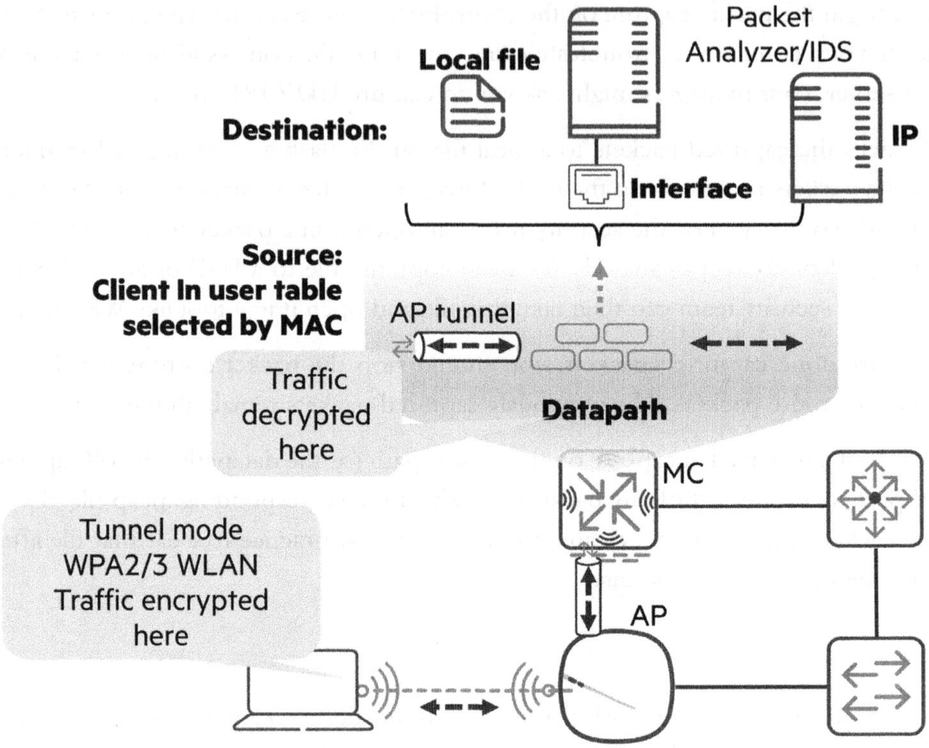

Figure 11-44: Datapath

Best practices for these packet captures are similar to those for the controlpath captures. Remember to turn them off and to clear out the pcap file when you no longer need the captured packets.

CLI Commands: Implementing Packet Captures on MCs

This section provides CLI commands for implementing packet captures with MCs.

Use the commands shown in Figure 11-45 when you are instructed to run a packet capture on a MC.

You can capture packets on the controlpath or datapath. The controlpath packets are always captured to the local filesystem. If you are setting up a capture on the datapath, find out whether you should capture the packets locally or send them to a system, either connected to a MC interface or reachable at an IP address. If you are sending the packets out an interface or to an IP address, make sure that the receiving system is set up to receive the packets. Then set the packet capture destination on the MC. Then enable the datapath packet capture.

To stop a packet capture, enter the same command with **no** at the beginning.

In enable context (/mm or /mynode)

Step	AOS command	Notes
Enable or disable up a packet capture on the controlpath.	[no] packet-capture controlpath <interprocess | other | sysmsg | tcp <all | *ports*> | udp <all | *ports*>>	sysmsg captures internal messaging and is used for debugging. Captured packets are stored in /var/log/oslog/filter.pcap.
Set the packet capture destination.	packet-capture destination <interface *slot/port* | ip-address <*IP*> | local-filesystem>	Local filesystem is the only option for controlpath captures.
Enable or disable a packet capture on a client in the datapath by MAC address.	[no] packet-capture datapath mac <*MAC*> <all | encrypted | decrypted>	Use colon delimiters in the MAC address. "ipsec" is an alternative to "mac," but IPSec is not covered in this course. If you set local-filesystem for the destination, captured packets are stored in /var/log/datapath.pcap

Figure 11-45: CLI Commands: Implementing Packet Captures on MCs

In Figure 11-46, you see commands for managing the packet capture on a MC. You can view enabled packet captures. You can also view the packets captured in the local filesystem. However, your managers or the security team members will probably want you to export the captured packets so that they can view them in a packet analyzer like Wireshark. The commands for doing so are shown here.

You also see the command for clearing the packets captured in the local filesystem.

In enable context (/mm or /mynode)

Step	AOS command	Notes
View which packet captures are enabled.	show packet-capture	
View packets captured in the local filesystem.	show packet-capture <controlpath-pcap | datapath-pcap>	
Copy a packet capture from the default location to a zip file in top level of flash.	packet-capture copy-to-flash <controlpath-pcap | datapath-pcap>	Destinations include ftp:, scp: and usb: Use ? to find the options for each.
Clear out the packets captured in the local filesystem.	packet-capture reset-pcap <controlpath-pcap | datapath-pcap>	If the capture is still enabled, the previously captured packets are cleared, and the device continues to capture new packets

Figure 11-46: CLI Commands: Managing Packet Captures on MCs

Configuring Wireshark to Receive the Capture

If you choose an IP address as the destination for a datapath capture, the MC sends the traffic with GRE encapsulation. Wireshark typically decodes GRE traffic by default. If you want to limit Wireshark to showing the traffic sent by the MC, configure Wireshark to filter for GRE traffic (**ip proto 47**) in the capture options, as shown in Figure 11-47.

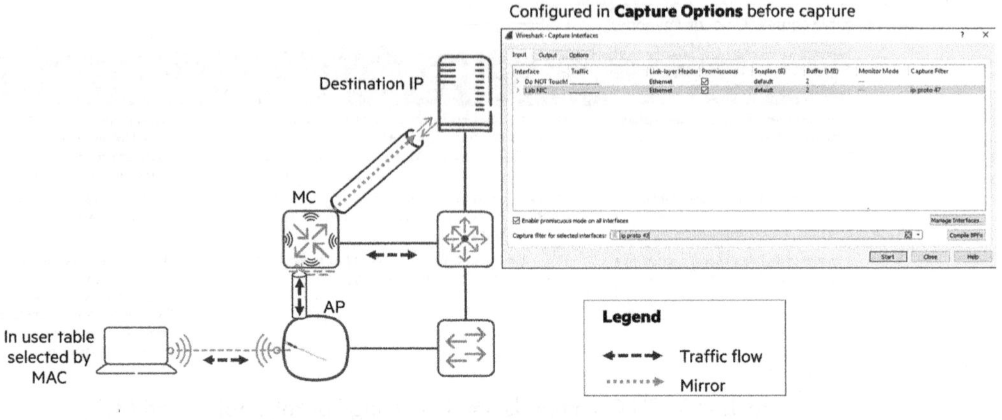

Figure 11-47: Configuring Wireshark to Receive the Capture

Implementing a Packet Capture on an HPE Aruba Networking AP

Sometimes your managers might ask you to capture traffic as seen in the air. For example, they might need to troubleshoot an issue by determining whether traffic transmitted through the wired network is actually being transmitted over the wireless network.

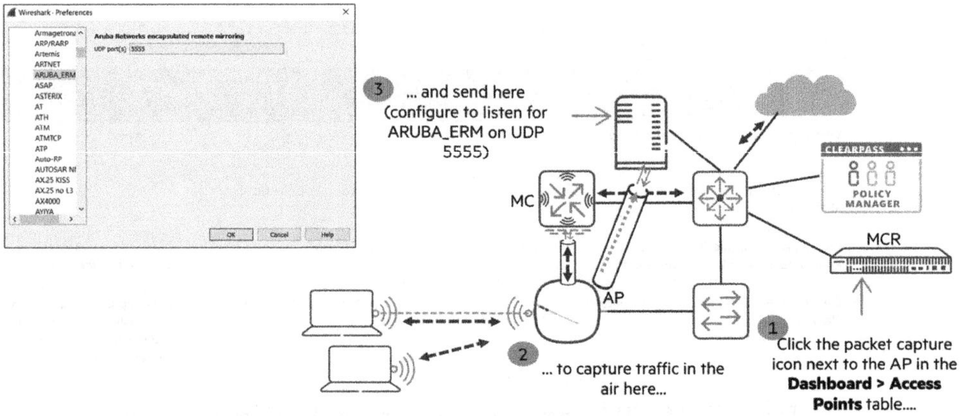

Figure 11-48: Implementing a Packet Capture on an HPE Aruba Networking AP

As shown in Figure 11-48, if you are using an MCR, you can implement a packet capture directly on a controlled AP. You simply go to the Access Points table in **Dashboard > Access Points**. Then click the packet capture icon next to the AP on which you want to capture traffic.

You can select options such as the frequency and WLANs on which to capture traffic and the destination to which to send the packets. Click **Start** to initiate the capture.

The AP uses a protocol called Aruba Encapsulated Remote Mirroring (ERM). You must set up Wireshark preferences to receive the traffic correctly, as shown in Figure 4-48. Select **File > Preferences**. Expand **Protocols** and find ARUBA_ERM in the list of protocols. Specify 5555 for the port. Wireshark then listens for traffic on UDP port 5555 and decapsulates the traffic for further analysis.

Implementing a Packet Capture for Traffic through an AOS-CX Switch

Similarly, your managers might direct you to capture packets in the wired infrastructure. For example, another security solution might have detected possibly suspicious activity from a wired client, and the security team wants to investigate that client's traffic more closely.

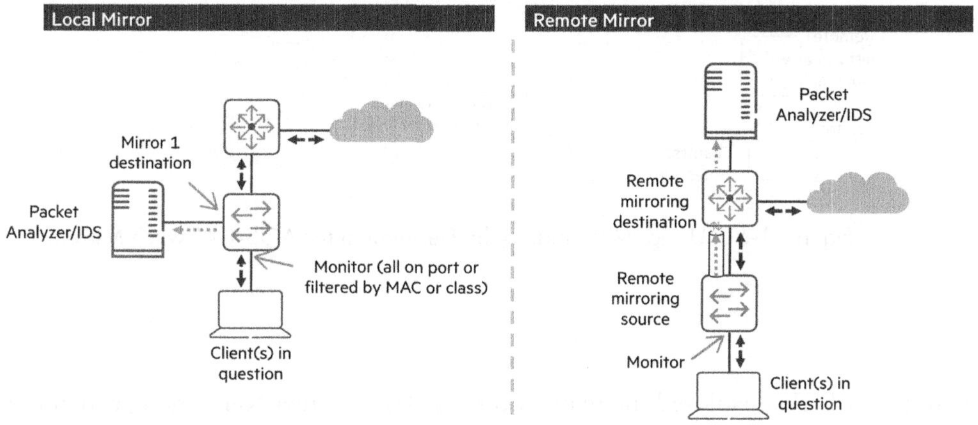

Figure 11-49: Implementing a Packet Capture for Traffic through an AOS-CX Switch

If you want to capture packets flowing through the wired infrastructure, you should set up a mirror on the switch as close to the source of the traffic in question as possible. As shown on the left of Figure 11-49, the mirror copies the packets, sending one copy toward the destination as usual and one copy out the mirror destination, which is another switch port when you are using local mirroring. As usual, you would connect a computer running a packet analyzer or IDS to that port.

You can capture all traffic inbound on one or more ports, outbound on the ports, or both. You can also filter the mirroring by MAC address or traffic class.

It is simplest to set up a local mirror, which directs the copied traffic out another interface on the same switch. Sometimes, though, it is not feasible to connect the packet analyzer or IDS to the switch where you want to capture the traffic. In that case, you can create a remote mirror, as shown on the right of Figure 11-49. The remote mirror encapsulates the mirrored traffic in a UDP-based tunnel (no protection) and sends it to another AOS-CX switch that is set up as the remote mirror destination.

Using Wireshark as the Destination for AOS-CX Switch Mirrors

You can also set up a device running Wireshark as the destination for a remote mirroring session. The AOS-CX remote mirroring tunnel uses GRE. So, you should configure Wireshark to filter for GRE traffic (**ip proto 47**) in the capture options, as shown in Figure 11-50.

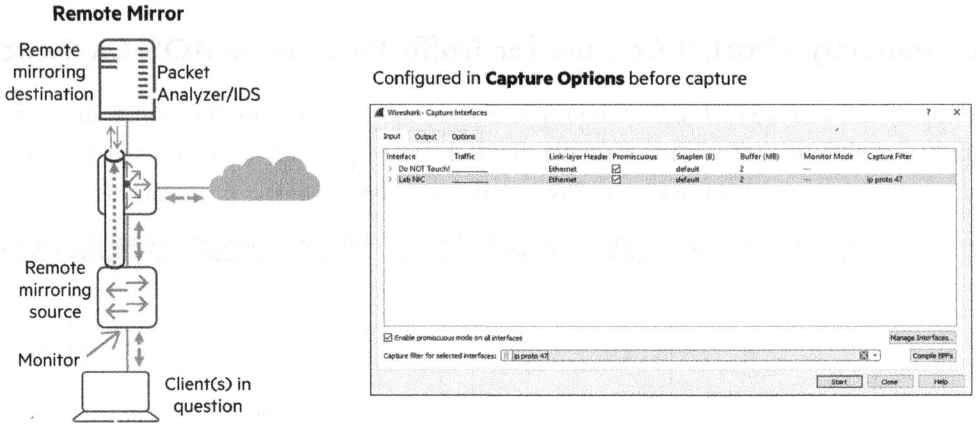

Figure 11-50: Using Wireshark as the Destination for AOS-CX Switch Mirrors

Summary

Congratulations! You explored the many tools that HPE Aruba Networking provides to help you monitor the network on an ongoing basis. You understand ClearPass Insight's reporting capabilities. You can integrate your network infrastructure devices into your company's SIEM solution. You also understand what information you can see in HPE Aruba Networking mobility dashboards and when you should look to AirWave and other tools for more extensive analysis and historical data.

You now have the tools that you need to troubleshoot authentication and related security issues. You can analyze information within CPPM Access Tracker and Event Viewer, supplementing that information with device logs as necessary, in order to find the root cause for users' issues.

Sometimes a security incident or a troubleshooting ticket require a more in depth view of traffic. You can now respond to your colleagues' requests for packet captures implemented on APs or MCs. You can also use mirroring on switches to direct traffic to a device running a packet analyzer or IDS.

Learning checks

1. Match each piece of information that you might want to find with the tool where you can find that information.

   Information that you want to find

   a. Reason that CPPM rejected an authentication request

   b. Report with analysis of authentication trends

   c. 802.1X related events that have occurred on a MC

   d. Global overview of issues related to association, authentication, DHCP, and DNS

   Found in this tool

   __ Process logs in MC UI

   __ ClearPass Insight

   __ WiFi Connectivity in HPE Aruba Networking Central

   __ CPPM Access Tracker

2. You have enabled logging to 10.1.10.10 on an AOS-CX switch. No other logging or debug commands are configured. Which logs does the switch send to 10.1.10.10?

   a. No debugs or logs

   b. Logs of info level and up

   c. Logs of error level and up

   d. Debugs but not logs

3. What could cause an AOS-CX switch an issue in downloading a user role from CPPM?

   a. The role name that CPPM sent doesn't match a role name configured on the switch.

   b. The switch lacks the root CA certificate for CPPM's HTTPS certificate.

   c. The switch does not have SFTP enabled on it.

   d. The switch does not have a local certificate of its own installed on it.

4. You set up a MC to enforce authentication to CPPM on a WPA3-Enterprise WLAN. When you test the connection, authentication fails. You check CPPM Access Tracker and cannot find a record associated with your attempt. What might be the problem?

   a. You entered the wrong username and password on the client.

   b. Your client does not support the right EAP types.

   c. Your client does not trust CPPM's certificate.

   d. The RADIUS secret configured on the MC is incorrect.

5. You have a controlled wireless solution. You want to capture unencrypted packets associated with a particular authenticated wireless user, who is connected to a WPA3-Enterprise WLAN in tunnel mode. Where should you implement the packet capture?

   a. On the user's AP

   b. On the switch connected to the user's AP

   c. On the MC's control plane

   d. On the MC's data plane

6. You have set up a packet capture from a CAP to a client running Wireshark. What should you do on the client to prepare to receive the traffic?

   a. Tag the client NIC for VLAN 4094.

   b. Bridge traffic between multiple NICs on the client.

   c. In Wireshark, set the Aruba_ERM protocol to UDP 5555.

   d. In Wireshark, filter for UDP 514.

7. Consider these questions about the purposes of, and best practices for, logging. Take notes on your answers:

   - Why is sending logs to a centralized solution on an ongoing basis important?
   - What options do you have for protecting the integrity and privacy of logs?
   - You can filter the logs that you forward in various ways. What arguments can you make for sending all logs, from informational level and up, or for sending relatively fewer logs?

Note

For answers to all learning checks, refer to Appendix B.

---

12 Threat Detection and Forensics

EXAM OBJECTIVES

- ✓ Identify a chain of custody for security events
- ✓ View, interpret, and acknowledge WIP events

Assumed Knowledge

- Basics of IP networking
- Basics of Ethernet switching
- Basics of 802.11 wireless networks

Overview

In this chapter, you will learn about digital forensics and best practices for responding to security events, compiling evidence, and maintaining a chain of custody. You will then apply some of these methods and look at proper ways to respond to and analyze HPE Aruba Networking WIP events. You will review this feature and then examine the Security Dashboard in detail, learning how to interpret what you see and respond to threats.

Forensics

You will now learn about proper digital forensics techniques for formal investigations of security incidents.

Digital Forensics

As defined by the National Institute of Standards and Technology (NIST), digital forensics is the "application of science to the identification, collection, examination, and analysis of data while preserving the integrity of the information and maintaining a strict chain of custody for the data" (https://nvlpubs.nist.gov/nistpubs/Legacy/SP/nistspecialpublication800-86.pdf).

While companies can use digital forensics for many purposes, this chapter focuses on using forensics as part of a security investigation, whether you are investigating a violation of company policy or of law. If your company plans to take legal action, your forensics policies must be even more strict. The guidelines provided in this chapter can give you a starting point, but you should not take them as legal advice. Consult with your organizations' legal department.

Chain of Custody

Maintaining a "chain of custody" as data passes through different forensic stages forms a critical component of forensics. To understand the concept of chain of custody, examine its origins in criminal investigations. Criminal investigators use a chain of custody to document where evidence has been and who has access to it so that they can verify that no one has tampered with the evidence and so that the evidence is admissible in court.

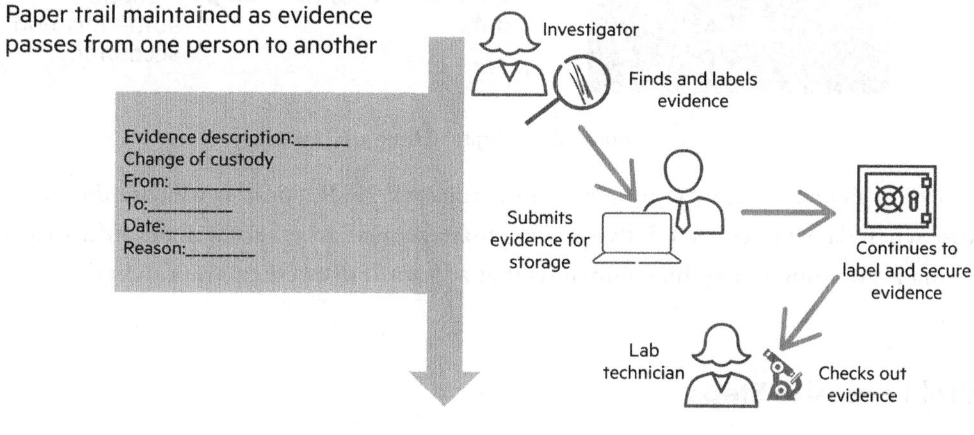

Figure 12-1: Chain of Custody

In Figure 12-1, you see an example. Investigators into a high-profile theft have found strands of hair at the crime scene. They hope to apply DNA sequencing to the hair and match the hair to a suspect. But for the DNA testing to be admissible as evidence, the investigators must be able to prove that the hair sample that ends up in the lab comes from the same hair that was found at the crime scene. The chain of custody provides this proof. The investigators log where they found the hair. When they turn the hair over to the department's evidence storage facility, they attest that the evidence has not left their keeping. Officials at the evidence storage facility must label the evidence and store it securely. When the officials turn the hair over to the DNA lab technicians, the lab technicians must sign for it. Again, the lab must store the evidence securely and take measures to label it properly. Any mix-ups could result in a mistrial. These measures ensure that everyone knows who is responsible for the evidence at each moment and that the evidence for each case remains valid.

Digital Chain of Custody

Similarly networking forensics requires a chain of custody. As data is collected and analyzed, you must establish when the data was collected, where the data was stored and replicated, and who had access to the data. Because the "investigators" collecting the raw data, and sometimes examining it as well, are typically infrastructure devices, you can see again the importance of having followed best practices to harden the devices.

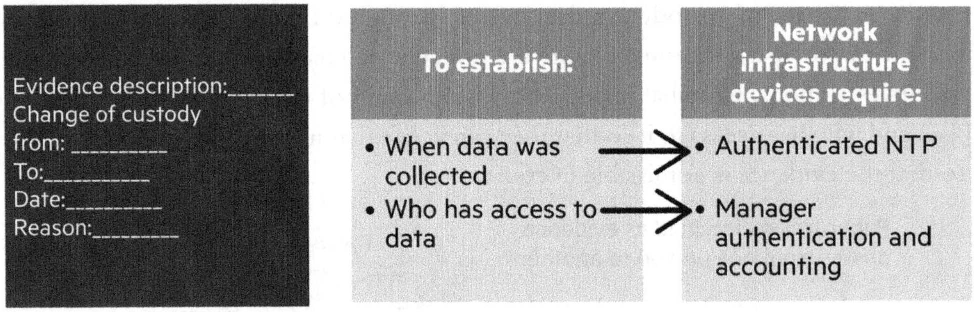

Figure 12-2: Digital Chain of Custody

As shown in Figure 12-2, by configuring authenticated NTP, you can validate the timestamps that indicate when data was collected. By securing management access to network infrastructure devices and using accounting to log the commands that administrators enter, you can verify who had access to logs.

Digital Forensics Steps

Now that you know about some general best practices, you will look at the forensics phases in more depth. As shown in Figure 12-3, NIST defines a model for digital forensics that helps companies properly collect and evaluate digital data so that they can conduct security investigations properly and assemble valid evidence. From collection to reporting, these steps take raw data to formal evidence.

The source for this model is:

https://nvlpubs.nist.gov/nistpubs/Legacy/SP/nistspecialpublication800-86.pdf).

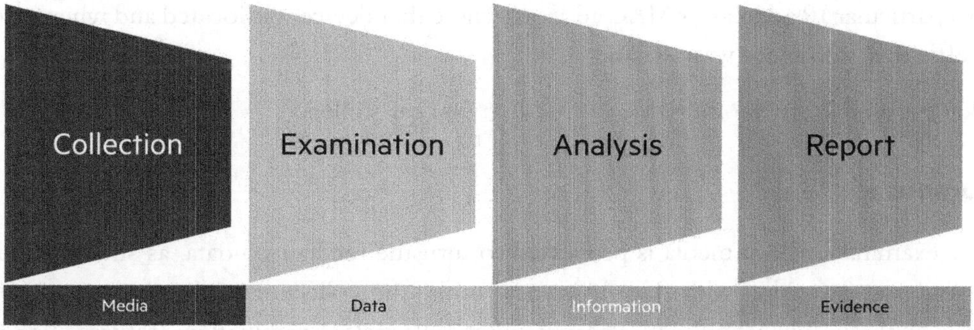

Figure 12-3: Digital Forensics Steps

Collection

As shown in Figure 12-4, the first step in forensics is the collection of data. For networking forensics, our focus, you collect a great deal of data before you know which data will be relevant for a particular investigation. Network device logs form one key source of information. In addition to storing these logs on individual devices, you can forward logs to a centralized Syslog server and incorporate the information into solutions like SIEM systems.

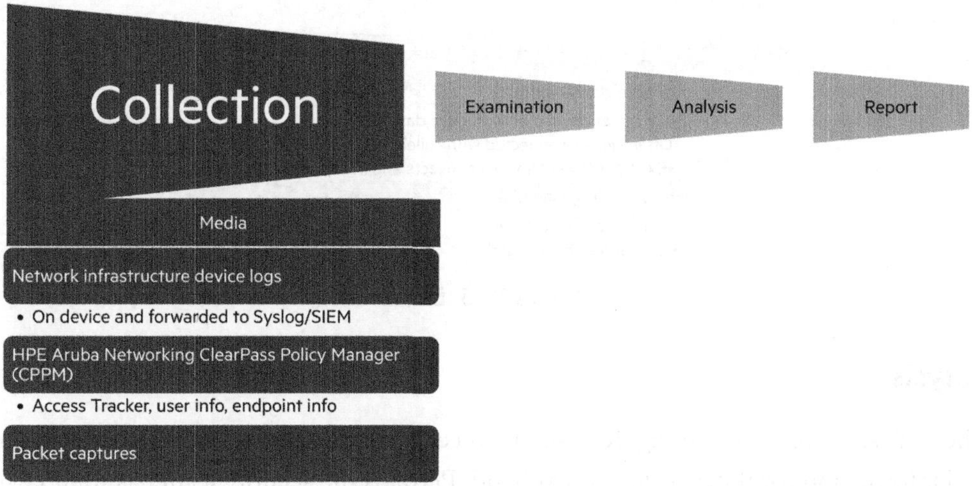

Figure 12-4: Collection

HPE Aruba Networking ClearPass forms another valuable source for data collection. With the ClearPass Policy Manager (CPPM) Access Tracker, you can track when each user and device connected to the network. ClearPass also tracks information about every user and device. In a security investigation, it can be very important that you are able to demonstrate who is using a device with a particular IP address or MAC address, where that device was located and when, and the types of traffic that the device was sending.

Packet captures form another type of data that you can collect.

Examination

In the examination step media is processed to turn the media into data, as shown in Figure 12-5. When you are working with digital forensic tools, the collection and examination phases often overlap with the collection tool also providing automatic examination features. For example, a packet analyzer like Wireshark both collects network packets as bits and bytes and transforms those bits into human readable data. Similarly, CPPM's Access Tracker logs RADIUS authentication events and also outputs them in an accessible way in the CPPM Web UI.

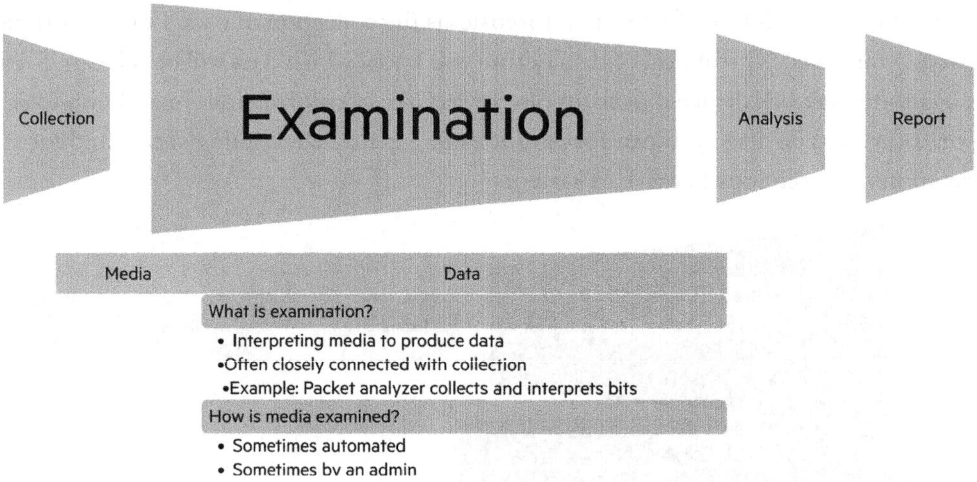

Figure 12-5: Examination

Analysis

At the analysis step, shown in Figure 12-6, the security team analyzes collected data and turns it into useful information for the investigation at hand. Perhaps the security team has found evidence that a security breach occurred sometime between Thursday and Monday. You want to know who connected to the network during that time window and whether any unusual access patterns occurred. CPPM's Access Tracker gives you a list of logged events—the data—but you need to analyze those events to find an answer your question.

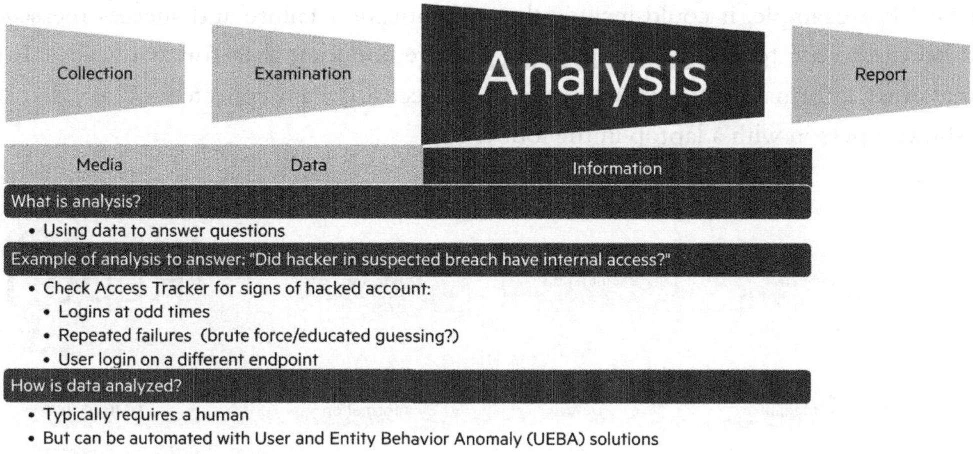

Figure 12-6: Analysis

You obtain a list of all the successful—and unsuccessful—authentications for the time in question, and start to analyze them. That's a lot of information, so you might start by looking for unusual times for someone to connect to the network—late at night for example—and for authentication failures. Perhaps you'll see a few failed authentication attempts for a username and then a success. This could be innocent, or it could be a hacker who was able to guess the employee's password because the employee used a variation of a compromised password. Maybe you'll note that the user logged in on a different endpoint from usual, and at an unusual time. You can note this as potentially suspicious information and follow up with the user to ask whether the user truly connected at that time or whether this could be the hacker's login.

The analysis step typically involves humans. However, a technology called User and Endpoint Behavior Analytics (UEBA) can automate the process. Using machine learning, UEBA can track normal user and device behavior such as when users normally log in, where they log in, and how they use the network. UEBA can then detect anomalies and generate alerts with a score as to how suspicious the behavior is. UEBA is particularly helpful because often you do not know about the security breach, so you do not start analyzing the information. You need a solution that is constantly analyzing the information so that you know when to start looking. You can attend higher level training to learn about UEBA.

Reporting

The final step in the forensics process is reporting, as shown in Figure 12-7. Reporting takes the information discovered in the analysis phase and transforms it into formal evidence in the form of reports. In the analysis phase, the security team probably went through vast amounts of information in attempts to find the relevant information. The final report omits the "hay" and keeps only the "needle." For example, it could include the authentication failure and success messages that the team believes relate to the security incident and the endpoint data for that access. It might also include other information that other parts of the security team collected—like video surveillance that shows a person with a laptop in the lobby.

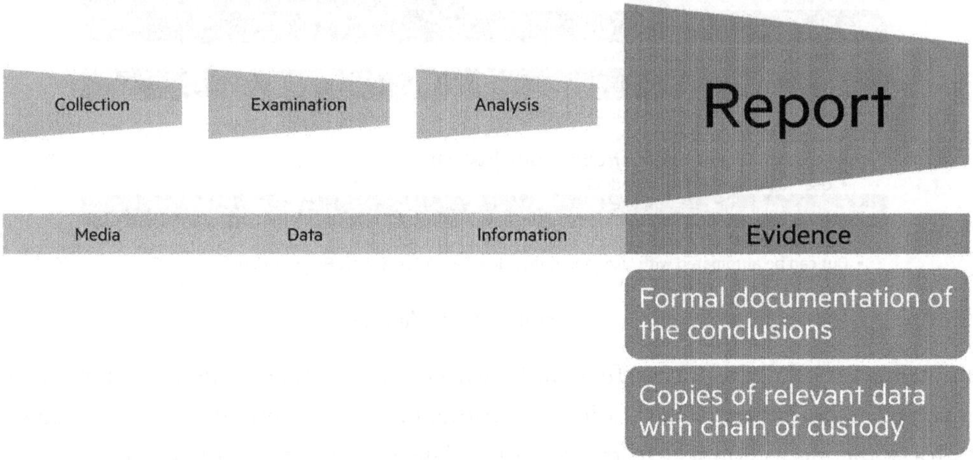

Figure 12-7: Reporting

The report should also include the chain of custody, documenting where and when each piece of included information was collected and who has had access to the information.

Forensic Policies

NIST recommends that organizations define formal policies of how network admins should implement forensics on a normal day-to-day basis and how they should implement forensics when security breaches occur. You should communicate with your company's security team to verify that you understand how to execute your company's policies.

A "day-to-day" policy might include steps such as setting up infrastructure devices to log particular events and archiving logs after a certain time. The logs are considered more trustworthy when they are stored on the device that originally collected them, so the company might have rules about documenting log exports and when logs can be cleared. Because, of course, you do not know when a security breach will occur, you must put monitoring, auditing, and logging in place to run consistently. You should work with the company's IT security managers so that you understand where logs and other monitoring data should be forwarded for analysis. The security team might perform some analysis; however, due to the vast amounts of data in question, most enterprises will require solutions that can analyze data and automatically detect anomalies for the security team to investigate.

When a breach or potential breach is detected, the company should have policies for responding and for implementing the proper forensics. A security breach policy might dictate that admins fill out specific forms documenting the incident and that they contact specific security team members for additional instructions. The company should define how admins document their processes as they investigate the incident. When collecting potential evidence, admins should make a copy of the data to analyze (for example, export logs). At the same time, they should carefully document when they collected the data and how they collected it.

All policies should also dictate proper ways for admins to use forensic tools so as to do their jobs, while also maintaining users' privacy as much as possible. For example, to troubleshoot an issue or investigate a possible attack, an admin might need to collect network traffic with packet captures. However, the network admin should not use those packet captures to start reading users' email. Forensic responses should begin with the least intrusive techniques, such as examining general patterns and traffic headers, before moving to more invasive techniques. Even if the company suspects an insider attack, the security team should be required to make a clear case against the user before taking actions that could invade the user's privacy. The company's legal team might need to formulate policies that explain under what situations IT can examine users' data and network traffic at a deep level and include these policies with other policies that employees sign or that guests agree to when they access the network through a portal.

Analyze Wireless Intrusion Prevention (WIP) Events

You will now turn your focus to a specific type of security investigation: investigating potential threats discovered by the HPE Aruba Networking Wireless Intrusion Prevention (WIP) feature.

Review Wireless Intrusion Prevention (WIP)

Begin with a brief review of this feature. WIP detects rogue APs, wireless DoS attacks, and online cracking attempts. It can also mitigate the threats. As shown in Figure 12-8, WIP includes four major phases.

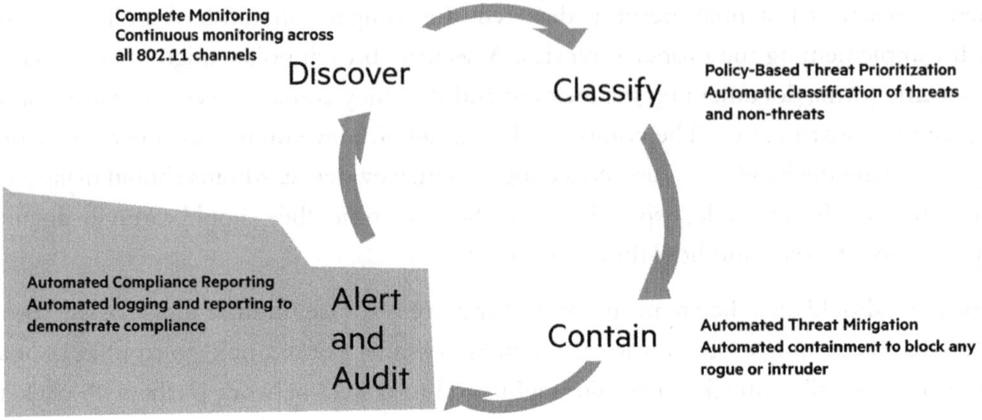

Figure 12-8: Review Wireless Intrusion Prevention (WIP)

In the discovery phase, WIP continuously monitors across all 802.11 channels to learn about wireless devices, wireless activity, and device configurations. Next, WIP automatically classifies discovered devices and what it has learned into various threats and non-threats. This increases your situational awareness, informing you of threat severity and potential risk.

If you are using only the intrusion detection functions, it is up to you to analyze the data, strategize, and take appropriate action. However, if you choose to use the intrusion prevention functionality, the HPE Aruba Networking solution can automatically implement threat containment. Containment actions include blocking anyone from using a rogue AP or denying intruder access.

Finally, alert and auditing functions enable ongoing analysis, reporting, and report distribution. You can use these reports to help the company prove compliance with company security policies and government regulations.

For this study guide you will be focusing on monitoring a solution that has already been configured, analyzing the collected information, and taking the appropriate next steps.

Review the Devices that Discover Threats

Take a moment to review the three modes in which HPE Aruba Networking APs can function to support discovery of wireless threats. (See Figure 12-9 and the sections below.)

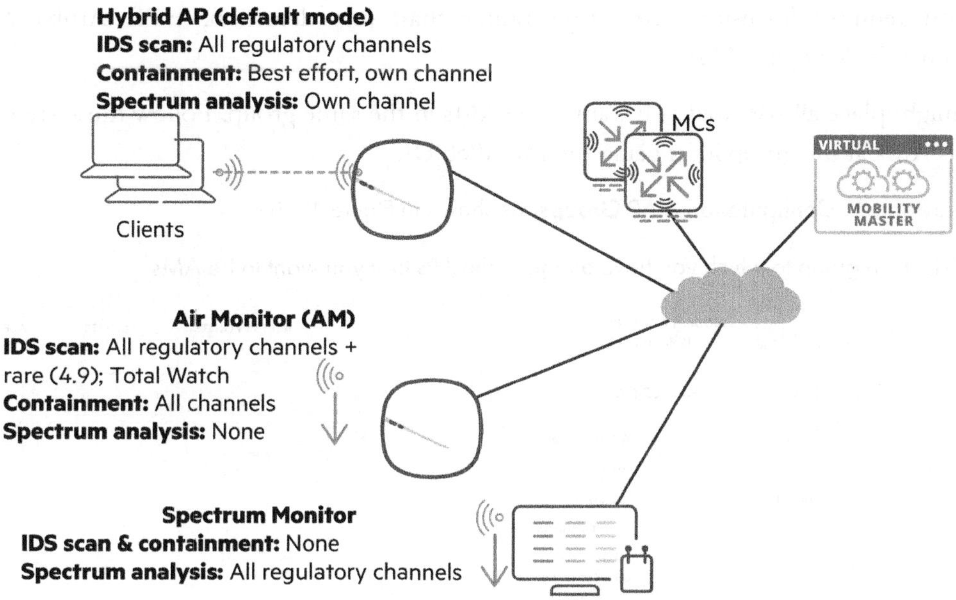

Figure 12-9: Review the Devices that Discover Threats

Hybrid Mode APs

APs are primarily responsible for serving WLAN clients. However, they can operate in hybrid mode and devote some of their time to detecting and mitigating threats as well. Hybrid mode APs go off channel every 10 seconds to scan one channel for about 100 ms. If they are serving clients, the APs back off the scans. Hybrid mode APs can typically sweep the regulatory channels and detect all wireless devices in the area within five to six minutes. Hybrid mode APs are capable of containing wireless devices that WIP has determined pose a risk, but only on their own channel and in a best effort manner. The hybrid mode APs also support spectrum analysis to help maintain RF health.

Hybrid mode operation is the default.

Air Monitors (AMs)

Also called Wireless Sensors, AMs are dedicated to scanning for rogue APs or other IDS events. They scan passively for Layer 2 802.11 frames. As passive devices, they listen for frames, but do not transmit RF and do not service clients. AMs move from channel to channel, listening for threats. They listen on all channels, not just channels that are permitted in the region, including rare ones in 4.9. Because they do not transmit RF, they can do so without violating the law. With the HPE Aruba Networking Total Watch feature, the AMs spend a longer scan time on active channels with detected wireless devices. AMs can contain wireless devices that pose a risk on any channel. If the solution requires intrusion prevention rather than just detection, HPE Aruba Networking recommends deploying AMs.

You might place all APs that you want to be AMs in the same group. Follow these steps to classify APs as AMs at the group level using the MC Web UI:

1. Navigate to **Configuration > AP Groups**, as shown in Figure 12-10.
2. Select the group to which you have assigned the APs that you want to be AMs.

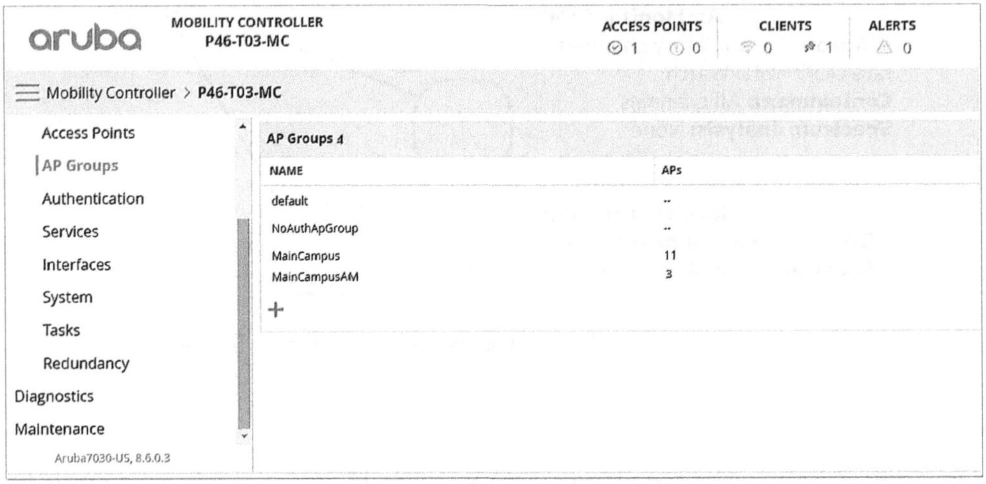

Figure 12-10: AOS UI (Configuration > AP Groups)

3. Click **Radio**, as shown in Figure 12-11.
4. Expand **Basic**.
5. Under both **2.4 GHz** and **5 GHz**, for **Radio mode**, select **am-mode**.
6. Click **Submit**.

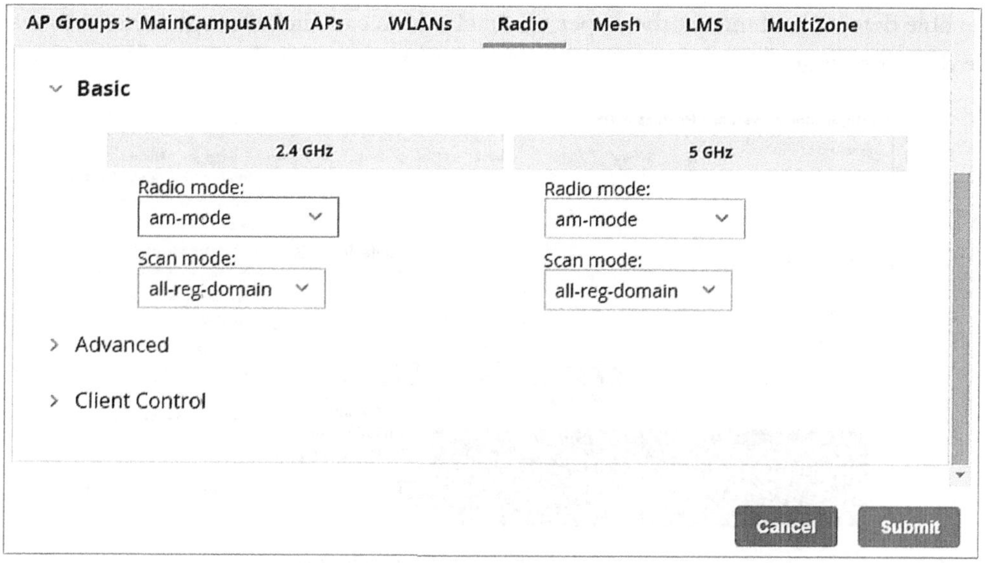

Figure 12-11: AOS UI (Configuration > AP Groups > *group name* > Radio)

7. Deploy the changes.

Spectrum Monitors (SMs)

Like AMs, Spectrum Monitors (SM) do not service clients. However, they examine Layer 1 RF data, rather than Layer 2 like AMs. By sending captured RF data to the Mobility Controller (MC) or Mobility Conductor (MCR) for spectrum analysis, SMs help companies detect RF noise and interference. Such interference could be introduced maliciously but is usually done so accidentally. Examples of devices that cause RF interference include non-802.11 devices like microwave ovens, baby monitors, video cameras, and Bluetooth devices; many other such devices exist.

You would typically deploy AMs to support your security efforts and SMs to monitor and maintain the environment's RF health.

IDS/WIP Profile

You can find the IDS profiles that your solution is using in the **Configuration > System > Profiles** page under **IDS**. In Figure 12-2, you see the default profile, which includes settings for detecting various types of attacks. For example, the IDS Unauthorized Device setting determines for which types of unauthorized devices the solution checks. As you see, by default, the solution is detecting and classifying rogue APs, among other unauthorized devices. If you click the IDS DOS settings, you can see that the default profile detects several types of DOS attacks against clients and that you can enable detection of many other types of attacks. You can click through other settings to explore more default settings.

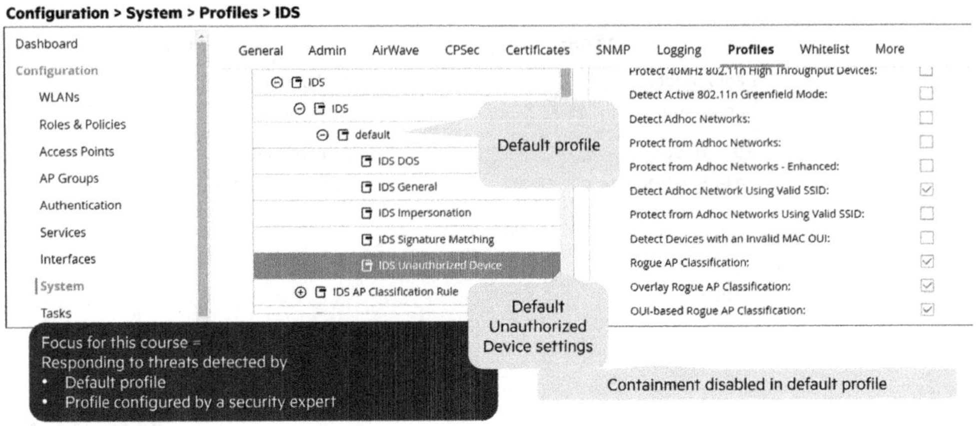

Figure 12-12: IDS/WIP Profile

Containment is *disabled* in the default profile, as automatic containment is a serious action that your company's legal team must approve.

Changing the settings and creating your own IDS profiles is an advanced topic beyond the scope of this study guide. The focus for this study guide is responding to threats detected by the default profile or by a profile configured by a security expert and activated on your solution.

HPE Aruba Networking Mobility Security Dashboard

The first place to look for those detected threats is in the HPE Aruba Networking Mobility Security Dashboard, available either from the MCR or a standalone MC (see Figure 12-13).

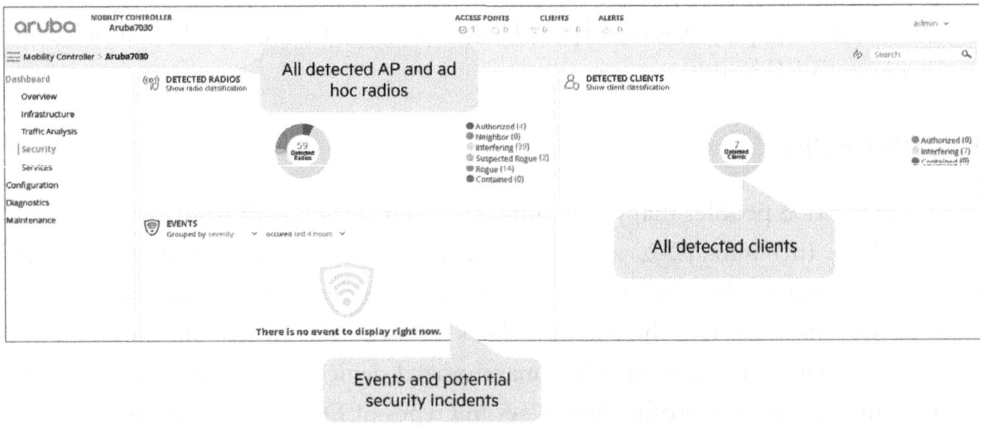

Figure 12-13: HPE Aruba Networking Mobility Security Dashboard

The detected radios section shows the AP radios that AMs or APs have discovered. The detected client section shows the clients, and the events section shows recent events, grouped by severity.

Interpreting AP Classifications

As you see in Figure 12-14, WIP classifies all detected AP radios. To interpret what you see, you need to understand what these classifications mean.

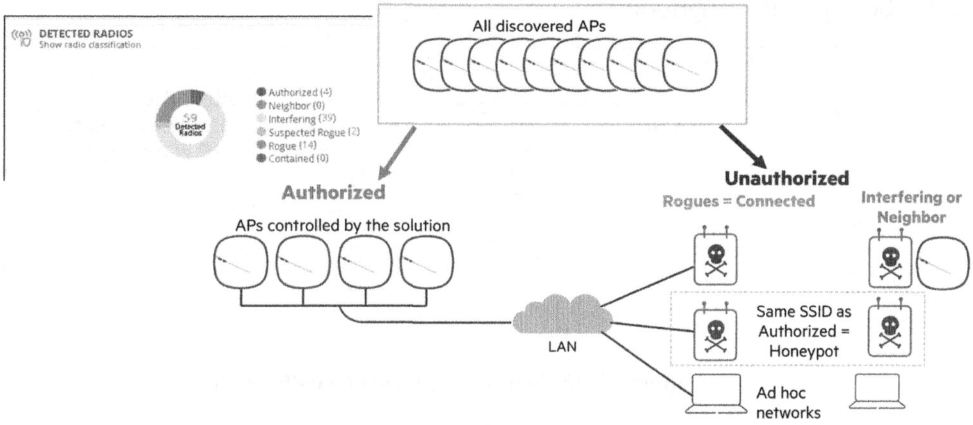

Figure 12-14: Interpreting AP Classifications

WIP first divides APs into Authorized APs and Unauthorized APs. Authorized APs are APs controlled by the solution. All other APs are unauthorized.

WIP further classifies unauthorized APs based on whether they connect to the corporate LAN. Unauthorized APs that are connected to the LAN are called Rogue APs. Unconnected, unauthorized APs are called either interfering APs or neighbor APs. Interfering APs are APs that are deployed on the company property, but are not connected to the LAN. Neighbor APs are legitimate APs that belong to neighboring organizations. When a corporation shares a building or even is just close to other buildings, some neighbor APs will be detected. By default, WIP classifies the unconnected APs as interfering. Admins can create rules, though, that define specific SSIDs as neighbor SSIDs. You can also manually re-classify an interfering AP as a neighbor.

If a rogue AP or interfering AP have the same SSID as authorized APs, WIP also calls them honeypot APs.

WIP also supports detecting ad hoc networks, which will be listed among detected radios. The 802.11 standard defines ad hoc networks as a way for stations to connect together as peers without the need for an AP. The client that hosts the ad hoc network could be connected to the LAN, which means that it is rogue, or not, which means that it is interfering or possibly neighbor.

Interpreting Client Classifications

You can also use the Security Dashboard to see a list of detected clients, as shown in Figure 12-15. WIP places clients in two categories: 1. authorized, also called valid, and 2. interfering. Authorized clients are clients that have authenticated to an authorized AP and passed encrypted traffic. Depending on how the solution is configured, the valid client status can age out or be permanent. All other clients are classified interfering. These clients might be unauthorized clients at the site, but could also belong to the neighbors.

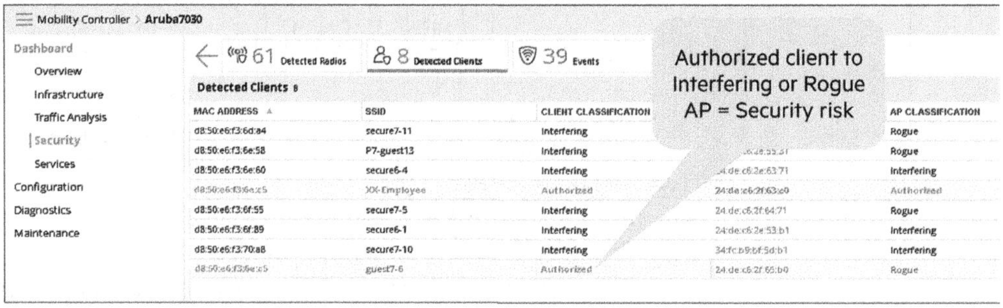

Figure 12-15: Interpreting Client Classifications

With clients, more than the client status itself, you are examining the client association. You want to look out for authorized clients connecting to rogue or interfering APs. This misassociation could pose a security risk for the client.

Understanding the Risks Posed by Unauthorized Devices

It is important that you understand the risks posed to the environment by the unauthorized wireless devices that WIP has detected.

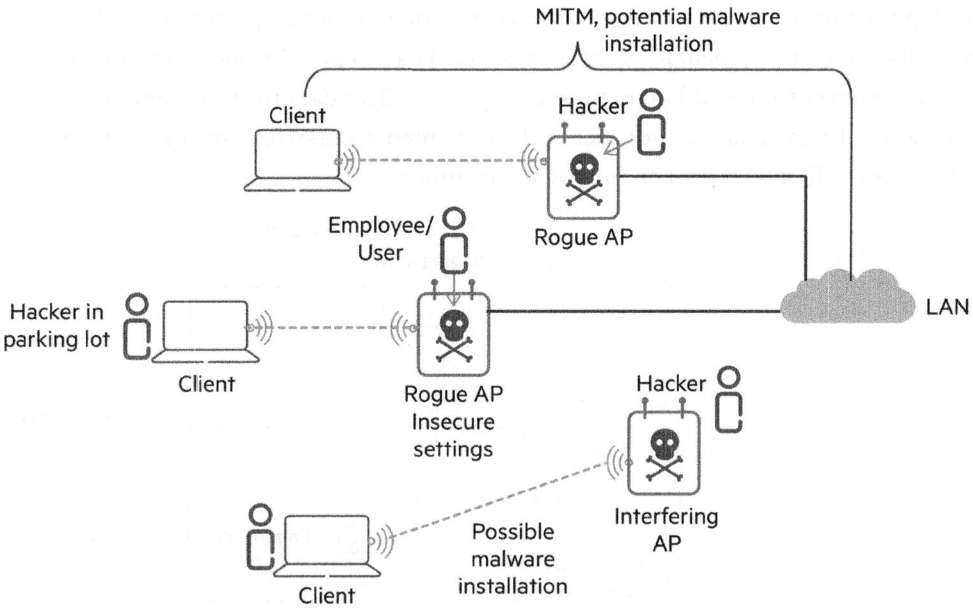

Figure 12-16: Understanding the Risks Posed by Unauthorized Devices

As shown in Figure 12-16, rogue APs can be introduced by hackers, in which case the hackers are usually attempting some form of an MITM attack. While the client might encrypt some traffic at the application layer—for example, if the client is communicating with a server over HTTPS—the hacker could try to attack the TLS tunnel. While TLS is largely secure, some ways that servers implement TLS are not. The hacker could also try to connect to the connected client and install malware on it.

Similarly a hacker who has introduced an interfering AP to which clients connect can try to install malware on the clients.

For both rogue and interfering APs, hackers can set the SSID the same as a prominent hotspot to try to get clients to connect to them automatically. Alternatively, they can set the SSID to the authorized SSID to try to escape detection or lure clients to connect.

Authorized users can also connect rogue APs to the network in an attempt to get wireless access in areas that are more convenient to them. Students are notorious for connecting rogue APs, but corporate users might do the same if the company has not given them wireless access where they want it. Even though the users do not mean harm, the rogue AP can create a huge security hole. The users might have the rogue AP advertise an open network or at best a WPA2-Personal network. Neither form of security is appropriate for enterprise environments, and hackers could use this weak spot to gain access from anywhere the AP's signal leaks. In effect, the users have installed an open port in the lobby or parking lot.

Other types of unauthorized wireless devices include computers participating in ad hoc networks. If an ad hoc network is open, it exposes sensitive data to eavesdroppers, as shown in Figure 12-17. If hackers connect to the ad hoc network, they can collect data from the devices and inject viruses and malware. Even if an ad hoc network implements password security and encryption, such security is not sufficient for an enterprise environment.

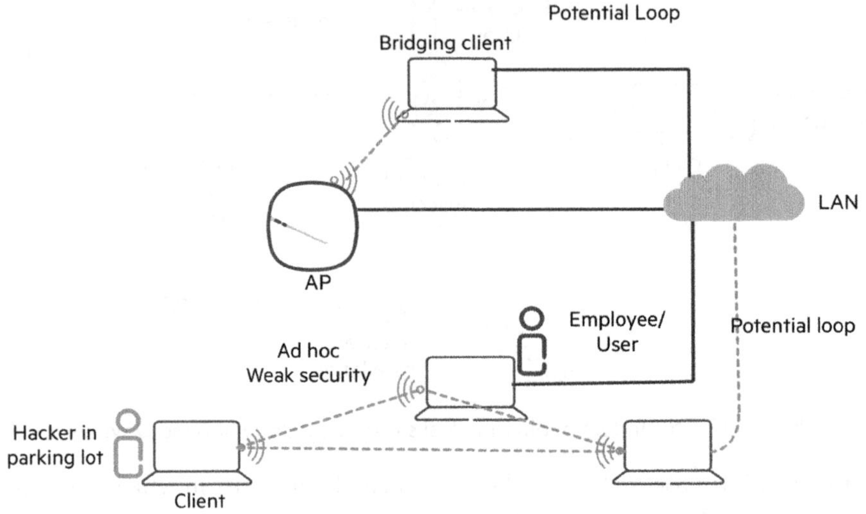

Figure 12-17: Understanding the Risks Posed by More Unauthorized Devices

As shown in Figure 12-17, if a computer in an ad hoc network connects to the corporate LAN and bridges between the wireless and wired network, like a rogue AP, it opens a backdoor into the network.

And if multiple bridging computers are participating in the ad hoc network and connected to the office, they could create a loop that causes a broadcast storm—essentially a DoS attack.

Even without an ad hoc network, a client in bridge mode might cause a loop when it is connected to both an AP and the wired network. WIP can detect bridging clients to help you find these issues.

Reviewing Some Ways to Mitigate Threats

Before you look at determining how to deal with detected alerts and WIP events, briefly review what containing an AP means. When WIP contains an AP, it prevents all clients from connecting to it. When WIP contains a client, it targets the client's association and prevents it from connecting to an AP.

Instead of completely containing a rogue AP or ad hoc network, you can also choose to protect valid clients. This feature prevents any valid clients from connecting to a non-authorized AP.

However, the APs or AMs allow non-valid clients to connect to the rogue APs. This approach can offer a good option for companies that want to protect their own users and devices from MITM attacks but want to avoid potentially interfering with neighbors.

HPE Aruba Networking WIP provides three containment methods.

As shown in Figure 12-18, wired AP containment poisons the AP's ARP table so that the AP's traffic does not reach the corporate default gateway. Some companies prefer to use this form of containment because it can only affect devices that are truly connected to your LAN; wired containment cannot accidentally contain a legitimate neighbor's AP.

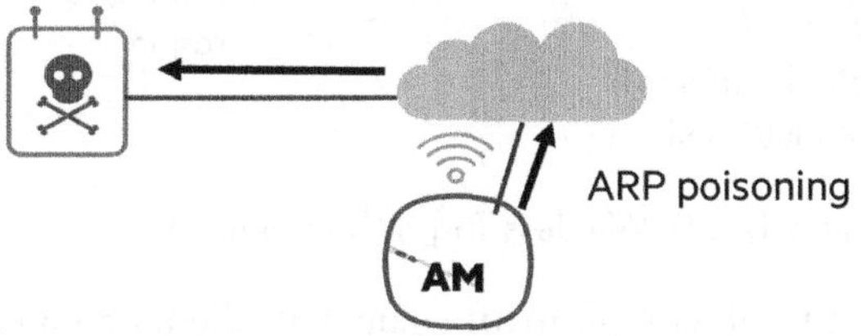

Figure 12-18: Wired Containment

When using the deauthentication method, APs or, more often, AMs send out deauthentication frames to disconnect clients from the contained AP (see Figure 12-19). The drawback of this method is that the AP or AM must continue to send the frames as clients usually continually try to reconnect to the rogue AP.

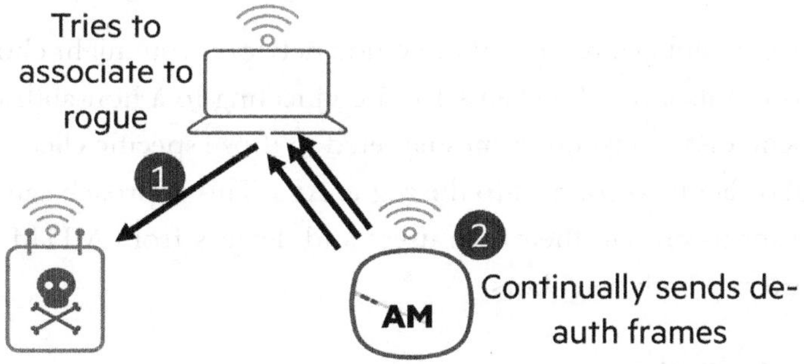

Figure 12-19: Wireless Deauthentication Containment

Tarpit containment, available with the RF Protect license, can be more effective than de-authentication containment. As shown in Figure 12-20, in addition to sending deauthentication messages, the AM spoofs probe responses and association responses on a different channel or BSSID from the rogue AP. Clients then connect to the AM instead. Once a client is associated with the AM, it stops trying to reconnect to the rogue AP. It is in a "tarpit." When all clients that try to connect to the rogue AP are in the tarpit, the AM no longer needs to send out frames, so this method imposes lower bandwidth requirements. The AM also does not need to spend as much time on the channel, so it can continue to scan for threats more effectively.

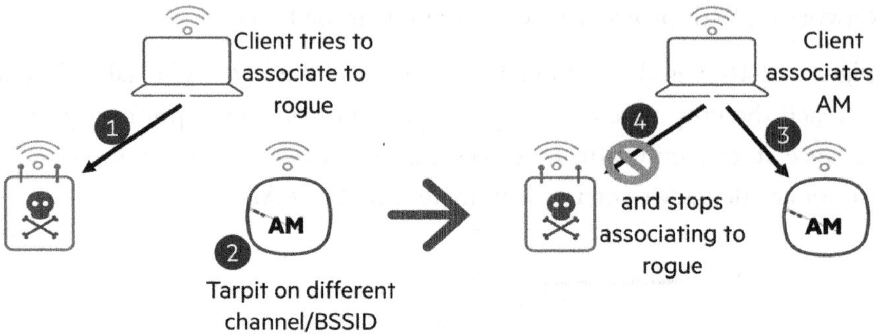

Figure 12-20: Wireless Tarpit Containment

Admins can choose to implement tarpit containment against all clients connected to a rogue AP or just non-valid clients. Since valid clients might accidentally connect to a rogue AP, the company might not want to punish them. The latter approach protects valid clients—the company's own clients—from being stuck in the tarpit and lets them connect to an authorized AP instead.

Wired AP containment works against rogue APs that are wireless routers. It poisons the AP's ARP table so that the AP's traffic does not reach the corporate default gateway. Some companies prefer to use this form of containment because it can only affect devices that are truly connected to your LAN; wired containment cannot accidentally contain a legitimate neighbor's AP.

Instead of completely containing a rogue AP or ad hoc network, you might choose to protect valid clients. This feature prevents any valid clients from connecting to a non-authorized AP by having hybrid APs or AMs send disassociation frames targeted to those specific clients. However, the APs or AMs allow non-valid clients to connect to the rogue APs. This approach can offer a good option for companies that want to protect their own users and devices from MITM attacks but want to avoid potentially interfering with neighbors.

Potential Follow Up Actions

To protect your network against these risks, you will need to take some follow up actions to investigate the potential threat, as well as document the threat and your actions. You might determine that an unauthorized device is not a threat. Maybe it belongs to the neighbors. You would then reclassify the device. Or you might determine that the device *is* a threat. In this case, you might want to contain the threat using the WIP capabilities, or deal with the threat yourself. You might also reclassify the AP if it has not been placed in the correct category. If you determine that some devices pose threats to your network, whether or not WIP is already containing them, you should attempt to find the device and eliminate it.

Understanding Unauthorized Device Policies

As you take the follow up actions, you should not be operating blindly. Instead you should follow your company's forensic policies around security incidents. If your company does not yet have such policies, you should work with the appropriate decision makers to define them.

An example policy, which might meet the needs of some companies, is shown in Figure 12-21.

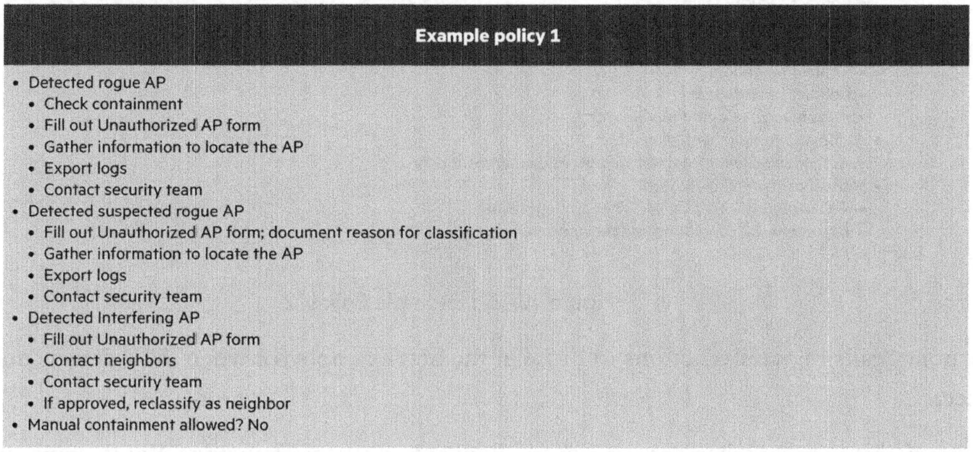

Figure 12-21: Example Policy 1

This company has chosen to implement automatic containment measures for rogue APs, and, as you see, the policy for when you detect a rogue AP tells you to check whether it has been contained. Containing a rogue AP can be a great first measure, but the ultimate goal is to remove the rogue AP from the environment, as well as investigate the incident. So this policy indicates that you should fill out an Unauthorized AP form, documenting the incident. It further directs you to gather information to locate the AP, export logs, and contact the security team.

If you see that WIP has detected a suspected rogue, in addition to completing similar steps as for a detected rogue, you might need to collect more information about where the AP is and why WIP defined it as a suspected rogue.

The policy might further provide guidelines for what you should do if you see that WIP has detected Interfering APs. You might need to fill out a similar form and contact neighbors to determine whether you should classify the AP as a neighbor.

Note that you can manually enable containment against an AP in the security dashboard, but your company might not want someone at your level to make this decision. You should discuss with your managers whether you are allowed to implement manual containment or not.

Another example policy is shown in Figure 12-22.

Example policy 2

- Detected rogue or suspected rogue AP
  - Fill out Unauthorized AP form; document reasons for classification
  - Gather information to locate the AP
  - Export logs
  - Contact security team
- Detected Interfering AP
  - Fill out Unauthorized AP form
  - Contact neighbors
  - Contact security team
  - If approved, reclassify as neighbor
- Automatic containment allowed?
  - Authorized client protection and wired containment only
- Manual containment allowed?
  - IF (conditions = XYZ) + Security team approval
  - Document the conditions and the approval

Figure 12-22: Example Policy 2

This policy calls for similar actions to those in the first example for when WIP detects unauthorized devices.

However, this company has chosen not to allow automatic containment. Some companies choose not to activate containment automatically because they do not want to take the risk of attacking a legitimate neighbor's wireless services, which can even be illegal in some regions (see the warning in Figure 12-23). While WIP takes steps to distinguish between rogue APs and interfering or neighbor APs, some companies simply do not want to assume the risks of automatic containment.

> Warning: The Federal Communications Commission ("FCC") and some third parties have alleged that, under certain circumstances, use of containment functionality violates 47 U.S.C. Section 333 and/or other FCC rules, regulations or policies. Before using any containment functionality, you should determine whether your intended use is allowed under the applicable rules, regulations and policies. Aruba shall not be liable for any claims, sanctions, or other direct, indirect, special, consequential or incidental damages related to your use of containment functionality.

Figure 12-23: Wireless Containment Warning

In this example, the company does permit manual containment. However, the company has specific policies that the security and legal teams have approved as for when you can enable containment manually. You might need to contact all neighboring businesses and verify what SSIDs they are using. You might need to document how WIP defined the AP as rogue and only take action based on specific sources of information. Or perhaps the AP must have a certain SNR. Understand what your company policies are and be ready to document how you fulfilled the policies before you start taking actions.

Viewing More Details About Interfering APs and Reclassifying Them

The next part of this chapter will give you some tips in how you can go about completing whatever tasks your policies require.

Assume that you are at the list of detected radios (see Figure 12-24).

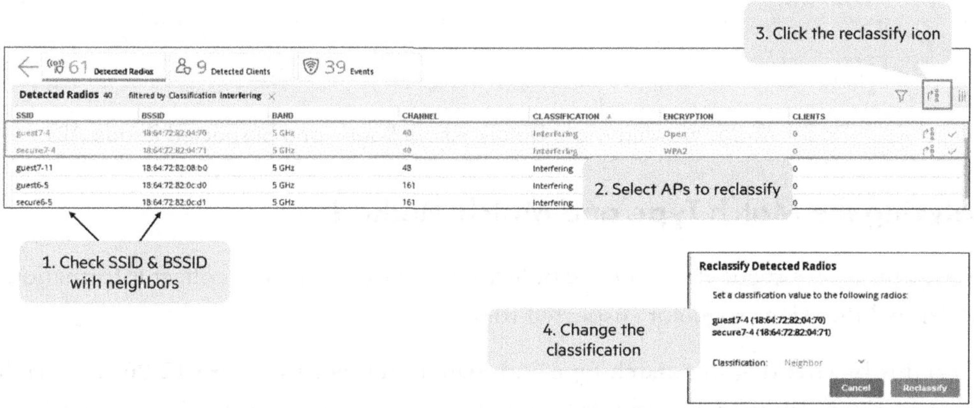

Figure 12-24: Viewing More Details About Interfering APs and Reclassifying Them

You can click on an AP type and see a list of the detected AP radios of that type. Here you see a list of interfering APs. If, after taking the steps dictated by your company's forensic policies, you determine that any of these interfering APs belong to neighbors, you can reclassify them as shown here. When you classify an AP as a neighbor, its BSSID becomes known, and WIP keeps its classification as neighbor. Remember that BSSIDs are specific to the AP, so, if the neighbor adds a new AP with the same SSID, you will once again need to classify that AP as a neighbor.

Any interfering APs that you cannot reclassify as neighbors represent a security risk. You will need to document these APs and make plans to contain and/or find them, as dictated by your company's policies.

Viewing More Details About Rogue and Suspected Rogue APs

You can also filter the list of detected radios to see rogue and suspected rogue APs. As you see in Figure 12-25, WIP provides you with a great deal of information about the APs, which you can collect for the documentation of the security incident.

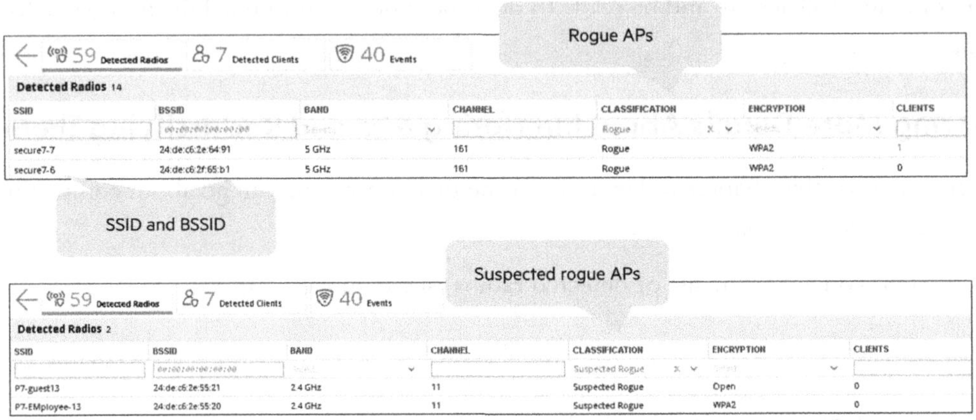

Figure 12-25: Viewing More Details About Rogue and Suspected Rogue APs

Analyzing the Match Type and Match Method

Your organization's rogue AP response policies might call on you to collect information about why WIP defined the AP as rogue or suspected rogue.

You do this by checking the match type and match method. In Figure 12-26, you see how to add these two columns to the table. In this example, you see several SSIDs with the match type as Base-BSSID-Override. WIP actually lists every BSSID on a radio separately. The Base-BSSID-Override match type means that WIP detected an unauthorized AP that has multiple BSSIDs because it is advertising multiple SSIDs. Once WIP decided that one of the BSSIDs is rogue, it automatically classified all the other BSSIDs on the AP as rogue. In this example, you can explain to the security team that you are looking for one AP, not two.

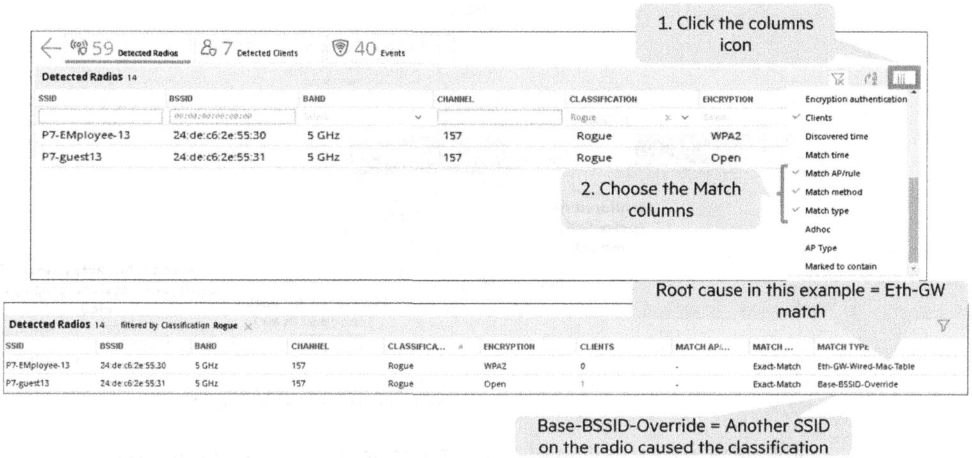

Figure 12-26: Analyzing the Match Type and Match Method

If you want to discover why the AP was classified as rogue, you must look for the BSSID with the matching initial characters that has a different match type listed. In the example shown in Figure 12-26, that match type is ETH-GW-Wired-Mac-Table.

Understanding an ETH-GW-Wired-Mac-Table Match

To understand what a match type like ETH-GW-Wired-Mac-Table means, you need to understand a bit about how WIP determines whether an unauthorized AP is connected to the LAN.

An AP, such as a soft AP, can act as a bridge, as shown in Figure 12-27. A bridging AP receives traffic from the wired LAN and transmits that traffic in the air with the source MAC address intact. If the AP is connected to your LAN, the AP's wireless traffic will have source MAC addresses that match MAC addresses of devices in your LAN.

Therefore, WIP (via hybrid APs and AMs) monitors the source MAC addresses in frames transmitted by unauthorized APs. It creates a table for each monitored AP with a list of these addresses. (The MAC addresses in this table are officially called "AP wired MACs;" these are source MAC addresses that were detected *in the air* but presumably bridged from the wired network.) WIP also maintains a table with MAC addresses that it knows are used in the company LAN. As discussed in more detail in a moment, Hybrid APs or AMs can learn these MACs in many ways, including snooping them, learning them from an MC, or being configured with them.

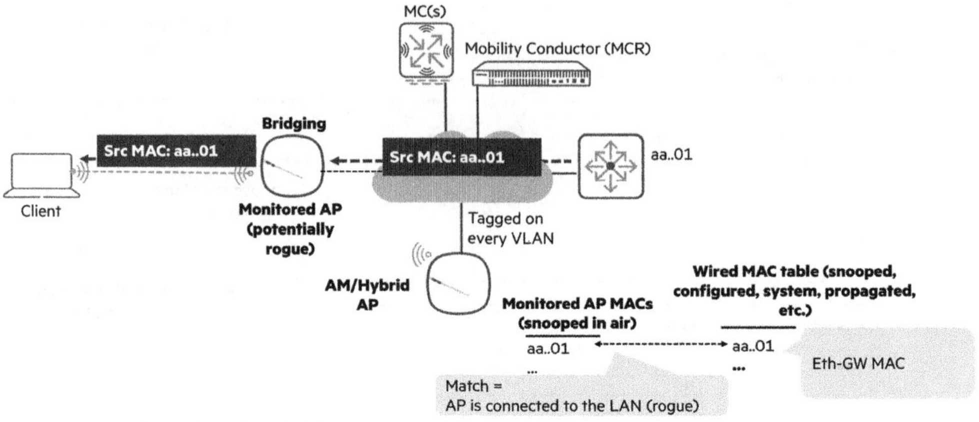

Figure 12-27: Understanding an ETH-GW-Wired-Mac-Table Match

WIP then checks the monitored AP MACs against a list of known wired MACs. If it finds a match, it knows the monitored AP is connected to the LAN. This type of match is called a Wired-Mac-Table match.

In the example shown in Figure 12-27, the monitored AP is transmitting frames with source MAC address aa…01 (MAC address abridged for simplicity). WIP knows that this is the MAC address of the LAN's default router, so it classifies this AP as a rogue. When the wired MAC address is the address of the default gateway, the match type is more specifically a ETH-GW-Wired-Mac-Table match.

This match type is common for bridging APs because most traffic destined to wireless clients is routed from elsewhere through the default gateway.

As a final topic, you should understand a bit more about how WIP builds the Wired MAC table. APs and AMs track the MAC addresses of the default gateways on their VLANs. Of course, an AP or AM knows the MAC address for its own gateway. If the device has multiple VLANs on its port, though, it can also learn the MAC addresses for the gateway on other VLANs. You should typically tag AMs' links for every VLAN used so that WIP can detect rogue APs connected to any of the company's VLANs.

Hybrid APs and AMs also snoop Ethernet traffic such as ARP requests so that they can build a list of MAC addresses in the corporate LAN for non-gateway devices. One of these wired MAC addresses could show up as the source in transmitted wireless traffic if a wireless client connected to the monitored AP is communicating with a wired device at Layer 2.

Understanding Other Eth-Wired-MAC Matches

You should also understand how WIP detects rogue APs that route, rather than bridge, traffic. Users installing their own AP are likely to simply grab a wireless router from their homes or a big box store, so some unauthorized APs act as wireless routers rather than bridges. When an AP is a wireless router, the source MAC address for transmitted wireless traffic does *not* come from the wired network. Instead, it is the MAC address that the AP advertises as part of its default gateway functions. As shown in Figure 12-28, this could be equivalent to the BSSID or a base wireless MAC from which the BSSIDs derive. To determine if this type of rogue AP is connected to the corporate LAN, WIP tries to determine if any of the MAC addresses in the corporate LAN seem to belong to the unauthorized AP.

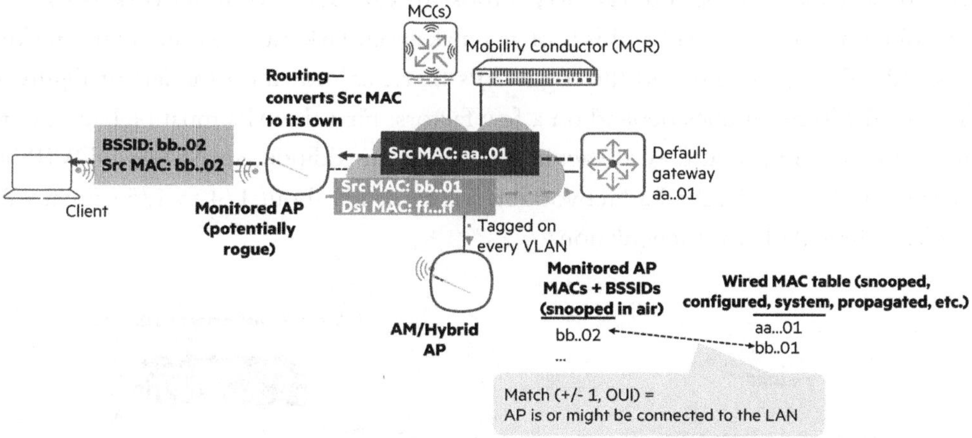

Figure 12-28: Understanding Other Eth-Wired-MAC Matches

WIP can discover wired MAC addresses in many ways. In addition to the Eth-GW MAC address, APs and AMs create lists of MAC addresses in traffic that they snoop on their Ethernet ports (ETH-MAC), as shown in Figure 12-28. They can also receive lists of MAC addresses snooped by the MC or propagated by the MC from other APs and AMs. Admins can also configure MAC addresses manually.

WIP takes advantage of the fact that many wireless routers assign sequential MAC addresses to their wireless and wired NICs and that BSSIDs derive from MAC addresses. Depending on various factors, including whether the AP BSSID or wired MAC matches, what type of wired MAC address it matches, and whether the match is exact or not, WIP classifies the AP as a rogue or suspected rogue with a different confidence level.

Other Match Types

You might also see a match type that indicates that a different method, rather than internal heuristics, made the identification. This could be AP-Rule, in which case WIP made the match based on rules that look at SSID, SNR, and number of detecting radios.

Note that some of your customers might have legacy AirWave still deployed in their environments. If AirWave makes a match and informs WIP, the match type is EMS.

Locating Detected Devices

If you and the other members of the security team have determined that some of the detected devices are potential threats, you want to get those threats out of your network as soon as possible. If your solution is using VisualRF this tool will make your task much easier. You can simply look at the VisualRF floor plan and find the rogue APs there, as shown on the left of Figure 12-29. The accuracy of the location does depend on a few factors. First the APs must be laid out on floorplan in the correct locations that they are actually installed. In addition, an AM-based WIP design needs to use enough AMs. HPE Aruba Networking recommends about 1/4 to 1/5 as many AMs as APs with at least three AMs for triangulation.

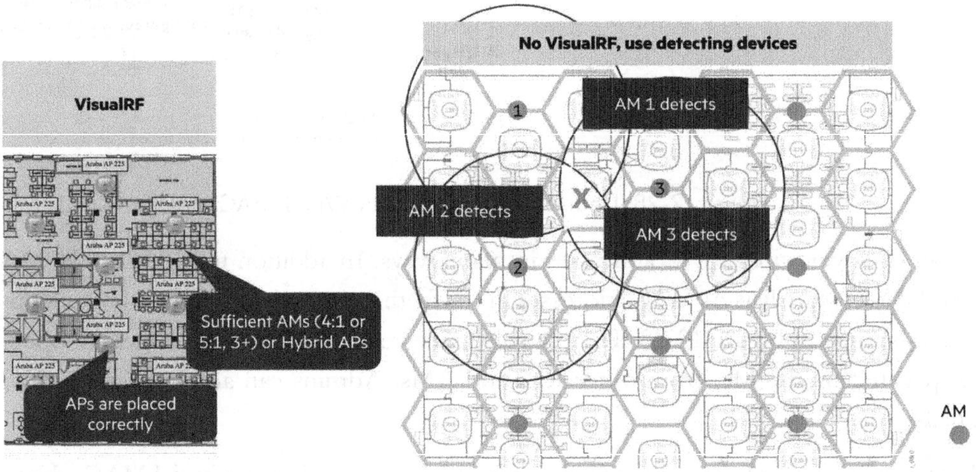

Figure 12-29: Locating Detected Devices

If the solution is not using VisualRF, you will need to hunt down the unauthorized devices yourself. As shown on the right of Figure 12-29, you can find which APs or AMs detected the unauthorized device in the logs for the events related to the device discovery. You can then use the identity of the detecting devices to triangulate the unauthorized device's location. In this example, the green circles indicate AMs. You see that AM 1, AM 2, and AM 3 detected a rogue AP. You should look in the area between those AMs as the likely location for that rogue device.

Viewing Events and Logs

You can view events in the Security Dashboard to find the name of the detector that detected an unauthorized device like a rogue AP (see Figure 12-30).

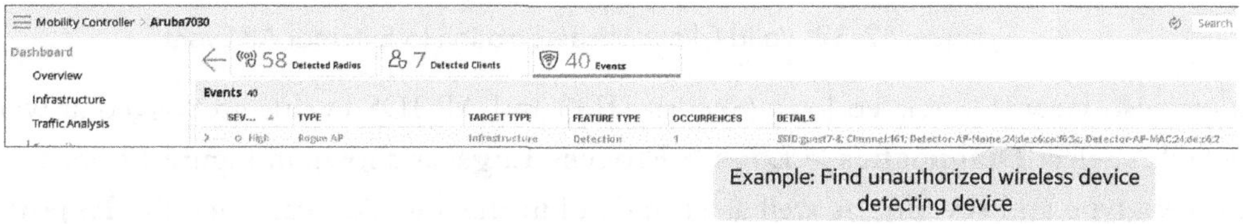

Figure 12-30: Viewing WIP Events

The Events view also lets you search for a wide array of threats. The dashboard categorizes events by the severity and their type. You can review details about the event. If the same event with the same details recurs, WIP increments the occurrence number rather than logging a new event.

Another example of an event is shown in Figure 12-31; WIP detected a Phony BSSID, which means that an unauthorized AP is using an SSID similar to authorized APs.

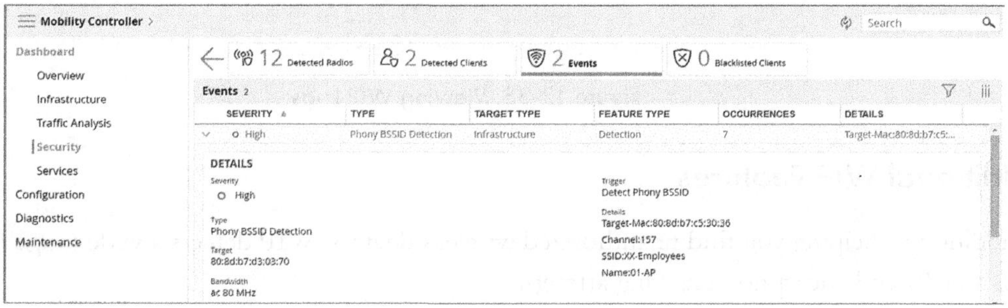

Figure 12-31: Phony BSSID Event

A third example is shown in Figure 12-32: Valid Client Misassociation to External AP. In other words, an authorized client associated to an interfering or rogue AP.

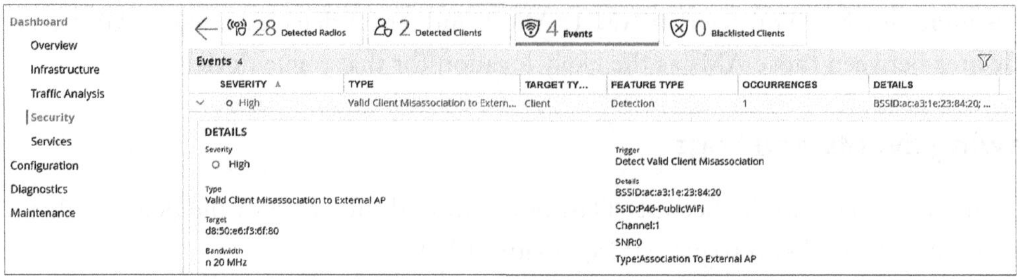

Figure 12-32: Valid Client Misassociation to External AP Event

You can also view IDS related logs. (Security IDS and AP-IDS events are logged at warning level by default.) Select **Diagnostics > Logs > Process Logs**, as shown in Figure 12-33. You can filter by process type and severity, as well as string within the log. You can click the **Export** button to save out the logs for analysis and archival outside the MC. Here you see how you have saved the logs with an explanation and data in the name.

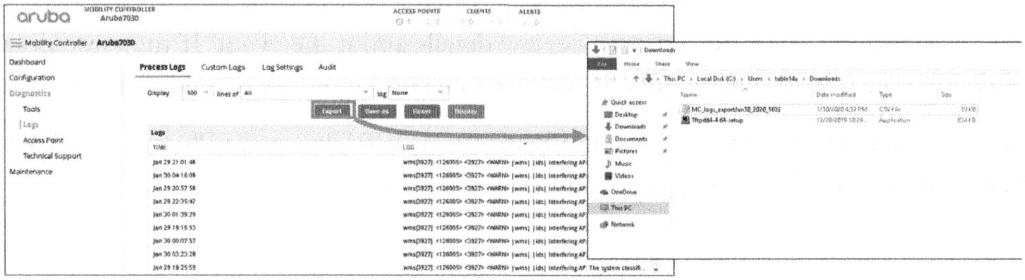

Figure 12-33: Viewing WIP Logs

Additional WIP Features

In addition to helping you find unauthorized wireless devices, WIP detects a wide range of wireless DoS attacks and encryption cracking attempts.

It is particularly important that you monitor for these events as WIP cannot contain them. Sending dissociation frames to a device that itself is sending such frames does not stop the device. Instead you and the security team must find the offender and stop the attack. Even though measures such as Protected Management Frames helps to prevent your clients from falling prey to such attacks, you should still deal with the offender, who might be causing other trouble as well.

Work with your company to obtain a policy for responding to such threats. (An example policy is shown in Figure 12-34.) You might need to fill out documentation and hunt for more information, documenting that information as you go. For example, the events can show you the MAC address for the offending device launching the DoS or other attack. You might be directed to gather more information. For example, while keeping in mind that the person launching the attack could be spoofing the device MAC address, you can check user table and Access Tracker for information about the potential offending device and what user has authenticated on that device. You might also need to notify the security team. And remember to locate and remove the offending device as possible.

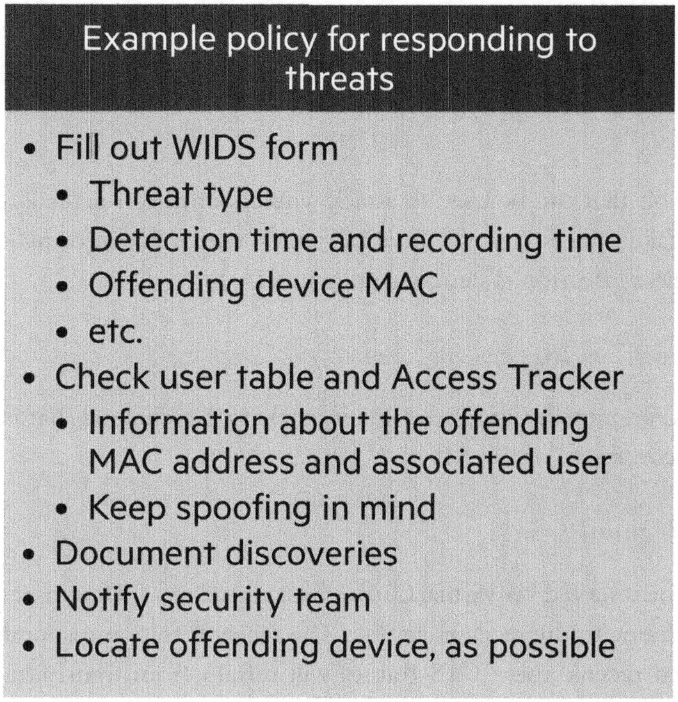

Figure 12-34: Example Policy for Threat Responses

Supplemental Information on IDS Events: Attacks on the Infrastructure

For your reference, types of attacks that WIP can detect are listed here in this section. Be aware, however, that non-malicious client and neighboring AP activity in busy environments might trigger some of these events. You should work with the IT security staff in your organization to track patterns and determine which events indicate true security issues in your environment. You can also take the *Implementing HPE Aruba Networking Security* course to learn more recommendations about which events to monitor (or read the *HPE Aruba Networking Certified Network Security Professional* study guide.)

For more information, read the sections below.

AP Flood Attack

Fake AP is a tool that was originally created to thwart wardrivers by flooding beacon frames containing hundreds of different addresses. This would appear to a wardriver as though there were hundreds of APs in the area, thus concealing the real AP. An attacker can use this tool to flood an enterprise or public hotspots with fake AP beacons to confuse legitimate users and to increase the amount of processing need on client operating systems.

Client Flood Attack

There are fake AP tools that can be used to attack wireless intrusion detection itself by generating a large number of fake clients that fill internal tables with fake information. If successful, it overwhelms the wireless intrusion system, resulting in a DoS.

Beacon Frame Spoofing Attack

In this type of attack, an intruder spoofs a beacon packet on a channel that is different from that advertised in the beacon frame of the AP.

RTS or CTS Rate Anomaly

The RF medium can be reserved via Virtual Carrier Sensing using a CTS transaction. The transmitter station sends a RTS frame to the receiver station. The receiver station responds with a CTS frame. All other stations that receive these CTS frames will refrain from transmitting over the wireless medium for an amount of time specified in the duration fields of these frames.

Attackers can exploit the Virtual Carrier Sensing mechanism to launch a DoS attack on the WLAN by transmitting numerous RTS and/or CTS frames. This causes other stations in the WLAN to defer transmission to the wireless medium. The attacker can essentially block the authorized stations in the WLAN with this attack.

Overflow EAPOL Key

Some wireless drivers used in access points do not correctly validate the EAPOL key fields. A malicious EAPOLKey packet with an invalid advertised length can trigger a DoS or possible code execution, achieved after a successful 802.11 association exchange.

Overflow IE Tags

Some wireless drivers used in access points do not correctly parse the vendor-specific IE tags. A malicious association request sent to the AP containing an IE with an inappropriate length (too long) can cause a DoS and potentially lead to code execution. The association request must be sent after a successful 802.11 authentication exchange.

Malformed Frame-Assoc Request

Some wireless drivers used in access points do not correctly parse the SSID information element tag contained in association request frames. A malicious association request with a null SSID (that is, zero length SSID) can trigger a DoS or potential code execution condition on the targeted device.

Malformed Frame-Auth

Malformed 802.11 authentication frames that do not conform to the specification can expose vulnerabilities in some drivers that have not implemented proper error checking. This feature checks for unexpected values in an Authentication frame.

Malformed Frame-HT IE

The IEEE 802.11n HT IE is used to convey information about the 802.11n network. An 802.11 management frame containing a malformed HT IE can crash some client implementations, potentially representing an exploitable condition when transmitted by a malicious attacker.

Malformed Frame-Large Duration

The virtual carrier-sense attack is implemented by modifying the 802.11 MAC layer implementation to allow random duration values to be sent periodically. This attack can be carried out on the ACK, data, RTS, and CTS frame types by using large duration values. This attack can prevent channel access to legitimate users.

AP Spoofing

An AP Spoofing attack involves an intruder sending forged frames that are made to look like they are from a legitimate AP. It is trivial for an attacker to do this, since tools are readily available to inject wireless frames with any MAC address that the user desires. Spoofing frames from a legitimate AP is the foundation of many wireless attacks.

Valid SSID Misuse

If an unauthorized AP (neighbor or interfering) is using the same SSID as an authorized network, a valid client may be tricked into connecting to the wrong network. If a client connects to a malicious network, security breaches or attacks can occur.

Wellenreiter

Wellenreiter is a passive wireless network discovery tool used to compile a list of APs along with their MAC address, SSID, channel, and security setting in the vicinity. It passively sniffs wireless traffic, and with certain versions (versions 1.4, 1.5, and 1.6), sends active probes that target known default SSIDs.

Invalid MAC OUI

The first three bytes of a MAC address, known as the MAC OUI, is assigned by the IEEE to known manufacturers. Often, clients using a spoofed MAC address do not use a valid OUI and instead use a randomly generated MAC address.

Invalid Address Combination

In this attack, an intruder can cause an AP to transmit deauthentication and disassociation frames to all of its clients. Triggers that can cause this condition include the use of broadcast or multicast MAC address in the source address field.

Bad WEP Initialization

This is the detection of WEP initialization vectors that are known to be weak. A primary means of cracking WEP keys is to capture 802.11 frames over an extended period of time and searching for such weak implementations that are still used by many legacy devices.

NetStumbler

NetStumbler is a popular wardriving application used to locate 802.11 networks. When used with certain NICs, NetStumbler generates a characteristic frame that can be detected. Version 3.3.0 of NetStumbler changed the characteristic frame slightly.

Wellenreiter

Wellenreiter is a passive wireless network discovery tool used to compile a list of APs along with their MAC address, SSID, channel, and security setting in the vicinity. It passively sniffs wireless traffic, and with certain versions (versions 1.4, 1.5, and 1.6), sends active probes that target known default SSIDs.

802.11n 40 MHz Intolerance Setting

When a client or AP sets the HT capability intolerant bit to indicate that it is unable to participate in a 40 MHz BSS, the AP must use lower data rates with all of its clients. Network administrators often want to know if there are devices that are advertising 40 MHz intolerance, as this can impact the performance of the network.

Active 802.11n Greenfield Mode

When 802.11 devices use the HT operating mode, they cannot share the same channel as 802.11a/b/g stations. Not only can they not communicate with legacy devices, but also the way the greenfield mode devices use the transmission medium is different. This can cause collisions, errors, and retransmissions. Again, admins want to know about the device because, even if not introduced maliciously, it can cause harm.

Supplemental Information on IDS Events: Attacks on Clients

WIP detects many attacks targeted against clients. Some of the attacks are intended to prevent the client from connecting to legitimate APs. Others might be intended to establish a man-in-the-middle (MITM) attack or an intrusion. As with attacks on the infrastructure, some attack checks can be sensitive, causing events when no true threat has occurred. Again, you can take the *Implementing HPE Aruba Networking Security* course to learn more recommendations about which events to monitor (or read the *HPE Aruba Networking Certified Network Security Professional* study guide.)

For more information, read the sections below.

Disconnect Station Attack

A disconnect attack can be launched in many ways; the end result is that the client is effectively and repeatedly disconnected from the AP.

Block ACK DoS

The Block ACK mechanism that was introduced in 802.11e, and enhanced in 802.11nD3.0, has a built-in DoS vulnerability. The Block ACK mechanism allows for a sender to use the ADDBA request frame to specify the sequence number window that the receiver should expect. The receiver will only accept frames in this window. An attacker can spoof the ADDBA request frame, causing the receiver to reset its sequence number window. The legitimate server will continue sending frames outside that range, so the receiver will drop them.

EAP Rate Anomaly

The hacker sends EAP-Failure or EAP-Success frames to the client, spoofing the AP, to disrupt the authentication state on the client. This confuses the client's state, causing it to drop the AP connection. By continuously sending EAP Success or Failure messages, an attacker can effectively prevent the client from authenticating with the APs in the WLAN.

FATA-Jack Attack

FATA-Jack is an 802.11 client DoS tool that tries to disconnect targeted stations using spoofed authentication frames that contain an invalid authentication algorithm number.

Meiners Power Save DoS Attack

To save on power, wireless clients will sleep periodically, during which they cannot transmit or receive. A client indicates its intention to sleep by sending frames to the AP with the Power Management bit ON. The AP then begins buffering traffic bound for that client until it indicates that it is awake. An intruder could exploit this mechanism by sending (spoofed) frames to the AP on behalf of the client to trick the AP into believing the client is asleep. This will cause the AP to buffer most, if not all, frames destined for the client.

Null Probe Response

A null probe response attack has the potential to crash or lock up the firmware of many 802.11 NICs. In this attack, a client probe-request frame will be answered by a probe response containing a null SSID. A number of popular NIC cards will lock up upon receiving such a probe response.

Rate Anomalies

Many DoS attacks flood an AP or multiple APs with 802.11 management frames. These can include authenticate or associate frames, which are designed to fill up the association table of an AP. Other management frame floods, such as probe request floods, can consume excess processing power on the AP.

Omerta Attack

Omerta is an 802.11 DoS tool that sends disassociation frames to all stations on a channel in response to data frames. The Omerta attack is characterized by disassociation frames with a reason code of 0x01. This reason code is unspecified and is not used under normal circumstances.

AirJack Attack

AirJack is a suite of device drivers for 802.11(a/b/g) raw frame injection and reception. It was intended to be used as a development tool for all 802.11 applications that need to access the raw protocol. However, one of the tools included allowing users to force all users off an AP.

Broadcast Deauthentication

A deauthentication broadcast attempts to disconnect all stations in range. Rather than sending a spoofed deauth to a specific MAC address, this attack sends the frame to a broadcast address.

Broadcast Disassociation

By sending disassociation frames to the broadcast address (FF:FF:FF:FF:FF:FF), an attacker can disconnect all stations on a network for a widespread DoS.

Hotspotter Attack

The Hotspotter attack is an evil-twin attack which attempts to lure a client to a malicious AP. Many enterprise employees use their laptop in Wi-Fi area hotspots at airports, cafes, malls, etc. They have SSIDs of their hotspot service providers configured on their laptops. The SSIDs used by different hotspot service providers are well known. This enables the attackers to set up APs with hotspot SSIDs in close proximity of the enterprise premises. When the enterprise laptop Client probes for hotspot SSIDs, these malicious APs respond and invite the client to connect to them. When the client connects to a malicious AP, a number of security attacks can be launched on the client. Airsnarf is a popular hacking tool used to launch these attacks.

ChopChop Attack

ChopChop attempt to recover plaintext in WEP encrypted networks. It exploits WEP's predictable CRC32 checksum and lack of replay detection. The hacker deletes one byte from a captured frame, which causes the ICV to no longer check out. But the hacker can predict the byte that needs to be changed in the ICV to make the ICV correct. The hacker guesses a value and replays the packet. Only if the guess was correct, the AP returns the packet. With only 256 possible values, the hacker keeps guessing until finding the correct value. Due to the predictable way that the ICV is constructed, the hacker can then derive the plaintext that was in the truncated byte, as well as derive that byte of the keystream. The hacker can continue chopping off another byte to decrypt more of the packet.

Networks that use WPA with AES are not susceptible to this attack.

TKIP Replay Attack

TKIP is vulnerable to replay (via WMM or QoS) and plain text discovery (via ChopChop). This affects all WPATKIP usage. By replaying a captured TKIP data frame on other QoS queues, an attacker can manipulate the RC4 data and checksum to derive the plain text at a rate of one byte per minute. By targeting an ARP frame and guessing the known payload, an attacker can extract the complete plain text and MIC checksum. With the extracted MIC checksum, an attacker can reverse the MIC AP to Station key and sign future messages as MIC compliant, opening the door for more advanced attacks.

Networks that use WPA with AES are not susceptible to this attack.

ASLEAP

ASLEAP is a tool created for Linux systems used to attack Cisco LEAP authentication protocol.

Review Licensing for WIP Features

The table in Figure 12-35 provides a reference of the features available in the base HPE Aruba Networking Wireless Operating System (AOS) software. (Those features are still available with an RF Protect license.)

Feature	AOS Base	RFProtect License
Air monitor (2.4 and 5GHz)	✓	✓
Wireless rogue scanning and identification	✓	✓
Wired rogue containment	✓	✓
Wireless rogue containment via de-authorization	✓	✓
Wi-Fi interference detection	✓	✓

Figure 12-35: Review Licensing for WIP Features

The table in Figure 12-36 shows the additional features that are only available with the RFP license. Every AP or AM requires one RFP license.

Feature	RFProtect License
Spectrum analysis (Hybrid and Spectrum monitor)	✓
Wi-Fi interference classification	✓
Wi-Fi interference visualization	✓
Wireless IDS attack signatures detection	✓
Security threat management visualization	✓
Wireless intrusion configuration wizard	✓
Total Watch enhanced air monitoring	✓
Air monitoring of all bands (2.4, 4.9 and 5GHz)	✓
Dynamic channel dwell times	✓
In-between channels rogue scanning	✓
Advanced wireless rogue containment via tarpitting	✓
Detect and contain Windows Bridge	✓
Security events correlation	✓

Figure 12-36: Review Licensing for More WIP Features

Summary

Congratulations! You can now implement best practices to keep your mobile infrastructure secure. You have examined digital forensics and how you can follow best practices to maintain a chain of custody as you collect, examine, analyze, and report on evidence in security incidents. In particular, you have looked at responding to security incidents detected by WIP. You can find unauthorized devices and other security events in the Security Dashboard. You can interpret device classifications and know how to respond to various wireless devices and threats.

Learning checks

1. What is a Chain of Custody?
    a. A plan for containing rogue and interfering APs
    b. A list of rogue and suspected rogue APs
    c. The stage of the kill chain at which the hacker gains control of the infected system
    d. Documentation of when evidence was collected, how it was stored, and who had responsibility for it

2. What does it mean if HPE Aruba Networking WIP has classified an AP as a suspected rogue?

    a. The AP is authorized by the HPE Aruba Networking solution, but it is configured with the wrong SSID or security settings

    b. The AP might connect to the corporate LAN, based on matches between the source MAC or BSSID in the AP's wireless traffic and MACs in the wired network

    c. WIP has detected the AP sending frames that might indicate DoS attacks but at a lower rate than the threshold for an alarm

    d. The AP has violated at least one anti-DoS policy on WIP or connected to an interfering client

3. True or false: The Security Dashboard shows the location of rogue APs on a floorplan.

    a. True

    b. False

Note

For answers to all learning checks, refer to Appendix B.

---

13 Practice Exam

Introduction

This exam validates candidates' knowledge, skills, and ability to describe common security threats and vulnerabilities. It tests a candidate's knowledge of device hardening, implementation of security at the edge with AAA, basic roles and firewall policies, dynamic segmentation, and endpoint classification. It covers basic threat detection technologies, collecting logs and alarms.

Minimum Qualifications

The candidate should be a network engineer or help desk engineer working in either a customer or partner environment. The candidate should have worked six months to a year in networking in both wired and wireless knowledge.

Exam Details

Below are details about the HPE6-A78 exam.

- Exam type: Proctored
- Types of exam items: multiple choice, single response and multiple responses
- Exam duration: 1 hour 30 minutes
- Exam length: 60 questions
- Passing score: 63%
- Delivery languages: English

To register for the exam, you need an HPE Learner ID and a Pearson VUE login and password.

No reference material is allowed at the testing site. During the exam, you can make comments about the exam items. We welcome these comments as part of our continuous improvement process.

Testing Objectives

This exam validates that you can successfully perform the objectives listed in Table 13-1. The objectives are grouped into three sections, and each section is given a weighting.

Table 13-1: Testing Objective

Weighting (% of exam)	Objectives
70%	**Protect and Defend** • Describe common security threats (MITM, DDOS, spoofing, zero day, and so on) and identify the difference between a threat and a vulnerability • Explain common security protocols and their use cases • Describe PKI components • Describe firewall (PEF), dynamic segmentation, RBAC, AppRF • Describe user roles and policy enforcement • Explain the purpose and methods of a packet capture • Explain social engineering and defense • Disable insecure protocols and follow best practices for implement secure management protocols such as SSH, HTTPS, authenticated NTP, and CPsec • Describe best practices to protect network infrastructure devices with passwords, physical security, and out-of-band management • Use external authentication for network managers • Compare and contrast wireless LAN methodologies • Deploy basic WLAN security • Describe and deploy basic user roles for wireless users • Compare and contrast wired LAN methodologies • Define and deploy basic user roles for wired users • Compare endpoint classifications methods
24%	**Analyze** • View and acknowledge WIPS and WIDS alarms • Explain attack stages and kill chain • Troubleshoot with ClearPass Access Tracker • Collect and monitor historical network pattern data

Weighting (% of exam)	Objectives
6%	**Investigate** • Collect logs • Identify chain of custody • Initiate an investigation

Advice to Help You Take This Exam

Complete the training and review all course materials and documents before you take the exam.

Exam items are based on expected knowledge acquired from job experience, an expected level of industry standard knowledge, and possibly other prerequisites (events, supplemental materials, etc.).

Successful completion of the course alone does not ensure you will pass the exam.

Read this HPE Exam Preparation Guide and follow its recommendations.

Visit HPE Press for additional reference materials, study guides, practice tests, and HPE books.

Practice Exam Items

1. A certificate includes a key. What is a possible purpose of that key?

    a. It is a public key in a public/private keypair owned by the certificate issuer. Other devices can use the public key to validate that this certificate is valid and trusted.

    b. It is a public key in a public/private keypair owned by the certificate subject. Other devices can use the public key to check data that the subject signed with the private key.

    c. It is a symmetric key, which the certificate subject and devices checking the certificate must agree on in advance. The devices must use the same key in order for the certificate to be valid.

    d. It is a symmetric key that the certificate subject and issuer must agree on in advance. The certificate itself only contains a hash of the key to protect it from unauthorized use.

2. Which responsibilities belong to the authenticator in the 802.1X process? (Select two.)

    a. Checking user credentials and determining whether they are valid

    b. Encapsulating EAP messages in RADIUS

    c. Using policies to determine whether a user is permitted access and at what level

d. Enforcing the access decision communicated by the authentication server

e. Checking the authentication server credentials and informing the supplicant whether they are valid

3. An AOS-CX switch is set up to use RADIUS to authenticate managers who access the switch with SSH. A manager uses SSH and logs in. The RADIUS server sends back an Access-Accept with Service-Type AVP set to 5. (The message does not include any VSAs.) The manager is denied access. What could be the problem?

   a. The switch is not configured to accept Service-Type AVPs.

   b. The switch cannot interpret Service-Type AVPs. The server must use the Aruba-User-Role VSA instead.

   c. The switch cannot interpret Service-Type AVPs. The server must use the HPE-Admin-Role VSA instead.

   d. The switch does not accept 5 as a valid Service-Type value; the value should be 6 or 7.

4. What is one active method of endpoint classification that is supported by HPE Aruba Networking ClearPass Policy Manager (CPPM)?

   a. DHCP fingerprinting

   b. HTTP User-Agent-based fingerprinting

   c. SNMP scanning

   d. Crowd sourcing

5. What is one security enhancement that WPA3-Personal provides over WPA2-Personal?

   a. It ensures that every client session uses a unique pairwise master key (PMK).

   b. It ensures that every client session uses a unique pairwise temporal key (PTK).

   c. It uses 802.1X authentication as opposed to passphrase authentication.

   d. It uses certificate authentication as opposed to passphrase authentication.

6. You want to apply application rules to users assigned to a specific role on an HPE Aruba Networking Mobility Controller (MC). The rules will filter traffic by application category. What is a prerequisite?

   a. The role is configured as the default role for the user's WLAN.

   b. The MC has sufficient RF Protect (RFP) licenses for the user count.

   c. Web Content Classification (WebCC) is enabled globally on the firewall and enabled on the role.

   d. Deep Packet Inspection (DPI) is enabled globally on the firewall and enabled on the role.

7. Refer to Figure 13-1.

```
ubt zone mc vrf default
  primary-controller ip 10.1.10.10
  papi-security-key ciphertext @A&E1378AOEUF;QJKXFEAULDAOE!
  enable
!
ip source-interface ubt interface vlan10 vrf default
!
ubt-client-vlan 4000
!
port-access role group1a
   gateway-zone zone mc gateway-role group1b
   session-timeout 1800
   reauth-period 7200
```

Figure 13-1: Exhibit for Item 7

Figure 13-1 shows relevant portions of the configuration on an AOS-CX switch. (Assume that all VLANs referenced in this portion of the configuration are configured on the switch.) The device with IP address 10.1.10.10 is an MC.

What correctly describes how the switch handles clients assigned to the "group1a" role?

a. The switch informs the MC that the clients' role name is group1a, but if group1a does not exist on the MC, the MC can assign them to group1b. The MC uses the role to assign the clients to a VLAN.

b. The switch tunnels the clients' traffic to the MC using VLAN 4000 as the forwarding VLAN. This VLAN must extend across the network infrastructure between the switch and MC.

c. The switch tunnels the clients' traffic to the MC using VLAN 4000 as a placeholder, and the MC will determine the correct VLAN for the clients.

d. The switch informs the MC that it should forward the clients' traffic in VLAN 4000, regardless of the VLAN specified in the clients' role on the MC.

8. Refer to Figure 13-2.

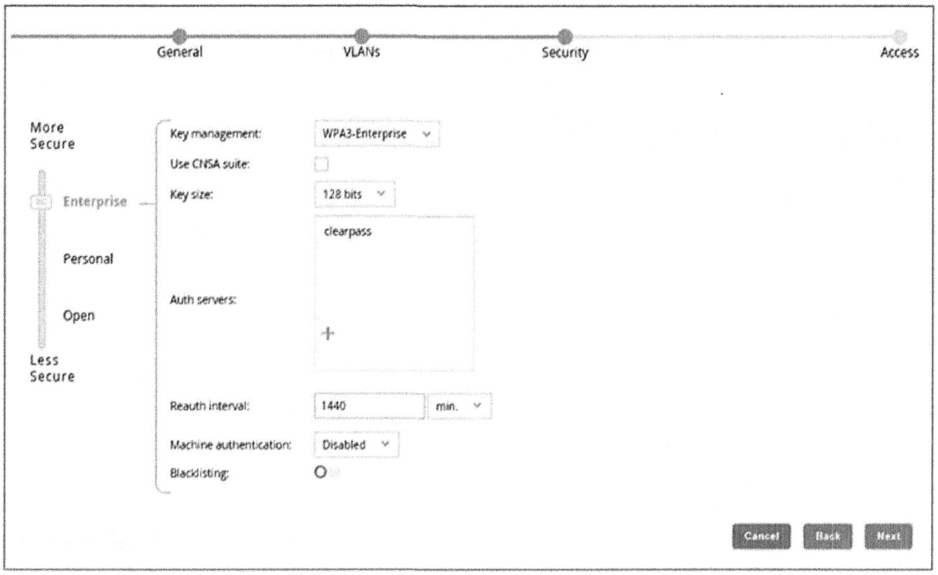

Figure 13-2: Exhibit for Item 8

What are correct settings if you want to permit both WPA2-capable and WPA3-capable clients to connect to the WLAN?

a. Leave the **Use CNSA suite** box clear and set the **Key size** to 128 bits.

b. Leave the **Use CNSA suite** box clear and set the **Key size** to 256 bits.

c. Select the **Use CNSA suite** box and set the **Key size** to 128 bits.

d. Select the **Use CNSA suite** box and set the **Key size** to 256 bits.

9. What is a difference between a controlpath and datapath packet capture on a MC?

a. A controlpath capture can mirror traffic to a destination IP address such as an Intrusion Detection System (IDS). A datapath capture can only collect traffic in a file.

b. A controlpath capture collects traffic passing through the MC and stops the traffic from flowing. A datapath capture collects traffic passing through the MC but also forwards the traffic to its destination.

c. A controlpath capture collects traffic to and from the MC's own IP address. A datapath capture collects traffic to and from a client connected through an MC.

d. A controlpath capture collects all traffic flowing through the MC. A datapath capture actually occurs on APs controlled by the MC and collects traffic in the air.

10. Refer to Figure 13-3 and Figure 13-4.

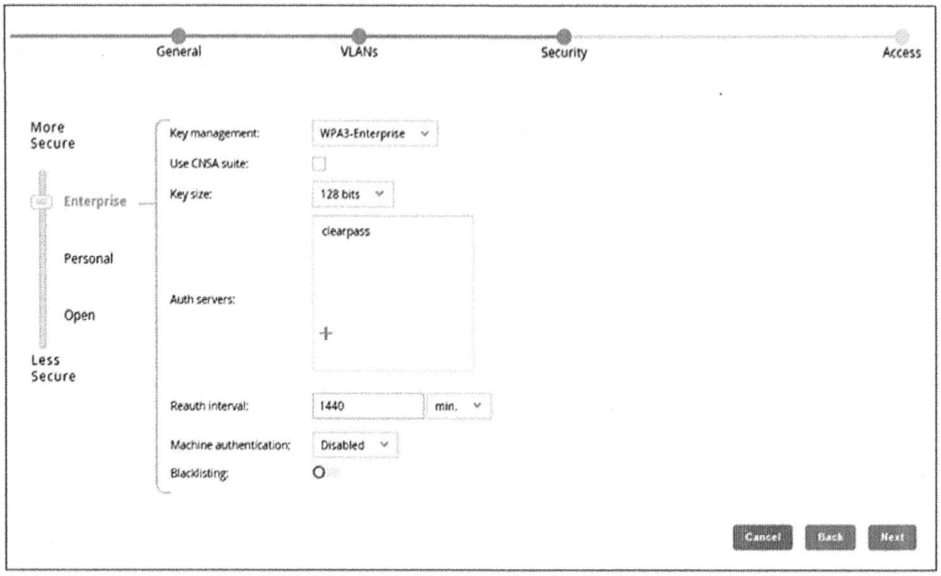

Figure 13-3: Exhibit 1 for Item 10

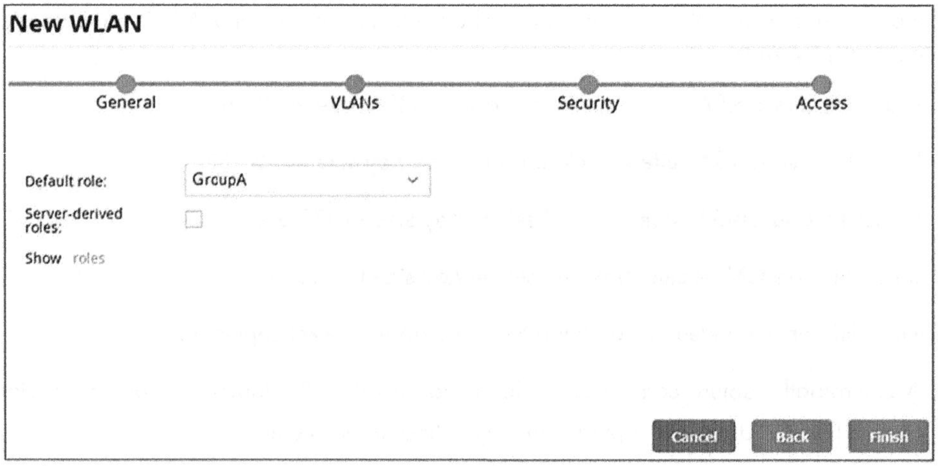

Figure 13-4: Exhibit 2 for Item 10

A company has a mobility solution that uses an MC and Campus APs (CAPs). A WLAN uses the Security and Access settings shown in Figure 13-3 and Figure 13-4. (The admin only used the WLAN wizard and did *not* change settings in the AAA profile.)

A user connects to this WLAN and undergoes authentication. The RADIUS server sends an Access-Accept with the Aruba-User-Role VSA set to "GroupB." What does the MC do?

a. It assigns the user to the "GroupA" role, whether or not the "GroupB" role exists on the MC.

b. It assigns the user to the "GroupB" role, whether or not the "GroupB" role exists on the MC.

c. It assigns the user to the "GroupB" role. However, if this role does not exist on the MC, it assigns the user to the "GroupA" role.

d. It assigns the user to the "GroupB" role. However, if this role does not exist on the MC, it disconnects the user from the WLAN.

11. What distinguishes spear-phishing from other types of phishing?

a. Spear-phishing is targeted to a specific user while, in general, phishing targets users more or less indiscriminately.

b. Spear-phishing involves spoofing email while other forms of phishing involve MAC or IP spoofing.

c. Spear-phishing is a zero-day threat while patches exist to protect users from other forms of phishing.

d. Spear-phishing uses multiple attacking devices as a distributed attack while phishing originates from a single hacker.

12. What is one way to help ensure that a network device's logs are useful and valid for security investigations?

a. Turn on debugging for all events and leave it running during normal operation.

b. Have the network device take its time from a trusted NTP server that uses authentication.

c. Configure the device to routinely log events with the highest severity level only; activate lower level logging if an incident occurs.

d. Ensure that the device uses UDP, rather than TCP, when forwarding logs to a central Syslog server.

13. Managers need to access an AOS-CX switch's CLI remotely over an IP network. What is one way to help protect managers from man-in-the-middle (MITM) attacks?

a. Set up SNMPv3 with an individual SNMPv3 user for each manager who needs to access the switch.

b. Set up SNMPv3 with a single SNMPv3 user. Make sure that this user has an auth and a priv key.

c. Set up SSH. Export the switch's SSH public key and install it on the managers' SSH clients.

d. Set up SSH. Make sure to use local authentication, rather than external authentication, to authenticate the managers.

14. What is one security benefit of a Trusted Platform Module (TPM) on an HPE Aruba Networking MC?

a. It provides a secure location for the storage of private keys associated with certificates.

b. It hosts an Ethernet port for establishing an out-of-band-management (OOBM) network.

c. It provides a default set of Wireless Intrusion Prevention (WIP) signatures for scanning wireless traffic.

d. It provides a repository of trusted and untrusted Web sites that the MC can use to filter wireless traffic.

15. You have been asked to set up an AOS-CX switch to send HTTP user-agent strings to ClearPass Policy Manager (CPPM) for clients connected to ports 1/1/1-1/1/24. The switch currently has no device fingerprinting settings configured on it. You enter these commands:

```
Switch(config)# client device-fingerprint profile myProfile
Switch(myProfile)# http user-agent
Switch(myProfile)# exit
Switch(config)# interface 1/1/1-1/1/1/24
Switch(config-if-1/1/1-1/1/24)# client device-fingerprint apply-
profile myProfile
```

What else should you do to ensure that the switch can send the information to CPPM?

a. Activate the precise HTTP attributes that you want to send in the "myProfile" device-fingerprint profile.

b. Configure an IF-MAP policy on the switch, specifying CPPM's IP address within the policy.

c. Make sure that its HTTPS server enabled on the default VRF (or the VRF to which ports 1-24 are associated.)

d. Make sure that CPPM is configured as a RADIUS accounting server on the switch and enable the VSA associated with device-fingerprint profiles.

16. Refer to Figure 13-5.

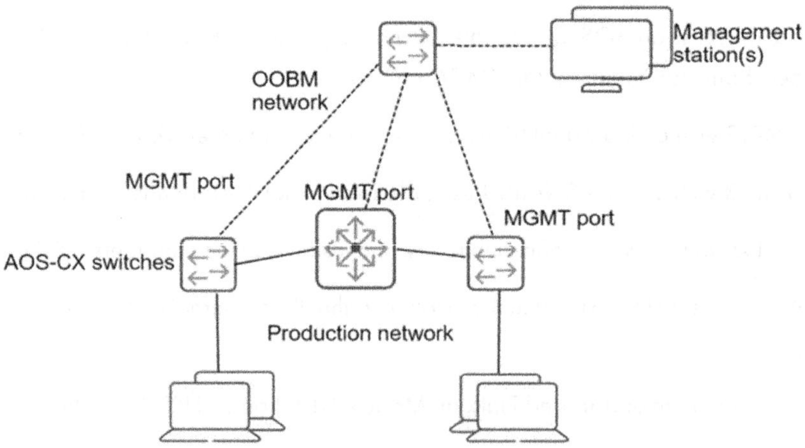

Figure 13-5: Exhibit for Item 16

A customer has asked you to establish an out-of-band management (OOBM) network for managing the customer's AOS-CX switches, as shown in Figure 13-5. (The network includes more access switches, which are connected in the same way as the three shown.) Managers should only be able to reach the switches' CLI through the OOBM network.

The switches are currently using their default settings for management protocols. What step should you take on each AOS-CX switch to fulfill this task?

a. Disable Telnet on the default VRF. Enable SSH on the default VRF.

b. Create custom control plane policing policies (CoPP) to deny user IP addresses and apply them on the default VRF.

c. Disable SSH on the default VRF and leave it enabled on the mgmt VRF only.

d. Create a CSR, obtain a CA-signed certificate, and install the certificate on the switch.

17. ClearPass Policy Manager (CPPM) has profiled a wireless client. It wants the Mobility Controller (MC) to which the client is connected to re-authenticate the client so that CPPM can give it a new role. What are two settings that the MC requires for this process to be successful? (Select two.)

a. A dynamic authorization (RFC-3576) server with the correct CPPM IP address and shared secret

b. An IF-MAP policy that specifies the CPPM IP address and credentials

c. A user rule set with DHCP option rules specified in the AAA profile associated with the client's WLAN

d. The dynamic authorization (RFC-3576) server with the correct IP address enabled in the AAA profile associated with the client's WLAN

e. Server derivation rules which specify Aruba-No-DHCP-Fingerprint VSA settings as criteria

18. What is a correct description of a stage in the Lockheed Martin kill chain?

a. During the exploitation stage, the hacker collects information about targets and probes for vulnerabilities.

b. During the reconnaissance stage, the hacker assesses the effects of the attack and how much information the hacker has exfiltrated.

c. During the delivery stage, the hacker typically exfiltrates data and sends it to a command and control server.

d. During the delivery stage, hackers often send spoofed emails with links to malware.

19. You have installed a new AOS-CX switch and configured it to enforce 802.1X authentication to ClearPass Policy Manager (CPPM) on edge ports. When you test the setup, you are not able to authenticate

successfully on any client. You suspect that you have the wrong RADIUS shared secret. What will admins see in the CPPM Access Tracker if your suspicion is true?

a. No record of the test authentication attempts

b. Records of the test authentication attempts that show Reject

c. Records of the test authentication attempts that show Error

d. Records of the test authentication attempts that show Timeout

20. You are examining the Client list in the Security Dashboard for an HPE Aruba Networking MC. What correctly describes what you should interpret in various scenarios?

a. You see an Unauthorized client connected to a Neighbor AP. This indicates a serious security threat against your neighbor.

b. You see an Authorized client connected to an Interfering AP. This indicates a potential security threat against one of your users.

c. You see an Authorized client connected to a Suspected Rogue AP. This indicates that the Suspected Rogue is probably really Authorized, and this is not a threat.

d. You see an Unauthorized client connected to a Rogue AP. This is not a potential threat; because the client is Unauthorized, the AP is likely actually a Neighbor.

21. You have been asked to collect and send low-level "radius" debug messages from an AOS-CX switch to a central SIEM server at 10.1.2.50. The only logging or debug command currently defined on this switch is:

```
logging 10.1.2.50
```

You enter this command:

```
debug destination syslog
```

What is an additional step that must you complete to send the debug messages to the SIEM server?

a. Set the global logging severity level to "debugging."

b. Enable debugging on the "radius" category.

c. Enable the switch's Network Analytics Engine (NAE).

d. Add the SIEM server's IP address to the default control plane policing (CoPP) policy.

22. What is a correct definition of the specified form of encryption?

    a. Asymmetric encryption acts as a one-way function; it creates a "digest" that cannot be reverted to the plaintext.

    b. Asymmetric encryption is often used to encrypt data in transit because it tends to be faster than symmetric encryption.

    c. Symmetric encryption is the foundation for digital certificates deployed in a Public Key Infrastructure (PKI).

    d. Symmetric encryption relies on a shared secret, which two parties must agree on before using symmetric encryption.

23. What is one difference between TACACS+ and RADIUS?

    a. TACACS+ provides sufficient security for AAA communications over untrusted networks while RADIUS does not.

    b. TACACS+ uses certificates to authenticate the TACACS+ client and TACACS+ server to each other while RADIUS uses shared secrets for this purpose.

    c. RADIUS can be used to authenticate users with 802.1X while TACACS+ cannot.

    d. RADIUS provides authentication, authorization, and accounting while TACACS+ provides only authentication and authorization.

24. You need to install a CA-signed certificate on an HPE Aruba Networking MC. What is one consideration for how to create the Certificate Signing Request (CSR)?

    a. The CSR must be created online on the MC if you want to install the certificate as a Server Certificate.

    b. Creating the CSR offline allows devices in a cluster to use the same certificate and private key.

    c. If you create the CSR online, you must remember to export the private key for archival.

    d. If you create the CSR online, you must install the CA-signed certificate in PKCS12 format.

25. Refer to Figure 13-6 and Figure 13-7.

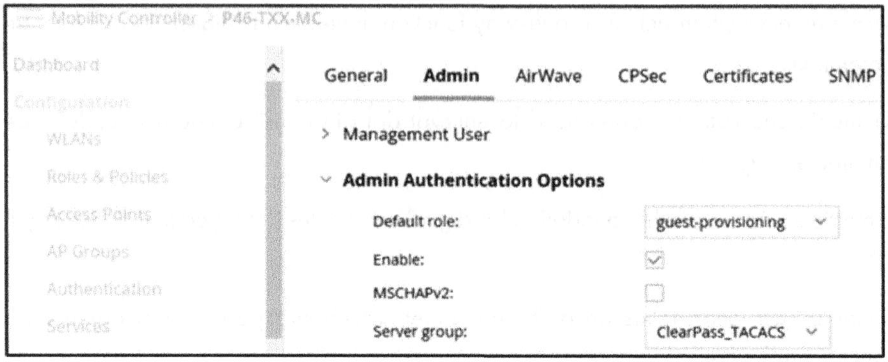

Figure 13-6: Exhibit 1 for Item 25

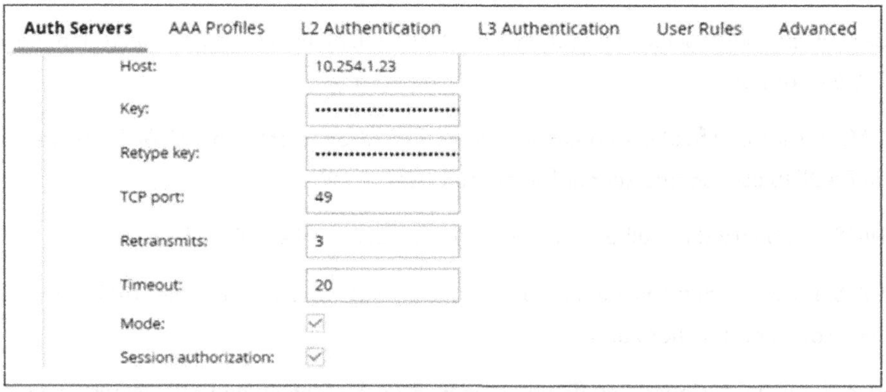

Figure 13-7: Exhibit 2 for Item 25

You have set up TACACS+ authentication for managers logging into an HPE Aruba Networking MC Web UI, as shown in Figure 13-6. In Figure 13-7, you can see the TACACS+ server in the "ClearPass_TACACS" group. This server is a ClearPass Policy Manager (CPPM) server.

CPPM is supposed to assign managers to roles such as "root" and "read-only." But when you test logging into the Web UI, you always receive the "guest-provisioning" role.

What is the likely problem?

a. The default role is incorrect and should be set to "root."

b. Your test credentials are incorrect, and you are failing authentication.

c. The TACACS+ key does not match on the MC and CPPM.

d. CPPM is not set up to send the proper values in Aruba-Admin-Role VSAs.

26. What is one practice that can help you to maintain a digital chain of custody in your network?

    a. Ensure that all HPE Aruba Networking Mobility Controllers have the proper PEF licensing.

    b. Enable automatic containment of rogue APs detected in the vicinity of the corporate LAN.

    c. Establish and enforce policies for secure collection and forwarding of logs.

    d. Ensure that network devices use the certificates installed in their TPMs at factory default settings.

27. A company wants to use 802.1X to authenticate wireless and wired users. What is one consideration for using EAP-TLS versus PEAP-MSCHAPv2?

    a. EAP-TLS requires the RADIUS/EAP authentication server to have a valid certificate while PEAP-MSCHAPv2 does not.

    b. PEAP-MSCHAPv2 is the only correct option if the company wants to use WPA2-Enterprise while EAP-TLS is the only correct option if the company wants to use WPA3-Enterprise.

    c. PEAP-MSCHAPv2 provides a higher level of security than EAP-TLS because PEAP-MSCHAPv2 includes two layers of authentication.

    d. EAP-TLS provides a higher level of security than PEAP-MSCHAPv2 because EAP-TLS enforces certificate authentication for both supplicants and the server.

28. Refer to Figure 13-8.

```
interface 1/1/1
    no shutdown
    no routing
    vlan access 1
    aaa authentication port-access dot1x authenticator
        enable
    exit
#same config on all other edge ports
aaa authentication port-access dot1x authenticator enable
```

Figure 13-8: Exhibit for Item 28

You are implementing 802.1X authentication on an AOS-CX switch. The company has these requirements:

- The switch authenticates users to a RADIUS server that does *not* send Aruba-User-Role or Aruba-CPPM-Role VSAs.
- All authenticated users should receive full access in VLAN 50.

- All unauthenticated users should be denied any access.

You have set up 802.1X, as shown in Figure 13-8. You have also set up the RADIUS server settings (not shown in Figure 13-8).

What additional steps should you take to fulfill the requirements of this scenario?

a. Make sure that the edge ports are configured in access mode and configure VLAN 50 as the access VLAN ID.

b. Create a port-access role that specifies 50 as the access VLAN ID and apply the role as an auth-role on edge ports.

c. Set edge ports' access VLAN ID to 50. Also apply the "denyall" role as the reject role on edge ports.

d. Disable user role authorization on the switch and configure VLAN 50 as the access VLAN ID on edge ports.

29. What correctly describes a way that HPE Aruba Networking Wireless Intrusion Prevention (WIP) can respond to wireless threats?

    a. Wired containment activates 802.1X on the ports that connect to unauthorized wireless devices and shuts them out of the corporate LAN.

    b. Tarpit containment uses a combination of wired and wireless frames to contain unauthorized wireless devices, providing greater reliability than other methods.

    c. Valid client protection sends disassociation frames to disconnect Unauthorized Clients from Unauthorized APs, but does not affect Authorized Clients.

    d. Tarpit containment disassociates clients from the contained AP and also sets up a decoy association to keep the clients from reconnecting to the AP.

30. Your company has AOS-CX switches at the edge. The company wants to prevent hackers from launching ARP poisoning attacks. Which two features should you configure on the switches?

    a. ARP inspection and DHCP snooping

    b. ARP inspection and BPDU guard (also called BPDU protection)

    c. ARP proxy and DHCP snooping

    d. ARP proxy and BPDU guard

Explanations

1. **B** is correct. A certificate always includes a public key in a public/private keypair owned by the certificate subject. The purpose of the certificate is distributing this key and tying it to the certificate subject's identity. Other devices can use the public key to check data signed the subject with the private key. They can also use the public key to encrypt data that they want to send to the subject.

   A is incorrect. The public key is part of a keypair that belongs to the certificate subject, not the issuer. The certificate *signature* is for validating the certificate. **C** and **D** are incorrect. The public key within a certificate is used with asymmetric encryption; it is *not* a symmetric key.

   For more information on this topic, refer to Chapter 2.

2. **B** and **D** are correct. The authenticator acts as a courier between the supplicant and authentication server; it encapsulates the supplicant's EAP messages in RADIUS and sends them to the authentication server. The authentication server makes access decisions, and the authenticator is responsible for enforcing those decisions (for example, by opening or closing a Layer 2 connection).

   A is incorrect. The authentication server is responsible for checking user credentials and determining whether they are valid. **C** is incorrect. The authentication server is responsible for using policies to determine whether a user is permitted access and at what level. **E** is incorrect. The supplicant checks authentication server credentials and authenticates the authentication server itself. The authenticator does check an authentication server's shared secret as part of accepting the server's RADIUS messages; however, it does not pass that information to the supplicant.

   For more information on this topic, refer to Chapter 6.

3. **D** is correct. The AOS-CX switch prefers Aruba-Admin-Role VSAs, but this RADIUS server is not sending VSAs. Therefore, the switch looks at the Access-Accept message's Service-Type AVP. This AVP must be 6 (for managers) or 7 (for operators). All other values are invalid, and if the switch receives one of these values, it denies the management user access. This behavior matches the description in the item, and it is a valid explanation of the probable issue.

   A is incorrect. AOS-CX switches do not need to be configured to accept Service-Type AVPs. Their behavior for management authentication is always to prefer Aruba-Admin-Role VSAs and then consider Service-Type AVPs. **B** and **C** are incorrect. AOS-CX *do* accept Service-Type AVPs, so these are not valid explanations of the problem. Also these answers list the wrong VSAs to use. (While using *Aruba-Admin-Role* VSAs is preferred, this RADIUS server might not support them.)

   For more information on this topic, refer to Chapter 3.

4. **C** is correct. Active classification methods involve CPPM sending out queries and receiving replies. SNMP scanning is one mechanism that CPPM can use for active classification.

A and **B** are incorrect. CPPM does support these classification methods; however, they are passive methods. In other words, with these methods, CPPM receives traffic that endpoints send as part of their normal operation.

D is incorrect. CPPM does not support this form of classification; ClearPass Device Insight does.

For more information on this topic, refer to Chapter 9.

5. **A** is correct. WPA3-Personal uses Simultaneous Authentication of Equals (SAE) to establish a unique shared secret between the client and AP/MC while also verifying that both the client and AP/MC know a passphrase. The unique shared secret becomes a unique PMK for the session. In WPA2-Personal, on the other hand, every client connected to the WLAN derives the same PMK based on a passphrase.

   B is incorrect. In both WPA3-Personal and WPA2-Personal, each client session has a unique PTK. Therefore, this is not a difference between the versions. **C** and **D** are incorrect. WPA3-Personal uses passphrase authentication. WPA2-*Enterprise* and WPA3-*Enterprise* use 802.1X authentication with certificate-based authentication as a possibility.

   For more information on this topic, refer to Chapter 6.

6. **D** is correct. Filtering by application category requires DPI. You must enable this feature globally on the firewall (and the MC must then be rebooted). The feature also needs to be enabled on the role.

   A is incorrect. How a user is assigned to the role does not affect whether the role can implement application rules. **B** is incorrect. RFP licenses enable Wireless Intrusion Prevention (WIP) features and do not affect the use of application rules. **C** is incorrect. WebCC licenses are required for applications rules that filter traffic by Web category and reputation, but not for application rules that filter traffic by application category.

   For more information on this topic, refer to Chapter 8.

7. **C** is correct. The ubt-client-vlan simply acts as a placeholder VLAN ID for tunneled traffic. After the MC receives the tunneled traffic, it determines the correct forwarding VLAN for the clients based on their role.

   A is incorrect. The "group1a" role name is the role to which clients are assigned on the *switch*. The switch informs the MC that the clients' role name is the gateway-role name, "group1b." **B** is incorrect. The switch tunnels the traffic using the UBT source interface. The ubt-client-vlan is simply a placeholder, and it should *not* exist in the network infrastructure. **D** is also incorrect. The MC uses the VLAN specified in the clients' role to determine the VLAN in which it bridges the clients' traffic into the LAN.

   For more information on this topic, refer to Chapter 8.

8. **A** is correct. WPA2 clients only support 128-bit keys, so the **Key size** must be set to 128 bits. 128-bit keys are only an option when the **Use CNSA suite** box is clear.

   B and **D** are incorrect. The key size must be 128 bits for this scenario. **C** is incorrect. 128 bits is not a valid setting when CNSA mode is enabled.

   For more information on this topic, refer to Chapter 7.

9. **C** is correct. A controlpath capture collects traffic to and from the MC's own IP address. A datapath capture collects traffic to and from a client connected through the MC.

   A is incorrect. A controlpath capture can only send traffic to a local file. A datapath capture can mirror traffic to a local file, port, or destination IP such as an IDS. **B** is incorrect. Both controlpath and datapath captures mirror traffic. That is, they collect the traffic for analysis but also let the traffic continue on its way. **D** is incorrect. Both controlpath and datapath captures, implemented on an MC, apply to traffic flowing through the MC.

   For more information on this topic, refer to Chapter 11.

10. **C** is correct. When a user authenticates successfully with 802.1X (the authentication method enforced in a WPA3-Enterprise WLAN), the MC uses the Aruba-User-Role VSA as the highest priority information for assigning the role. Therefore, the MC assigns the user to "GroupB" *if the "GroupB" role exists on the MC.* However, if the role is invalid—it does not exist on the MC—the MC cannot assign it. The MC would next use server derivation rules; however, this WLAN is not using those rules, as shown in the exhibit. Therefore, the MC would assign the user to the default role, which is "GroupA" in this scenario.

    A is incorrect. The Aruba-User-Role VSA takes precedence over the default authentication role. **B** is incorrect. The MC cannot assign a user to a role that does not exist on it. **D** is incorrect. If the role assigned by the Aruba-User-Role VSA is invalid, the default role is used, as explained above.

    For more information on this topic, refer to Chapter 8.

11. **A** is correct. Spear-phishing targets a specific user, often using information collected from social networking.

    B is incorrect. Many types of phishing use spoofed email. **C** is incorrect. Spear-phishing and phishing use social engineering to lure users to click links or perform other actions. The hacker *might* then proceed to exploit vulnerabilities, for which patches *might or might not* exist. However, none of these characteristics are what define spear-phishing and phishing as threats. **D** is incorrect. How many devices launch an attack does not distinguish spear-phishing from phishing.

    For more information on this topic, refer to Chapter 1.

12. **B** is correct. Valid, useful logs require an accurate timestamp. A good way to ensure accurate timestamps is to have devices take their time from an NTP server. Authentication provides extra assurance that the NTP server is trusted.

    A is incorrect. Debugging is resource intensive and should typically be used on specific events for temporary periods for the purposes of troubleshooting.

    C is incorrect. If you log only the highest severity events routinely, you will collect very few logs. You need to collect logs before an incident occurs to obtain the most useful logs. Different companies will have different policies for the precise severity level at which to log routinely. However, the highest level is too high.

D is incorrect. Depending on the network device, you might be able to use UDP or TCP to forward logs. TCP establishes a session between the network device and the Syslog server while UDP simply has the device send logs without a session. UDP is not inherently more reliable than TCP. Neither mechanism provides encryption on its own.

For more information on this topic, refer to Chapters 11 and 12.

13. **C** is correct. SSH is recommended for accessing a switch's CLI remotely over an IP network, as SSH provides encryption and mutual authentication. Mutual authentication (both the client and the switch authenticate) helps to prevent MITM attacks. The switch uses its SSH public/private keypair to authenticate. Installing the public key on manager clients in advance enables the clients to check the switch's authentication. In this way, managers ensure that they are establishing an SSH session with the legitimate switch and not a man-in-the-middle.

    A and **B** are incorrect. SNMPv3 is the recommended version for SNMP. And it is recommended that the SNMPv3 user has an auth and a priv key to provide data integrity and encryption. However, SNMPv3 does not provide access to the CLI, so it is irrelevant for this scenario. **D** is incorrect. External authentication to a TACACS+ or RADIUS server is generally recommended for use with SSH.

    For more information on this topic, refer to Chapter 3.

14. **A** is correct. The TPM is a secure hardware component in which HPE Aruba Networking devices store the private keys associated with their certificates. The TPM also stores a cryptographically secure fingerprint for the OS, which empowers the secure boot feature.

    B, **C**, and **D** incorrect. These are not features of a TPM.

    For more information on this topic, refer to Chapter 4.

15. **D** is correct. The AOS-CX switch sends the information collected by the device-fingerprint profile to CPPM using a "dfp-client-info" VSA in RADIUS accounting messages. The switch must use CPPM as a RADIUS accounting server and send interim messages to it. You must also enter a command to enable the "dfp-client-info" VSA.

    A is incorrect. You do not need to activate any other HTTP attributes in the device-fingerprint profile. **B** is incorrect. The AOS-CX switch does not have IF-MAP policies. **C** is also incorrect. The switch itself does not have to act as an HTTPS server for it to collect the HTTP user-agent strings.

    For more information on this topic, refer to Chapter 9.

16. **C** is correct. By default, SSH is enabled on both the default VRF and the mgmt VRF. Disabling SSH on the default VRF prevents users from using SSH to contact switches' CLI over the production network. Leaving SSH enabled on the mgmt VRF enables managers to use SSH to contact switches over the OOBM network. This behavior meets the company's specifications. Telnet is insecure and should be disabled. However, the scenario indicates that the switch is using default settings for it, which means that it *is* disabled.

A is incorrect. SSH *is* recommended over Telnet. However, to meet the needs of this scenario, you should not leave SSH enabled on the default VRF.

B is incorrect. The CoPP policies apply rate limits to control plane traffic; they are intended to prevent DoS attacks, not to limit users' access to the switch CLI through specific protocols. A control plane *ACL* could serve the purpose of denying SSH traffic to the switch from specific IP addresses. However, for this scenario, it is best to simply disable SSH on the default VRF and have it enabled on the mgmt VRF, which is associated with the OOBM network.

D is incorrect. A CA-signed certificate is not necessary for SSH, and installing one does not address the requirements in this scenario.

For more information on this topic, refer to Chapter 3.

17. **A** and **D** are correct. CPPM sends a change of authorization (CoA) message to the MC to tell the MC to re-authenticate the client. For the MC to accept this message, CPPM's IP address must be specified as a dynamic authorization (RFC-3576) server. In addition, the AAA profile associated with the client's WLAN must have that dynamic authorization server specified in it.

    B is incorrect. An MC uses an IF-MAP policy to send HTTP Agent-Strings to CPPM. Such a policy is not required for the MC to accept CoA messages. **C** is incorrect. The MC would use user rules with DHCP options to profile clients on its own. In this scenario, the MC only needs to accept CoA messages from CPPM.

    E is incorrect. The MC uses server derivation rules to determine the correct role to which to assign an authenticated client (if the server did not send a valid Aruba-User-Role VSA). These rules do not relate to the MC accepting CoA messages, which is what is required in this scenario.

    For more information on this topic, refer to Chapter 9.

18. **D** is correct. The delivery stage is when hackers deliver malware to a targeted user's device. Many delivery mechanisms exist, but email is a common one.

    A is incorrect. The hacker collects information about targets and probes for vulnerabilities in the reconnaissance stage. In the exploitation stage, malware delivered to a device exploits a vulnerability. **B** is incorrect. The reconnaissance stage occurs early in the attack when hackers collect information about targets and probe for vulnerabilities. **C** is incorrect. The hacker typically exfiltrates data in the Actions on objective stage. In the delivery stage, the hacker delivers malware to the target.

    For more information on this topic, refer to Chapter 1.

19. **A** is correct. When CPPM receives an authentication request with an incorrect shared secret, it does *not* create a record for that request in the Access Tracker. However, you and other ClearPass admins can see the error in CPPM's Event Viewer, which is where you should check if you suspect shared secret issues.

    B, **C**, and **D** are incorrect. If the shared secret on the switch is incorrect, CPPM will not create a record for the authentication request.

For more information on this topic, refer to Chapter 11.

20. **B** is correct. An Authorized client is a client that has connected to an Authorized AP and passed encrypted traffic; in other words, it is typically a client that belongs to the organization. An Interfering AP is an Unauthorized AP that is not connected to your corporate LAN (or has not yet been detected as connected). Even if the Unauthorized AP is not connected to your corporate LAN, it can still launch attacks against a client that is connected to it. Therefore, this could be a potential security threat, which you should investigate.

    A is incorrect. An Unauthorized client is any client that has never connected to your APs and passed encrypted traffic. Your neighbor's clients will show up as Unauthorized Clients, and an Unauthorized Client connected to a Neighbor AP is likely just your neighbor's own client connected to your neighbor's wireless network. This is not a serious nor certain indication of a threat.

    C is incorrect. A Suspected Rogue AP is an Unauthorized AP that might or might not be connected to the corporate LAN. In either case, it is *not* an Authorized AP. An Authorized Client connecting to the Suspected Rogue AP does *not* indicate that the AP should be Authorized. Instead the AP could be luring one of your users to connect to it so that it can install malware or launch another attack on it. In addition, if it turns out that the Unauthorized AP *is* actually connected to the corporate LAN, it could be providing unauthorized access into the LAN. No matter what, you should investigate and follow up on the potential threat.

    D is incorrect. A Rogue AP is an Unauthorized AP that is connected to your corporate LAN, not a Neighbor AP. A Rogue AP always indicates a potential security issue, when either Authorized and Unauthorized Clients connect to it.

    For more information on this topic, refer to Chapter 12.

21. **B** is correct. Debugging is not enabled by default. You must enable it on the specific category that you want to debug.

    A is incorrect. You do not need to set a global logging severity level. **C** is incorrect. NAE is a separate feature from debugging. **D** is incorrect. CoPP policies rate limit traffic of particular protocols, destined to the switch control plane. The switch can use the default policies and send logging traffic without issue.

    For more information on this topic, see Chapter 11.

22. **D** is correct. In symmetric encryption, the same key encrypts plaintext and decrypts the resulting ciphertext. When two parties want to use symmetric encryption to encrypt their communications, they must agree on that key, or shared secret, in advance. Agreeing on the shared key securely is one of the challenges of symmetric encryption, which is addressed by technologies such as Diffie-Hellman and TLS.

    A is incorrect. Asymmetric encryption creates ciphertext that *can* be reverted to plaintext with the correct key. In asymmetric encryption, keys are pairs of public and private keys. The private key can decrypt what the public key encrypts and vice versa. A hash function is what creates a "hash" or "digest" that cannot be reverted to the original plaintext.

B is incorrect. Asymmetric encryption is usually slower and more processor intensive than symmetric encryption. For this reason, a protocol like TLS uses asymmetric encryption for the initial stage to enable two parties to agree on a shared secret securely. The parties then use that shared secret with symmetric encryption to secure future communications.

C is incorrect. Digital certificates and a PKI are used to distribute public keys, a part of *asymmetric* encryption.

For more information on this topic, see Chapter 2.

23. **C** is correct. 802.1X uses RADIUS for authentication; TACACS+ is not supported by 802.1X. But you can use RADIUS or TACACS+ to authenticate managers attempting to log in to a network infrastructure device Web UI or CLI.

    A is incorrect. TACACS+ provides encryption, but this encryption is not sufficient to provide true protection over an untrusted network. If you need to run AAA communications over an untrusted network, you should use RadSec.

    B is incorrect. Both TACACS+ and RADIUS use shared secrets. RadSec, which carries RADIUS communications over TLS, does have the RADIUS client and server authenticate with certificates.

    D is incorrect. Both TACACS+ and RADIUS support authentication, authorization, and accounting.

    For more information on this topic, see Chapter 2.

24. **B** is correct. When you create the CSR offline, you have control over the private key. You can then bundle the CA-signed certificate and the private key into a PFX or PKCS12 file and install the certificate and private key on multiple devices in a cluster.

    A is incorrect. You can install a Server Certificate that was created from a CSR generated online *or* offline. If you (or someone else) created the CSR offline, you simply need to bundle the signed certificate with the private key in a PFX or PKCS12 file. You can then install that file as a Server Certificate.

    C is incorrect. When you create the CSR online, the private key remains on the MC and cannot be exported.

    D is incorrect. A PKCS12 file is typically used to bundle a certificate and the associated private key. When you create the CSR online, the MC has the private key on it already. You can import the CA-signed certificate using one of several formats, including PEM.

    For more information on this topic, see Chapter 4.

25. **D** is correct. The scenario states that CPPM is supposed to assign managers to roles. It does so by sending values in Aruba-Admin-Role VSAs within Access-Accept messages. If the Access-Accept does not have an Aruba-Admin-Role VSA with a valid role name, the MC assigns managers to the default role, which is what is happening in this scenario. You should check with the ClearPass admins and have them configure CPPM to send the proper Aruba-Admin-Role VSAs.

A is incorrect. The problem is with CPPM assigning the roles, not with the default role, which is the role assigned to managers for whom CPPM does not assign a valid role. Changing the default role will not meet the requirements of the scenario, which is that CPPM assigns the roles. Changing the default role to "root" could also open a security hole.

B is incorrect. If the credentials were incorrect, the authentication would fail, and you would not be able to log in to the UI at all. Similarly, **C** is incorrect. If the keys did not match, CPPM would drop the authentication request, and you would not be able to log in to the UI at all.

For more information on this topic, see Chapter 4.

26. **C** is correct. Maintaining a digital chain of custody refers to using the proper procedures to collect, examine, and analyze logs and other data in order to use it as evidence in a security investigation. You and your company should establish policies related to secure collect and forwarding of logs. For example, you could establish policies for how logs are exported and stored. You could use TLS or IPsec to encrypt forwarded logs if they need to cross an untrusted network. You should also ensure that devices have correct clock settings so that logs have the correct timestamps. You can also help to ensure a proper chain of custody by securing management access to devices and establishing auditing to see what managers do on a device.

    A is incorrect. Installing PEF licenses on MCs enables the MCs to use the role-based AOS firewall. However, this step is not directly related to maintaining a chain of custody. **B** is incorrect. Dealing with rogue APs is a best security practice, but it is not directly related to maintaining a chain of custody. In addition, some companies avoid containing rogue APs with wireless disassociated frames, and that is a valid choice. **D** is incorrect. A device using its factory-installed certificate, rather than a CA-signed certificate, does not improve the chain of custody.

    For more information on this topic, refer to Chapter 12.

27. **D** is correct. EAP-TLS provides a higher level of security than PEAP because it enforces certificate authentication for both supplicants and the server. PEAP supports a variety of authentication types for supplicants; with PEAP-MSCHAPv2 supplicants authenticate with usernames and passwords. The industry considers certificate-based authentication as more secure than password-based authentication for a variety of reasons. Certificates are more resistant to dictionary attacks and less likely to be leaked. In addition, MSCHAPv2 has known vulnerabilities.

    A is incorrect. Both of these authentication types require the RADIUS/EAP authentication server to have a valid certificate. **B** is incorrect. Both WPA2-Enterprise and WPA3-Enterprise require 802.1X and can support a variety of EAP methods, including EAP-TLS and PEAP-MSCHAPv2. **C** is incorrect. PEAP-MSCHAPv2 does feature an inner method and an outer method, but this does not make it more secure than EAP-TLS.

    For more information on this topic, see Chapter 6.

28. **B** is correct. If an AOS-CX switch does not receive a valid VSA in the Access-Accept, informing the switch of the authenticated user's proper role, the switch applies the port's auth-role. You should create a port-access role that specifies 50 as the access VLAN ID. You must then apply that role on edge ports as the auth-role. The switch will then assign this role to authenticated users and forward those users' traffic in VLAN 50 (regardless of the edge ports' configured access VLAN ID). The reject role, as well as the pre-auth and critical roles, are deny all traffic by default, so unauthenticated users are already denied access.

    A is incorrect. Simply making VLAN 50 the access VLAN on the edge ports will not fulfill the scenario requirements. The switch will assign authenticated users to the "denyall" role and drop their traffic. **C** is incorrect for the same reason. Also note that the "denyall" role is already the reject role by default. **D** is incorrect. User role authorization is always enabled on AOS-CX switches.

    For more information, see Chapter 7.

29. **D** is correct. Tarpit containment disassociates clients from the contained AP and also sets up a decoy association to keep the clients from reconnecting to the AP.

    A is incorrect. HPE Aruba Networking WIP does support wired containment. However, this feature involves Air Monitors (AMs) or hybrid APs using ARP poisoning against the contained device. **B** is incorrect. Tarpit containment does not use wired frames. **C** is incorrect. Valid client protection disassociates only *Authorized Clients* from Unauthorized APs, protecting those clients from accidentally connecting to a hacker's device.

    For more information, see Chapter 12.

30. **A** is correct. ARP inspection enables switches to monitor ARP responses and only permit ones with valid IP-to-MAC bindings. In this way, the switches prevent ARP poisoning attacks. To learn valid IP-to-MAC bindings for clients with dynamic IP addresses (which most have), the switch uses DHCP snooping.

    B is incorrect. BPDU guard prevents unauthorized devices from joining the spanning tree and is not related to ARP inspection. **C** and **D** are incorrect. ARP inspection, not ARP proxy, is the correct feature for this use case.

    For more information, see Chapter 5.

Appendix A Glossary

Access Control Entry (ACE)

An ACE is an entry in an access control list (ACL), which selects traffic based on criteria, such as packet headers, and applies actions such as "permit" or "deny." An ACE is often called a rule.

Access Control List (ACL)

An ACL is an ordered list of ACEs, or rules, which filter traffic that passes through a firewall.

Access Point (AP)

An AP includes one or more radios, on which it advertises SSIDs and associates with clients to establish WLANs. The AP also acts as a bridge to corporate LAN. HPE Aruba Networking offers several types of APs, including Instant APs (IAPs), which form controllerless clusters, Campus APs (CAPs), which are controlled by Mobility Controllers (MCs), and Remote APs, which are deployed at home offices or small branch offices and controlled by MCs.

Accounting

Accounting tracks how users or devices use network services.

Active Directory (AD)

AD is the Microsoft Windows directory solution. AD stores the directory of users and devices in a Windows domain, manages domain policies, and makes directory information available to authorized users and applications.

Add Block Acknowledecment (ADDBA)

802.11n and 802.11ac require acknowledgements (ACKs). 802.11e introduced the concept of block ACKs to decrease the number of ACKs required. Both the AP and client must agree to use block ACKs. An ADDBA frame requests the use of block ACKs.

Address Resolution Protocol (ARP)

ARP is a protocol for resolving IP addresses. A device broadcasts an ARP Request to learn the MAC address that is associated with an IP address within the device's own subnet. The device that owns the requested IP address sends an ARP Response with its MAC address.

Advanced Encryption Standard (AES)

The National Institute of Standards and Technology (NIST) created AES, a symmetric block cipher, to protect sensitive information. It has been in use since 2002. AES supports 128-bit, 192-bit, and 256-bit keys.

Adware

Adware is a Potentially Unwanted Program (PUP) that displays advertisements.

Air Monitor (AM)

An HPE Aruba Networking AM is an AP that has been configured in AM mode. AMs do not service clients. They are dedicated to scanning for Layer 2 802.11 frames in order to detect unauthorized wireless devices, such as rogue APs, and other potential threats. They can also take actions to contain threats.

Algorithm

In the context of cryptography, an algorithm is a set of instructions for transforming data with a key. An encryption algorithm is sometimes called a cipher.

Amazon Web Services (AWS)

AWS is Amazon's public cloud service, which provides companies with on-demand compute and storage resources through a subscription payment model.

Act on the Protection of Personal Information (APPI)

The APPI is data protection legislation passed in 2003 by the Japanese government "to protect the rights and interests of individuals while taking consideration of the usefulness of personal information."

Application Layer Gateway (ALG)

An ALG is a firewall component that handles traffic based on application. It helps the firewall to correctly handle applications that use dynamic ports, which are not known in advance.

Application Programming Interface (API)

An API is a programmatic interface that exposes information, features, and functions of one application to other applications. APIs enable scripts or third-party tools to configure settings on the device that hosts the API through calls to the API.

Artificial Intelligence (AI)

Sometimes called machine intelligence, AI refers to the ability of devices or applications to take action based on environmental factors and inputs. In the strictest sense, AI refers to the ability of devices or applications to actually learn and exhibit cognitive behavior.

Aruba-User-Role Vendor Specific Attribute (VSA)

RADIUS servers such as ClearPass Policy Manager (CPPM) can send this attribute in an Access-Accept message for a client's authentication. The AOS device, or AOS-CX switch, that receives the message uses the value in the attribute to assign the client to a role.

Application-Specific Integrated Circuit (ASIC)

ASICs are purpose-designed circuits that are meant to execute only one function, for example, holding hardware forwarding tables. The specificity of their design enables them to execute that function very quickly. ASICs form the basis for high-speed fabrics on network infrastructure devices.

ASLEAP

ASLEAP is a tool created for Linux systems and used to attack Cisco LEAP authentication protocol.

Attribute Value Pair (AVP)

In the context of RADIUS and TACACS+ packets, an AVP is a field that defines the value for a specific attribute. For example, a RADIUS packet could have an AVP that specifies "Alice" as the value for the User-Name attribute.

Authenticated Encryption with Associated Data (AEAD)

An AEAD algorithm both encrypts data and ensures the data integrity. In addition to ensuring the integrity of the ciphertext, the algorithm ensures the integrity of "associated data" that is not encrypted.

Authentication

Authentication verifies a user or device's identity.

Authentication and Key Management (AKM)

As defined by 802.11i and implemented by WPA/WPA2/WPA3, AKM defines the mechanism by which a client authenticates to the WLAN. It further defines how the AP and client will establish the hierarchy of keys used to secure the connection. The AKM is how APs using various version of WPA advertise whether the WLAN operates in Personal mode (pre-shared key plus a 4-way handshake) or Enterprise mode (802.1X plus a 4-way handshake).

Authentication Authorization Accounting (AAA)

AAA is a framework for securing access to network services. It requires users to authenticate, or verify their identity, then receive authorization, which determines what users can do, and then be subject to accounting, which records users' actions. Examples of AAA protocols are RADIUS and TACACS+.

Authentication, Authorization, and Accounting (AAA) profile

AOS devices use AAA profiles to specify AAA-related settings for clients. Among other settings, the AAA profile specifies the enabled authentication types, the servers used for authentication, and default roles for various authentication methods. Each WLAN has an AAA profile attached to it.

Authenticity

Data authenticity refers to ensuring that data was truly transmitted by the device that claims to have sent it. It is related to data integrity, which ensures that data is not altered in transit.

Authorization

Authorization determines which rights a user or device has.

Availability

Availability ensures that network services are available for legitimate users.

Base Service Set Identifier (BSSID)

The BSSID is the MAC address of an AP on an SSID. The BSSID is included in all 802.11 frames sent between the AP and clients connected to the SSID.

Border Gateway Protocol (BGP)

BGP is a routing protocol used to route traffic between autonomous systems (ASes). While BGP is most often used between Internet Service Providers (ISPs), some companies also use BGP internally.

Bot and Botnet

A bot is a device that has been infected with malware that places it under the control of a hacker. A botnet is a network of many bots.

Branch Gateway (BGW)

A component of an HPE Aruba Networking EdgeConnect SD-Branch, an HPE Aruba Networking BGW links a branch LAN to the corporate datacenter and other outside resources. A BGW works with HPE Aruba Networking Central to provide SD-WAN capabilities, as well as AOS firewall policies and other features to protect the branch LAN.

Bridge Protocol Data Unit (BPDU)

BPDUs are data frames used in Spanning Tree Protocol (STP), Rapid Spanning Tree Protocol (RSTP), and Multiple Spanning Tree Protocol (MSTP). BPDUs advertise information about the spanning tree such as the bridge ID of the root and the cost for reaching the root from the switch transmitting the BPDU.

Bring Your Own Device (BYOD)

BYOD is the trend of employees using their own devices at work. Smartphones are the most common example of BYOD, but employees might also take their own tablets, laptops, and other types of devices to work.

Broadcast domain

A broadcast domain defines the extent over which broadcasts are propagated. In Ethernet, a virtual LAN (VLAN) typically defines the broadcast domain. Devices in the same VLAN receive each other's broadcasts and communicate at Layer 2. Routing is required to permit communications between devices in different broadcast domains.

Campus AP (CAP)

An HPE Aruba Networking CAP is an AP that is controlled by a Mobility Controller (MC). HPE Aruba Networking offers Unified APs (UAPs), which can be configured to operate in CAP or IAP mode.

Captive portal authentication

Captive portal authentication is a form of Layer 3 authentication. When a client initially connects, the client is provided only enough network access to be redirected to a Web server. The Web server hosts a portal through which users can log in. ClearPass Guest and ClearPass Policy Manager (CPPM) can work together to deliver captive portal authentication.

Cardholder Data Environment (CDE)

A CDE is a system that processes, stores, and transmits sensitive payment authentication data.

Central Processing Unit (CPU)

A CPU executes processes that allow applications to run. A network infrastructure device's CPU runs the device software, which supports the device's management and control planes.

Certificate

A digital certificate binds a subject's identity with the subject's public key. Certificates can be used for purposes such as authentication.

Certificate Signing Request

A CSR is a request for a digital certificate. It includes an entity's subject name and public key and sometimes other fields such as Subject Alternative Name (SAN). You submit a CSR to a Certification Authority (CA) to be transformed into a certificate and signed.

Certification Authority (CA)

A CA vouches for the authenticity of the identity of the certificate holder. All valid certificates must trace their verification back to a root CA. Root CAs can be public CAs, which are well-known CAs that are widely believed to be trustworthy, such as Verisign and DigiCert, or private CAs, which are maintained and trusted within a private organization. Root CAs sign certificates for subordinate or intermediate CAs. Intermediate CAs sign end-entity certificates.

Chain of custody

Chain of custody refers to the tracking of how data is collected and stored in order to preserve its integrity.

Challenge Handshake Authentication Protocol (CHAP)

CHAP is an authentication protocol that authenticates a client with a username and password. To prove that it knows a password, the client transmits a hash of the password and a challenge value unique to the exchange. To validate the password, the authentication sever must have access to plaintext passwords stored with reversible encryption. RADIUS CHAP is not considered a highly secure protocol, so it should be run over trusted network connections or protected with RadSec.

Change of Authorization (CoA)

A RADIUS server sends a CoA message to a Network Access Server (NAS) to tell the NAS to change the authorization status of a client. For example, a CoA can be a disconnect message (DM), which tells the NAS to terminate the client's session.

Chief Financial Officer (CFO)

A CFO is responsible for ensuring the financial well-being of a company through financial planning, risk management, and other means.

Cipher

A cipher is an algorithm for transforming data from plaintext to ciphertext.

Cipher Block Chaining Message Authentication Code (CBC-MAC or CMAC)

CBC is a mode of operation for block ciphers. It uses each ciphertext block as an additional input for encrypting the next plaintext block. CBC MAC uses the final ciphertext block as a Message

Authentication Code (MAC), which a receiving device with the same key can use to validate the data's authenticity and integrity.

Ciphertext

Ciphertext is data that has been encrypted.

Cisco Discovery Protocol (CDP)

CDP is a Layer 2 protocol that enables networking applications to obtain information about neighboring devices, such as system name and IP address.

Clear to Send (CTS)

As an optional part of the 802.11 protocol, Request to Send (RTS) and Clear to Send (CTS) frames can reduce collisions in circumstances in which clients connected to the same AP cannot hear each other. When a client connected to an AP wants to transmit, it sends an RTS. If no other client is using the air, the AP sends a CTS.

ClearPass Device Insight (CPDI)

CPDI provides an advanced solution for profiling and categorizing endpoints. Collectors, deployed on-prem, can use a variety of passive and active scanning methods. They submit metadata to cloud-based Device Insight Analyzer, which can apply machine learning and supervisor crowd-sourcing to further identity devices.

ClearPass Guest

ClearPass Guest can be deployed with ClearPass Policy Manager (CPPM) to provide more sophisticated authentication services for guests. ClearPass Guest provides a login portal for guests and many other services.

ClearPass Insight

ClearPass Insight provides customizable reports for a ClearPass environment.

ClearPass Onboard

ClearPass Onboard provides an easy-to-use portal from which users can install an agent, which automatically configures the users' clients with the correct authentication settings and enrolls the clients for any necessary certificates.

ClearPass Policy Manager (CPPM)

A central component of the HPE Aruba Networking security strategy, HPE Aruba Networking CPPM provides sophisticated, policy-based access control for users and devices. Among many other features, CPPM acts as a RADIUS server and TACACS+ server. It also provides device profiling capabilities.

Cloud Access Security Broker (CASB)

CASB stands as a "broker" between users and SaaS applications. It can then enforce security controls, such as inspecting traffic and restricting certain actions (such as file transfers)—all based on the organization's security policies.

Command Line Interface (CLI)

A CLI provides a text-based interface for admins to interact with a device. Admins can enter commands to view and alter the device's settings and configuration.

Commercial National Security Algorithm (CNSA) Suite

The CSNA Suite includes the algorithms that NIST recommends for ensuring security as the world transitions toward quantum computing.

Computer Reseller News (CRN)

CRN is a publication that offers market analysis and updates regarding the technology industry.

Confidentiality

Data confidentiality refers to ensuring that only authorized parties can read the data. It is sometimes called data privacy.

Control plane

The control plane of a network infrastructure device runs protocols. It builds routing tables and MAC forwarding tables, and it processes traffic that is destined for the device itself.

Control Plane Policing (CoPP)

Control Plane Policing (CoPP) helps to protect a network device's control plane from excessive traffic, which could overwhelm the device and prevent it from forwarding traffic or permitting management access. CoPP monitors traffic incoming into the control plane and applies rate limits to that traffic based on class.

Control Plane Security (CPsec)

CPsec provides security for the PAPI channels between APs and Mobility Controllers (MCs). An AP and its MC use certificates to mutually authenticate each other and then construct a secure IPsec tunnel, which provides data encryption and integrity for their communications.

Counter Mode (CTR)

CTR mode is a mode of operation for block algorithms, which in some ways imitates a stream cipher. For the first block, it encrypts an Initialization Vector (IV), the counter, and the key with the block algorithm. It then XORs the result with the plaintext data block to produce the ciphertext. The next block uses the previous counter plus 1.

Counter Mode with CMAC

CCM combines Counter Mode (CTR) encryption with Cipher Block Chaining-MAC (CMAC) to provide data integrity. CMAC first authenticates the complete packet, including data that is not encrypted (such as MAC addresses and IP addresses). Then, CTR mode encrypts the payload and the MAC.

Counter Mode with CMAC Protocol

CCMP is an encryption protocol for wireless security. It uses a MIC key and Cipher Block Chaining-MAC (CMAC) to calculate a Message Integrity Code (MIC). The MIC is a single block that acts a fingerprint for the frame to prevent tampering with the data. AES CTR mode then encrypts the packet payload and the MIC with a key called the temporal key (TK).

Cyclic Redundancy Check 32 (CRC32)

CRC32 is a mathematical operation performed to validate the integrity of data being transmitted. Wired Equivalent Privacy (WEP) uses CRC32, but CRC32 is not a secure way to ensure integrity.

Data Loss Prevention (DLP)

DLP monitors users' communications with applications. It detects file transfers and blocks those transfers if they include unauthorized data. For example, the organization might not permit transfers that include customers' personal data.

Data plane

A network infrastructure device's data plane consists of the paths over which data traffic flows within the device. It is sometimes called the fabric or forwarding plane.

Decryption

Decryption is the act of transforming text from ciphertext back to plaintext.

Deep Packet Inspection (DPI)

DPI enables devices to examine packets (rather than just packet headers) and filter packets, based on data in the packet payload. The AOS firewall can use DPI to implement application rules, which control traffic based on the type of application such as gaming or social networking.

Demilitarized Zone (DMZ)

In an IT context, a DMZ stands between an untrusted external network and a more trusted internal network and has a firewall on each side of it. The DMZ often hosts more public facing services with highly secured servers, enabling the company to minimize external access to the internal network.

Denial of Service (Dos)

A DoS is an attack that interferes with legitimate users receiving network services.

Department of Defense (DoD)

The DoD is an executive branch department of the United States responsible for managing agencies that are charged with maintaining national security, including the National Security Agency (NSA).

Differentiated Services Code Point (DSCP)

A DSCP is an IP packet header value used for Differentiated Services (DiffServ)-based quality of service (QoS). A DSCP can be used to request a specific priority or type of handling for traffic.

Diffie-Hellman (DH)

Two parties can use a Diffie-Hellman key exchange to securely create a shared secret over a network that might have eavesdroppers. The parties can then use that secret for symmetric encryption. The mathematics for generating the values exchanged and the secret key involve prime modulus or elliptic curve operations. Unauthenticated Diffie-Hellman is susceptible to MITM attacks. However, authenticated Diffie-Hellman, in which parties use asymmetric encrypt to "sign" exchanged values, is not.

Digital Encryption Standard (DES)

DES is a symmetric block algorithm. It is considered to be insecure.

Digital Experience Monitoring (DEM)

HPE Aruba Networking DEM is part of HPE Aruba Networking SSE. It continuously monitors sessions, granting admins insights. It helps admins detect performance issues and resolve problems more quickly.

Digital Signature Algorithm (DSA)

Developed by the National Institute of Security Technologies (NIST), DSA is an asymmetric algorithm. It uses a private key to encrypt or "sign" data; other entities can then use the corresponding public key to validate the signature.

Disconnect Message (DM)

A DM is a type of RADIUS Change of Authorization (CoA) message, which tells the Network Access Server (NAS) to disconnect a client from the network.

Distributed Denial of Service (DDoS)

DDoS attacks amplify the effects of a DoS attack by using a large number of devices to launch the attack. The devices involved in a DDoS attack are often botnets.

Domain Name System (DNS)

Many applications specify destinations by memorable and consistent domain names or fully qualified domain names (FQDNs), as opposed to IP addresses. DNS enables clients to query servers to discover the IP address that is currently associated with a domain name.

Downloadable User Role (DUR)

DURs are user roles that a Network Access Device (NAD) (such as an AOS MC, IAP, or AOS-CX switch) downloads from ClearPass Policy Manager (CPPM). Using DURs moves role configuration from the network admins to the ClearPass admins.

Drive by download

A drive by download occurs when software is downloaded and installed on a device without the user's knowledge. Hackers can conduct drive by downloads by exploiting OS or application vulnerabilities.

Dynamic Host Configuration Protocol

DHCP enables DHCP servers to dynamically assign client devices valid IP addresses, subnet masks, and other settings such as default gateway address and DNS server address.

Dynamic Path Selection (DPS)

In an HPE Aruba Networking EdgeConnect SD-Branch solution, DPS chooses the best path for traffic based on information from path quality monitoring (PQM). DPS could mean, for example, that the branch gateway (BGW) chooses to route real-time voice and video on the path identified by PQM as having the lowest latency and jitter, while routing bulk traffic on the path with most bandwidth.

EdgeConnect Software-Defined Branch (SD-Branch)

HPE Aruba Networking EdgeConnect SD-Branch is an HPE Aruba Networking solution that combines the benefits of a software-defined WAN with a software-defined LAN, all with rigorous security.

Software-Defined Local Area Network (SD-LAN) SD-LAN allows IT to automate the establishment of branch LANs. HPE Aruba Networking branch LANs benefit from the same role-based security as HPE Aruba Networking-based campus LANs.

EdgeConnect Software-Defined Wide Area Network (SD-WAN)

HPE Aruba Networking EdgeConnect SD-WAN is the lead HPE Aruba Networking SD-WAN solution. (See also "Software-Defined Wide Area Network WAN.)

Elliptic Curve Cryptography (ECC)

ECC utilizes elliptic curves (plane algebraic curves that have no cusps or self-intersections) as the basis for asymmetric encryption, also called public-key cryptography. As compared to finite field cryptography (FCC)-based asymmetric encryption, ECC can achieve the same security with shorter keys.

Elliptic Curve Diffie-Hellman

ECDH is the creation of a shared secret as outlined by Diffie-Hellman using elliptic-curve cryptography (ECC).

Elliptic Curve Digital Signature Algorithm

ECDSA is a variation of the Digital Signature Algorithm that uses elliptic curve cryptography (ECC).

Encapsulated Remote Mirroring (ERM)

ERM is the protocol that a network infrastructure device uses for remote mirroring. The device copies traffic flowing through it, encapsulates the traffic with UDP, and sends the encapsulated traffic to a destination IP address.

Encryption

Encryption is the act of transforming text from plaintext to ciphertext.

Endpoint Detection and Response (EDR)

An EDR system usually uses software agents on endpoints to continuously monitor those devices. The resulting data is stored and processed in a central location and if an issue is detected, an alert or automatic response may be created.

Enterprise Mobility Management (EMM)

EMM solutions allow companies to monitor and manage the use of mobile devices within a company's network, as well as to secure the company's data on those devices.

Ethernet

Ethernet is a Layer 2 technology often used in local area networks (LANs). Ethernet enables devices to connect together over copper or fiber optics cables. In a typical Ethernet network, endpoints connect to switches which place the endpoints in the appropriate broadcast domains or virtual local area networks (VLANs). Switches can connect to other switches to extend VLANs over a campus or segment of a data center. A routing switch or router is required to route traffic between devices in different VLANs; a router might also connect the Ethernet LAN to the Internet or to other sites in a wide area network (WAN).

European Union (EU)

The EU is a political, social, and economic confederation of European states, headquartered in Brussels.

Exclusive OR (XOR) cipher

The XOR cipher is a symmetric stream algorithm. Each bit in plaintext is compared to a bit in a key. If the bits are the same, the bit for the ciphertext becomes 0. If the bits are different, the bit for the ciphertext becomes 1.

Extensible Authentication Protocol (EAP)

EAP is a RADIUS authentication protocol used with 802.1X authentication. EAP is a framework, and many EAP methods exist.

Extensible Authentication Protocol over LAN (EAPOL)

EAPOL is the protocol that transports EAP messages over Ethernet or 802.11.

Extensible Authentication Protocol-Transport Layer Security (EAP-TLS)

Supported on many supplicants, EAP-TLS is the most secure option for EAP. EAP-TLS provides mutual authentication using digital certificates on both sides.

Extensible Authentication Protocol-Tunneled Extensible Authentication Protocol (EAP-TEAP)

Supported in Windows 10 and ClearPass Policy Manager (CPPM) 6.9, EAP-TEAP uses a Tunneled Layer Security (TLS) handshake to authenticate the server and create a secure TLS tunnel. An inner

EAP method then authenticates the user. EAP-TEAP supports chaining, in which an authentication request can include a token from a former authentication.

Extensible Authentication Protocol-Tunneled Transport Layer Security (EAP-TTLS)

EAP-TTLS uses a Tunneled Layer Security (TLS) handshake to authenticate the server and create a secure TLS tunnel. An inner EAP method then authenticates the user.

Fabric

A network infrastructure device's fabric consists of the paths over which data traffic flows within the device. It is sometimes called the data plane or forwarding plane.

FATA-Jack

FATA-Jack is an 802.11 client DoS tool that tries to disconnect targeted stations using spoofed authentication frames that contain an invalid authentication algorithm number.

File Transfer Protocol (FTP)

FTP is a standard protocol for transferring files between a client and a server over a network.

Finite Field Cryptographic (FFC)

FFC is a form of cryptography that uses finite fields. A finite field, also called a Galois field, is a finite set of numbers for which mathematical operations such as addition, subtraction, multiplication, and division function as for real and rational numbers. RSA is an example of FFC. In RSA a finite field is created by multiplying two large prime numbers. That field is then used for calculating a public and private keypair and for encrypting/decrypting data with those keys.

Firewall

A firewall is a device that stands inline with traffic and filters traffic that flows through it, often based on ACLs. The AOS firewall is role-based and stateful. It provides capabilities such as application layer gateways (ALGs) and deep packet inspection (DPI).

Firewall as a Service (FWaaS)

An FWaaS is a firewall that is hosted in the cloud. It filters traffic as a service.

Forensics or digital forensics

Digital forensics refers to the act of collecting and analyzing data in a way that preserves the integrity of the information and permits the data to be used as evidence in a security investigation.

Fully Qualified Domain Name

The FQDN is a domain name for a specific computer or host that details the host's exact location in the tree hierarchy of the Domain Name System. It consists of a hostname and a domain name.

Galois Counter Mode (GCM)

GCM is a mode of operation for block algorithms. It takes as inputs an IV, a key, plaintext that needs to be encrypted, and data that needs to be authenticated but not encrypted. It outputs ciphertext, created with CTR mode encryption, and a MAC, created with Galois field multiplication. The most processor-intensive operations for generating this MAC can be parallelized so that GCM can secure data at higher speeds than CCM.

General Data Protection Regulation (GDPR)

GDPR is an EU regulation that defines personal data and how companies can collect and store that data. All companies that collect information about EU citizens are subject to GDPR.

Generic Routing Encapsulation (GRE)

GRE is a standard for "tunneling" traffic through an underlying network. GRE encapsulates packets with a GRE header and a delivery IP header destined to the other side of the tunnel. The network forwards the traffic to the delivery destination, as it would a normal IP packet. The other side of the tunnel decapsulates the packet—removes the delivery and GRE headers. It then forwards the packet based on the original destination. In this way, a GRE tunnel enables traffic to move from one part of a network to another without the underlying network needing to be aware of, or able to provide forwarding for, the original source and destination addresses. A GRE tunnel should not be confused with a secure tunnel; it does not provide encryption.

If GRE operates at Layer 2, it retains the traffic's Layer 2 header before adding the GRE and delivery IP headers. For example, HPE Aruba Networking CAPs and MCs can encapsulate 802.11 frames and send them over GRE tunnels. This form of operation enables the extension of the same broadcast domain across an underlying network that uses routing.

Gigabit Ethernet (GbE)

Defined in the IEEE 802.3-2008 standard, Gigabit Ethernet (GbE or 1 GigE) enables the transmission of Ethernet frames at a rate of a gigabit per second (1,000,000,000) bits per second.

Gigabit per Second (Gbps)

Gbps is a unit of measurement that describes how many billions of bits are transmitted per second.

Google Cloud Platform (GCP)

GCP is Google's public cloud platform. It provides companies with IT services such as compute and storage on a pay-per-use basis.

Group Master Key (GMK)

Within WPA/WPA2/WPA3, a GMK is the key used to derive a group temporal key (GTK), which encrypts and decrypts broadcast and multicast traffic.

Group Policy Object (GPO)

A Windows domain uses GPOs to control settings for users and devices in the domain from a central location.

Groupwise Temporal Key (GTK)

Within WPA/WPA2/WPA3, a GTK is the key that encrypts and decrypts broadcast and multicast traffic.

Hacker

A hacker is a bad actor who introduces threats, seeks to disrupt systems, and seeks to gain access to resources to which the hacker is not authorized.

An "ethical hacker" could be a security researcher, who uses similar tools as hackers in order to discover vulnerabilities and responsibly inform vendors.

Hash

A hash is data of a fixed size generated by a hash algorithm from data of any size. The hash algorithm should be one-way; the original data cannot be derived from the hash. In addition, a change to the message should change the hash in an unpredictable way, and it should be infeasible to predict two messages that will produce the same hash.

Hash algorithm

A hash algorithm is an algorithm that is used for generating a hash from a message. Examples of hash algorithms include MD5, SHA1, SHA2, and SHA3.

Hash-based Message Authentication

HMAC applies a hash algorithm to a key and data to produce a secure MAC, which is transmitted with the original data. The receiver can use the same key to check the MAC, verifying that the data comes from a trusted source (defined as knowing the key) and that data has not been altered in transit. Keys used for this purpose are sometimes called integrity or authentication keys.

Headquarters (HQ)

The central location of an organization, from which branch locations can be managed.

Health Information Technology for Economic and Clinical Health Act (HITECH)

The American federal government passed this law in 2009 to encourage the adoption and continued use of health information technology.

Health Insurance Portability and Accountability Act (HIPAA)

HIPAA is an American law that protects the privacy of individuals' personal health information from fraud and theft. It regulates many aspects of how organizations are allowed to collect, store, use, and communicate individuals' personal health information.

High Throughput (HT)

In 802.11 standards, HT refers to 802.11n. A wireless device can indicate that it is 802.11n capable by including the HT Capabilities Information Element (IE) in management frames like probe requests. An AP can indicate the cell operates in HT mode with the HT Operation IE in management frames like beacons and probe responses.

HyperText Transfer Protocol (HTTP)

Carried by TCP/IP, HTTPS is a protocol for transmitting hypermedia between clients and servers. Clients use HTTP, or the similar HTTPS, to request data from Web servers. Most communications across the Internet, and many internal network communications as well, use HTTP or HTTPS.

HyperText Transfer Protocol Secure (HTTPS)

HTTPS offers the same functionality as HTTP but with greater security by establishing a Transport Layer Security (TLS) tunnel over which to send the HTTP messages.

Information Element (IE)

APs include IEs in their beacons and probe responses to advertise capabilities. Similarly, clients include IEs in probe requests and association requests.

Information Technology (IT)

IT involves the use of personal computing devices, servers, supporting infrastructure such as storage and networking devices (both wired and wireless), and applications. IT enables users to create, store, transfer, and manipulate data.

Initialization Vector (IV)

An IV is a number used as the first value for an iterated process, particularly for data encryption. Many encryption algorithms use an IV with a key to prevent the same plaintext and key from always producing the same ciphertext.

Instant AP (IAP)

HPE Aruba Networking IAPs provide a controllerless solution. They automatically form a cluster which hosts a single point of management. Configuration changes are propagated across the cluster. IAPs support many of the same AOS features as controller-based solutions, but each IAP implements those features, and forwards wireless client traffic into the LAN, individually.

HPE Aruba Networking delivers APs as Unified APs, which can be configured in IAP or campus AP (CAP) mode.

Institute of Electrical and Electronics Engineers (IEEE)

The IEEE describes itself as "the world's largest technical professional society—promoting the development and application of electrotechnology and allied sciences for the benefit of humanity, the advancement of the profession, and the well-being of our members." It defines many of the standards that regulate communications across computer networks.

Integrity

Data integrity refers to ensuring that data is not altered between the transmitter and the receiver.

Integrity Check Value (ICV)

Wired Equivalent Privacy (WEP) used Cyclic Redundancy Check 32 (CRC32) to create ICVs to verify the integrity of packets. However, CRC32 and the ICV were not secure.

Interface for Metadata Access Point (IF-MAP)

IF-MAP provides an open framework for orchestrating actions across multiple devices. It defines a standard by which metadata access points can publish data to a database server. AOS devices support IF-MAP for sending HTTP User-Agent string metadata to ClearPass Policy Manager (CPPM).

Internet of Things (IoT)

The Internet of Things (IoT) consists of things that are not primarily computing devices but are also connected to the Internet. These things include machinery, HVAC systems, consumer goods, and many more.

Internet Protocol (IP)

IP is a protocol used to send data from one computer to another on the Internet. Each computer (known as a host) on the Internet has at least one IP address that uniquely identifies it from all other computers on the Internet.

Intrusion Detection System (IDS)

An IDS helps to detect threats to the network and to alert administrators to those threats. It is often deployed in a central location and has traffic directed to it for examination, although it can also be located within the traffic flow. An IDS might use signatures to detect specific patterns indicative of an attack, or it might use various methods to detect anomalies or suspicious behavior indicative of an attack.

An IDS is often combined with an Intrusion Prevention System (IPS).

Intrusion Prevention System (IPS)

An IPS is a system that detects threats and takes action to mitigate those threats. The IPS might be installed within the traffic flow, or other devices might send traffic to the IPS for examination. In the latter case, the IPS would need some way to interface with other devices in order to mitigate the threat.

IP Address Manager (IPAM)

IPAM is a form of software used to assign and dictate the use of IP addresses on a given network. It might manage DHCP and DNS services in a network.

IP Security (IPsec)

IPsec is an industry standard for establishing virtual private network (VPN) tunnels, which securely deliver data across untrusted networks. IPsec can use Authentication Header (AH) or Encapsulating Security Protocol (ESP); ESP uses shared keys to provide privacy and authentication for data transmitted over the IPsec tunnel.

IP subnet

A campus or data center LAN typically consists of multiple IP subnetworks or subnets. Devices within the same subnet can communicate at Layer 2 and send broadcasts to each other. Routing is required for devices in different subnets to communicate. IP subnets come in various sizes, as defined by the prefix length of the network address. In an Ethernet LAN, each subnet is more or less contiguous with a virtual local area network (VLAN).

Keccak Message Authentication Code

KMAC is a keyed hash function (though it can also be used without a key as a regular hash function) that is approved by NIST for message authentication.

Key Confirmation Key (KCK)

Part of the WPA key hierarchy, the KCK is one of the keys derived from the Pairwise Transient Key (PTK). The client and AP (or its controller) use the KCK to authenticate messages transmitted during the 4-way handshake and to implicitly demonstrate that they know the Pairwise Master Key (PMK), which is the ultimate source of the KCK.

Key Encryption Key (KEK)

Part of the WPA key hierarchy, the KEK is one of the keys derived from the Pairwise Transient Key (PTK). The AP (or its controller) uses the KEK to encrypt the Groupwise Transient Key (GTK) transmitted during the 4-way handshake.

Kill chain

Lockheed Martin developed a "kill chain" model to describe targeted attacks. The stages in the kill chain include: Reconnaissance, Weaponization, Delivery, Exploitation, Installation, Command and Control, and Actions on Objective.

Layer 2 authentication

Layer 2 authentication requires a client to authenticate before the client's Layer 2 connection comes up.

Layer 3 authentication

Layer 3 authentication provides an unauthenticated client with limited network access such as DHCP for receiving IP settings and DNS for resolving domain names. The client must authenticate to receive more complete access.

Lightweight Directory Access Protocol (LDAP)

LDAP is a protocol used for accessing directory services. AOS devices can implement some forms of authentication by querying an LDAP server. ClearPass Policy Manager (CPPM) can also use LDAP to use an external directory to authenticate clients.

Link Layer Discovery Protocol (LLDP)

LLDP allows network devices to advertise information about themselves to peers on a network, usually an Ethernet LAN.

Local Area Network (LAN)

A LAN is a network of devices, which might include personal computing devices, printers, servers, and Internet of Things (IoT) devices, at a single location such as a home, office building, or campus. Enterprise and SMB LANs typically use Ethernet, and the LAN might include multiple subnets through the use of virtual LAN (VLAN) technology. Access points (APs) often integrate mobile devices into the LAN as well.

Log

A log is a record of an event. Network infrastructure devices create logs of many events that occur on them. They can store the logs locally and also forward them to a Syslog server.

MAC Authentication (MAC-Auth)

MAC authentication is an example of Layer 2 authentication. It authenticates devices based on their source MAC address and requires no special capabilities or configuration on the client. However, MAC-Auth is easily spoofed and not highly secure.

Malvertising

In malvertising, hackers create ads with embedded malware and use ad servers to deliver that malware to many web sites.

Malware

Malware is a malicious code that executes actions unwanted by the owner of the device running the malware. Malware seeks to disrupt a system or to use the system in unauthorized ways.

Management Frame Protection (MFP)

MFP increases the security of management frames sent over a wireless 802.11 network. It encrypts and authenticates unicast management frames with the Temporal Key (TK) and authenticates broadcast management frames with the Groupwise Transient Key (GTK).

Management plane

The management plane of a network infrastructure device holds the configuration. It also hosts the CLI, Web UI, and other interfaces used to monitor, manage, and configure the device.

Man-in-the-Middle Attack (MITM)

In an MITM attack, a hacker inserts himself or herself between two communicating devices. The hacker might simply eavesdrop on communications or also tamper with the communications.

Media Access Control (MAC) address

A MAC address, sometimes called a hardware address, is the Layer 2 address assigned to an Ethernet or wireless network interface card (NIC). A NIC's MAC address is typically assigned at the factory, although it is possible to change a NIC's MAC address manually. Layer 2 devices such as switches use traffic's destination MAC address to determine how to forward the traffic.

For IP communications over Ethernet, devices use IP addresses to address traffic for the correct destination, and devices in different subnets have no awareness of each other's MAC addresses. When the traffic reaches the IP subnet in which the destination device resides, infrastructure devices use Address Resolution Protocol (ARP) to find the MAC address associated with the destination IP address and deliver the traffic accordingly.

Megahertz (MHz)

The hertz is a measure of frequency, and one hertz is defined as one cycle per second; the megahertz is one million hertz, that is, one million cycles per second. 802.11 channels are 20 MHz, but channels can be bonded into 40 MHz, 80 MHz, or even wider channels.

Message Authentication Code (MAC)

A MAC is securely generated data that is transmitted with a message to authenticate the message and indicate whether it has been changed. Generating and validating a MAC requires knowledge of a key. Hash-Based Message Authentication Code (HMAC) is one method for generating MACs.

Message Digest 5 (MD5)

MD5 is an example of a hash function; however, MD5 suffers from various shortcomings and its use is deprecated.

Microsoft Challenge Handshake Authentication Protocol (MSCHAP)

MSCHAP is a RADIUS authentication protocol, which provides mutual authentication. Credentials are a username and password. The peers prove knowledge of a password hash by using it to encrypt data exchanged in the MCSHAP process. AOS devices can use MSCHAP for RADIUS authentication of managers, as well as for MAC-Auth or captive portal authentication. While more secure than PAP or CHAP, MSCHAP has known vulnerabilities and should be implemented over trusted network connections or a secure tunnel.

Mobile Device Management (MDM)

MDM refers to the administration of mobile devices. An MDM solution performs functions such as deploying configurations to mobile devices, distributing applications to them, monitoring how the devices are used, and enforcing other policies.

Mobility Controller (MC)

An HPE Aruba Networking MC controls Campus APs (CAPs). The MC provides the main management interface for configuring the mobility solution. The MC pushes the appropriate configuration to managed CAPs. CAPs also typically tunnel wireless client traffic to their MC (based on the forwarding mode for the WLAN). The MC then processes the wireless client traffic, authenticates clients with an internal server or to an external authentication server such as ClearPass Policy Manager (CPPM), assigns clients to roles, applies role-based firewall policies, among other features, and finally forwards the traffic into the LAN. MCs can also function as branch gateways (BGWs) and VPN concentrators.

When an organization has multiple MCs, the MCs can, in turn, be managed by a Mobility Conductor (MCR). In that case, the hosts the main management interfaces.

Mobility Conductor (MCR)

An HPE Aruba Networking MCR provides single-pane-of-glass management for multiple Mobility Controllers (MCs), or managed devices (MDs).

multicast DNS (mDNS)

mDNS is a variant of DNS that resolves hostnames to IP addresses without requiring a DNS server, although it can work in conjunction with DNS servers. A host transmits a multicast message, asking the device that owns a DNS name to respond with its IP address.

Multiple Spanning Tree Protocol (MSTP)

MSTP is a version of spanning tree protocol (STP). It uses multiple spanning tree instances, each of which is mapped to a different set of VLANs. Each instance operates like a unique spanning tree using Rapid Spanning Tree Protocol (RSTP). Because different instances can block different ports, MSTP enables VLAN-based load sharing of traffic over redundant links.

Multi-Protocol Label Switching (MPLS)

Often used by service providers, MPLS enables the provider network to transport traffic from multiple customers over stable and efficient label switched paths (LSPs). With an MPLS-based WAN connection, a customer can connect to the MPLS provider over any media that supports IP. The service provider can grant the customer various amounts of bandwidth and quality of service (QoS) based on service level agreements (SLAs). MPLS connections tend to be relatively expensive, as compared to broadband Internet connections. However, they do not require adding a protocol such as IP Security (IPsec) to establish a virtual private network (VPN).

National Institute of Standards and Technologies (NIST)

NIST is a US agency that has laboratory programs in many areas. NIST develops and investigates various forms of encryption, as well as other IT security measures.

National Institute of Standards and Technologies (NIST) Suite B

NIST Suite B, which was originally released in 2005 and has had multiple updates, consists of algorithms approved by the US National Security Agency (NSA) for securing various levels of information from non-classified to top secret. Private entities can follow similar guidelines for securing their sensitive data.

National Security Agency (NSA)

The NSA is responsible for ensuring the cryptographic safety of the United States and its allies, particularly as related to signals and intelligence (SIGINT) and cybersecurity.

Negative Acknowledgement (NAK)

A NAK message acknowledges that a request has been received but cannot be satisfied as desired by the requester.

Network Access Device (NAD)

An NAD is another name for a Network Access Server (NAS). In the AAA framework, a NAD is a device that offers access to a network service to another device. Examples of HPE Aruba

Networking devices that can act as NADs are Mobility Controllers (MCs), Instant APs (IAPs), and AOS-CX switches.

Network Access Server (NAS)

In the AAA framework, a NAS is a device that offers access to a network service to another device. Examples of HPE Aruba Networking devices that can act as NASes are Mobility Controllers (MCs), Instant APs (IAPs), and AOS-CX switches.

Network Address Translation (NAT)

NAT is the process by which network routers or other routing devices change an address found in the IP header of a frame to make that address compatible with another address space. Source NAT translates source IP addresses while destination NAT translates destination IP addresses. NAT is often used between a private network and the Internet to translate private IP addresses to a public IP address that is valid in the Internet.

Network Interface Card (NIC)

A NIC is the component within a computing device that enables the device to connect to a wired or wireless network. The NIC generates the signal to transmit Layer 2 frames and also receives and processes incoming signals.

Network Time Protocol (NTP)

Carried by UDP (well-known port 123), NTP is a protocol used to synchronize the clocks of endpoints and network infrastructure devices with the clock of a single reference source, called the NTP server.

Nmap

Nmap is an open source network mapping tool. It sends out pings to identify the IP addresses for active devices. It then scans each device for open ports by sending TCP requests. It can further send TCP requests and analyze the response to fingerprint a device's operating system (OS).

Number used once (nonce)

A nonce is a number that is used once and only once. An initialization vector (IV) for encryption typically must be a nonce to prevent the exact same key from being re-used.

One-Time Pad (OTP)

An OTP is a single-use key shared between communicating parties in advance of transmitting data across an insecure channel. The OTP must be as long as or longer than the data to be encrypted with the OTP. When the XOR algorithm uses an OTP, the algorithm is mathematically proven to be uncrackable. However, because the OTP must be shared in advance and can only be used once, using OTPs has serious, pragmatic limitations.

Open Shortest Path First (OSPF)

OSPF is a dynamic routing protocol that routers and routing switches use to find the best path to route packets from one network to another and deliver them to their final destination. OSPF routing devices use link state advertisements (LSAs) to advertise information to neighbors. Through LSA propagation, the OSPF routing devices can create a topology of the network and use that topology to calculate routes. LSA updates communicate changes to the network topology.

Open Systems Interconnections (OSI)

The OSI model categorizes the various components of network communication as belonging to one of seven, distinct layers. For example, Layer 1 is the physical layer, which is created with technologies such as Ethernet; Layer 4 is the transport layer, which encompasses technologies such as TCP and UDP that are used to transfer data from a server to a client.

Operating System (OS)

The OS is the underlying software that manages a computer's hardware, enables the functionality of other software, and provides interactions with peripherals.

Opportunistic Wireless Encryption (OWE)

OWE provides enhanced security for open 802.11 WLANs. It enables users to connect to the WLAN without submitting a passphrase or credentials; however, encryption and authentication is provided for the communications. OWE uses an unauthenticated Diffie-Hellman exchange between the client and AP to generate the pairwise master key (PMK) from which the data encryption and authentication keys are derived.

Organizationally Unique Identifier (OUI)

An OUI is a number composed of 24 bits that identify an organization (such as a vendor or manufacturer). The first three octets of a MAC address are an OUI.

Out-of-Band Management (OOBM)

OOBM refers to managing network infrastructure devices by connecting directly into the management plane, rather than reaching the management plane through the same data plane that provides connectivity to endpoints. Traditionally OOBM used console ports. However, many AOS-CX switches also support Ethernet OOBM or Mgmt ports, which permit the establishment of an Ethernet OOBM network.

Packet capture

A packet capture creates a copy of traffic flowing through a network infrastructure device and presents that traffic for analysis.

Pairwise Master Key (PMK)

A PMK forms the root of the WPA key hierarchy and is used to derive the Pairwise Transient Key (PTK). The source of the PMK depends on the WPA version and mode. WPA2-Personal derives the PMK from a pre-shared key or passphrase, so the PMK is identical for all clients in the WLAN. WPA3-Personal derives the PMK using Simultaneous Authentication of Equals (SAE), and the PMK is unique for every client session. WPA2-Enterprise and WPA3-Enterprise derive the PMK from the TLS master secret established during 802.1X EAP authentication, and the PMK is unique for every client session. Opportunistic Wireless Encryption (OWE) derives the PMK from a Diffie-Hellman exchange, and the PMK is unique for every client session.

Password Authentication Protocol (PAP)

PAP is a RADIUS authentication protocol, which requires a client to submit a username and password to authenticate. RADIUS PAP encrypts the password with the RADIUS shared secret, but the encryption is not secure and can expose both the password and the shared secret to attack. HPE Aruba Networking devices can use PAP for authenticating managers to an external RADIUS or TACACS+ server or for MAC-Auth or captive portal authentication. RADIUS PAP and TACACS+ PAP are considered insecure and should be used only over trusted network connections or a secure tunnel.

Path Quality Monitoring (PQM)

PQM, provided by HPE Aruba Networking branch gateways (BGWs), actively and passively monitors paths for latency, jitter, packet loss, and throughput.

Payment Card Industry Data Security Standard (PCI DSS)

PCI DSS applies to organizations that collect, store, or transmit credit cardholder data. Compliance comprises 12 requirements and covers the complete cardholder data environment (CDE), which includes anywhere that cardholder or account data is stored, processed, or transmitted.

Penetration tester

A penetration tester investigates an organization's network for vulnerabilities that hackers could exploit. Penetration testers use many of the same tools as hackers, but for ethical reasons.

Personal Computer (PC)

In contrast to a server, a PC is designed to be used by one end-user at a time.

Personal Identification Number (PIN)

A PIN is a unique numeric or alphanumeric code with which users identify themselves.

Personal Information Protection and Electronic Documents Act (PIPEDA)

PIPEDA is a Canadian federal privacy law for private-sector organizations, which "sets out the ground rules for how businesses must handle personal information in the course of their commercial activity."

Per-VLAN Spanning Tree Plus (PVST)

PVST+ is a Cisco proprietary protocol that is similar to spanning tree. However, rather than create a single spanning tree, PVST creates one spanning tree per VLAN.

Phishing

Phishing is a type of social engineering attack in which hackers spoof emails to lure users to click links. The link might install malware, or it might be a spoofed login page that collects the user credentials.

Plaintext

Plaintext is unencrypted data, which can be read by anyone.

Potentially Unwanted Program (PUP)

A PUP is an application that does not pose a threat to the system, but could be unwanted by the user.

Pre-shared Key (PSK)

A PSK is a shared secret that is shared over a secure channel between two or more parties before it must be used.

Primary Account Number (PAN)

A PAN is usually a 14-16 digit number that uniquely identifies an account. If the PAN is for a credit card's account, the first digit is the "major industry identifier," which associates the account with either a major credit card company (Visa cards start with a 4) or an industry (petroleum company cards start with a 7).

Privacy

Data privacy refers to ensuring that only authorized parties can read the data. It is sometimes called data confidentiality.

Private IP address

IPv4 defines several ranges of IP addresses that companies can use in their corporate LAN. These ranges include 192.168.1.0-192.168.1.255, 172.16.0.0-172.31.255.255, and 10.0.0.0-10.255.255.255. Because many companies might be using the same IP addresses, traffic with a private IP source must be translated to a public IP address before entering the Internet.

Private key

In asymmetric encryption, a private key exists as part of a pair with a public key. It encrypts data that the public key decrypts and vice versa. It must be kept private and stored securely.

Process Application Programming Interface (PAPI)

PAPI provides a control channel between HPE Aruba Networking APs and Mobility Controllers (MCs) or between HPE Aruba Networking switches and MCs.

Protected EAP (PEAP)

PEAP is an EAP method that establishes a Transport Layer Security (TLS) tunnel to carry messages for another EAP method, called the inner method. PEAP provides mutual authentication. The authentication server authenticates with a digital certificate during the TLS tunnel establishment. The client authenticates using the inner method, which is usually a less secure method that uses usernames and passwords for credentials. A common form of PEAP is PEAP-MSCHAP, in which EAP-MSCHAP is the inner method.

Protected Management Frames (PMF)

PMF is another term for Management Frame Protection (MFP). PMF increases the security of management frames sent over a wireless 802.11 network. It encrypts and authenticates unicast management frames with the Temporal Key (TK) and authenticates broadcast management frames with the Groupwise Transient Key (GTK).

Public key

In asymmetric encryption, a public key exists as part of a pair with a private key. It encrypts data that the private key decrypts and vice versa. It can be distributed freely.

Public key cryptography

Public key cryptography is another term for asymmetric encryption and the concepts related to it.

Public Key Infrastructure (PKI)

Using asymmetric encryption to authenticate communications relies on being able to reliably associate an identity with a public/private keypair. PKI encompasses all of the components that enable computing systems to bind their identities to a private/public keypair reliably. Within a PKI, end-entity digital certificates are signed by Certification Authorities (CAs), which vouch for the certificate's authenticity. A PKI typically includes a root Certification Authority (CA), which has a self-signed certificate and which signs subordinate CA certificates. Subordinate CAs sign end-user certificates. The PKI also includes the processes and mechanisms for issuing, distributing, and revoking certificates.

Quality of Service (QoS)

On the Internet or a company's private network, QoS mechanisms can be applied to certain types of traffic to ensure that traffic receives priority handling. QoS mechanism can reduce latency and jitter, improving the performance of sensitive applications such as voice and video.

Radio Frequency (RF)

RF is the medium for wireless transmissions. A radio transmits waves at a specific frequency and modulates the waves in specific ways to encode data. In this course, the RF transmissions discussed are 802.11 transmissions, which enable users to connect to a backend local area network (LAN) through a wireless connection. The 802.11 standard uses the 2.4GHz and 5GHz frequency bands.

Ransomware

Ransomware is malware that locks a system (typically by encrypting it). The hacker demands money in exchange for unlocking the system.

Rapid Spanning Tree Protocol (RSTP)

RSTP is a version of Spanning Tree Protocol (STP). Like STP it ensures loop-free topology. "Rapid" refers to the protocol's ability to provide faster convergence than its predecessor when a topology change occurs.

Remote Administrator Tool/Remote Access Trojan (RAT)

A RAT gives a remote user full access to the system on which the RAT runs. If it is installed by a hacker to gain unauthorized access, it is malware.

Remote AP (RAP)

An RAP connects to a Mobility Controller (MC) over the Internet, rather than a more traditional LAN connection. The RAP and MC establish an IP Security (IPsec) tunnel to ensure security.

Remote Authentication Dial-In User Service (RADIUS)

RADIUS is an Authentication, Authorization, and Accounting (AAA) protocol, which authenticates users, authorizes them for access to a network service, and tracks what the users do with the service. RADIUS uses UDP for transport (well-known port 1812 for authentication and authorization and well-known port 1813 for accounting).

Remote Browser Isolation (RBI)

RBI hosts users' browsers in the cloud, potentially protecting the users' own devices from malware and other threats.

Representational State Transfer (REST)

REST is a simple stateless architecture that runs over HTTP/HTTPS. A RESTful application program interface (API) allows clients to use HTTP requests to execute GET, PUT, POST, and DELETE operations on the system that hosts the API.

Request to Send (RTS)

As an optional part of the 802.11 protocol, Request to Send (RTS) and Clear to Send (CTS) frames can reduce collisions in circumstances in which clients connected to the same AP cannot hear each other. When a client connected to an AP wants to transmit, it sends an RTS. If no other client is using the air, the AP sends a CTS.

Rivest Cipher 4 (RC4)

RC4 is a symmetric stream algorithm, which offers speed and simplicity. However, RC4 has several known vulnerabilities and is now deprecated.

Rivest Shamir Adleman (RSA)

RSA is an asymmetric encryption, or public-key, algorithm. RSA remains recommended by NIST with the proper key lengths.

Robust Security Network (RSN)

RSN, defined by 802.11i and used in WPA2 and WPA3, enables an AP and client to negotiate the encryption and authentication algorithms that will secure 802.11 communications. An AP's beacons and probe responses list the sets of algorithms that are supported for that SSID in an RSN element.

Role

AOS assigns all clients that connect on an untrusted interface (such as wireless clients or tunneled node clients) to a role. The role dictates settings such as which firewall policies and bandwidth contracts (rate limits) apply to the client's traffic.

Role-based firewall

A role-based firewall can take clients' role into account when determining which rules to apply to their traffic. The AOS firewall is role-based.

Secure Access Service Edge (SASE)

Combining networking and security functions, SASE securely connects users or devices anywhere—branch offices, home offices, or others—to applications that run anywhere—from the private data center to the public cloud. The two key components of SASE or SD-WAN and SSE.

Secure File Transfer Protocol (SFTP)

SFTP is a secure file transfer protocol that encrypts and authenticates transfers using Secure Shell (SSH).

Secure Hash Algorithm (SHA)

SHA is a cryptographic hash function. SHA has many variants, including SHA-1, SHA-2—which contains six variants—and SHA-3.

Secure Service Edge (SSE)

SSE consists of cloud-based services for securing user access to data and applications.

Secure Shell (SSH)

SSH is a TCP/IP protocol for securing communications between a client and server over an insecure network. Network infrastructure devices can act as SSH servers, offering managers remote access to the devices' CLI over an IP network. SSH uses a public/private keypair to authenticate the server; the client can authenticate with a username and password or a public/private keypair. SSH encrypts and authenticates data that passes between the client and server.

Secure Web Gateway (SWG)

SWG provides security for users' Web browsing. It can perform actions such as blocking users from accessing untrustworthy sites or sites prohibited by the organizations' policies. It can also inspect traffic for malware or other threats.

Security Association (SA) Query

Defined in 802.11w, SA Queries and SA Responses are intended to prevent hackers from replaying association requests for already associated clients.

Security Information and Event Management

A SIEM solution aggregates security information and events submitted by devices and applications across the network. The solution stores, processes, and interprets that information.

Server derivation rule

AOS devices support server derivation rules. The criteria for a rule can include attribute value pairs (AVPs) or vendor specific attributes (VSAs). If the AVPs/VSAs in the Access-Accept for a client's authentication match the criteria in a rule, the AOS device assigns the client the role specified in that rule. However, an Aruba-User-Role VSA with a valid role always overrides a server derivation rule.

Service Set Identifier (SSID)

An SSID is essentially the name of a wireless network, and it is included in frames like beacons, probe requests, and probe responses. SSIDs can be anywhere from zero to 32 octets long and are usually expressed in a natural language (such as English). An SSID can extend across multiple APs, each with its own BSSID, and clients can roam between those APs.

Service Set Identifier (SSID) profile

AOS devices use SSID profiles to define settings such as the security options for a WLAN.

Shared secret

A shared secret is agreed on by two or more parties in a secure manner. Shared secrets are often used as keys for symmetric encryption.

Signal to Noise Ratio (SNR)

The SNR measures the strength of a signal compared to the background noise that might interfere with that signal. It is measured by dividing the power of the signal by the power of the noise.

Simple Network Management Protocol (SNMP)

SNMP is an application layer protocol that helps network administrators manage and monitor their network. SNMP agents run on network devices. They respond to queries from a managing device called an SNMP server or Network Management Station (NMS). Queries can be to get information (read-only) or to change a device's settings (read-write). Queries are directed to objects within a Management Information Base (MIB). For example, a property for a switch interface can be a MIB.

SNMP can use SNMPv1/v2c or SNMPv3.

Simple Network Management Protocol version 1 and version v2c (SNMP v1/v2c)

SNMPv1/v2c versions use communities to regulate access to read-only and read-write functions. A server must include the proper community string in its request. However, community strings are plaintext, so they do not offer true authentication, and the use of SNMPv1/v2c is deprecated.

Simple Network Management Protocol version 3 (SNMPv3)

SNMPv3 regulates access to SNMP objects with users. An SNMPv3 user can have an auth key, which is used to authenticate data, and a priv key, which is used to encrypt data. SNMP servers and network devices are set up with matching user configurations. The network device might assign particular MIBs to which the user has read-write or read-only access.

Simultaneous Authentication of Equals (SAE)

Based on RFC 7664, SAE is a passphrase-based form of authentication that can be used securely over an insecure channel. It enables two entities to prove to each other that they each know the passphrase without revealing information about the passphrase that a hacker could use to derive the passphrase. SAE uses some of the same cryptographic concepts as Diffie-Hellman and can use either Finite Field Cryptography (FFC) or Elliptic Curve Cryptography (ECC).

SAE is used by WPA3-Personal.

Social engineering

Any attack that relies on exploiting human behavior is social engineering. Phishing involves social engineering by using emails to convince users to click unsafe links or send sensitive information to a hacker. Other forms of social engineering exist as well.

Soft Access Point (AP)

A soft AP is software that enables the device running the software to act as an AP.

Software as a Service (SaaS)

SaaS, or "on-demand software," is a model of licensing and providing software via subscription, in exchange for regular payments. Netflix, Workday, and Salesforce are all examples of SaaS.

Software-Defined Wide Area Network (SD-WAN)

SD-WAN uses automation and orchestration to establish WAN connections and regulate how traffic flows over those connections. In an SD-WAN, a branch could connect to headquarters over multiple connections, including an IPsec tunnel over broadband Internet and an MPLS network. The branch gateway can dynamically steer traffic over different connections based on policies and current conditions.

Spanning Tree Protocol (STP)

STP is a Layer 2 network protocol that prevents broadcast storms in Ethernet networks with redundant links. By blocking some links, STP establishes a loopless topology that resembles a tree with its base at the root bridge. If the topology changes, as when a link fails, the topology re-converges, opening blocked ports and blocking new ports, as necessary. STP switches use bridge protocol data units (BPDUs) to elect the root and maintain the tree.

Spear phishing

Spear phishing is a variant of phishing, in which hackers use data collected from sources such as social media to create a convincing email for a particular target.

Spectrum Monitor (SM)

SMs are HPE Aruba Networking APs that have been converted to SM mode. They collect Layer 1 RF data, which they can submit to a Mobility Controller (MC) or Mobility Conductor (MCR) for spectrum analysis. In this way, SMs help companies find sources of RF noise and interference such as microwave ovens, baby monitors, video cameras, Bluetooth devices, and more.

Spoofing

In an IT context, spoofing consists of faking the source of communications in some way. MAC spoofing and IP spoofing fake the source MAC or IP address of traffic. Email spoofing fakes the source of the email.

Spyware

Spyware is malware that collects information from the infected system. A key logger, for example, can collect keystrokes to discover sensitive information like passwords.

Stateful firewall

A stateful firewall handles traffic flows at a session level. This allows it, for example, to automatically permit return traffic for a session that was previously permitted.

Synchronize (SYN), Synchronize-Acknowledge (SYN-ACK), and Acknowledge (ACK)

A TCP 3-way handshake consists of these three messages. The client sends a SYN to the server. The server sends a SYN ACK. The client sends an ACK, and the connection is established.

Syslog

Syslog is a protocol for collecting logs from multiple devices on a central server.

Tamper Evident Label (TEL)

TELs secure products against fraud and tampering as they become irreversibly altered when, for example, someone breaks the label to access a console port.

Temporal Key Integrity Protocol (TKIP)

TKIP is a legacy wireless security protocol, used by the original version of WPA. TKIP uses key mixing of an Initialization Vector (IV) and the Temporal Key (TK) to create a per-frame key, and it uses the RC4 algorithm to encrypt data with that key. It also uses a MIC key to create a Message Integrity Code (MIC) to authenticate the frame. TKIP has known vulnerabilities and is deprecated.

Terminal Access Controller Access-Control System Plus (TACACS+)

TACACS+ is an Authentication, Authorization, and Accounting (AAA) protocol, developed by Cisco, based on TACACS. While TACACS+ is sometimes called TACACS, the two protocols are different. TACACS+ authenticates users, authorizes them for access to a network service, and tracks what the users do with the service. TACACS+ uses TCP for transport (well-known port 49) and provides basic packet encryption.

Transmission Control Protocol (TCP)

TCP is a Layer 4 protocol carried by IP. It provides the transport for higher level applications such as HyperText Transfer Protocol (HTTP), HTTP Secure (HTTPS), Secure Shell (SSH), and many more. TCP handles functions such as grouping traffic into a session, ordering packets, and redelivering lost packets.

Transport Layer Security (TLS)

TLS enables a client and server to establish a secure tunnel for transmitting data over an insecure network. It features a handshake in which the client and server determine the algorithms that will secure future communications, the server authenticates with a digital certificate, and the client optionally authenticates with a certificate. The handshake further enables the server and client to generate a shared key with a method such as authenticated Diffie-Hellman.

A TLS tunnel can carry many types of protocols, including HTTP (HTTPS) and RADIUS (RadSec).

Trivial File Transfer Protocol (TFTP)

TFTP is a simple protocol for transferring files between clients and servers. TFTP does not encrypt data.

Trojan

A Trojan is malware that is included in seemingly innocent software that a user installs intentionally.

Unified Communications (UC)

UC refers to the integration of communication tools, such as email, chat, voice, and video, into a single solution.

Unified Communications and Collaboration (UCC)

Unified communications and collaboration (UCC) adds collaboration tools to the communications tools available in UC. Collaboration tools go beyond helping people communicate to enabling teamwork and working together. Microsoft Skype for Business is an example of a UCC program.

Universal Serial Bus (USB)

A USB is a plug-and-play interface between a computer and add-on devices. USB devices include flash storage, media players, keyboards, mouse device, telephones, digital cameras, scanners, joysticks, and printers.

User and Endpoint Behavior Analytics (UEBA)

UEBA detects and mitigates threats based on detecting anomalies, rather than using signatures for attack patterns. A UEBA solution aggregates and analyzes data from many sources. It might use AI to create a baseline of expected behavior for each user and endpoint. It then detects deviations from the baseline that could indicate potential threats, such as devices compromised with malware. It then alerts admins and possibly works with other solutions to mitigate the threat automatically.

User Datagram Protocol

UDP is Layer 4 protocol, carried by IP. It provides the transport for higher level applications. However, it does not provide any session management such as acknowledges or recovery for lost packets. UDP is mainly used for establishing low-latency and loss-tolerating communications.

User Interface (UI)

A UI presents users with a convenient way to interact with a computer, application, or program. A UI is sometimes called a Graphical User Interface (GUI).

User-Derivation Rule (UDR)

AOS devices support UDRs, which are applied in AAA profiles. A UDR can include criteria such as MAC address or DHCP option; it further specifies a role. If a client matches the criteria, the AOS device assigns it to that role. UDRs with Layer 2 criteria such as MAC address are executed before a client is authenticated and can be overridden by a role assigned during authentication. UDRs with Layer 3 criteria such as DHCP option (fingerprint) apply after authentication and override the role assigned during authentication—the exception is if the authentication server sent an Aruba-No-DHCP-Fingerprint VSA.

Validated Reference Design (VRD)

An HPE Aruba Networking VRD is a technology focused guide that includes descriptions of HPE Aruba Networking technology, recommendations for product selections, network design, configuration, and deployment best practices.

Vendor Specific Attribute (VSA)

Within a protocol such as RADIUS or TACACS+, Attribute Value Pairs (AVPs) communicate the value for various attributes in this session. A VSA is a special attribute defined by vendors like HPE Aruba Networking to communicate settings important for that vendor's solutions. A key example of an HPE Aruba Networking VSA is Aruba-User-Role.

Virtual AP (VAP) profile

A VAP profile defines a WLAN on AOS devices. The VAP profile also specifies an SSID profile and an AAA profile.

Virtual LAN (VLAN)

A VLAN is a logical group of devices that have been placed in the same broadcast domain. A VLAN typically corresponds to an IP subnet. Network admins can use VLANs to logically segregate endpoints on the same switch. They can also extend VLANs over multiple switches to group endpoints by criteria other than location.

Virtual Machine (VM)

A VM is an isolated software container, including an operating system (OS) and its application, which runs on a physical host. A host has a hypervisor, which it uses to host multiple VMs and provide hardware resources to them.

Virtual Private Network (VPN)

A VPN establishes a secure tunnel between devices connected over an insecure network. This tunnel can provide data authenticity, data integrity, and data privacy. Types of VPNs include Secure Sockets Layer (SSL) VPNs, which use Transport Layer Security (TLS), and IP Security (IPsec) VPNs.

Virtual Private Network Concentrator (VPNC)

Within an SD-Branch solution, an HPE Aruba Networking VPNC, also called a headend gateway, acts as a hub that allows branches to connect to a network. HPE Aruba Networking Branch Gateways (BGWs) establish IPsec tunnels with VPNCs to connect edge devices at the branch to resources at the headquarters.

Virtual Routing and Forwarding (VRF)

VRF is a technology used to create multiple, virtual routing tables that exist on the same router. On a VRF-capable device such as an AOS-CX switch, every Layer 3 interface belongs to a single VRF instance. The VRF instance has its own routing table, consisting of the local subnets configured on interfaces in that instance, static routes created in that instance, and dynamic routes learned by routing protocols running on interfaces in that instance. Traffic that arrives on an interface within a specific VRF is routed by that VRF's specific routing table, maintaining complete separation between subnets in different VRFs.

A VRF instance is sometimes called, more colloquially, a VRF.

AOS-CX switches have two VRF instances at factory default settings: "mgmt" VRF, which includes the Mgmt port and can only include this port, and the "default" VRF, which includes all the data plane ports.

Virus

A virus is malware that replicates itself and seeks to spread to other systems. A virus requires a user to take an action (like click a link or install software) to infect a system.

Voice over IP (VoIP)

VoIP enables voice calls over IP networks such as campus LANs and the Internet. VoIP phones might be hardware phones that look and function much like traditional phones or "soft" phones that run as applications on user devices.

Vulnerability

A vulnerability is anything that a hacker can exploit to disrupt a system or gain unauthorized rights to resources. A vulnerability has the potential to be exploited, but does not necessarily mean that a threat has already occurred.

Web Content Classification (WebCC)

The AOS firewall supports WebCC (with WebCC licenses) to help companies better control which Web services clients access. WebCC analyzes clients' Web traffic and determines to which URLs it is destined.

When WebCC is enabled, an AOS application rule can specify permitted or denied Web categories. The rule can further specify a reputation, which refers to how trusted the Web site is.

Wide Area Network (WAN)

A WAN connects computing devices at multiple sites over wider geographical areas.

A WAN is a network of interconnected devices that are dispersed at different geographic locations, as opposed to a LAN, which connects devices at a single site.

WiFi Protected Access (WPA)

WPA/WPA2/WPA3 define mechanisms for securing 802.11 communications.

All versions of WPA operate in one of two modes. Personal Mode authenticates clients with a passphrase shared by all users authorized to connect to the WLAN. Enterprise Mode authenticates clients individually with 802.1X.

Available in 2003, WPA was an intermediate measure between WEP and WPA2. It used Temporal Key Integrity Protocol (TKIP) to provide stronger security measures for wireless communications, as compared to Wired Equivalent Privacy (WEP). WPA2 complied with 802.11i and provided stronger data protection based on Advanced Encryption Standard (AES) Counter mode with CBC-MAC Protocol (CCMP). Until WPA3 was released in January 2018, WPA2 was considered the only viable option for protecting enterprise wireless access.

WPA3 was designed to remedy the security deficiencies of WPA2. It institutes Simultaneous Authentication of Equals (SAE) for WPA3-Personal. It also adds new supported algorithms such as AES with Galois Counter Mode with a 256-bit key (GCM-256). WPA3-Enterprise can operate in a CNSA mode, which mandates use of certain EAP-TLS options, GCM-256, and other secure algorithms.

Windows Management Instrumentation (WMI)

Devices with Microsoft operating systems use WMI to advertise information about themselves. ClearPass Policy Manager (CPPM) can use WMI to discover a device's MAC and IP addresses, OS information, hostnames, services, processes and applications, local users defined in the OS, and open ports.

Wired Equivalent Privacy (WEP)

WEP is an outdated security algorithm that was previously the standard method for encrypting data on a wireless network. It used Rivest Cipher 4 (RC4) to encrypt data and maintained the integrity of that data with Cyclic Redundancy Check 32 (CRC32). Static WEP derived a key from a passphrase while dynamic WEP derived the key during the 802.1X EAP process. Both forms of WEP have many known vulnerabilities, and WEP should not be used in modern networks.

Wireless Intrusion Detection System/Wireless Intrusion Prevention System (WIDS/WIPS)

WIDS monitors the network for unauthorized APs and clients, as well as for wireless Denial of Service (DoS) attacks. WIDS also logs information and generates reports based on those logs. WIPS adds intrusion protection features such as wired and wireless containment.

AOS WIP detects unauthorized wireless devices such as rogue APs, wireless DoS attacks, and online cracking attempts. It can also mitigate the threats.

Wireless Local Area Network (WLAN)

A WLAN gives mobile users access to a local area network (LAN) through a wireless (radio) connection. A WLAN is defined by a service set Identifier (SSID) and various security options associated with that SSID.

Wireless Network Management (WNM)

Defined in 802.11v, WNM adds frames that all APs to configure clients.

Worm

A worm is malware that is similar to a virus. It replicates and seeks to spread to other systems. However, it can spread through the network and infect vulnerable systems without user interaction.

Zero day exploit

A zero day exploit is a threat that exploits a vulnerability that has not yet been discovered or patched by the vendor.

Zero day vulnerability

A zero day vulnerability is a vulnerability that has not yet been discovered or patched by the vendor.

Zero Touch Provisioning (ZTP)

ZTP allows a less experienced IT staff member to connect HPE Aruba Networking devices, which are then automatically provisioned in a secure manner. The devices automatically contact HPE Aruba Networking Activate over HTTPS. Activate directs the devices to HPE Aruba Networking Central, from which they obtain their configurations.

Zero Trust Network Access (ZTNA)

ZTNA replaces traditional client-to-site VPNs. Rather than tunnel a remote user into an internal network, ZTNA brokers access for each authenticated user to the precise private applications to which that user is authorized.

Appendix B
Learning Check Answers

Chapter 1

1. What are ways to protect yourself from phishing attempts?

<u>You can look for signs of phishing such as:</u>

- <u>Link that does not go where it claims (hover over)</u>
- <u>Misspellings in domain name or email sender name</u>
- <u>Unsolicited email</u>
- <u>Threats such as loss of account access or services without immediate action</u>

<u>Steps to protect yourself include:</u>

- <u>Stop and think before you click</u>
- <u>Do not click links; navigate to the site yourself</u>
- <u>Follow up with requests in person or on the phone</u>

2. What is a primary security risk introduced by the fact that ARP has no built-in authentication and integrity mechanisms?

   a. Hackers can add malicious code to ARP messages to install viruses on user machines.

   b. Hackers can send malformed ARP messages that cause user devices to blue screen.

   c. <u>Hackers can poison devices' ARP tables in order to implement MITM attacks.</u>

   d. Hackers can use ARP to change the MAC address on their machines and gain unauthorized network access.

3. What is the correct order for the chain? (Put the steps below in order.)

    _3_ Delivery

    _5_ Installation

    _1_ Reconnaissance

    _4_ Exploitation

    _2_ Weaponization

    _7_ Actions on Objective

    _6_ Command and Control

4. What is one way an HPE Aruba Networking solution supports micro-segmentation?

    a. It replaces third-party security solutions with a single network-based solution.

    b. It installs a host-based firewall on every device that connects to the network.

    c. It applies granular access controls as external traffic flows into the internal network.

    d. It identifies users and devices at network connection and tunnels their traffic to a firewall that implements role-based access control.

Chapter 2

1. To which organizations does PCI DSS 3.2 apply?

    a. Any US federal agency or contractor

    b. Any organization that collects and stores data about EU citizens

    c. Any organization that handles users' personal health data

    d. Any organization that transmits or stores credit cardholder data

2. Match each concept with the correct description.

    Concepts

    a. Stream algorithm

    b. Symmetric encryption

    c. Block algorithm

d. Initialization vector (IV)

Descriptions

_d_ A value used within the encryption process to help prevent identical key reuse

_c_ Encryption algorithms such as AES, which encrypt sets of data that are the same size as the key

_b_ The same key encrypts and decrypts data

_a_ A form of encryption that encrypts data bit by bit

3. True or false: MD5 is recommended for key-based hashing over SHA2.

   a. True

   b. False

4. Alice hashes her message, encrypts the hash with her private key, and sends the message plus the hash to Bob. What key should Bob use to validate the message?

   a. His own private key

   b. His own public key

   c. Alice's private key

   d. Alice's public key

5. Match each entity with the correct description.

   Concepts

   a. Root CA

   b. Intermediate/subordinate CA

   c. End entity (leaf)

   Descriptions

   _b_ Entity that issues certificates to end entities

   _c_ Device that obtains certificates, but does not issue certificates

   _a_ Entity believed to be trustworthy

6. What are some of the purposes of TLS? (Select two.)

   a. Ensuring that users select good passwords that are resistant to dictionary attack

   b. Establishing encrypted and authenticated communications for upper layer protocols

   c. Submitting CSRs to a CA and obtaining signed certificates

   d. Preventing malicious users from exploiting ARP

   e. Authenticating a server and optionally a client with certificates

7. What level of security does a RADIUS shared secret provide?

   a. It encrypts the RADIUS message but does not ensure the message's integrity and authenticity.

   b. It validates the message's integrity and authenticity but does not provide encryption.

   c. It both validates the message's integrity and authenticity and provides encryption.

   d. It enforces pre-authentication for the NAS to the RADIUS server but does not provide message integrity or privacy.

8. Make notes on what you have learned about certificates. Consider questions such as these:

   - What values do certificates offer for authenticating devices' identity and messages?
   - What does a chain of trust mean?
   - What weak points can exist in certificate deployments?
   - How can you use Windows tools to find a certificate's subject name and SAN?

You might have noted ideas such as these:

- Certificates let users and devices authenticate themselves (tie their identity to a public/private key pair used for signing data) and establish secure communications.
- Certificates exist in a hierarchy. To trust a certificate, a device must trust the entity that signed that certificate. To trust the root CA, the device must have the root CA certificate installed on it as trusted.
- If users click past certificate warning, the certificates cannot serve their purpose to authenticate. Also, if the CA is not truly trustworthy or protected, security for the entire PKI is compromised.
- You can open the certificate with the default Windows application. The CN of the subject name is shown on the General tab. You can see the full subject name in the Details tab. You can also scroll through the Details tab to see the SAN.

Chapter 3

1. You have set up an Ethernet out-of-band management network and connected AOS-CX switch Mgmt ports to it. What additional step should you take to ensure that users in the production network cannot access the management interfaces (Web UI or CLI)?
   a. Apply a control plane ACL on the mgmt VRF.
   b. Enable management protocols on the mgmt VRF and disable them on the default VRF.
   c. Create a list of authorized management IP addresses.
   d. Enable and disable SSH/HTTPS access on specific physical ports.
2. You are configuring SSH authentication to a RADIUS server on an AOS-CX switch. You want Bob and Alice to be assigned admin level status. What should you tell the RADIUS server manager about the AVP to send when they authenticate?
   a. Configure the RADIUS server to send the Service-Type set to 1.
   b. Configure the RADIUS server to send the Service-Type set to 15.
   c. Configure the RADIUS server to send the Aruba-Admin-Role VSA set to "administrators."
   d. Configure the RADIUS server to send the Aruba-Priv-Admin-User VSA set to 1.
3. Which of the following is a list of the secure protocols that HPE Aruba Networking recommends using?
   a. Telnet, TFTP, HTTP, SNMPv1/v2c
   b. SSH, SFTP, HTTP, SNMPv3
   c. SSH, SFTP, HTTPS, SNMPv3
   d. Telnet, SFTP, HTTPS, SNMPv1/v2c
4. You are using SSH to access your switches. What is one recommendation for configuring the management station to authenticate the switch?
   a. **View the switch's public SSH key in the CLI and copy the key into the management station SSH client's known hosts file**.
   b. Configure the switch's username and password in the known hosts file for the management station's SSH client.
   c. Configure the management station to communicate with a RADIUS server to check the switch's credentials.

d. Configure the management station to communicate with a TACACS server to check the switch's credentials.

Chapter 4

1. True or false: If the console port on an MC or AP is unprotected, attackers can easily gain access using the password recovery procedure.

   a. True

   b. False

2. True or false: HPE Aruba Networking recommends entirely disabling local authentication

   a. True

   b. False

3. Which of these are reasons to avoid using SNMPv1/v2c? (Select all that apply.)

   a. Passwords are sent in plaintext

   b. Critical data is sent in clear text

   c. Hackers can easily send packets with valid community strings

   d. The protocols place a strain on compute resources

   e. The protocols are difficult to configure

4. What security is provided by implementing CPsec between APs and MCs?

   a. CPsec encrypts and authenticates the PAPI control channel and in the GRE tunnels that carry user traffic.

   b. CPsec authenticates the PAPI control channel but does not encrypt it.

   c. CPsec encrypts and authenticates the PAPI control channel.

   d. CPsec encrypts and authenticates the GRE tunnels that carry user traffic and authenticate the PAPI control channel.

5. Consider these questions related to securing access to an MC's control plane. Take notes on your thoughts.

   - Through what protocols can users and management solution access the MC's data and configuration settings?

- What are the risks of unauthorized users accessing the MC through particular protocols?
- What strategies would you take toward locking down access to the correct users and devices?

Your ideas might include:

- Devices can access the MC management plane with SSH, HTTPS, SNMP, RADIUS, IPsec, and many other protocols.
- It is critical to protect the MC from unauthorized access. Hackers who gain unauthorized management access can cause DoS attacks, perform network reconnaissance, open backdoors, capture traffic, and alter firewall policies.
- While password protections are good, it is sometimes better to prevent hackers from ever reaching the login page.

Chapter 5

1. Match each feature with its description.

   Features

   a. BPDU guard

   b. Root guard

   c. CoPP on STP

   Descriptions

   _c_ Prevents strain on CPU

   _b_ Prevents an unauthorized device from becoming root

   _a_ Prevents rogue spanning tree devices from connecting

2. True or false: DHCP includes native protections to enable clients to authenticate the DHCP server as legitimate.

   a. True

   b. False

3. How does an AOS-CX switch learn dynamic MAC-to-IP bindings for implementing ARP protection?

   a. MAC authentication

   b. SNMP

   c. <u>DHCP snooping</u>

   d. OSPF

4. What basic protections can you implement to prevent unauthorized devices from altering your networks' routing tables? (Select two.)

   a. ARP snooping

   b. <u>OSPF or BGP authentication</u>

   c. <u>OSPF passive interface</u>

   d. ARP protection

   e. BPDU guard

Chapter 6

1. Match each role to the actions associated with that role.

   Name of role

   a. Supplicant

   b. Authenticator

   c. Authentication server

   Actions associated with role

   _b_ Grants access to the network

   _c_ Determines whether or not to grant access

   _a_ Requests access to the network

2. What is the main flaw of WPA2-Enterprise security?

   a. <u>The TLS handshake that generates the PMK can be implemented with insecure algorithms.</u>

   b. Users tend to choose easily guessed passwords and leave the passwords lying around.

c. AES encryption is not valid for securing data in modern networks.

d. All users share the same passphrase.

3. Match the security enhancement to the security mode to which it applies.

Security enhancement

a. Unauthenticated Diffie-Hellman to negotiate a PMK

b. Simultaneous authentication of Equals (SAE), which protects weak passwords from attack

c. CNSA mode to ensure secure algorithms

Security mode

_c_ WPA3-Enterprise

_b_ WPA3-Personal

_a_ OWE

4. Consider these questions about implementing 802.1X with EAP-TLS.

- Assume that decision makers at your company have asked you about certificate-based versus password-based authentication for the wireless and/or wired networks. They want to know if certificates are really more secure. What would you explain?
- Assume that your company has decided to implement certificate-based authentication with EAP-TLS. What would you take into account as you make a plan for implementing the new security method?

You might have listed ideas such as these:

- Certificates provide stronger authentication than passwords.
- Users cannot easily choose a bad certificate the way that they can easily choose a bad password.
- Certificates are less subject to dictionary attacks and cracking attempts.
- But deploying certificates takes planning and input from multiple groups. You need to plan ways to deploy the proper certificates to the proper devices. You also need to make plans for automatically updating certificates.
- With certificates, the work on the part of admins can actually ease user experience; for example, with certificates, users do not have to change passwords periodically.

Chapter 7

1. True or false: Campus APs (CAPs) in Tunnel mode forward authenticated user traffic into the VLAN assigned to the WLAN.

   a. True

   b. <u>False</u>

2. You have campus APs and an MC. How can you control where an SSID is deployed?

   a. Assign location tags to APs. Specify the location tag within the New WLAN wizard.

   b. <u>Assign APs in different locations to different groups. In the New WLAN wizard, choose the specific AP Groups for the WLAN.</u>

   c. Within the SSID profile, choose the names of the APs that should broadcast the SSID.

   d. Create a location tag for areas of your network. Within the WLAN's AAA profile, select the location tag.

3. Which are steps for enabling 802.1X on an AOS-CX switch? (Select multiple options.)

   a. Creating a tunnel VLAN

   b. Setting a non-default admin password

   c. <u>Enabling 802.1X on a per-interface basis</u>

   d. <u>Enabling 802.1X globally</u>

4. What additional setup do you need to perform if you want to use RadSec between a network infrastructure device like an MC or switch and CPPM? (Take notes on this open-ended question.)

<u>The network infrastructure device requires a certificate that it can use for client authentication. That certificate must have a root CA that CPPM trusts. It might also require a specific subject name. The network infrastructure device must also have installed on it the root CA certificate associated with CPPM's RadSec certificate. You must also make sure that the hostname or IP address that you configure for CPPM on the network infrastructure device matches up with CPPM's certificate's subject name.</u>

<u>On MCs you must specify these certificates and enable RadSec from the CLI.</u>

Chapter 8

1. True or false: A role assigned with an Aruba-User-Role VSA takes precedence over a role derived from a server derivation role.

    a. True

    b. False

2. What is one feature of the AOS stateful firewall?

    a. It establishes sessions and can automatically permit return traffic in a session.

    b. It ensures that all users' traffic is controlled in the same way.

    c. It maintains good performance by offloading packet processing to a third-party firewall such as CPPM.

    d. It adds encryption to unencrypted traffic before forwarding it on the LAN.

3. An AOS-CX switch has a UBT zone profile configured on it. How does the switch decide when to tunnel a user's traffic?

    a. Based on whether the initial role is bound to tunneling globally

    b. Based on whether that zone is specified in the user role assigned to the user

    c. Based on whether CPPM sends an VSA called "Aruba-Tunnel"

    d. Based on whether UBT is enabled on the user's port

4. Consider these open-ended questions about role assignment and take notes on your answers:

    - Assume that you are using ClearPass Policy Manager (CPPM) as the authentication server, and you have a network with an MCR, several MCs, and many CAPs. Which component should be the decision maker for which roles users are assigned?
    - What approaches can companies take toward dividing users into different roles? When should users have the same role? When should they have different roles?

CPPM should generally determine the role.

Companies can take many flexible approaches for defining the role. Companies might create roles for different departments or even for different jobs within departments. Whenever users require different access rights, they need to be assigned to different roles.

5. Consider these open-ended questions about defining policies on the AOS firewall. Take notes on your answers.

    - Describe allow list versus deny list approaches to defining firewall policies. Discuss advantages and disadvantages.
    - What strategies can you take toward defining the services that users need?
    - How can you distinguish between and control different types of applications that use HTTP/HTTPS?

An allow list denies all access by default and permits access only to the specified resources. A deny list permits all access by default and denies access only to the specified resources. An allow list is the more restrictive and secure approach.

You can define the services to which users need access based on destination IP addresses, protocol and protocol port (such as TCP 80 for HTTP). You can further define permitted services at an application level, such as social networking. To determine to which services users need access, you might need to conduct surveys, as well as discuss with stakeholders and security teams.

Many applications use HTTP/HTTPS. The AOS firewall lets you distinguish between these applications using AppRF 2.0 and DPI.

6. What are the benefits of roles, policies, and aliases for the AOS firewall? Take notes on your answers.

- With roles, you do not typically have to worry about the source IP for access control rules. Rules within a role automatically apply to any traffic to and from a client assigned to that role.
- Policies let you create building blocks that you apply to multiple roles.
- Aliases let you use intuitive names for IP addresses and services, which you can update.

7. You are defining a WLAN that uses WPA3-Enterprise security and authenticates users to CPPM. CPPM sends the roles for authenticated users in Aruba-User-Role VSAs. What should you set for the default role in the WLAN's Access settings?

When CPPM is sending roles, the 802.1X default role should grant restricted access so that an error (like a misspelled role) does not grant elevated access. A typical choice is "denyall."

8. This learning check refers to the policies listed in Table 8-1 and Table 8-2.

    a. Will contractors be able to send an HTTPS request to these IP addresses:

    - 10.254.1.15 **Yes**
    - 10.254.1.21 **No**

- 203.0.13.3 **Yes**

b. Will this traffic will be permitted:

- HTTPS response from 10.254.1.15 after an HTTPS request from a contractor client **Yes**
- HTTPS request from 10.254.1.15 (a server outside the firewall) **No**
- DNS response from 10.254.1.21 after a DNS request from a contractor client **Yes**

Chapter 9

1. What is a difference between passive and active endpoint classification?
   a. Passive classification collects data for admins to examine while active classification uses device types to assign roles.
   b. Passive classification uses traffic sent by endpoints on their own while active classification involves sending out probes or requests.
   c. Passive classification always occurs on-prem while active classification involves sending data to the cloud.
   d. Passive classification relies on agents installed on devices before they connect to the network while active classification occurs post authentication.
2. What should you configure in the network infrastructure to help CPPM perform DHCP fingerprinting?
   a. Configure CPPM as an additional IP helper address in the VLANs with endpoints.
   b. Configure MCs and switches to mirror all traffic to CPPM.
   c. Configure active scans with SNMP and DHCP on MCs.
   d. Configure user derivation rules with DHCP options on MCs.
3. Which purpose does an IF-MAP profile on an MC serve?
   a. It enables the MC to use Nmap to perform active classification on endpoints.
   b. It allows you to configure whether you want the MC to prioritize DHCP fingerprint roles from CPPM or the MC.
   c. It maps user roles sent by CPPM to user roles configured on the MC firewall.
   d. It sends HTTP User-Agent strings from the MC to CPPM.

4. Assign numbers to each criterion to indicate its precedence for assigning roles on MCs. The criterion with highest precedence should be 1.

   __4__ User Derivation Rule with L2 criteria

   __3__ Aruba-User-Role VSA only

   __2__ User Derivation Rule with L3 (DHCP option) criteria

   __1__ Aruba-User-Role VSA + Aruba-No-DHCP-Fingerprint VSA

5. What is one difference between profiling capabilities in CPDI and CPPM?

   a. CPDI supports active classification methods, and CPPM only supports passive classification.

   b. <u>CPDI can use packet inspection, analysis of traffic patterns, and machine learning to identify IoT devices more precisely than CPPM.</u>

   c. CPDI can use device categories as conditions for assigning roles while CPPM on its own cannot.

   d. CPDI runs as a module on MCs while CPPM is a standalone solution.

6. Consider these questions and take notes on your thoughts:

   - How can the profiling capabilities about which you learned in this chapter help companies improve visibility into their network?
   - How can the profiling capabilities about which you learned in this chapter help companies improve security?
   - What are two ways that you could use these capabilities in your or your customers' networks?
   - Assume that you are working with a company that already uses CPPM to authenticate users, but does not use device profiling. Now ClearPass admins want to add profiling. What would you discuss with the admins to help the implementation go smoothly?

<u>Profiling helps companies understand the devices in their network. It gives them clarity into their inventory. Profiling can help HPE Aruba Networking infrastructure to place devices that do not support authentication in the proper roles anyway. Profiling can let you limit users' access when they are using more vulnerable devices.</u>

<u>Before implementing profiling, make sure to coordinate roles between CPPM and the infrastructure. Also make sure that dynamic authorization is enabled on the devices. Also make sure that the infrastructure is set up to help CPPM obtain necessary information. For example, set up CPPM as another helper address on DHCP relays.</u>

Chapter 10

1. What are two of the trends that make it difficult for companies to close security gaps? (Select two.)

    a. Most companies lack perimeter security solutions such as firewalls.

    b. A majority of the workforce now works remotely at least some of the time.

    c. Most user clients lack support for advanced technologies such as 802.1X.

    d. Users are moving away from mobile connections toward Ethernet connections.

    e. The number of IoT devices is rapidly growing.

2. Which HPE Aruba Networking solution provides cloud access security broker (CABS) functions?

    a. HPE Aruba Networking EdgeConnect SD-WAN

    b. HPE Aruba Networking EdgeConnect SD-Branch

    c. HPE Aruba Networking EdgeConnect Microbranch

    d. HPE Aruba Networking SSE

Chapter 11

1. Match each piece of information that you might want to find with the tool where you can find that information.

    Information that you want to find

    a. Reason that CPPM rejected an authentication request

    b. Report with analysis of authentication trends

    c. 802.1X related events that have occurred on a MC

    d. Global overview of issues related to association, authentication, DHCP, and DNS

    Found in this tool

    _c_ Process logs in MC UI

    _b_ ClearPass Insight

    _d_ WiFi Connectivity in HPE Aruba Networking Central

    _a_ CPPM Access Tracker

2. You have enabled logging to 10.1.10.10 on an AOS-CX switch. No other logging or debug commands are configured. Which logs does the switch send to 10.1.10.10?

   a. No debugs or logs

   b. Logs of info level and up

   c. Logs of error level and up

   d. Debugs but not logs

3. What could cause an AOS-CX switch an issue in downloading a user role from CPPM?

   a. The role name that CPPM sent doesn't match a role name configured on the switch.

   b. The switch lacks the root CA certificate for CPPM's HTTPS certificate.

   c. The switch does not have SFTP enabled on it.

   d. The switch does not have a local certificate of its own installed on it.

4. You set up a MC to enforce authentication to CPPM on a WPA3-Enterprise WLAN. When you test the connection, authentication fails. You check CPPM Access Tracker and cannot find a record associated with your attempt. What might be the problem?

   a. You entered the wrong username and password on the client.

   b. Your client does not support the right EAP types.

   c. Your client does not trust CPPM's certificate.

   d. The RADIUS secret configured on the MC is incorrect.

5. You have a controlled wireless solution. You want to capture unencrypted packets associated with a particular authenticated wireless user, who is connected to a WPA3-Enterprise WLAN in tunnel mode. Where should you implement the packet capture?

   a. On the user's AP

   b. On the switch connected to the user's AP

   c. On the MC's control plane

   d. On the MC's data plane

6. You have set up a packet capture from a CAP to a client running Wireshark. What should you do on the client to prepare to receive the traffic?

   a. Tag the client NIC for VLAN 4094.

   b. Bridge traffic between multiple NICs on the client.

   c. <u>In Wireshark, set the Aruba_ERM protocol to UDP 5555.</u>

   d. In Wireshark, filter for UDP 514.

7. Consider these questions about the purposes of, and best practices for, logging. Take notes on your answers:

   - Why is sending logs to a centralized solution on an ongoing basis important?
   - What options do you have for protecting the integrity and privacy of logs?
   - You can filter the logs that you forward in various ways. What arguments can you make for sending all logs, from informational level and up, or for sending relatively fewer logs?

<u>Your ideas might have included ones such as these:</u>

- <u>You need to collect logs on an ongoing basis to have the information when you need it. Otherwise, once you know you need the information, it is too late.</u>
- <u>SIEM solutions rely on having a great deal of information from many sources.</u>
- <u>Logging to a remote device preserves logs (in case of deletion or reboot).</u>
- <u>You can use TLS or IPsec to protect Syslog messages and avoid revealing information about your network.</u>

Chapter 12

1. What is a Chain of Custody?

   a. A plan for containing rogue and interfering APs

   b. A list of rogue and suspected rogue APs

   c. The stage of the kill chain at which the hacker gains control of the infected system

   d. <u>Documentation of when evidence was collected, how it was stored, and who had responsibility for it</u>

2. What does it mean if HPE Aruba Networking WIP has classified an AP as a suspected rogue?

   a. The AP is authorized by the HPE Aruba Networking solution, but it is configured with the wrong SSID or security settings

   b. The AP might connect to the corporate LAN, based on matches between the source MAC or BSSID in the AP's wireless traffic and MACs in the wired network

   c. WIP has detected the AP sending frames that might indicate DoS attacks but at a lower rate than the threshold for an alarm

   d. The AP has violated at least one anti-DoS policy on WIP or connected to an interfering client

3. True or false: The Security Dashboard shows the location of rogue APs on a floorplan.

   a. True

   b. False

Note: VisualRF shows rogue APs on a floorplan; the Security Dashboard lists rogue APs.

Appendix C Supplemental Material

The information throughout Appendix C is provided for your reference. The HPE6-A78 exam does not test you on the information included in this appendix.

Chapter 2: Security Technologies

This section includes supplemental information about TLS.

Example TLS 1.2 Handshake

An example of the TLS 1.2 handshake is shown in Figure C-1. This handshake is used in other technologies like EAP-TLS.

The process occurs as follows:

1. The client sends a hello to the server with information such as a session ID, TLS version, and the cipher suites that the client supports. The cipher suite specifies the key exchange algorithm, as well as the encryption and data integrity algorithms that will be used to secure the session after the handshake.
2. The server sends several TLS "records" in one or more messages:
    a. Server Hello with similar information, including the cipher suite that the server has selected
    b. Its certificate. Some key exchange algorithms do not require the server to authenticate with a certificate, but typically the server does. Of course, hackers could try to steal a certificate and submit it as their own, but during the course of the rest of the exchange, the server will prove that it has the private key associated with the public key in the certificate.
    c. ServerKeyExchange. This message is required if this handshake is using ephemeral Diffie-Hellman for the key exchange. This means that every session uses new values for the Diffie-Hellman exchange, which provides "forward secrecy." In other words, if a hacker later compromises the server's private

key, the hacker cannot decrypt all the old sessions. The message includes the DH exchange values, the values required for initiating Diffie-Hellman and the server's public value. The server "signs" these values (hashes them and encrypts them with its private key). This validates that the server really owns the certificate and that the DH public value is its own. It prevents a MITM attack.

d. CertificateRequest—only if client authentication is required

e. Hello Done

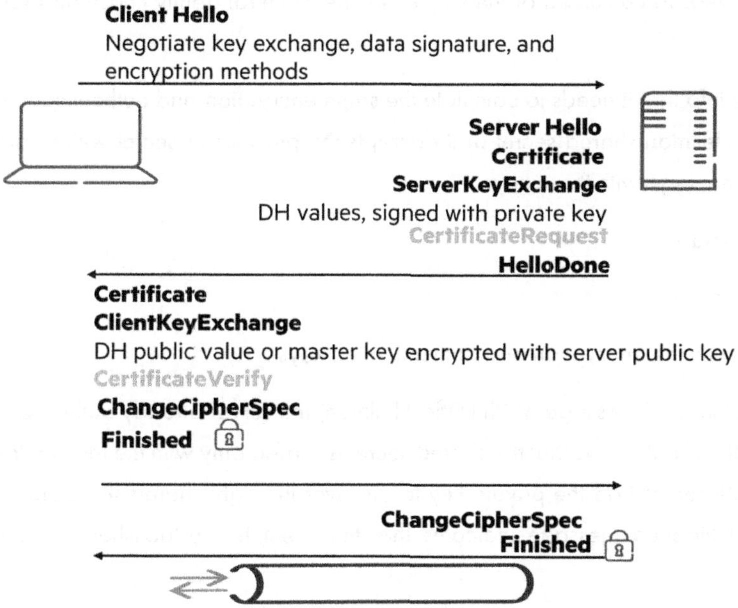

Figure C-1: Example TLS 1.2 Handshake

3. The client validates the certificate.

4. The client sends a few more records:

   a. Its certificate—only if requested by the server

   b. ClientKeyExchange—for ephemeral Diffie-Hellman, this message includes the client's public value. An alternative key exchange method is RSA , in which case the client generates a pre-master secret, which it encrypts with the server's public key so that only the server can read it. RSA does not provide forward secrecy; if a hacker captures the TLS handshake and obtains a server's private key later, the hacker can obtain the pre-master secret.

   c. CertificateVerify—only if it sent a certificate. This is a hash of the entire exchange that the client signs with its private key. The server will decrypt with the client's public key and compare to its own hash of the conversation. In this way, the client proves that it has the private key and verifies the integrity of the exchange.

d. A Change Cipher Spec message, indicating that the client is going to start using encryption and authentication.

e. A Finished message that is secured with symmetric encryption and authentication keys. These keys come from the pre-master secret, which is either the Diffie-Hellman shared secret (which the client has all the info it needs to calculate) or the secret that the client randomly chose and sent encrypted to the server.

5. The server now has the info that it needs to calculate the same encryption and authentication keys (either it calculates the Diffie-Hellman shared secret or it decrypts the pre-master secret with its private key). It validates the Finished message with these keys.

6. It then sends its own messages:

   a. ChangeCipherSpec

   b. Finished, secured with the encryption and authentication keys

7. The client validates the Finished message. With Diffie-Hellman, the server already authenticated its public value using its key, so the client knows that the shared secret is shared only with the legitimate server. With RSA, only the legitimate server had the private key to discover the right shared secret. In either case, a successfully encrypted Finished message indicates that the client has established a session with the legitimate server.

The client and server can now exchange data that is encrypted and authenticated. The data sent with this security is called tunneled data, and the connection established between the client and server is called a tunnel.

Chapter 5: Secure LAN Protocols

This section provides supplemental information about spanning tree protocols.

Review Spanning Tree

Redundant links help to ensure that a path continues to exist across the network even if one link or even one switch fails. For example, in the topology shown in Figure C-2, Access-1 could continue to reach Core-1 if its direct link to Core-1 failed. It could also continue to reach the core if Core-1 failed entirely. However, adding redundant Layer 2 links without a protocol to manage the links results in network loops. These loops, in turn, create broadcast storms and mis-learned MAC addresses, making the network inaccessible. To function properly, an Ethernet network must have only *one* active pathway between any two devices.

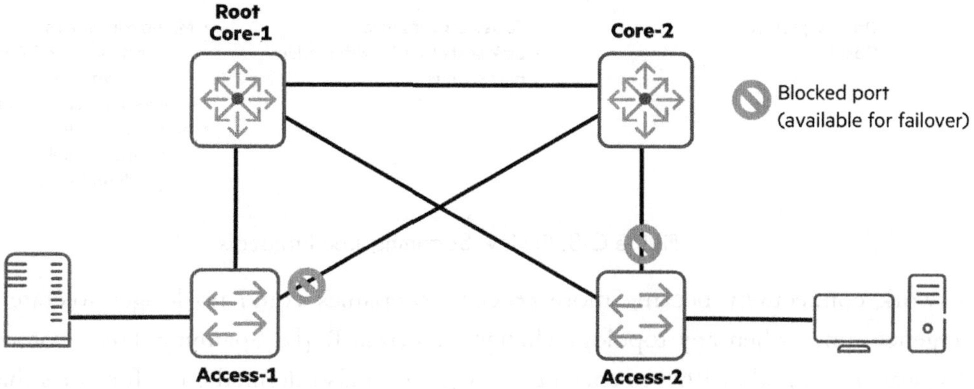

Figure C-2: Review Spanning Tree

A spanning tree protocol provides the traditional solution for adding redundant links to a Layer 2 network without causing loops. The spanning tree consists of bridges, or switches. One switch is the root. In the example shown in Figure C-2, Core-1 is the root.

Every other spanning tree bridge has a lowest cost path to the root. The links in the lowest cost paths forward traffic. Other links are blocked. As you see, the network resembles a tree, rooted at the root bridge, branching to other switches, and finally branching to the endpoints on switch edge ports. Because links that are not part of this tree are blocked, the network has no loops.

The blocked links remain available for use if a change occurs in the topology. For example, if the link between Access-1 and Core-1 fails, the blocked link between Access-1 and Core-2 will transition to forwarding.

Review Spanning Tree Protocols

Several spanning tree protocols exist, as shown in Figure C-3. The original standard, Spanning Tree Protocol (STP), is defined in 802.1D. This standard defined the basic behaviors still used in spanning tree such as electing a root bridge, calculating the best path to that bridge, and blocking alternate paths. However, STP was designed in a time when a minute or so for convergence was acceptable. It requires ports to move through several port states, each with a relatively long (10 to 20 second) wait time, before the ports can start forwarding. Any change in the topology can force ports to progress slowly through these states again.

STP	RSTP	MSTP
802.1D pre-2004	802.1w merged into 802.1D 2004	802.1s merged into 802.1Q 2005

Original protocol
Obsolete
	• Faster convergence
	• Link costs that fit with modern port speeds
		• Multiple instances
		  • Instance = RSTP tree with its own root
		• VLANs mapped to instances
		• Better use of links
		  • Different links active in different instances

Figure C-3: Review Spanning Tree Protocols

As network connectivity became more crucial, companies could no longer tolerate such long convergence times when any topology change occurred. Rapid Spanning Tree Protocol (RSTP), 802.1w, was developed in 1998 to speed convergence. It also defined costs for ports that better fit with modern port speeds. In 2004 RSTP was merged into the 802.1D, superseding STP.

Multiple Spanning Tree Protocol (MSTP) operates in the same basic way as RSTP but supports multiple instances. A different set of VLANs are mapped to each instance. Each instance has its own topology with its own root. This allows switches to use their redundant links more efficiently. A link can be inactive in one instance, but active in another. In this way, switches load share traffic across redundant links on a VLAN by VLAN basis.

Review BPDUs, Bridge IDs, and Root Election

Spanning tree switches elect the root based on bridge ID. Every switch has a bridge ID, which consists of a 2-byte priority value followed by the switch's MAC address. The switch with the *lowest* bridge ID becomes root. Because the priority comes before the MAC address in the bridge ID, the priority value will always act as the most important criteria for root election while the MAC address acts as a tie breaker if two switches have the same lowest priority value. However, you should configure priorities from preventing the need for a tie breaker as it leads to unexpected roots.

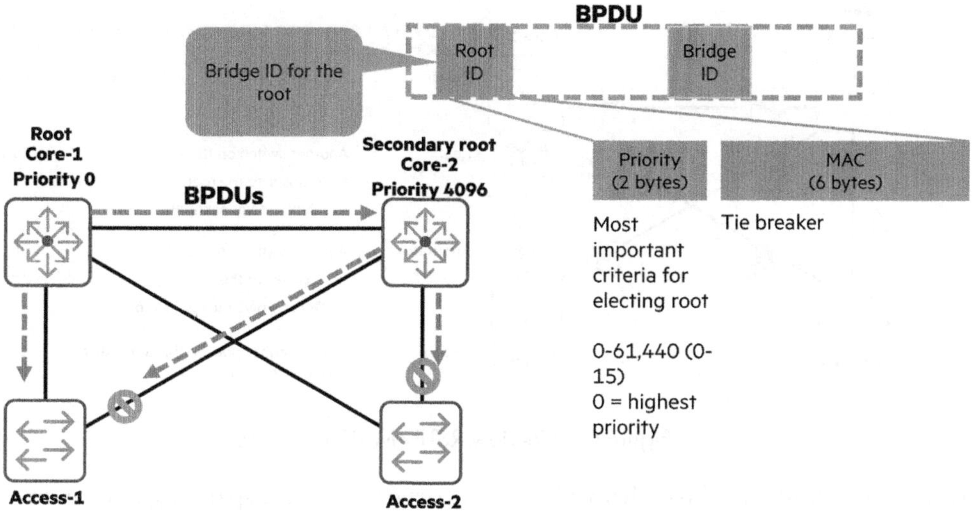

Figure C-4: Review BPDUs, Bridge IDs, and Root Election

In the example shown in Figure C-4, Core-1 has priority 0, so it is elected root. Core-2 has priority 4096 while the access switches have the default priority of 32,768. If Core-1 fails, Core-2 will become root. (Some switches specify priority with 16 values from 0 to 15, each value representing a step of 4096; in this case, Core-2's priority would be 1.)

The election actually occurs through Bridge Protocol Data Units (BPDUs), which are spanning tree protocol frames that spanning tree switches use to maintain the spanning tree. Each BPDU includes a field for the switch's bridge ID and for the root ID, which is the bridge ID of the switch that the switch believes is root. When you first enable spanning tree on a switch, the switch believes that it is root. It sends out BPDUs on all of its links with its own bridge ID in the root ID field. When the switch receives a "superior BPDU" with a lower value root ID, the switch accepts that root as the root for the spanning tree and changes the root ID in its BPDUs accordingly.

As the BPDUs propagate, the switches agree on a single root. They can then determine which ports to block, as discussed in a moment. In the stable, "converged" state, BPDUs flow from the root down through the spanning tree, as shown above, and all BPDUs have the same root ID.

Review RSTP/MSTP Port Roles

RSTP switches do not have a complete view of the topology. The same BPDUs that elect the root also help switches to calculate the correct port role for each of their ports.

Switches that run RSTP or MSTP determine whether a link should forward traffic or be blocked based on the port's role.

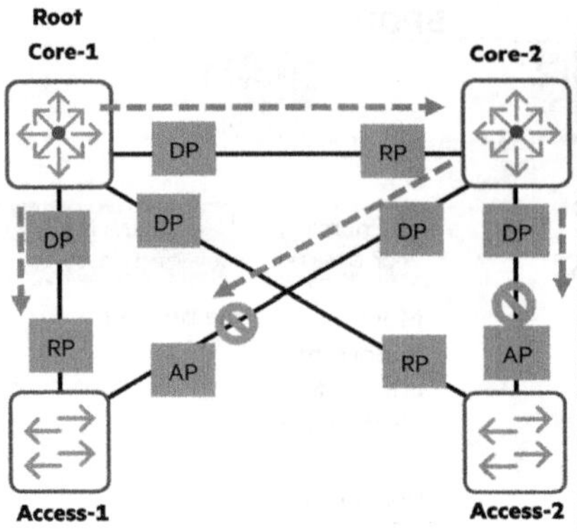

Port role	What it means	Final state
Designated	I'm closest to the root on this link	Forwarding Transmits BPDUs
Root	Another switch on this link: • Is closer to the root • Gives me my best path	Forwarding
Alternate	Another switch on this link: • Is closer to the root • Doesn't give me my best path	Discarding (blocking)
Backup	I have another port on this link that's closer to the root	Discarding (blocking)

Figure C-5: Review RSTP/MSTP Port Roles

In Figure C-5, consider the links that will become part of the active spanning tree, or, in other words, part of the switches' set of lowest cost paths to the root. For each of these links:

- The port on one side is Designated. This is the port that is closer to the root (or is, in fact, on the root).
- The port on the other side of the link is the root port.

In other words, the root port is the port that gives a non-root switch its best path to the root. A switch can only have *one* root port because it can have only one best path.

The port role affects the states that the port is allowed to pass through. In normal operation, designated and root ports transition to forwarding state. In this state, ports learn MAC addresses and forward traffic. Because both the designated port and root port are in a forwarding state, the link is active. In addition to forwarding traffic, Designated ports transmit BPDUs to maintain the spanning tree.

Now look at the links that the spanning tree blocks. For each of these links:

- The port on one side is Designated. Again, this is the port that is closer to the root.
- The port one the other side of the link is either:
- Alternate (it is connected to a different switch)
- Backup (it is connected to the same switch)

Typically, you will see alternate ports. A backup port occurs if a switch port is accidentally looped to a port on the same switch. It also occurs if a switch port is connected to downstream switches that do not support spanning tree and that introduce a loop. From the point of view of spanning tree, those switches are transparent, so the spanning tree switch appears to be connected to itself.

Alternate and backup ports remain in the discarding state. The discarding state is sometimes called the blocking state. In this state, ports drop incoming traffic, do not learn MAC addresses from incoming traffic, and do not forward traffic. They continue to receive and process BPDUs from the designated port on the other side so that the switch knows that the link is still available for failover.

Note that ports that connect to endpoints are still part of the spanning tree. These switch ports have a Designated role so that they can forward traffic to the endpoints. The ports can also be defined as edge ports.

Chapter 6: Network Authentication and Encryption Technologies

This section provides information about PEAP. This information is provided for your reference.

PEAP-MSCHAPv2 Process

PEAP begins very much like EAP-TLS. PEAP is called an outer method because it sets up a tunnel for a less secure authentication method, called the inner authentication method.

1. As shown in Figure C-6, the process begins, as always, with an EAP Request/Identity message. The supplicant responds with an identity. This identity might be anonymous to protect the username from hackers. This keeps hackers from sending wrong password attempts to lock users out of their accounts, for example.
2. The RADIUS/EAP server starts PEAP.
3. If the supplicant supports PEAP, the supplicant responds with a PEAP response, which begins the TLS handshake. The server continues the TLS handshake much as with EAP-TLS, sending its certificate to authenticate. However, the server does not request a supplicant certificate.

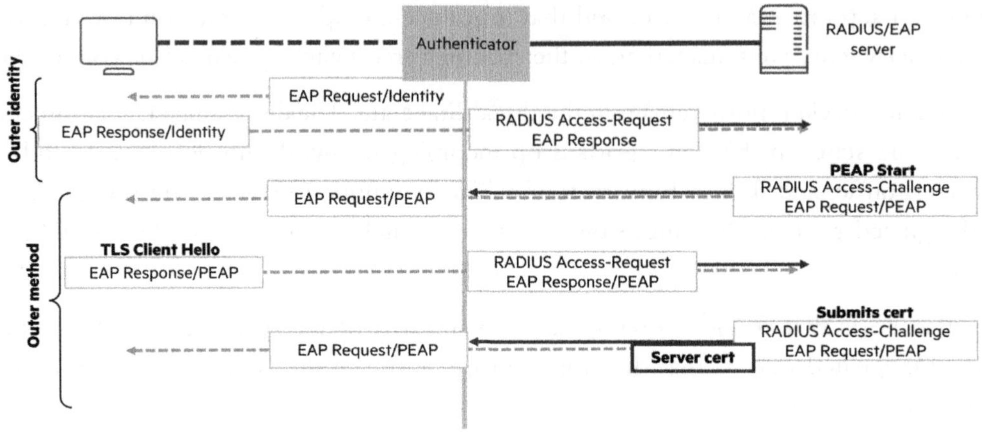

Figure C-6: PEAP-MSCHAPv2 Process (Steps 1-3)

4. As shown in Figure C-7, the supplicant validates the server's certificate. It then finishes its side of the TLS handshake.

5. The server finishes the TLS handshake too, encrypting the message using the TLS shared secret. Its ability to generate the TLS secret proves that it really owns the private key associated with its certificate.

6. The supplicant validates the server's message and sends a success message for the outer authentication method.

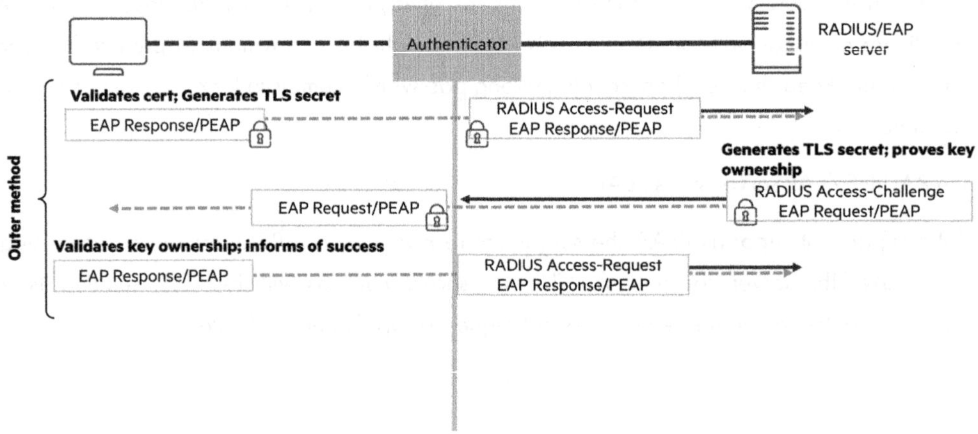

Figure C-7: PEAP-MSCHAPv2 Process (Steps 4-6)

7. As shown in Figure C-8, inside the TLS tunnel, the server starts EAP over. It sends a new EAP request/identity.

8. The supplicant responds with its inner identity, which might not match the outer identity if the outer identity was anonymized.

9. The server then initiates the inner authentication method. A common inner method is EAP-MSCHAPv2. EAP-MSCHAPv2 begins with a server challenge

10. EAP-MSCHAPv2 continues with a supplicant response, in which the supplicant proves that it knows the password hash (by encrypting challenge values with it).

11. EAP-MSCHAPv2 ends with a server response, in which the server also proves that it knows the password hash. While MSCHAPv2 is more secure than PAP or CHAP, it is still susceptible to attack for reasons such as that it uses outdated DES encryption. This is why it needs to be carried inside the TLS tunnel.

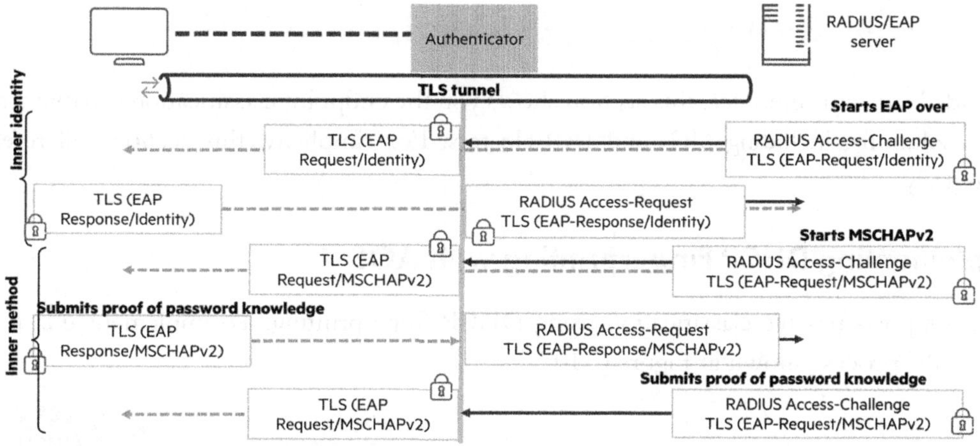

Figure C-8: PEAP-MSCHAPv2 Process (Steps 7-11)

12. As shown in Figure C-9, after the server and supplicant have authenticated each with MSCHAPv2, the supplicant sends a message indicating that the inner method was successful.

13. The server responds with its inner method success message.

14. The client then sends a PEAP message concluding the PEAP exchange.

15. The server sends a RADIUS Access-Accept to the authenticator.

16. The authenticator sends an EAP Success to the supplicant, and the connection opens.

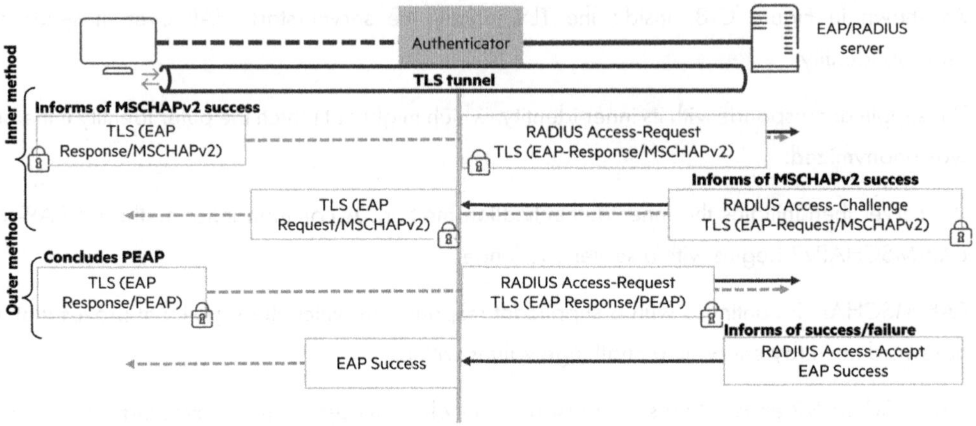

Figure C-9: PEAP-MSCHAPv2 Process (Steps 12-16)

Chapter 9: Identify and Classify Endpoints

Provided for your reference, this section shows you the endpoint classification capabilities built into AOS solutions, including MCs and IAP clusters. For simplicity, this section will refer to AOS generically.

Implementing DHCP Fingerprinting with AOS

AOS supports passive classification using DHCP fingerprinting. To implement this feature, you follow the process shown in Figure C-10.

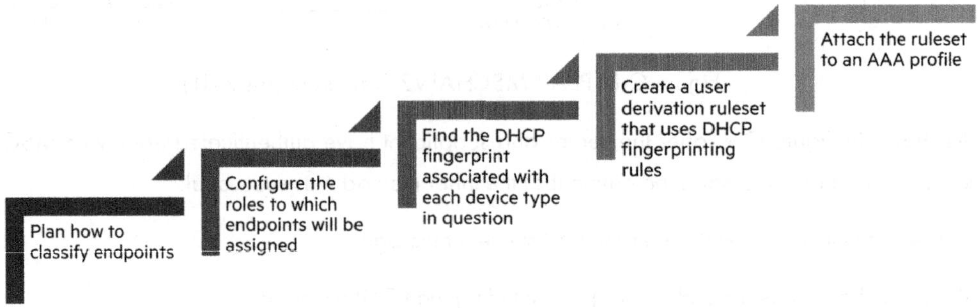

Figure C-10: Implementing DHCP Fingerprinting with AOS

First, you must plan how you want to classify endpoints. You then configure the roles to which you want to assign devices on the MC. You must also find the DHCP fingerprint associated with each device type that you want to classify.

You can then create a ruleset of user derivation rules. Each rule in the set will map a DHCP fingerprint to a role. Finally, you will attach that ruleset to an AAA profile.

Example Endpoint Classification Plans

Begin by planning roles for devices. Of course, hundreds, or even thousands, of device types exist, and you do not necessarily have to categorize every one. You need to understand what the customer is trying to achieve with endpoint classification.

Refer to Figure C-11 for some examples. Company A might want to assign all users on smartphones to a smartphone role with limited network access, regardless of the users' identity. Customer B might want to identify VoIP phones and security cameras and assign them to voice and camera roles while leaving other devices uncategorized.

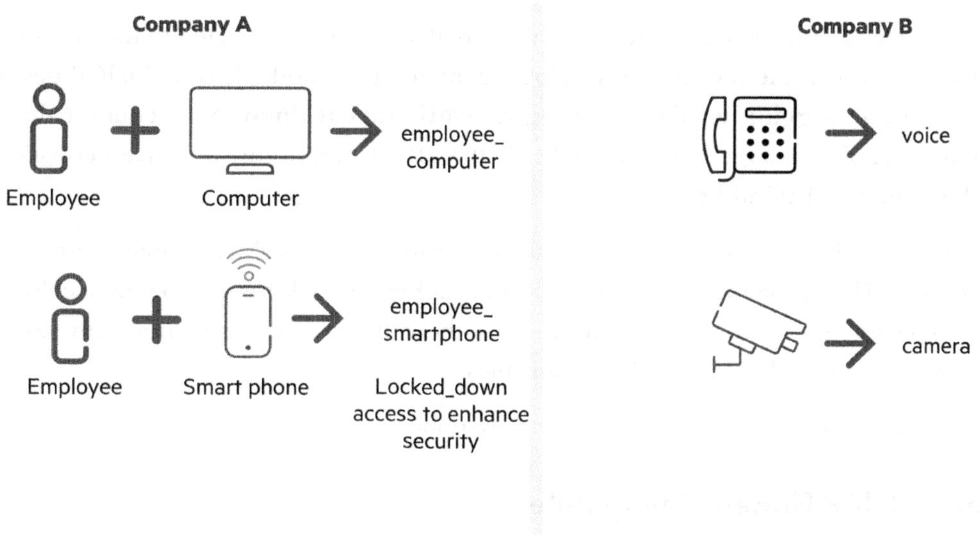

Figure C-11: Example Endpoint Classification Plans

The complexity of planning for large numbers of devices is one reason that CPPM or CPDI, which have many built in classification rules, is preferred.

Find Devices' DHCP Fingerprints

If you are setting up user derivation rules on an MC to classify devices, you can find the correct DHCP fingerprint for a particular device by using the network logs, as shown in Figure C-12.

Through the MC

Obtain device in question; make sure it is disconnected

```
MC ^[mynode] (config) logging level debugging network subcat dhcp
```

Connect the device

```
MC ^[mynode] (config) show log network all
Jan 24 01:06:44  dhcpdwrap[4033]: <202541> <4033> <DBUG> |dhcpdwrap| |dhcp|
Received DHCP packet from Datapath, Flags 0x100440, Opcode 0x5a, Vlan 102, Ing
ress tunnel 9, Egress vlan 102, SMAC d8:50:e6:f3:6e:c5
Jan 24 01:06:44  dhcpdwrap[4033]: <202534> <4033> <DBUG> |dhcpdwrap| |dhcp|
Datapath vlan102: DISCOVER d8:50:e6:f3:6e:c5 Transaction ID:0x333102f1 Options
 3c:616c64726f69642d646863702d3130 37:0103060f1a1c333a3b2b
```

Copy and take out the colon

Figure C-12: Find Devices' DHCP Fingerprints

Obtain a device of the correct type and make sure that it is not connected to the network. Establish a CLI session with the MC and go to global configuration mode. Enable DHCP debugging with this command: **logging level debugging network subcat dhcp**. Now connect the device and look for DHCP messages in the MC CLI. If the MC is supporting an active network, you might need to filter by MAC address.

Next look for the most relevant options for fingerprinting, typically beginning with the Parameter List option. The debug messages display options in hex, so look for a 37 before a colon. Then find the hex characters after the 37. The 37 plus the other characters without any colons are the fingerprint that you should include in your rules.

Remember to turn off debugging when you are done.

Create DHCP Fingerprinting Rules

You are now ready to create user rules that map the fingerprints to roles. You create the rules in the **Configuration > Authentication > User Rules** window, as shown in Figure C-13.

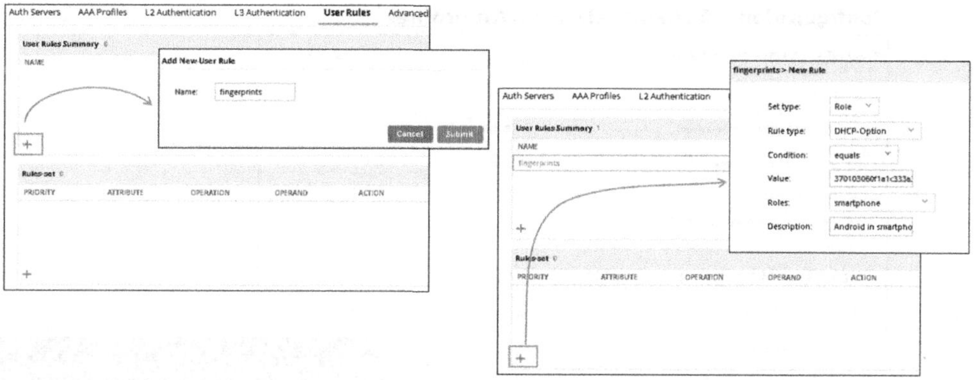

Figure C-13: Create DHCP Fingerprinting Rules

First add a user rule (actually set of rules). In this example, the rule is named "fingerprints." Then add one or more rules to it. As you see in this example, the set type is role, which means that clients that match this rule will be assigned to the specified role. In the example shown in Figure C-13, that role is smartphone. The rule type is DHCP-Option and you have set the value to the value that you found in the DHCP Discover message in the MC's network logs.

Attaching the User Rule Set to a AAA Profile

In Figure C-14, you see how you can attach the user rule set to an AAA profile—in this example, the profile associated with the company's employee WLAN. Remember that user derivation rules can include L2 criteria and L3 criteria. Any rules with L2 criteria take effect before authentication, and the client is re-assigned to a new role at authentication. However, rules with L3 criteria, *which include DHCP fingerprinting rules*, assign clients roles *after* authentication.

Configuration > Authentication > AAA profiles

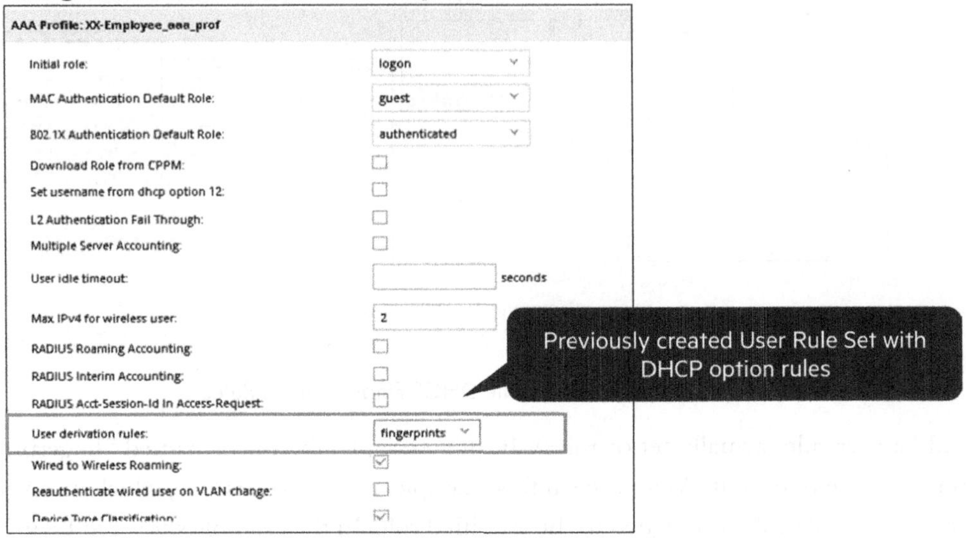

Figure C-14: Attaching the User Rule Set to a AAA Profile